Progress in Botany 67

67 PROGRESS IN BOTANY

Genetics
Physiology
Systematics
Ecology

Edited by

K. Esser, Bochum
U. Lüttge, Darmstadt
W. Beyschlag, Bielefeld
J. Murata, Tokyo

Springer

With 57 Figures

ISSN 0340-4773
ISBN 3-540-27997-0 Springer-Verlag Berlin Heidelberg New York
ISBN 03978-3-540-27997-6

The Library of Congress Card Number 33-15850

Springer is a part of Springer Sciences+Business Media
springeronline.com

© Springer-Verlag Berlin Heidelberg 2006
Printed in Germany

Cover design: Design & Production, Heidelberg
Typesetting: SPI Publisher Services
31/3150 - 5 4 3 2 1 0 - Printed on acid-free paper

Contents

Function of genetic material:
From genomics to functional markers in maize 53
Chun Shi, Gerhard Wenzel, Ursula Frei, Thomas Lübberstedt

Extranuclear inheritance:
Gene transfer out of plastids 75
Ralph Bock

Plant Breeding:
MADS ways of memorizing winter: vernalization
in weed and wheat
Günter Theißen

Biotechnology:
Engineered male sterility in plant hybrid breeding
Kerstin Stockmeyer and Frank Kempken

Physiology

New insight into auxin perception, signal
transduction and transport 219
May Christian, Daniel Schenck, Michael Böttger,
Bianka Steffens, Hartwig Lüthen

Molecular chaperones—holding and folding. 315
Christoph Forreiter

Systematics

Recent progress in floristic research in Korea 345
Chong-Wook Park

Ecology

Structural determinants of leaf light-harvesting capacity and photosynthetic potentials

Ülo Niinemets and Lawren Sack

**Biodiversity experiments – artificial constructions
or heuristic tools?** 486
Carl Beierkuhnlein and Carsten Nesshöver

List of Editors

Professor Dr.Dr.h.c. mult. K. Esser
Lehrstuhl für Allgemeine Botanik, Ruhr Universität
Postfach 10 21 48
44780 Bochum, Germany

Phone: +49-234-32-22211; Fax: +49-234-32-14211
e-mail: karl.esser@ruhr-uni-bochum.de

Professor Dr. U. Lüttge
TU Darmstadt, Institut für Botanik, FB Biologie (10)
Schnittspahnstraße 3-5
64287 Darmstadt, Germany

Phone: +49-6151-163200; Fax: +49-6151-164630
e-mail: luettge@bio.tu-darmstadt.de

Professor Dr. W. Beyschlag
Fakultät für Biologie
Lehrstuhl für Experimentelle Ökologie
und Ökosystembiologie
Universität Bielefeld, Universitätsstraße 25
33615 Bielefeld, Germany

Phone: +49-521-106-5573; Fax: +49-521-106-6038
e-mail: w.beyschlag@biologie.uni-bielefeld.de

Professor Dr. Jin Murata
Botanical Gardens
Graduate School of Science
University of Tokyo
3-7-1 Hakusan
Bunkyo-ku, Tokyo 112-0001
Japan

Phone: +81-3-3814-2625; Fax: +81-3-3814-0139
e-mail: murata@ns.bg.s.u-tokyo.ac.jp

Curriculum Vitae

Diter H. von Wettstein, born 20, September 1929, Göttingen, Germany
Dep. of Crop and Soil Sciences, Washington State University
267 Johnson Hall, Pullman, WA 99164-6420, USA
Tel:+1-509-3353635; Fax:+1-509-3358674; E-mail: diter@wsu.edu

Education:
Tübingen University: Dr. rer. nat.(Ph.D.) 1953; (Biology, Biochemistry.)
Stockholm University: Fil.Lic. (Ph.D.) 1953; Genetics
Stockholm University: Fil. Dr. (D.Sc.) 1957; Genetics
Honors: Foreign Associate of the National Academy of Sciences USA; Member of the following: Royal Danish Academy of Sciences; Royal Physiographical Society, Lund; European Molecular Biology Organization; Deutsche Akademie der Naturforscher Leopoldina; Royal Swedish Academy of Sciences, Academy of Technical Sciences, Copenhagen, Academia Europaea, Acadeémie Royale des Sciences de Belgique Österreichische Akademie der Wissenschaften, Nordrhein-Westfälische Akademie der Wissenschaften. Honorary Member of the Swedish Seed Association, Svalöf; Awarded the Lillö-Stiftelsens Prize for Genetic Research, Gregor Mendel Medal & Kurt Mothes Gold Medal, Leopoldina, Dr.agro.h.c. Copenhagen.

Professional experience:
1 October 1996 R.A. Nilan Distinguished Professor, Dep. of Crop and Soil Sciences & School of Molecular Biosciences, Washington State University
1972-September 1996 Professor and Head, Department of Physiology, Carlsberg Laboratory, Copenhagen
1975-1988 Director of Carlsberg Plant Breeding
1962-1975 Professor and Head, Institute of Genetics, University of Copenhagen
1966, 1972, 1974 Visiting Professor, University of California, Davis
1969 Visiting Professor, Washington State University
1958 Rockefeller fellow, California Institute of Technology, Pasadena & Carnegie Institution of Washington, Cold Spring Harbor and Stanford
1957 Assistant Professor in Genetics, Stockholm University Has published 332 papers in genetics, plant breeding, developmental physiology, cell biology, plant biochemistry and molecular biology.

Honorary Offices:
Member and Chair, Scientific Advisory Board Friedrich Miescher Institute, Basel, 1980-91; Chair Scientific and Technical Advisory Committee UNDP/World Bank/WHO Programme for Research and Training in Tropical Diseases, 1985-89; Member of Sainsbury Laboratory Council, Norwich, 1987-95; Member Fachbeirat, Max-Planck-Institut für Züchtungsforschung, Köln 1992-1996; Chair Fachbeirat, Institut für Pflanzengenetik & Kulturpflanzenforschung, Gatersleben, 1992-95; Chair International Advisory Board, Graduate School, Experimental Plant Sciences. Wageningen Agricultural University, 1993-95.

Fascination with Chloroplasts and Chromosome Pairing

Diter von Wettstein

During the second half of the twentieth century biological research could be characterized as a period of strong convergence. Genetics, physiology, biochemistry and other sub-disciplines of biology were joined in the common goal of clarifying the molecular processes behind the function of organelles, cells, organs and organisms. The whole chain from the information contained in the genome to the properties and function of an organism was and is analysed with sophisticated methods.

It has been a pleasure and privilege to contribute to these ventures and at the outset I would like to mention and thank my mentors in the different disciplines. They taught me to carry out research and to ask important questions: Erwin Bünning and Adolph Butenandt in Tübingen, Jacob Seiler and Albert Frey-Wyssling in Zürich, Åke Gustafsson in Stockholm, Frank Stahl and Salvador Luria at Cold Spring Harbor, and Paul Stumpf at Davis and Mogens Westergaard in Copenhagen. But the results could likewise not have been achieved without the imaginative and enthusiastic efforts of co-workers, students, postdoctoral fellows and visiting scientists. They include 54 students who completed their master's degree and 65 their PhD, and I will try to review some of their work here. During my time at the Carlsberg Laboratory, the Department of Physiology hosted 115 postdoctoral fellows and visiting scientists; they provided much of the inspiration that guided innovation and progress.

In this review I would like to discuss two areas of my interests:

1. Biosynthesis of the photosynthetic membrane and chloroplast biogenesis.
2. Chromosome pairing, the mechanism of crossing-over and genome analysis.

1 Biosynthesis of the photosynthetic membrane and chloroplast biogenesis

My interest in the development of chloroplasts and chlorophyll biosynthesis began when I became an assistant to Åke Gustafsson in Stockholm in 1951 and joined the multidisciplinary Swedish Group of Mutation Research he had created and was leading with great success. One of my tasks was to analyse mutation rates, and spectra, in the M_2 generation of barley grains treated with various ionising radiation and chemical mutagens. This was done by a test he had devised in the 1930s and consisted of counting the white, yellow, light-green or tiger-striped lethal seedling mutants emerging from thousands of spikes planted in the greenhouse during the winter season. These tests were done to find the most efficient treatments for inducing mutants suitable for barley breeding programs. Interestingly, now there is hardly a cultivar that does not contain an induced mutant among its ancestors, but at the time it was considered that all induced mutations were detrimental and therefore useless in plant breeding, a view propagated by Herman J. Muller and L. J. Stadler. Due to the tireless efforts of Åke Gustafsson and a few others like Bob Nilan in Pullman it was shown that induced mutations could yield improved cultivars – Muller and Stadler overlooked the fact that the majority of spontaneous mutations were also detrimental, and that mutations are still a major factor in the evolution of genes to organisms. The discussions for and against were not unlike the present discussions concerning the use of transgenic plants in breeding. As history repeats itself, the time will come when transformed cultivars will be as accepted and considered as "traditional" as crop plants containing induced mutations.

Encouraged by the successful efforts of Beadle and Tatum in analysing metabolic pathways by knock-out mutations, it seemed to me that all these hundreds of interesting mutants should be useful for a detailed analysis of the development of chloroplasts and pigment biosynthesis. I thus started to collect representatives of the different types of mutants and to conduct crosses to determine allelic relationships by complementation tests. At that time electron microscopy of thin sections started to reveal the ultrastructure of animal and plant cells. So I took my mutants and spent a few days every week at Arne Tiselius's Biochemistry Institute at Uppsala University, where Håkan Leyon had constructed a microtome and developed embedding procedures, and where I could use the third electron microscope built by Siemens in 1940. It had been acquired by The Svedberg and was installed next to his ultracentrifuges. The mutants turned out to be very useful for characterizing the development of chloroplast structure as presented in a summary (von Wettstein 1959). In higher plants, chloroplasts develop from proplastids in the light or via the etioplast pathway after an initial dark period. The primary thylakoid layers are formed by alignment of vesicles budded from the inner membrane of the plastid envelope. In contrast to the in depth knowledge obtained since then of the organization of the photosynthetic membrane and the import of the protein components into the chloroplast and their targeting to the thylakoids, progress in learning how the lipid bilayer membranes are formed is less apparent (von Wettstein 2001). This may change with the discovery by Kroll et al. (2001) and Westphal et al. (2001) of a function of the vesicle-inducing protein in plastids (VIPP).

In pea chloroplasts the 37-kDa VIPP protein is located both in the vicinity of the chloroplast envelope and the thylakoid membranes and was considered by Li et al. (1994) as a candidate for the transfer of galactolipids from their site of synthesis at the chloroplast envelope to the thylakoids. Daniella Kroll and co-workers (Kroll et al. 2001) studied a recessive *Arabidopsis* T-DNA insertion mutant with severe disturbances in the photosynthetic electron transport chain and the formation of the thylakoids. The insertion was identified in the gene encoding VIPP and the mutant could be rescued by transformation with the *VIPP* cDNA. The cause for the disturbed development or maintenance of the thylakoids was the failure of the mutant to bud the vesicles from the inner chloroplast envelope membrane, which transfer lipids from the inner envelope to the thylakoid membranes. In the transformants the process of vesicle budding was re-established and the thylakoid organization normalized. The companion paper by Sabine Westphal and co-workers (2001) identifies *VIPP 1* genes in the genomes of *Synechocystis*, *Anabaena*, *Synechococcus* and *Nostoc*. In these cyanobacteria, the protein is located in the plasma membrane, but its disruption in *Synechocystis* by insertion mutagenesis with a kanamycin cassette prevents ordered thylakoid formation and light-dependent oxygen evolution.

The photosynthetic membrane in barley and other higher plants converts solar energy into chemical energy, and as we now know, it uses six larger protein complexes for this purpose (Simpson and von Wettstein 1989; von Wettstein et al. 1995) (Fig. 1). They are called the reaction centres of photosystem I and II, the two light harvesting complexes of photosystem I and II, the cytochrome b6/f complex and the chloroplast coupling factor (synthesizing ATP). The polypeptides in these complexes bind and orient chlorophyll and carotenoid molecules and the different metals and molecules which are required for energy channelling and electron transport. Circa one half of the ~57 membrane proteins are encoded by genes in the nucleus and the other half in the chloroplast's own DNA genome. This cooperation between the two genomes in the plant cell also takes place in the assembly of the CO_2-fixing enzyme, Rubisco, that is made up of two, different-sized subunits, of which the larger one is encoded in chloroplast DNA and the smaller one in the nuclear chromosomal DNA. The following results of our research are of special significance.

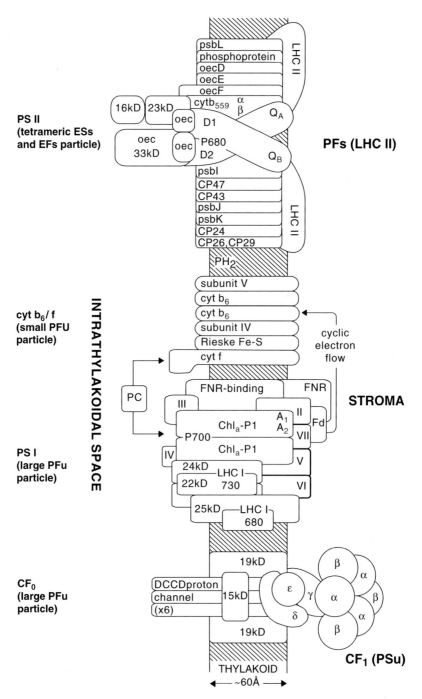

Fig. 1a, b. Model of the photosynthetic membrane showing the polypeptide components of the major complexes. The site of coding is indicated by shading (chloroplast DNA) or is unshaded (nuclear gene). (Modified from Simpson and von Wettstein 1989)

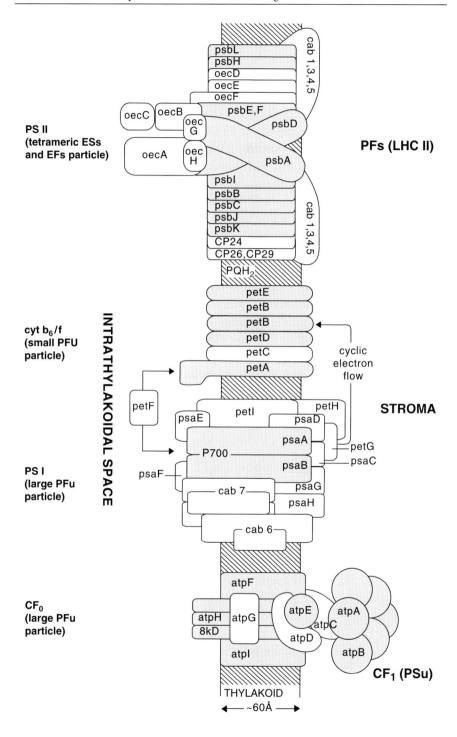

In cooperation with the Biological Laboratories of Harvard University the nucleotide sequence of the first plant gene, the structural gene for the large subunit of Rubisco, was determined in 1980 (Mc Intosh et al. 1980; von Wettstein 1981). Carsten Poulsen, Anthony Holder, Brian Martin and Ib Svendsen had produced peptide maps of the large subunit of Rubisco of barley and the genus *Oenothera* and obtained partial amino acid sequences (von Wettstein et al. 1978; Holder 1978; Poulsen et al. 1979). Lawrie Bogorad called me one day in 1977 to say that he had heard that we had amino acid sequences of the large subunit and to ask if I would share them with him, since he wanted to sequence the maize gene and this would be of great help to his project. "Sure," I said, "and I would also like to send you Carsten Poulsen with his Carlsberg fellowship to help with the sequencing." After supplying Carsten with a large supply of liquorice, he and Lee completed the task. Peptide mapping of the large and small Rubisco subunits also led to the identification of the pomato, the first somatic hybrid between potato and tomato produced by Georg Melchers in Tübingen (Melchers et al. 1978; Poulsen et al. 1980; von Wettstein 1983). The most interesting aspect of the analysis of these generic hybrids was the finding that they only retained the tomato or potato chloroplast genome at an equal frequency, but not both. We still do not know how this happens.

Over the years 357 barley mutants with defects in chloroplast development and chlorophyll synthesis have been assigned to 105 gene loci. Together with Albert Kahn, Ole Frederik Nielsen, Simon Gough and Naomi Avivi-Bleiser (von Wettstein et al. 1974; Kahn et al. 1976) structural and regulatory genes of chlorophyll synthesis were identified. Knud Henningsen, John Boynton, David Simpson, Otto Machold, Gunilla Høyer-Hansen, Roberto Bassi, Bob Smillie and Torsten Fester analysed the different categories of the mutants with regard to their ultrastructure, pigment levels, thylakoid polypeptide composition and photosynthetic capacity (Henningsen et al. 1993; Simpson and von Wettstein 1980; Simpson et al. 1985; Smillie et al. 1978).

The mutants were used to localize the macromolecular photosynthesis complexes, as recognized by freeze-fracture particles, to the different domains of the chloroplast membranes (e.g. Simpson et al. 1989; Simpson and von Wettstein 1989). Birger Lindberg Møller analysed the function of the grana and stroma membranes by isolating and purifying these membrane types, by separating the membrane polypeptides and reconstituting them to give photosynthetically active membranes (e.g. Henry et al. 1982; Møller and Høj 1983; Møller 1985). The gene family encoding the light-harvesting proteins of photosystem I was also identified (e.g. Knoetzel et al. 1992). The first transcription map of a chloroplast genome was established for barley (Poulsen 1983) and alternative transcription was demonstrated for the gene encoding the large subunit of Rubisco. The longer transcript is used by the plant when a large amount of protein is synthesized in the light (Poulsen 1984).

A single molecule of chlorophyll and haem is synthesized from eight molecules of 5-aminolevulinic acid. In 1975 it was shown by isotope labelling that higher plants, in contrast to animals and humans, synthesize this non-protein amino acid from the intact carbon skeleton of glutamate (Beale et al. 1975). Gamini Kannangara, Simon Gough, postdoctoral fellows, students and visiting scientists have elucidated this three-step pathway at the biochemical and molecular level over a period of 19 years (cf. Kannangara et al. 1994; von Wettstein et al. 1995; von Wettstein 2000a, b). This pathway is used by higher plants, algae, cyanobacteria, *Escherichia coli* (not recognized for over 30 years), as well as a number of other bacteria. Animals and humans, yeast and photosynthetic bacteria form 5-aminolevulinate by condensation of glycin and succinate. Entirely surprising was the discovery that the glutamic acid has to be activated by ligation to a glutamyl tRNA before it can be reduced to glutamate-semialdehyde and thereafter transaminated by an aminomutase to 5-aminolevulinic acid (Schön et al. 1986). It is so far the only known case in which a tRNA participates in the conversion of a low molecular weight compound. In higher plants this tRNA is encoded in chloroplast DNA and also has to serve for the translation of mRNA on chloroplast ribosomes. The three enzymes are encoded in nuclear DNA, and have to be translated on cytosolic ribosomes and imported into the chloroplast.

The importance of the pathway for chlorophyll synthesis is demonstrated by transgenic tobacco plants expressing an antisense gene for the glutamine semialdehyde aminotransferase (Höfgen et al. 1994). The barley enzyme that requires the glutamyl tRNA as substrate was purified and a partial amino acid sequence obtained (Pontoppidan and Kannangara 1994). This work identified the structural gene for this enzyme as the *HemA* gene, already cloned and sequenced in many organisms but not recognized as encoding glutamyl RNA^{Glu} reductase. Finally this interesting enzyme was expressed as a fusion protein in *E. coli* (Vothknecht et al. 1996, 1998). It turned out that haem, a prominent inhibitor of chlorophyll synthesis, binds to the N-terminal extension of the protein that is characteristic for plant enzymes, but absent in bacteria.

In 1994 Lucien Gibson, Ph.D. student with Neil Hunter, University of Sheffield, arrived and brought with him plasmids that contained the *bchH*, *bchD* and *bchI* genes from *Rhodobacter spheroides*. Lucien, Robert Willows and Gamini Kannangara expressed the proteins of these three genes in *E. coli* and demonstrated for the first time that the association of these three proteins *in vitro* inserts the Mg atom into protoporphyrin IX (Gibson et al. 1995; Willows et al. 1996). Reconstitution of Mg chelatase activity required only ATP, Mg^{2+} and protoporphyrin. This opened the way to learn more about how the metal ion is incorporated into the porphyrin ring. The insertion of Mg^{2+} into protoporphyrin IX proceeds in two stages. In the first stage subunits BchD (70 kDa) and BchI (40 kDa) undergo activation by complex formation in the presence of ATP and Mg^{2+}. The protein–protein interaction of these two subunits was subsequently confirmed for the tobacco subunits with the yeast two-hybrid system (Gräfe et al. 1999). Thereafter Mg^{2+} is inserted into the protoporphyrin IX substrate that is bound to the large subunit BchH (140 kDa).

The information of the nucleotide sequence of the *Rhodobacter* genes permitted the identification, cloning and molecular characterization of the corresponding barley and other higher plant genes and their mutants (Jensen et al. 1996; Kannangara et al. 1997; cf. von Wettstein 2000b). That three different gene products are required for the insertion of Mg^{2+} into protoporphyrin IX was originally found with xantha mutants at three gene loci in barley that accumulate protoporphyrin IX when fed 5-aminolevulinate (Gough 1972; von Wettstein et al. 1974; cf. von Wettstein 2000a). They belong to the first mutants isolated and analysed in 1953. Gene *Xantha-f* corresponds to *bchH*, *Xantha-g* to *bchD* and *Xantha-h* to *bchI*.

One of the post-genomic challenges is to determine the function of the genes discovered in genome sequencing projects. Usually > 50% of the genes uncovered in the sequenced genomes have no significant matches to proteins or cloned genes in the databases for other organisms. Furthermore, while such matches can hint at similar functions they do not prove the function of the gene in question. To determine the precise function of a gene its cloning is required, frequently carried out by positional cloning. While this is expedient with small genomes like that of *Arabidopsis* it is difficult with large sequenced or un-sequenced genomes like those of small grain cereals. Due to the availability of the transcript-deficient barley mutant *xantha-h*[57] we were able to develop the microarray method for transcript-based cloning of genes only known through their mutant phenotype (Zakhrabekova et al. 2002).

Libraries of genomic clones or cDNA clones or expressed sequence tag clones representing several thousand genes are microarrayed on glass slides. Each clone occupies a round spot on the slide. cDNAs made from the mRNAs of the transcript-deficient mutant and its wild type is differentially labelled with green and red fluorescing nucleotides, respectively, and hybridized in equal amounts to the microarrayed clones. Because of the absence of the mutant transcripts, pure red fluorescence from a spot will result from wild type DNA and identify the gene sought This technique also worked with the xantha-f 27 and xantha-f 40 mutants, which display non-sense-mediated mRNA decay, a surveillance system developed by organisms to reduce the abundance of mRNA with nonsense codons (Gadjieva et al. 2004). It can be exploited to clone genes through mutants with reduced transcript abundance. This then will allow functional identification of a majority of the ca. 1,000 barley genes for which > 8,322 mutants have been identified through Åke Gustafsson's and Udda Lundqvist's efforts and are conserved in the Nordic Gene Bank.

The three-dimensional structure of the ATP-binding subunit BchI of *Rhodobacter capsulatus* solved at a resolution of 2.1 Å by Michel Fodje and Salam Al-Karadaghi in combination with the molecular genetic analyses of Mats and Andreas Hansson has allowed remarkable insights into the molecular basis of the insertion of Mg^{2+} into protoporphyrin IX (Fodje et al. 2001). It provides the starting point for clarifying the mechanism by which Mg^{2+} is inserted into the chlorophyll molecule.

BchI belongs to the chaperone-like "ATPase associated with a variety of cellular activities" (AAA) family of ATPases. Its structure could be compared with those of other members of this protein family, such as the heat shock protein 100 of *E. coli*, the delta-prime subunit of DNA polymerase III clamp loader complex and the hexamerization domain D2 of the *N*-methylmaleimide-sensitive membrane vesicle fusion protein. The domains of these proteins are highly conserved, but are located in different ways in the overall structure. BchI also contains loop structures forming a deep positively charged groove that might be involved in interaction with the other subunits of Mg-chelatase. Electron microscopy of BchI in solution in the presence of ATP revealed that it forms in the same way as hexameric ring structures of other AAA proteins. The primary structure of the BchD subunit consists of an AAA module at the N-terminal portion and an integrin I domain in the C-terminal half. An acidic, proline-rich region links the two domains and is predestined to bind to the positively charged cleft of BchI. Both BchI and BchH (the protoporphyrin-binding subunit) contain integrin I domain-binding amino acid sequences. Most likely the hexamer ring of BchI is connected to a hexameric ring of the BchD-AAA module via the proline-rich domain. The integrin BchI domains bind to BchH linking porphyrin metallation by BchH to ATP hydrolysis by BchI.

Among the seven mutant alleles of the barley *xantha*-h gene encoding the smallest subunit of magnesium chelatase (corresponding to BchI) four are recessive and three are semi-dominant. The homozygous mutants are yellow, because of a lack of chlorophyll. The heterozygotes of the recessive mutants are fully green whereas the heterozygotes carrying the semidominant allele are pale to yellow-green. The recessive mutations prevent transcription of the gene (Jensen et al. 1996), while the semidominant alleles are mis-sense mutations leading to single amino acid substitutions (Hansson et al. 1999). Identification of the mutated residues in the BchI three-dimensional structure located all three of them in the interface between two neighbouring subunits in the AAA$^+$ hexamer and close to the region forming the ATP-binding site. The three amino acid changes were made by site-directed mutagenesis in the BchI gene of *R. capsulatus* and the subunits expressed in *E. coli*. Combination of wild type BchD and BchH subunits with modified BchI subunits were deficient in ATP hydrolysis and Mg-chelatase activity. However mixtures of the mutated and wild type BchI subunits could form oligomeric complexes with the D and H subunits. The oligomerization is ATP dependent but results in complexes lacking Mg-chelatase activity. Furthermore the presence of mutant BchI subunits in the oligomer did not inhibit the ATPase activity of the wild type subunits but prevented the insertion of Mg^{2+} into prototoporphyrin IX. It is suggested that a small amount of hexamers consisting only of wild type subunits rescues the heterozygous plants. It remains to be seen if disruption of ATP hydrolysis in the mixed hexamers prevents the conformational change expected to permit chelation of Mg^{2+}.

2 Chromosome pairing, mechanism of crossing-over and genome analysis

The ascomycete *Neottiella rutilans* turned out to be an excellent object with which to study the assembly and disassembly of the synaptonemal complex, the 200-nm-wide ribbon between the paired pachytene chromosomes, by electron microscopy of serial sections. Jane Mink Rossen and Mogens Westergaard had shown that the DNA replication in this organism before meiotic prophase occurs in the crozier nuclei prior to karyogamy, which laid to rest the textbook theory of chromosome pairing in connection with a DNA replication at meiotic prophase. In this ascomycete the chromosomes are always at a condensed chromatin stage, also during mitosis and meiosis, which makes it a highly favourable subject for ultrastructural studies (Westergaard and von Wettstein 1966). In a study of all stages of meiosis of *Neottiella* (Westergaard and von Wettstein 1970; von Wettstein 1971, 1977) it was demonstrated that after a rough alignment of the homologous chromosomes to within 300 nm, the lateral components (protein and RNA) are laid down between the two sister chromatids of each chromosome (Fig. 2). This causes the appearance of the leptotene chromosome as undivided in the light microscope. At the same time the central region pre-assembles in the nucleolus and is then transported together with recombination nodules into

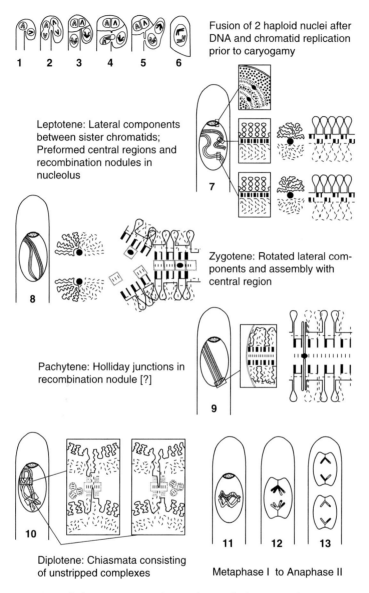

Fusion of 2 haploid nuclei after DNA and chromatid replication prior to caryogamy

Leptotene: Lateral components between sister chromatids; Preformed central regions and recombination nodules in nucleolus

Zygotene: Rotated lateral components and assembly with central region

Pachytene: Holliday junctions in recombination nodule [?]

Diplotene: Chiasmata consisting of unstripped complexes

Metaphase I to Anaphase II

Fig. 2. Formation of the synaptonemal complex and chiasmata during meiosis in an ascomycete (*Neottiella*)

the space between the roughly aligned homologues. The two sister chromatids relocate, so that that the lateral components are positioned lateral to the chromatin of the chromosome. In the pairing fork the central region material is organized alternately on one or the other lateral component, and the synaptonemal complex is completed by attachment of the free lateral component of the homologue (Fig. 3). It was also demonstrated that at diplotene short pieces of the complex remain and constitute the chiasmata. It was concluded that the molecular pairing of the DNA of two non-sister chromatids and recombination takes place inside the syaptonemal complex at the domains, which subsequently are retained as chiasmata at diplotene (Westergaard and von Wettstein 1972).

Thereafter, Søren Rasmussen and Preben Bach Holm (1984) together with postdoctoral and visiting scientists set out to investigate in various species and special cytogenetic situations the concept that the synaptonemal complex is a vector for chromosome pairing and disjunction as evidenced by: (1) the universality of its occurrence in eukaryotic organisms displaying four-strand crossing-over, (2) the evolutionary stability of its structural organization, and (3) its role in the formation of chiasmata, the microscopic counterpart of crossing-over. They also sought and found some solutions to the question: how could it be that crossing-over and gene conversion can occur in principle between or within any genes along the giant DNA double helix spanning from one telomer to the other, but in the individual bivalent

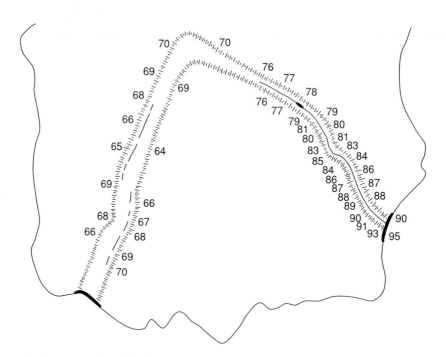

Fig. 3. Reconstruction of bivalent 22 of *Neottiella* at synapsis from electron micrographs of a serially sectioned zygotene nucleus. (Modified from von Wettstein 1977)

of one meiocyte, there are only one, two or, more rarely, three to six crossover events?

By three-dimensional reconstruction from electron micrographs of serially sectioned meiotic nuclei or by spreading of the synaptonemal complexes on an air-water interphase and subsequent contrasting with phospho-tungstic acid or ammoniacal silver ions, meiosis was analysed in the human male and female, the male mouse, in the silkworm *Bombyx mori*, in *Drosophila melanogaster*, in the higher plants *Lilium longiflorum*, maize, barley, wheat and *Lolium*, in the ascomycetes *Neurospora* and *Sordaria*, in the basidiomycetes *Coprinus* and *Schizophyllum*, and in the Phycomycetes *Allomyces* and *Blastocladiella*. Both standard genotypes, autopolyploids and chromosome structural rearrangements were studied. The following information was obtained (von Wettstein 1984; Rasmussen and Holm 1984):

1. The concepts of the assembly and disassembly of the synaptonemal complex as a vector for pairing and disjunction were confirmed in these species.

2. In *Sordaria*, *Lilium* and maize a complete lateral component is assembled at leptotene before the formation of the synaptonemal complex ensues. In *Coprinus*, *Bombyx* and human spermatocytes and oocytes the lateral components are first laid down close to the telomeres and pieces of the synaptonemal complex are formed in these regions before the lateral components are completed. In *Sordaria* the rough alignment of all homologues into pairs takes place after karyogamy and prior to the formation of the synaptonemal complex. In most other organisms with long/and or many chromosomes studied up until now, the alignment of homologues takes place progressively, segment by segment, during leptotene and zygotene (e.g. *Bombyx*, maize, wheat, lily and human meiosis). The reconstructions have shown, that in the majority of diplont organisms, the homologous chromosomes at premeiotic interphase are widely separated within the nucleus. The biochemical apparatus required for the alignment of the homologous chromosomes to within a distance of 300 nm during zygotene is still under intensive investigation as is that for the analogous phenomenon of somatic pairing present in some groups of insects.

3. The three dimensional reconstructions of zygotene nuclei established that the joining of chromosome segments with the synaptonemal complex at this stage requires genetic homology, and that precise site-to-site matching with the synaptonemal complex can take place interstitially at several independent places within a long pairing bivalent. The reconstruction of an early zygotene nucleus of lily by Preben Holm (1977) with 12 partially synapsed bivalents revealed between five and 36 independent initiation sites.

The long stretches of unpaired lateral components between the synaptone-mal complex segments were not aligned and their homologous regions were separated by distances of up to 30 μm, i.e. a distance almost equivalent to the diameter of the nucleus. The close agreement of the length of two lateral components in each of the 12 partially paired bivalents showed that the lateral component, and thus the chromosome length is determined before the site-to-site synapsis is initiated. From the large number of reconstructions of zygotene nuclei in several species, it became evident that the length of a particular chromosome pair varies in different meiocytes of the same individual, while the lateral components of homologues within the nucleus are of equal or nearly equal length. Extensive interstitial initiation of synaptonemal complex assembly was also found in maize, wheat, rye and human oocytes, while in human spermatocytes, *Bombyx, Coprinus* and *Sordaria* there were few interstitial initiation sites. It was demonstrated extensively that multivalent formation with the synaptonemal complex is unavoidable in polyploids and translocation heterozygotes at zygotene in all organisms studied and that the synaptonemal complex can be used for the identification of structural chromosome rearrangements, where classic light microscopy failed.

4. In classic cytological analyses, chromosome and bivalent interlocking was considered a rare accident that could be increased in frequency by treating meiocytes with various physical and chemical agents; in hindsight this was an erroneous conclusion derived from studies of diplotene to metaphase I stages. If zygotene is analysed by three-dimensional reconstruction from serial sections, both chromosome and bivalent interlockings are frequently encountered with a frequency of four per nucleus. They are resolved by chromosome breakage and precise reunion of the broken ends prior to pachytene. This has been documented for all species investigated, including human spermatocytes and oocytes. It can be stated that every plant and every human being has developed from gametes, in which one or several chromosomes had been broken and precisely repaired. Unfortunately this mechanism does not protect against radiation-induced chromosome breakage.

5. Non-homologous pairing was recognized by McClintock (1933) in pachytene nuclei containing unbalanced or structurally heterozygous chromosome complements, and was also shown to occur in normal sporocytes. Synaptonemal complexes of normal size have indeed been found early on in non-homologously paired chromosomes or chromosome segments in haploid tomato, petunia, snapdragon, barley and wheat or in foldback paired univalent chromosome segments. Such non-homologously paired regions do not give rise, however, to chiasmata or translocations. This puzzle was solved by investigation of diploid, triploid and autotetraploid *Bombyx* females. These lack crossing-over and somatic pairing but disjoin their 28

bivalents by retaining the synaptonemal complex in amplified form until the beginning anaphase movement when the complexes are left behind in the metaphase plane. This was described at the end of the last century as elimination chromatin. I studied such preparations under the light microscope during my student days with Jakob Seiler at the Institute of Technology at Zürich. Remembering this I thought that the elimination chromatin might be discarded synaptonemal complexes and discussed the matter with the eminent Russian silkworm geneticist Boris Astaurov at a meeting in Dushambe. This led to a visit of Søren Rasmussen in Moscow, where appropriate material was made available for electron microscopic studies, and to the proof that indeed the "elimination chromatin" consists of discarded synaptonemal complexes. But this also opened the way to study homologous and non-homologous chromosome pairing without the complication that crossing-over might introduce. In triploid females typical trivalents are formed with the synaptonemal complex and exchange of pairing partners, but at early pachytene the trivalents are reorganized into 28 bivalents completely paired with synaptonemal complexes and 28 univalents which now display fold back pairing and non-homologous association with synaptonemal complex formation. It thus became clear that at early zygotene strict homology is required for pairing with the synaptonemal complex, but at the transition from zygotene to pachytene synaptonemal complex formation no longer requires DNA homology and permits extensive rearrangement of the synaptonemal complex in multivalent associations leading to an optimization of bivalent formation. This is directly demonstrated by the analysis of autotetraploid *Bombyx* females. At zygotene, quadrivalents, trivalents and univalents are frequent, whereas at pachytene, nuclei with 56 bivalents paired with the synaptonemal complex from telomere to telomere are found and the eggs of these females are fertile. In Bombyx males meiosis occurs with crossing-over and disjunction with the aid of chiasmata; the synaptonemal complexes are shed in a normal way at diplotene. Analysis of autotetraploid males revealed an average of 13.3 quadrivalents and 25.1 bivalents at zygotene, while at pachytene the mean number of quadrivalents was reduced to 8.7 and that of bivalents correspondingly increased to 37.1. This and other observations revealed that an established crossing-over efficiently prevents conversion of multivalents into bivalents, but the correction mechanism is effective also in chiasmatic meiosis. In humans it was demonstrated by the analysis of early and late pachytene stages of a man with a balanced translocation 46,XY, t(5p–;22p+). It has to be pointed out that these rearrangements play a crucial role in auto- and allopolyploid plants, as they permit correction of multivalents into bivalents, the mechanism responsible for disomic inheritance in polyploids.

Ernie Sears had discovered that a major or two controlling genes for chromosome pairing in wheat are located in the middle of the long arm of chromosome 5B. In the absence of chromosome 5B or its long arm, or in karyotypes with extra copies of the chromosome, the disomic inheritance breaks down due to crossing-over and chiasma formation between the homoeologues in hexaploid wheat or in hybrids between wheat and related species. Plant breeders use such hybrids in order to transfer disease-resistant genes from related species into wheat. Palle Hobolth, Preben Holm, Glyn Jenkins, Bente Wischmann and Xingshi Wang analysed chromosome pairing and chiasma formation at diplotene in euploid wheat, in lines nullisomic and monosomic for chromosome 5B, and in lines monoisosomic, diisosomic and triisosomic for the long arm of chromosome 5B. Also analysed were trihaploid wheat with and without chromosome 5B and wheat-rye hybrids nullisomic for 5B (Hobolth 1981; Jenkins 1983; Holm 1986; Wischmann 1986; Wang 1988; Wang and Holm 1988; Holm and Wang 1988). Analyses of euploid wheat revealed that chromosome pairing and synaptonemal complex formation at zygotene primarily occurred between homologues, but in most nuclei one or more quadrivalents, pentavalents or hexavalents had formed due to pairing partner exchange. Nearly all multiple associations were corrected before pachytene. Among lines with zero to six copies of the long arm of chromosome 5B, only in lines with two copies was pairing of bivalents with the synaptonemal complex achieved to 97%, while it was reduced from 90 to 25% in the karyotypes deviating from the diploid number of chromosome 5B. The arrest in completion of pairing at the various levels during zygotene still permits primarily homologous crossing-over and chiasma formation, but due to the lack of correction of multiple associations homoeologous crossing-over occurs.

6. Recombination nodules attach to the synaptonemal complex from zygotene to pachytene. In *Bombyx* males and in *Coprinus*, as in *Neottiella*, recombination nodules associated with the synaptonemal complex at pachytene are converted into the chromatin chiasmata at diplotene, diakinesis and metaphase I. Thus, a recombination nodule at pachytene and a retained segment of synaptonemal complex at diplotene can mark the site of a reciprocal exchange, a crossing-over. At the end of zygotene, about twice as many recombination nodules are present as there are chiasmata found at diplotene. Quantitative determinations in combination with computer modelling by Søren Rasmussen, revealed that the nodules at zygotene are distributed at random, leaving many bivalent arms without a nodule. A subsequent redistribution of the recombination nodules minimizes the number of bivalents and bivalent arms without a nodule. Two sources for positive chiasma interference, as observed in a majority of organisms with four-

strand crossing-over, are suggested: the availability of a limited number of recombination nodules restricts the number of reciprocal exchanges in a bivalent arm. Preferential attachment of recombination nodules to certain domains of the synaptonemal complex along the bivalent arm, coupled with the mechanism that after zygotene direct nodules to domains devoid of recombination nodules, will reduce the chance of adjacent crossovers. The number and chromosomal distributions of recombination nodules are in good agreement with crossing-over frequencies in Neurospora and *Drosophila*, considering the transitory nature of the nodules in the latter species.

Dramatic differences in the length of the synaptonemal complex were revealed in *Homo sapiens*, where the length of the female complement from zygotene to the end of pachytene exceeds that of the male by a factor of 2. In this case the difference is not accompanied by a comparable difference in chiasma frequencies (as has been found in special lines of maize). The number of crossovers in the male has been estimated to be about 70 by cumulating the number of recombination nodules throughout pachytene, while the mean number of chiasmata at diakinesis totals 50. The average number of chiasmata at diakinesis in oocytes is 44, while the number of recombination nodules at pachytene is 60 compared to 75 in spermatocytes (Bojko 1985).

Gene conversion has been found in all organisms in which appropriate genetic fine structure analysis has been possible. It varies in frequency depending on the chromosome and species investigated, but it can be more frequent than reciprocal exchanges. It was therefore suggested that the randomly distributed recombination nodules at the end of zygotene effect gene conversions, but abort when reciprocal exchange and chiasma formation do not follow. This dual function of the recombination nodule would be in line with the present concept that crossovers and gene conversions arise in association with hybrid DNA by a common mechanism.

Much of the work referred to here has been made possible by grants awarded by Euratom of the European Community. By 1988 these programs had come to an end and the research arm of the European Community asked me to participate in the yeast genome-sequencing project. While reluctant at first, because it would defer biochemical studies of meiosis, I felt that it would provide the Carlsberg Laboratory with the opportunity to establish a highly efficient nucleotide sequencing team under the leadership of Søren Rasmussen and Jean Sage, who had been the laboratory's champion for serial sectioning of meiotic nuclei. We thus participated in establishing the complete nucleotide sequence of chromosomes III, XI and X in yeast (Oliver et al. 1992; Dujon et al. 1994; Galibert et al. 1996). The gene density

variations and the regular alternations of high and low G+C contents discovered along the chromosome arms may well form a basis for the preferred
domains at which crossing-overs are positioned as found in the analyses of
meiotic prophase. The many new genes discovered by the cooperative spirit
of more than 40 laboratories have served plant molecular biology and genetics of yeast well. This work also spearheaded the sequencing of other eukaryotic genomes, including that of *Homo sapiens.*

In the following I would like to provide a brief up-to-date review of the
recent progress in the molecular analysis of chromosome pairing and crossing-over, primarily achieved by cloning genes from meiotic mutants in yeast
and by fluorescent in situ hybridizations (FISH) with tagged antibodies for
synaptonemal complex proteins and DNA recombination enzymes. Shirleen
Roeder (1997) and Denise Zickler and Nancy Kleckner (1998, 1999) have
presented the results in comprehensive reviews.

The Zip1 protein from *Saccharomyces cerevisiae* and the Scp1/Syn1 proteins of rat, hamster, mouse and human with a primary structure of
875–1,000 amino acids and a coil–coil domain form the transverse filaments
that synapses the lateral components into the synaptonemal complex. Null
mutants assemble only lateral components. Zip2 protein co-localizes with
Zip1 at sites where central regions polymerize. Zip3 co-localizes with the
other two Zip proteins and interacts with the proteins involved in DNA
recombination (Agarwal and Roeder 2000). It is a candidate constituent of
recombination nodules inserted during polymerization of the central
region. Electron microscopic analyses have shown that half of the central
region with the central component attaches first to one lateral component in
the pairing fork. Then the free surface of the central region associates with
the homologous segment of the other lateral component (von Wettstein
1977). Further studies can now determine how the Zip proteins accomplish
the assembly of the complex between the homologous lateral components.

Proteins of the lateral component Cor1 in hamster and ScP3 in rat carry
coiled coil domains. They localize to the lateral components of unsynapsed
chromosomes and remain with chromosome cores until metaphase I. In
yeast Red1 is associated with unsynapsed and synapsed lateral elements but
in a discontinuous pattern. It is required for the formation of the lateral elements and dissociates from the chromosomes, when the synaptonemal complex is disassembled.

Double strand breaks have been identified at recombination hotspots in
S. cerevisae (Sun et al. 1989) and led to support the double strand-break
repair model of meiotic recombination (Szostak et al. 1983; Sun et al. 1991).
Exonucleolytic digestion at the double strand break exposes single stranded
tails with 3′ termini, which invade an homologous DNA double strand.

Repair synthesis followed by branch migration produces two Holliday junctions. Resolution of the junctions in opposite direction results in a reciprocal crossing-over between markers that flank the region of strand exchange. A non-crossover (gene conversion) is formed when the strands are cleaved between the two Holliday junctions. In yeast many of the intermediates postulated by the model have been demonstrated physically, and mutants blocking the different steps in the repair process have been identified [cf. Fig. 6 in Roeder (1997) and references cited therein]. The enzyme that cleaves the double strand DNA is Spo11, a type II topoisomerase that causes the break by a trans-esterification reaction, which covalently links the 5′ termini to the Spo11 protein. This reaction provides the possibility for reversal of the double strand break, if a suitable homologous DNA invasion partner for recombination is not found (Keeney et al. 1997). Apart from Spo11, mutations in eight different genes (Rad50, Mre11, Xrs2, Mer2, Mei4, Rec102/104/114) lead to failed induction of double strand breaks. Mutations in the genes Rad50, Mre11 and Com1/Sae2 prevent the exonucleolytic 5′ to 3′ digestion to yield single stranded tails by their covalent linkage to protein.

In yeast, four genes (*RAD51*, *RAD55*, *RAD57* and *DMC1*) encode DNA strand exchange enzymes with homology to the bacterial RecA enzymes. Mutations in all four of them lead to defects in the repair of the single strand tails after their invasion into a homologous double strand molecule. Isolated Rad51 protein accomplishes strand exchange *in vitro* when supplied with a single stranded DNA binding protein and the Rad55 and Rad57 protein. Rad52 and Rad54/Tid1 promote the annealing of complementary single strands. These repair processes would lead to the formation of a joint molecule with two Holliday junctions one on each side of the repaired DNA strand exchange. Such joint molecules have been isolated by two-dimensional gel electrophoresis (Collins and Newlon 1994; Schwacha and Kleckner 1994, 1995). Cleavage of the molecules with Holliday junction-cleaving enzymes of bacteria yielded duplex DNA molecules, half of which were recombinant for the flanking markers as might be expected for random resolution of the two Holliday junctions.

The single strand tails formed by the meiotic recombination pathway in budding yeast starting with double strand breaks are involved in a genome-wide search for homology leading to chromosome pairing as revealed by detection of sequences at ectopic locations (e.g. Lichten et al. 1987). This is in contrast to observations made using FISH that demonstrate homologue pairing in the absence of meiotic recombination (e.g. Loidl et al. 1994). In this context the analysis of the Hop2 protein is of relevance (Tsubouchi and Roeder 2003). Mutation or deletion of this protein causes synaptonemal complex formation between non-homologous chromosomes and accumulation of complexes of organized central region material. Also double strand breaks are not repaired. The strand-exchange enzymes Rad51 and Dmc1 accumulate to aberrantly high levels on the chromosome. Disruption of the Rad51 and Dmc1 genes suppresses the homologue-pairing defect of *hop2* mutations. Additionally over-expression of Rad51 can suppress the deficiency of double strand-break repair caused by mutation in the Dmc1 gene. The conclusion is that double strand breaks and recombination are required for homologous pairing and that non-crossover and axial association leading to crossing-over can be obtained by two pathways, one involving Dmc1+Rad51 followed by Hop2+Mnd1 and one by Rad51 alone.

The *SPO11* gene of *Sordaria macrospora* has been identified recently (Storlazzi et al. 2003). A knockout mutant and a mutation changing the active site tyrosine of the double strand-breaking transesterase were molecularly constructed, as was a transformant expressing green fluorescent protein (GFP)-tagged enzyme (Spo11-GFP). The enzyme appears on the chromosomes during karyogamy of the two haploid nuclei and is present as numerous foci on the leptotene chromosomes during presynaptic alignment as well as during synapsis. Then the number of foci decreases and at mid to late pachytene Spo11-GFP is spread over the entire nucleus. Antibodies against Rad51 mark 60–50 foci from late leptotene to early zygotene, considered as markers for double strand breaks. Some of these are processed into crossing-overs seen as ~21 chiasmata at diplotene. This then is reminiscent of the transient appearance and random location of recombination nodules at zygotene in *Bombyx* males and human meiocytes and their reduction, relocation and conversion into chiasmata from pachytene to diplotene. The *Sordaria spo11* knockout deletion displays only a few Rad51 foci and a ~500-fold reduction in the number of chiasmata. Crossing-over between two spore colour markers is reduced from 44 to 2%. The lateral elements of the synaptonemal complexes are formed normally, but presynaptic co-alignment, so characteristic for *Sordaria* homologues and synaptonemal complex formation is absent. Most remarkably, exposure of young fruiting bodies of the knockout mutant to 200 Gy of γ-rays led to the appearance of many Rad51 foci at leptotene, restoration of homologous synapsis and formation of chiasmata. Thus γ-ray-induced double strand breaks can substitute for the double strand breaks formed by the transesterification reaction of the Spo11 topoisomerase. It is likely that the homologous pairing and synaptonemal complex formation in Bombyx females and polyploid plants also is initiated at leptotene and zygotene by topoisomerase-formed double strand breaks, but that these organisms have evolved mechanisms to prevent the consequences of multivalent formation at diplotene by untimely resolutions into reciprocal crossing-overs. This is clearly demonstrated by the reorganization of multivalent associations with the synaptonemal complex into optimal bivalent formations at pachytene. With the cloned genes available it will now be possible to determine whether these repair processes involve reversal of the trans-esterification reaction or inhibition/elimination of enzymes converting the double strand breaks into reciprocal crossing-overs.

3 Perspectives

1. Regarding the **biosynthesis of the photosynthetic membrane and chloroplast biogenesis**, many exciting new discoveries are being built on the knowledge I have briefly reviewed. Especially interesting are advances made by Kenneth Hoober, who has uncovered with the aid of loop forming, synthetic peptides, the fifth ligand that has to be provided to the Mg atom of chlorophyll by amino acid side chains or backbones in the reaction centres of photosystem I and II and in the light-harvesting proteins of algae and plants. We are beginning to understand why the different chlorophyll species *a*, *b*, *c* and *d* have evolved.

The protein domains with diverse functions in microbes, animals and humans that turned up in the complex inserting Mg into protoporphyrin IX during chlorophyll synthesis reminds us that there are only ~30,000 genes in the genomes of *Arabidopsis*, rice, barley, cows and humans. Thus the domains of this limited number of genes are "re-used" for many different functions. Clearly DNA homologies from genome sequencing projects provide an initial indication of possible functions, but detailed biochemical, physiological and crystallographic analyses are required more than ever to really understand the function of individual genes and proteins in the context of the cell biology of a species.

2. The research results of Storlazzi and her co-workers on **chromosome pairing and the mechanism of crossing-over** in *Sordaria* provides a new link in the DNA double strand-break-repair process of meiotic recombination together with the cell biology results of chromosome synapsis with the synaptonemal complex and formation of chiasmata from nodules effecting cross-overs. Intriguing questions will be solved using recombinant proteins made from the cloned genes of the components of the synaptonemal complex which are now available. Examples of this are:

1. How is the equal length of the lateral components of a pair of homologues established in a nucleus?
2. What proteins are involved in moving the double strand breaks to find homologous partners and how are these attached to the recombination nodules or to the lateral components of the synaptonemal complex?

3. How does the synaptonemal complex requiring DNA homology at zygotene differ in its molecular structure from that reorganized at later stages to optimize "bivalent" associations, even if these combine non-homologous chromosome segments?
4. At what stage is the repair process or resolution of the two Holliday junctions halted or aborted to allow delay until pachytene in order to ensure that a bivalent does not get more than one or two crossing-overs per chromosome arm but also that every pair at least obtains one to ensure disjunction. Three-dimensional crystal structures and use of additional organisms will be helpful in these endeavours.

An Apology and Special Thanks

I would like to apologize to those students, colleagues, postdoctoral collaborators and friends, including my wife Penny, whose inspiration, efforts and results during the years in Stockholm, Copenhagen and Pullman I could not include and mention. The space provided for the present publication prohibited this. It will have to be done on another occasion.

Special thanks and appreciation are due to Inge Sommer, Lisbeth Svarth, and Inger Braase who through their administrative and writing skills made life easy for me, and also to the many guest researchers and students with whom I have worked. I must also mention Nina Rasmussen for her excellent graphics and the talented photogapher Ann-Sofi Steinholz. Equally important for the scientific achievements discussed were the dedicated and enthusiastic laboratory technicians. I would like to mention on this occasion Ulla Edén, Sven Møller, Klaus Barr, Kirsten Kristiansen, Bibi Stampe Sørensen and Bent Hansen.

References

Agarwal S, Roeder GS (2000) Zip3 provides a link between recombination enzymes and synaptonemal complex proteins. Cell 102:245–255

Beale SI, Gough SP, Granick S (1975) Biosynthesis of d-aminolevulinic acid from the intact carbon skeleton of glutamic acid in greening barley. Proc Natl Acad Sci USA 72:2719–2723

Bojko M (1985) Human meiosis IX. Crossing-over and chiasma formation in oocytes. Carlsberg Res Commun 50:43–72

Collins I, Newlon CS (1994) Meiosis-specific formation of joint DNA molecules containing sequences from homologous chromosomes. Cell 76:65–75

Dujon B, Rasmussen SW, Wettstein D von, et al. (1994) Complete sequence of yeast chromosome XI. Nature 369:371–378

Fodje MN, Hansson A, Hansson M, Olsen JG, Gough S, Willows, RD, Al-Karadaghi S (2001) Interplay between an AAA module and an integrin I domain may regulate the function of magnesium chelatase. J Mol Biol 311:111–122

Gadjieva R, Axelson E, Olsson U, Vallon-Christersson J, Hansson M (2004) Nonsense-mediated mRNA decay in barley mutants allows the cloning of mutated genes by a microarray approach. Plant Physiol Biochem 42:681–684

Galibert F, Rasmussen SW, Wettstein D von, 54 other authors (1996) Complete nucleotide sequence of *Saccharomyces cerevisiae* chromosome X. EMBO J 15:2031–2049

Gibson LCD, Willows RD Kannangara CG, Wettstein D von, Hunter (1995) Magnesium-protoporphyrin chelatase of *Rhodobacter sphaeroides*: reconstitution of activity by combining the products of the *bchH, -I* and -*D* genes expressed in *Escherichia coli*. Proc Natl Acad Sci USA 92:1941–1944

Gough S (1972) Defective synthesis of porphyrins in barley plastids caused by mutation in nuclear genes. Biochim Biophys Acta 286:36–54

Gräfe S, Saluz H-P, Grimm B, Hanel F (1999) Mg-chelatase of tobacco: the role of the subunit CHLD in the chelation step of protoporphyrin IX. Proc Natl Acad Sci USA 87:4169–4173

Hansson A, Kannangara CG, Wettstein D von, Hansson M (1999) Molecular basis for semidominance of missense mutations in the XANTHA-H (42kDa) subunit of magnesium chelatase. Proc Natl Acad Sci USA 96:1744–1749

Henningsen KW, Boynton JE, Wettstein D von (1993) Mutants at *Xantha* and *albina* loci in relation to chloroplast biogenesis in barley (*Hordeum vulgare* L.). Biol Skr 42:1–349

Henry LEA, Møller BL, Anderson B, Åkerlund HE (1982) Reactivation of photosynthetic oxygen evolution in TRIS-inactivated inside-out photosystem II vesicles from spinach. Carlsberg Res Commun 47:187–198

Hobolth P (1981) Chromosome pairing in allohexaploid wheat var. Chinese spring. Transformation of multivalents into bivalents, a mechanism for exclusive bivalent formation. Carlsberg Res Commun 46:129–173

Höfgen R, Axelsen KB, Kannangara CG, Schüttke, I, Pohlenz H-D, Willmitzer L, Grimm B, Wettstein D von (1994) A visible marker for antisense mRNA expression in plants: inhibition of chlorophyll synthesis with a glutamate-1-semialdehyde aminotransferase antisense gene. Proc Natl Acad Sci USA 91:1726–1730

Holder AA (1978) Peptide mapping of the ribulose bisphosphate carboxylase large subunit from the genus *Oenothera*. Carlsberg Res Commun 43:391–399

Holm PB (1977) Three-dimensional reconstruction of chromosome pairing during the zygotene stage of meiosis in *Lilium longiflorum* (Thunb.). Carlsberg Res Commun 42:103–151

Holm PB (1986) Chromosome pairing and chiasma formation in allohexaploid wheat, *Triticum aestivum* analyzed by spreading of meiotic nuclei. Carlsberg Res Commun 51:239–294

Holm PB, Wang X (1988) The effect of chromosome 5B on synapsis and chiasma formation in wheat, *Triticum aestivum* cv. Chinese spring. Carlsberg Res Commun 53:191–208

Jenkins G (1983) Chromosome pairing in *Triticum aestivum* cv. Chinese spring. Carlsberg Res Commun 48:255–283

Jensen PE, Willows RD, Larsen Petersen B, Vothknecht UC, Stummann BM, Kannangara CG, Wettstein D von, Henningsen KW (1996) Structural genes for Mg-chelatase subunits in barley: *Xantha-f, -g* and -*h*. Mol Gen Genet 250:383–394

Kahn A, Avivi-Bleiser N, Wettstein D von (1976) Genetic regulation of chlorophyll synthesis analyzed with double mutants in barley. In: Bücher T., et al. (eds) Genetics and biogenesis of chloroplasts and mitochondria. Elsevier, North Holland Biomedical Press, Amsterdam pp 119–131

Kannangara CG, Andersen RV, Pontoppidan BR Willows RD, Wettstein D von (1994) Enzymic and mechanistic studies on the conversion of glutamate to 5-aminolaevulinate. In: The Biosynthesis of the Tetrapyrrole Pigments. Ciba Foundation Symposium no. 180. Wiley, Chichester, pp 3–25

Kannangara CG, Vothknecht UC, Hansson M, Wettstein D von (1997) Magnesium chelatase: association with ribosomes and mutant complementation studies identify barley subunit

Xantha-G as a functional counterpart of *Rhodobacter* subunit BchD. Mol Gen Genet 254: 85–92

Keeney S, Giroux CN, Kleckner N (1997) Meiosis-specific DNA double strand breaks are catalyzed by Spo11, a member of a widely conserved gene family. Cell 88:375–384

Knoetzel J, Svendsen I, Simpson DJ (1992) Identification of the photosystem I antenna polypeptides in barley. Eur J Biochem 206:209–215

Kroll D, Meierhoff K, Bechtold N, Kinoshita M, Westphal S, Vothknecht UC, Soll J, Westhoff P (2001) Vipp1, a nuclear gene of *Arabidopsis thaliana* essential for thylakoid membrane formation. Proc Natl Acad Sci USA 98:4238–4242

Li H, Kaneko Y, Keegstra K (1994) Molecular cloning of a chloroplastic protein associated with both the envelope and thylakoid membranes. Plant Mol Biol 25:619–632

Lichten M, Borts RH, Haber JE (1987) Meiotic gene conversion and crossing over between dispersed homologous sequences occurs frequently in *Saccharomyces cerevisiae*. Genetics 115:233–246

Loidl J, Klein F, Shertan H (1994) Homologous pairing is reduced but not abolished in asynaptic mutants of yeast. J Cell Biol 125:1191–1200

McClintock B (1933) The association of non-homologous parts of chromosomes in the mid-prophase of meiosis in *Zea mays*. Z Zellforsch Mikrosk Anat 19:191–237

McIntosh L, Poulsen C, Bogorad L (1980) Chloroplast gene sequence for the large subunit of ribulose bisphosphate carboxylase of maize. Nature 288:556–560

Melchers G, Sacristán MD, Holder AA (1978) Somatic hybrid plants of potato and tomato regenerated from fused protoplasts. Carlsberg Res Commun 43:203–218

Møller BL (1985) The photosynthetic membrane. In: Sixteen Research Reports by the Niels Bohr Fellows of the Royal Danish Academy of Sciences and Letters. K Dan Vidensk Selsk Biol Skr 25:123–172

Møller BL, PB Høj (1983) A thylakoid polypeptide involved in the reconstitution of photosynthetic oxygen evolution. Carlsberg Res Commun 48:161–165

Oliver SG, Rasmussen SW, Wettstein D von, et al. (1992) The complete DNA sequence of yeast chromosome III. Nature 357:38–46

Pontoppidan B, Kannangara CG (1994) Purification and partial characterisation of barley glutamyl-tRNA^{Glu} reductase, the enzyme that directs glutamate to chlorophyll biosynthesis. Eur J Biochem 225:529–537

Poulsen C (1983) The barley chloroplast genome. Carlsberg Res Commun 48:57–80

Poulsen C (1984) Two mRNA species differing by 258 nucleotides at the 5¢ end are formed from the barley chloroplast rbcL gene. Carlsberg Res Commun 49:89–104

Poulsen C, Martin B, Svendsen I (1979) Partial amino acid sequence of ribulose bisphosphate carboxylase from barley. Carlsberg Res Commun 44:191–199

Poulsen C, Porath D, Sacristán MD, Melchers G (1980) Peptide mapping of the ribulose bisphosphate carboxylase small subunit from the somatic hybrid of tomato and potato. Carlsberg Res Commun 45:249–267

Rasmussen SW, Holm PB (1984) The synaptonemal complex, recombination nodules and chiasmata in human spermatocytes. Symp Soc Exp Biol 38:271–292

Roeder GS (1997) Meiotic chromosomes: it takes two to tango. Genes Dev 11:2600–2621

Schön A, Krupp G, Gough S, Berry-Lowe S, Kannangara CG, Soll D (1986) The RNA required in the first step of chlorophyll biosynthesis is a chloroplast tRNA. Nature 332:281–284

Schwacha A, Kleckner N (1994) Identification of joint molecules that form frequently between homologs but rarely between sister chromatids during yeast meiosis. Cell 76:51–63

Schwacha A, Kleckner N (1995) Identification of double Holliday junctions as intermediates in meiotic recombination. Cell 83:783–791

Simpson DJ, Wettstein D von (1980) Macromolecular physiology of plastids XIV. Viridis mutants in barley: genetic, fluoroscopic and ultrastructural characterization. Carlsberg Res Commun 45:283–314

Simpson DJ, Wettstein D von (1989) The structure and function of the thylakoid membrane. Carlsberg Res Commun 54:55–65

Simpson DJ, Machold O, Høyer-Hansen G, Wettstein D von (1985) Chlorina mutants of barley (*Hordeum vulgare* L.). Carlsberg Res Commun 50: 223–238

Simpson DJ, Vallon O, Wettstein D von (1989) Freeze-fracture studies on barley plastid membranes VIII. In viridis[115], a mutant completely lacking photosystem II, oxygen evolution enhancer 1 (OEE1) and the a-subunit of cytochrome b559 accumulate in appressed thylakoids. Biochim Biophys Acta 975:164–174

Smillie RM, Henningsen KW, Bain JA, Critchley C, Fester T, Wettstein D von (1978) Mutants of barley heat-sensitive for chloroplast development. Carlsberg Res Commun 43:351–364

Storlazzi A, Tessé, S, Gargano S, James F, Kleckner N, Zickler D (2003) Meiotic double strand breaks at the interface of chromosome movement, chromosome remodeling, and reductional division. Genes Dev 17:2675–2687

Sun H, Treco D, Schultes NP, Szostak JW (1989) Double strand breaks at an initiation site for meiotic gene conversion. Nature 338:87–90

Sun H, Treco D, Szostak JW (1991) Extensive 3′-overhanging, single-stranded DNA associated with the meiosis specific double strand breaks at the *ARG4* recombination initiation site. Cell 64:1155–1161

Szostak JW, Orr-Weaver TL, Rothstein RJ, Stahl FW (1983) The double strand-break repair model for recombination. Cell 33:25–35

Tsubouchi H, Roeder GS (2003) The importance of genetic recombination for fidelity of chromosome pairing in meiosis. Dev Cell 5:915–925

Vothknecht UC, Kannangara CG, Wettstein D von (1996) Expression of catalytically active barley glutamyl tRNA[Glu] reductase in *Escherichia coli* as a fusion protein with glutathione S-transferase. Proc Natl Acad Sci USA 93:9287–9291

Vothknecht UC, Kannangara CG, Wettstein D von (1998) Barley glutamyl tRNA[Glu] reductase: mutations affecting haem inhibition and enzyme activity. Phytochemistry 47:513–519

Wang X (1988) Chromosome pairing analysis in haploid wheat by spreading of meiotic nuclei. Carlsberg Res Commun 53:135–166

Wang X, Holm PB (1988) Chromosome pairing and synaptonemal complex formation in wheat-rye hybrids. Carlsberg Res Commun 53:167–190

Westergaard M, Wettstein D von (1966) Mechanism of crossing over. III. On the ultrastructure of the chromosomes in *Neottiella rutilans* (Fr.) Dennis. C R Trav Lab Carlsberg 35:233–286

Westergaard M, Wettstein D von (1970) Studies on the mechanism of crossing-over. IV. The molecular organization of the synaptinemal complex in *Neottiella* (Cooke) *Saccardo* (ascomycetes). C R Trav Lab Carlsberg 37:239–268

Westergaard M, Wettstein D von (1972) The synaptinemal complex. Annu Rev Genet 6:71–110

Westphal S, Heins L, Soll J, Vothknecht UC (2001) *Vipp1* deletion mutant of *Synechocystis*: a connection between bacterial phage shock and thylakoid biogenesis. Proc Natl Acad Sci USA 98:4243–4248

Wettstein D von (1959) The effect of genetic factors on the submicroscopic structures of the chloroplast. J Ultrastruct Res 3: 234–240

Wettstein D von (1971) The synaptinemal complex and four strand crossing-over. Proc Natl Acad Sci USA 68: 851–855

Wettstein D von (1977) The assembly of the synaptinemal complex. Phil Trans R Soc Lond B 277:235–243

Wettstein D von (1981) The Emil Heitz lecture. Chloroplast and nucleus: concerted interplay between genomes of different cell organelles. In: Schweiger HG (ed) International cell biology 1980–1981. Springer, Berlin Heidelberg New York, pp 250–272

Wettstein D von (1983) Genetic engineering in the adaptation of plants to evolving human needs. Experientia 39:687–713

Wettstein D von (1984) The synaptonemal complex and genetic segregation. Symp Soc Exp Biol 38:195–231

Wettstein D von (2000a) Chlorophyll biosynthesis. I. From analysis of mutants to genetic engineering of the pathway. In: Kung S-D, Yang S-F (eds) Discoveries in plant biology, vol 3. World Scientific, Singapore, pp 75–93

Wettstein D von (2000b) Chlorophyll biosynthesis. II. Adventures with native and recombinant enzymes. In: Kung S-D, Yang S-F (eds) Discoveries in plant biology, vol 3 World Scientific, Singapore, pp 95–139

Wettstein D von (2001) Commentary: discovery of a protein required for photosynthetic membrane assembly. Proc Natl Acad Sci USA 98:3633–3635

Wettstein D von, Kahn A, Nielsen OF, Gough S (1974) Genetic regulation of chlorophyll synthesis analyzed with mutants in barley. Science 184:800–802

Wettstein D von, Poulsen C, Holder AA (1978) Ribulose-1,5-bisphosphate carboxylase as a nuclear and chloroplast marker. Theor Appl Genet 53:193–197

Wettstein D von, Gough S, Kannangara CG (1995) Chlorophyll biosynthesis. Plant Cell 7:1039–1057

Willows RD, Gibson LCD, Kannangara CG, Hunter CN, Wettstein D von (1996) Three separate proteins constitute the magnesium chelatase of *Rhodobacter sphaeroides*. Eur J Biochem 235:438–443

Wischmann B (1986) Chromosome pairing and chiasma formation in wheat plants triisosomic for the long arm of chromosome 5B. Carlsberg Res Commun 51:1–25

Zakhrabekova S, Kannangara, CG, Wettstein D von, Hansson M (2002) A microarray approach for identifying mutated genes. Plant Physiol Biochem 40:189–197

Zickler D, Kleckner N (1998) The leptotene-zygotene transition of meiosis. Annu Rev Genet 32:619–697

Zickler D, Kleckner N (1999) Meiotic chromosomes: integrating structure and function. Annu Rev Genet 33:603–754

Diter von Wettstein
Department of Crop & Soil Sciences
Washington State University
Pullman
WA 99164-6420
USA
Tel.: +1-509-3353635
Fax: +1-509-3358674
e-mail: diter@wsu.edu

Genetics

Recombination:
Cytoplasmic male sterility and fertility restoration in higher plants

Renate Horn

1 Introduction

Cytoplasmic male sterility (CMS) is a widespread maternally inherited trait in higher plants that results from the expression of novel, often chimeric genes located in the mitochondrial genome (Schnabel and Wise 1998). In many cases, specific dominant nuclear genes, termed restorers of fertility (Rf), have been identified that suppress the male sterility phenotype and restore fertility to plants carrying CMS mitochondrial genomes. While the mitochondrial genes that are associated with male sterility have been identified for a number of CMS systems, we have only now started to learn something about the molecular features of restorer genes and the proteins encoded by them (Schnable and Wise 1998; Hanson and Bentolila 2004).

For CMS/fertility restorer systems, two basic systems can be differentiated: sporophytic and gametophytic (Tang et al. 1999). In the sporophytic system, fertility restoration genes manifest their effects in the sporophytic tissues, such as in tapetal cells or pollen mother cells. Fertility restoration can result in the production of viable haploid pollen even though individual gametes may not carry the fertility restorer gene. In contrast, gametophytic sterility is expressed at the post-meiotic haploid stage, and viability of the gamete is determined by the genotype of the gamete, hence the presence of the restorer allele in the gamete.

Cytoplasmic male sterility and fertility restoration represent important agronomic traits in crops such as maize, sunflower, rice and rapeseed, which are essential for the production of hybrid seeds on a commercial scale. Hybrid breeding allows exploitation of heterosis and by this leads to higher yields and more yield stability. In addition to facilitating the commercial exploitation of CMS-Rf systems, detailed studies of CMS and Rf genes provide us with information that can increase our understanding of nuclear-cytoplasmic interactions (Budar 1998).

Progress in Botany, Vol. 67
© Springer-Verlag Berlin Heidelberg 2006

2 Mechanism of cytoplasmic male sterility

2.1 Open reading frames identified as cause of male sterility

Open reading frames associated with cytoplasmic male sterility often represent chimeric genes that seem to originate from multiple recombination events involving frequently known mitochondrial genes as well as their 5'- and 3'-flanking region in addition to sequences of unknown origin.

In maize, the chimeric *T-urf13* gene of the T-cytoplasm evolved through several recombination events and consists of coding and 3'-flanking region of the 26S rRNA gene as well as nine amino acids of unknown origin (Dewey et al. 1987). The recombinations have placed the *T-urf13* adjacent to the 5'-flanking region of the *atp6* gene, which enables the transcription of *T-urf13*. The *T-urf13* gene is located upstream of a conserved mitochondrial gene originally termed *orf221*, which has recently been demonstrated to be the membrane-bound gene product of the *atp4* gene (Heazlewood et al. 2003). In the CMS-S, two adjacent mitochondrial open reading frames *orf355* and *orf77* are associated with cytoplasmic male sterility (Zabala et al 1997). The 1.6 kb co-transcript of *orf355* and *orf77* is present in microspores of CMS-S plants and absent in microspores of male-fertile plants recovered from CMS-S plants by mitochondrial mutation. The *orf77* contains three segments derived from the mitochondrial ATP synthase subunit 9 (*atp9*) locus (Zabala et al. 1997). Editing of the cotranscript and the *atp9* gene was investigated (Gallagher et al. 2002).

In *Sorghum bicolor*, the CMS-specific open reading frame of the 9E cytoplasm also represents a chimeric gene. At least two recombination events lead to the enlargement of the *coxI* gene in the 5'-as well as in the 3'-region in this CMS type (Bailey-Serres et al. 1986a,b).

In chive, the CMS1 configuration is twice present in the mitochondrial genome, but with different 5'-regions (Engelke and Tatglioglu 2004). The two sequences are designated CMS_1-1 and CMS_1-2. The CMS1 configuration is derived in part from sequences of the essential genes *atp9* and *atp6*. Three open reading frames are predicted *orf780*, *orf744* and *orf501* using the three possible start codons, which encode predicted proteins of 29 kDa, 27.5 kDa and 19 kDa, respectively. The gene product of *orf501* would correspond to the size of the 18 kDa protein (based on the electrophoretic mobility) associated with the CMS_1 phenotype (Potz and Tatglioglu 1993).

In petunia, the *pcf* gene consists of 5'-flanking and 5'-coding region of the *atp9* gene, coding sequences of the *coxII* gene as well as the sequences of an open reading frame *urfS* (Young and Hanson 1987).

In rapeseed, for four CMS types (Polima, Ogura, Kosena and Tournefortii-Stiewe) open reading frames associated with male sterility have been characterized. In the Polima cytoplasm, the CMS-phenotype is caused

by a new open reading frame *orf224*, which required at least three recombinations events (Handa et al. 1995). This *orf224* consists of the 5' flanking
region of *orf158*, also known as *orfB* (Hiesel et al. 1987), which could recently
be demonstrated to be the subunit 8 of F_1F_0-ATP synthase (Sabar et al.
2003), a part of the exon 1 of the ribosomal protein S3 (*rps3* gene) and
unknown sequences (Handa et al. 1995). In the Ogura cytoplasm in rapeseed, the CMS phenotype was correlated with a 2.5 kb *NcoI* fragment
(Bonhomme et al. 1992). On this fragment, three open reading frames
orf158, *trnfM* and *orf138* were identified. The *orf138* was finally associated
with the male sterility (Bonhomme et al. 1992). The *orf125* correlated with
the CMS phenotype of the Kosena cytoplasm is very similar to the *orf138*
(Krishnasamy and Makaroff 1993; Bellaoui et al. 1999). These two open
reading frames only differ by two amino acid substitutions and a 39-bp deletion (Iwabuchi et al. 1999). For the Tournefortii-Stiewe CMS, a chimeric
gene, *orf193*, which encodes a predicted 22.7-kDa protein, exhibits partial
sequence identity to the *atp6* gene and is regarded as candidate gene for male
sterility (Dieterich et al. 2003). However, *orf193* is cotranscribed with one of
the newly identified *atp9* genes and might also be translated uninterrupted
into a chimeric 30.2 kDa protein.

In *Phaseolus vulgaris*, the unique *pvs*-region of the CMS Sprite contains
three open reading frames, *orf98*, *orf97* and *orf239* (Chase and Ortega 1992;
Johns et al. 1992). The *pvs*-region, which is flanked by the *atpA* gene and
sequences of the *cob* gene, shows no homology to nuclear DNA, but to the
intron of the plastidal tRNA alanine and a short part of the 5-kb repeat in
maize. The development of the *pvs*-region is probably the result of several
recombination events.

In sunflower CMS PET1, the CMS-associated open reading frame
orfH522, which is localized in the 3'-region of the *atpA* gene, consists of the
first 57 bp of *orfB* (*atp8*) and of unknown sequences for the remaining part
(Köhler et al. 1991). The *orfH522* is cotranscribed with the *atpA* gene on an
additional larger transcript (Köhler et al. 1991). For the PEF1 cytoplasm, de
la Canal et al. (2001) identified an insertion of 0.5 kb of unknown sequences
in the 3'-region of the *atp9* gene as cause for male sterility. Multiple recombination events involving known mitochondrial genes as well as sequences
of unknown origin seem to create the open reading frame associated with
inducing male sterility in higher plants.

2.2 CMS-specific proteins and possible functions

For a number of CMS systems, CMS-associated proteins could be identified
for the corresponding open reading frames correlated to male sterility. In

other cases, only the presence or absence of a protein in the in organello translation products of CMS systems compared to the fertile cytoplasm was observed without any knowledge about the corresponding open reading frame.

In maize, Dewey et al. (1987) identified an additional protein of 13 kDa to be expressed in the T-cytoplasm, which is encoded by the unique coding region *T-urf13*. The 13-kDa protein represents a prominent mitochondrially encoded protein, which confers sensitivity to the T-toxin. The T-toxin produced by the T race of *Cochliobolus heterostrophus*, causing the southern corn leaf blight, interacts with the protein, which is located in the inner mitochondrial membrane in an oligomeric arrangement. This induces the formation of a pore that makes the membrane leaky (Dewey et al. 1987; Korth et al. 1991). The binding of the pathotoxin renders the mitochondria incapable of performing oxidative phosphorylation (Kaspi and Siedow 1993; Rhoads et al. 1995). For the disruption of pollen development, it is assumed that the 13-kDa protein might interact with an anther-specific substance in a way similar to that observed with the T-toxin (Flavell 1974). For the CMS-C in maize, an additional 17.5 kDa protein was identified. For CMS-S, complex changes were observed with eight additional proteins that were all larger in size than 42 kDa (Forde and Leaver 1980).

In *Sorghum bicolor*, Bailey-Serres et al. (1986b) investigated the in organello translation products of a number of CMS systems. For CMS Milo, they identified an additional 65 kDa protein, for CMS 9E an additional 42 kDa protein and the absence of a 38 kDa protein (Dixon and Leaver 1981). For CMS IS II2 and MS M35-1(B), additional proteins of 12 kDa and 82 kDa, respectively, were detected (Bailey-Serres et al. 1986b).

In petunia, Nivison and Hanson (1989) identified the gene product of the CMS-associated *pcf* gene by using antibodies produced against a synthetic oligopeptide. The 43 kDa protein encoded by *pcf* is post-translationally processed at the N-terminus to give a 25-kDa protein (Nivison et al. 1994).

In rapeseed, a 19-kDa protein was identified as the product of the CMS-associated *orf138* of the Ogura cytoplasm (Grelon et al. 1994), and for the Juncea CMS type a 32 kDa protein was related to the male sterile phenotype (Landgren et al. 1996). The *orf125* of the Kosena CMS encodes a 17 kDa protein (Iwabuchi et al. 1999). An antibody produced against ORF125 was used to demonstrate that the accumulation is reduced in fertility restored hybrids (Koizuka et al. 2000).

Also in *Phaseolus vulgaris*, antibody allowed to detect the postulated gene product of the CMS-associated *pvs-orf239*. Apart from the predicted protein of 27.5 kDa, a second protein of 21 kDa could be identified, which might be a degradation product. However, different to all other CMS systems, in which the CMS-specific proteins are also expressed in the vegetative tissue, the 27.5 kDa protein in *Phaseolus vulgaris* can only be detected in the pollen mother cells and the developing microspores (Abad et al.1995).

In sunflower, a new open reading frame *orfH522* in the 3'-flanking region of the *atpA* gene could be associated with the CMS phenotype PET1 (Köhler et al. 1991; Laver et al. 1991). Using specific antibodies against the gene product of *orfH522* it was demonstrated that *orfH522* encodes the 16-kDa protein (Monéger et al. 1994; Horn et al. 1996) which represents the only difference between the in organello translation products of fertile and male-sterile lines (Horn et al. 1991). The 16-kDa protein is membrane-bound (Horn et al. 1996) and its expression is specifically reduced in the anthers of fertility restored hybrids (Monéger et al. 1994). The 16-kDa protein seems to be involved in initiating premature programmed cell death in tapetum cells via release of cytochrome C by the mitochondria (Balk and Leaver 2001). Comparing the mitochondrially encoded proteins of 28 CMS sources in sunflower, nine additional CMS sources could be identified that also have the same CMS mechanism as PET1 (Horn et al. 1996). This was a surprise, as these PET1-like CMS sources had different origins (Horn and Friedt 1999).

According to Serieys (1996), these cytoplasmic male sterile germplasms had been produced by either different interspecific crosses involving *H. argophyllus* (ARG1), *H. neglectus* (NEG1), *H. exilis* (EXI2), *H. anomalus* (ANO1), and two subspecies of *H. praecox* (PRR1, PRH1), or by mutagenesis of two maintainer lines for the PET1 cytoplasm (MUT1 and MUT2). In addition, one of the CMS types that arose spontaneously (ANN10) expressed the 16-kDa protein. All these PET1-like CMS cytoplasms showed the same organization at the *atpA* locus (Horn and Friedt 1999).

Apart from these PET1-like CMS sources, other groups of CMS cytoplasms could be identified which expressed new proteins (Horn and Friedt 1999). ARG3 and RIG1 showed an additional 16.9-kDa protein but missed a 17.5-kDa protein common to the other cytoplasms. ANN1 and ANN3 expressed three specific proteins of 34.0, 16.9 and 16.3 kDa in common. A protein of 12.4 kDa was unique for PET2 and GIG1. Although no specific open reading frame has been identified for those proteins yet, investigations of the mitochondrial DNA level as well as the differential fertility restoration pattern support the grouping of the CMS sources (Horn 2002, Horn et al. 2002).

CMS-associated proteins were also identified in *Allium schoenoprasum* (Potz and Tatglioglu 1993), *Beta vulgaris* (Boutry et al. 1984), *Nicotiana tabacum* (Håkansson et al. 1988), *Triticum aestivum* (Boutry et al. 1984), *Daucus carota* (Scheike et al. 1992) and *Vicia faba* (Boutry et al. 1984).

The reported molecular weights of proteins associated with cytoplasmic male sterility range from 12 kDa to 82 kDa. For some of the identified proteins, their presence or absence is only characterized by the comparison to the fertile cytoplasm but no direct correlation or verification of their involvement was yet performed by looking at the effects of the restorer genes on their expression.

3 Mechanism of fertility restoration

3.1 Genetics and functions of fertility restorer genes

The effect of restorer genes on transcripts and proteins of CMS-associated open reading frames has been studied intensively. Table 1 gives an overview about the functions described for restorer genes, so far. Most of the restorer genes seems to act on the RNA level or even translation or post-translational level. In addition, different marker systems (RAPD, AFLP, RFLP, SSR) have been applied to map restorer genes, to obtain markers for marker-assisted breeding and markers for positional cloning approaches in order to isolate restorer genes. Table 2 gives an overview about the marker analysis performed with regard to restorer genes. For some species the localization of different restorer genes in the genome is known.

In maize, fertility restoration of the T-cytoplasm requires two dominant, complementary nuclear encoded restorer genes, *Rf1* and *Rf2* (Levings and Dewey 1988). Approaches to clone the restorer genes *Rf1* and *Rf2* were made via map-based cloning as well as transposon-tagging. First, both genes could be mapped with closely linked RFLP-markers in five mapping populations (Wise and Schnable 1994). The *Rf1* gene mapped on chromosome 3 between the molecular markers umc97 and umc92, *Rf2* was located between the markers umc153 and sus1 on chromosome 9.

In parallel to the marker-assisted approach, the restorer gene *Rf2* was successfully identified by transposon-tagging (Schnable and Wise 1994). As a first step, 178,300 plants carrying the transposon families Cy and Spm were screened for a mutated *Rf2* allele (*Rf2-m*). As mutated *Rf2* alleles lost the ability to restore fertility, a functional product of the *Rf2* genes seems to be required. The seven *Rf2-m* alleles obtained by transposon-tagging are independent events according to the corresponding RFLP-analyses.

The cloned *Rf2* gene represents an aldehyde dehydrogenase, which acts by detoxifying toxic substances in the tapetum and by this allows the production of functional pollen (Cui et al. 1996; Liu et al. 2001). Complete (or

Table 1. Overview of the function described for restorer genes

Species	CMS	Restorer gene	Function	Reference
Bean	CMS-Sprite	*Fr*	Loss of CMS-specific region	Janska and Mackenzie (1993)
		Fr2	Post-transcriptional	Abad et al. (1995)
Maize	CMS-T	*Rf1*	RNA-processing or post-transcriptional	Kennell and Pring (1989)
		Rf2	AlDH, physiological function	Cui et al. (1996)
Petunia	RM	*Rf*	Reduced transcript amounts	Pruitt and Hanson (1991)
Rapeseed	Polima	*Rfp1/Rfp2*	RNA-processing	Handa and Nakajima (1992)
	Ogura	*Rfo*	Post-transcriptional	Krishnasamy and Makaroff (1994)
Rice	CMS-BT	*Rf-1*	RNA-processing	Iwabuchi et al. (1993)
Sunflower	PET1	*Rf1*	RNA stability	Monéger et al. (1994)

partial) restoration of fertility in presence of the T-cytoplasm is only possible if the *Rf2* gene is combined with one of three other restorer genes: *Rf1*, *Rf8* or *Rf**. Each of them results in a specific change of the processing of the T-*urf13* transcript (Wise et al. 1999). Also for the *Rf1* gene in maize, the identification of transposon-tagged mutated alleles was successful (Wise et al. 1996). All four *rf1-m* alleles in the male sterile plants cosegregated with an increased steady-state accumulation of the 1.6- and 0.6-kb-transcripts of the *T-urf13*. These transcripts represent processing derivates of the 2.0 or 1.8 kb transcripts, which are characteristic for the T-cytoplasm (Dewey et al. 1987). A functional gene product of the *Rf1* gene seems to be necessary, as in the case of the *Rf2* gene, to change the transcript pattern of the *T-urf13* gene and to reduce the CMS-specific polypeptide. *Rf8* leads to the accumulation of two transcripts of 1.42 and 0.42 kb, *Rf** of T-*urf13*-transcripts with 1.4 and 0.4 kb (Dill et al. 1997). The 5'-ends of the two transcript groups are only 22 nucleotides apart and show a conserved sequence motif 5'-CNACANNU-3'.

In the S-cytoplasm of maize, apart from the transcripts of 2.8 and 1.6 kb, which are typical for the CMS-specific region *orf355-orf77*, new additional

Table 2. Mapping activities and marker systems used for restorer genes

Plant species	CMS system	Restorer gene	Marker system	Reference
Beta vulgaris ssp. maritima	CMS H	*R1H*	RAPD, RFLP	Laporte et al. (1998)
Brassica napus	CMS polima	*Rfp1/Rfp2*	RAPD, RFLP	Jean et al. (1997)
	CMS ogura	*Rfo*	RAPD, RFLP	Delourme et al. (1998)
Gossypium hirsutum	CMS-D2	*Rf1*	RAPD, SSR	Liu et al. (2003)
	CMS-D8	*Rf2*	RAPD, STS	Zhang and Stewart (2004)
Helianthus annuus	PET1	*Rf1*	RFLP	Gentzbittel et al. (1995, 1999)
			RFLP, RAPD, AFLP	Berry et al. (1995), Horn et al. (2003), Kusterer et al. (2005)
	PEF1	*RF1-PEF1*	RAPD, isoenzyme	Quillet et al. (1995)
Hordeum vulgare	msm1	*Rfm1a*	RAPD, STS	Matsui et al. (2001)
Petunia hybrida	RM	*Rf*	RAPD, AFLP	Bentolila et al. (1998), Bentolila and Hanson (2001)
Phaseolus vulgaris	CMS-Sprite	*Fr, Fr2*	RAPD	He et al. (1995b)
Oryza sativa	CMS-BT	*Rf-1*	RFLP	Kurata et al. (1994), Ichikawa et al. (1997)
			PCR-marker	Komori et al. (2003)
	CMS-HL	*Rf5*	SSR	Huang et al. (2000)
		Rf6(t)	SSR, STS	Lui et al. (2004)
	CMS-L	*Rf2*	RFLP	Shinjyo and Sato (1994)

Table 2. *Continued*

Plant species	CMS system	Restorer gene	Marker system	Reference
	CMS-WA	*Rf3*	RAPD, RFLP	Zhang et al. (1997)
		Rf4	SSLP, RFLP	Yao et al. (1997), Jing et al. (2001), Zhang et al. (2002)
Secale cereale	CMS G	*Rfg1*	RAPD, RFLP	Börner et al. (1998)
	CMS P	*Rfp1/Rfp2*	AFLP, SCAR	Stracke et al. (2003)
Sorgum bicolor	A1 (milo)	*Rf1*	AFLP, SSR	Klein et al. (2001)
	A3	*Rf4*	AFLP	Wen et al. (2002)
Zea mays	CMS-T	*Rf1/Rf2*	RFLP	Wise and Schnable (1994)
		Rf8	AFLP	Wise et al. (1999)
	CMS-S	*Rf3*	RFLP	Kamps and Chase (1997)

transcripts of this region of 0.75, 1.1 and 2.1 kb were observed in developing microspores in the presence of the restorer gene *Rf3* (Wen and Chase 1999). The *Rf3* gene seems to have an influence on the processing of the CMS-specific transcripts. The *Rf3* gene responsible for restoring fertility in the presence of the S-cytoplasm in maize was mapped on the long arm of chromosome 2 in a distance of 4.3 cM from the *whp* Locus and 6.4 cM proximal to the RFLP-marker bnl17.14 (Kamps and Chase 1997).

In rice, so far six restorer genes for CMS cytoplasms have been localized in the genome (Shen et al. 1998; Huang et al. 2000). The restorer gene *Rf-1* is responsible for fertility restoration of the gametophytic cms-BT (also known as cms-bo) system (Shinjyo 1984). *Rf-1* mapped close to the marker G4003, and 3.7 cM from ORS33 on chromosome 10 (Akagi et al. 1996). In between, a candidate gene was isolated for *Rf-1* (Kazama and Toriyama 2003; Komori et al. 2004). The restorer gene *Rf2* of the CMS-L system is on chromosome 2 (Shinjyo and Sato 1994). Genetic studies of the sporophytic

cms-WA system in rice showed that two restorer genes are necessary for fertility restoration: *Rf3* on chromosome 1 and *Rf4* on chromosome 10 (Yao
et al. 1997; Zhang et al. 1997, 2002). *Rf4* mapped close to the S10019 region,
and 0.9 cM from the marker Y3-8. For the gametophytic cms-HL system, the
restorer genes, *Rf5* and *Rf6(t)*, could be localized on chromosome 10 in the
lines MY23 and 93-11, respectively. Linkage analysis revealed that these
genes are cosegregating with the SSR-markers, RM3150 and RM5373,
respectively and therefore are localized in some distance from *Rf1* and *Rf4*,
which are also on chromosome 10 (Liu et al. 2004).

A 105-kb BAC-clone, which carries the *Rf6(t)* locus, was isolated from the rice BAC library.
The candidate gene region was delimited to a 66 kb region by a combination of physical fine
mapping and BLASTX searches of the marker sequences in the genomic database
(http://btn.genomics.org.cn.rice).

In rye, the restorer gene *Rfg1* of the male sterility inducing G-cytoplasm
was mapped on chromosome 4RL distal of three RFLP markers (Xpsr119,
Xprs167, Xpsr899) and four RAPD markers (XP01, XAP05, XR11, XS10)
(Börner et al. 1998). *Rfg1* might be allelic to the gene, which is responsible
for fertility restoration of the P-cytoplasm and to *Rfc4*, which restores in rye
addition lines of chromosome 4RL the fertility in hexaploid wheat with the
T. timopheevi cytoplasm (Börner et al. 1998).

In *Phaseolus vulgaris*, two restorer genes, *Fr* and *Fr2*, allow fertility restoration of CMS-Sprite by two different mechanisms. *Fr2* suppresses the expression of the CMS-associated *pvs*-region (Abad et al. 1995), whereas *Fr* leads
to an irreversible elimination of this region (Janska and Mackenzie 1993). In
presence of both genes, the expression of the *pvs*-region is reduced, however
the region is not eliminated, as the gene product of the *pvs*-region seems to
be necessary for this (He et al. 1995a). In case of a spontaneous reversion to
fertility the precursor molecule, which contains *pvs-orf239*, remains present
in sub-stoichometric amounts in the reverted genome (Janska et al. 1998).
Using "bulked-segregant" analyses (Michelmore et al. 1991), four RAPD
markers closely linked to the restorer gene *Fr* could be identified (He *et al.*
1995b). These markers were integrated into the existing map for bean
(Vallejos et al. 1992). Apart from the *Fr*- and the *Fr2* gene two additional
restorer genes, $Fr_{PI207228}$ and Fr_{XR235}, showing the same function as the *Fr2*
gene, were identified (Jia et al. 1997). All four genes mapped on the same
linkage group, so that *Fr2*, $Fr_{PI207228}$ and Fr_{XR235} are probably allelic.

In petunia, an AFLP-marker ECCA/MCAT and the RAPD markers OP51,
OP704, OP413 as well as OP605 were identified as closely linked to the *Rf*
gene (Bentolila et al. 1998). OP704, CT24 and ECCA/MCAT hybridized with
the same *Mlu*I fragment, which had a size of 650 kb. As the genetic distance

between the markers OP704, CT24 and ECCA/MCAT was estimated to be 1.6 cM the ratio between physical and genetic map was calculated with 400 kb/cM for the genome region around the *Rf* gene in petunia. The restorer gene has been isolated in between (Bentolila et al. 2002).

In rapeseed, the two restorer genes, *Rfp1* and *Rfp2*, involved in fertility restoration of the CMS Polima, are localized at the same gene locus on chromosome 18. It is assumed that they represent alleles of this locus (Jean *et al.* 1997). For the Ogura CMS-system, Delourme et al. (1998) identified 30 markers (RFLP and RAPD), which cosegregated with the restorer gene *Rfo* in the cross R40 × Yuda. The *Rfo* gene has also been cloned, recently (Brown et al. 2003; Desloire et al. 2003).

In *Beta vulgaris* ssp. *maritima*, RFLP and RAPD markers could be identified, which were linked with the restorer gene R1H of the H-type of mitochondria. The closest RFLP marker pKP753 maps at a distance of 1.7 cM to the gene, the next RAPD marker $K11_{1000}$ on the other side with 5.2 cM (Laporte et al. 1998).

In sunflower, two restorer genes, *Rf1* and *Rf2*, seem to be necessary for fertility restoration of the PET1 cytoplasm (Serieys 1996). However, one of the restorer genes is also present in most of the maintainer lines, so that in most cases only *Rf1* is introduced into the hybrids by the restorer line (Leclercq 1984). The restorer gene *Rf1* for the PET1 cytoplasm specifically reduces the cotranscript of *atpA* and *orfH522* in the anthers of fertility restored hybrids. The degree of polyadenylation seems to play a role for the degradation of the mRNA by the mitochondrial RNase 2 (Gagliardi and Leaver 1999). One of the two restorer genes in sunflower may control the polyadenylation of the *atpA-orfH522* cotranscript (Gagliardi and Leaver 1999). The restorer gene *Rf1* mapped on linkage group 13 of the sunflower general genetic map (Tang et al. 2003; Kusterer et al. 2005). Three AFLP markers, E32M36-155, E42M76-125 and E44M70-275, and three RAPD markers, OP-K13_454, OP-Y10_740 and OP-H13_337, were closely linked to the restorer gene *Rf1*.

The RAPD markers were successfully converted into two STS markers and one CAPS marker, which are now available and used in marker-assisted breeding (Horn et al. 2003; Kusterer et al. 2005). A sunflower BAC library (Özdemir et al. 2002, 2004) has been used to identify BAC clones around the restorer gene *Rf1*. For other CMS cytoplasms such as PEF1, PET2 or ANN4 (Horn and Friedt 1997), restorer lines could be identified but the restorer genes have not yet been mapped on the sunflower general genetic map.

The investigations on the location of restorer genes in genomes of different plant species indicate that restorer genes for different CMS-types seem to be distributed over the whole genome, but in some cases they are also located on the same chromosome or may be alleles of one gene.

3.2 PPR genes function as restorer genes

All up-to-date isolated restorer genes, with one exception, the *Rf2* gene of maize (Cui et al. 1996), belong to a family of genes that contain a pentatricopeptide repeat (PPR) motif (Table 3). PPR motif carrying proteins show a characteristic tandem array of repeats of a degenerated motif of 35 amino acids. PPR-containing proteins are encoded by a gene family in *Arabidopsis thaliana* that contains over 450 members, with the number of repeats varying from 2 to 26 (Aubourg et al. 2000). The majority of these proteins are predicted to be targeted either to mitochondria or to chloroplasts. It is assumed that PPR proteins specifically interact with RNA in organelles and play a role in RNA processing or translation (Small and Peeters 2000).

The first restorer gene of the PPR-type was cloned from Petunia (Bentolila et al. 2002). Applying molecular marker techniques the genomic region containing the restorer gene was delimited to a BIBAC of 37.5 kb (Bentolila et al. 1998; Bentolila and Hanson 2001). Transgenic approaches were then used to identify the restorer gene from the candidate genes. The restorer gene encodes 592 amino acids and is targeted to mitochondria via a signal sequence. This *Rf-PPR592* restored fertility in transgenic plants carrying the Petunia CMS cytoplasm. The abundance of the CMS-associated PCF protein was considerably reduced in these plants. *Rf-PPR592* contains 14 copies of the pentatricopeptide motif, representing 87% of the coding region. The *Rf* locus in Petunia shows a complex genomic structure and contains a second PPR gene (*Rf-PPR591*) coding for a protein of 591 amino acids with an unknown function. The homolog gene in *rf/rf* lines shows a

Table 3. Overview of the up to date cloned restorer genes of the PPR-type

Species	CMS-type	Restorer gene	PPR motifs	Reference
Petunia hybrida	RM	*Rf*	14 Repeats	Bentolila and Hanson (2001), Bentolila et al. (2002)
Oryza sativa	cms-BT	*Rf-1*	18 Repeats	Kazama and Toriyama (2003), Akagi et al. (2004), Komori et al. (2004)
Raphanus sativus	Kosena	*Rfk1*	16 Repeats	Imai et al. (2003), Koizuka et al. (2003)
Raphanus sativus	Ogura	*Rfo (g26)*	16 Repeats	Brown et al. (2003), Desloire et al. (2003)

deletion in the promoter region compared to *Rf-PPR592* and differences in the predicted amino acid sequence (Bentolila et al. 2002).

In rice, cms-BT (cms-bo), male sterility is caused by an aberrant *atp6* gene (*B-atp6*) in the mitochondrial genome. This *B-atp6* is transcribed into a 2.0 kb transcript that contains apart from the *atp6* gene a unique sequence of *orf79* located in the 3' region of *atp6*. In presence of the restorer gene *Rf-1* two transcripts of 1.5 kb and 0.45 kb are formed by processing the 2.0 kb transcript (Iwabuchi et al. 1993, Akagi et al. 1994). Sequence analysis of the cDNA demonstrated that the processed B-*atp6*-RNA is as effectively edited as the N-*atp6*-RNA. The *Rf-1* gene seems to be involved in processing, which also has an effect on the post-transriptional editing process of *atp6* (Iwabuchi *et al.* 1993). Several research groups were able to clone the *Rf-1* gene by combining a map-based cloning strategy and a candidate gene approach (Kazama and Toriyama 2003; Akagi et al. 2004; Komori et al. 2004). Kazama and Toriyama (2003) were the first to identify *PPR8-1*, coding for a PPR-protein, as candidate gene for *Rf-1*. Using a transgenic approach, they could demonstrate that this gene participates in processing of the transcripts of *atp6*. The protein encoded by *PPR8-1* promotes the formation of the 0.45 kb RNA from the transcript of the *B-atp6* gene in the same way as *Rf-1* is supposed to act. However, this group did not investigate whether the introduction of *PPR8-1* results in fertile regenerated plants. *PPR791* (*Rf-1*), which was cloned by Komori et al. (2004) following a fine mapping strategy (Komori et al. 2003), is identical to *PPR8-1*. The *Rf-1* gene was predicted to encode a 791 aa protein, containing 16 PPR-motifs, 14 of which are in a tandem array (Komori et al. 2004). The recessive allele (*rf-1*) encodes a truncated protein of 266 amino acids because a 1-bp deletion in the putative coding region leads to a frame shift and a premature stop codon. The 574 bp deletion located in the 3'-region of the coding sequence probably does not play a role for the function of the restorer gene.

Komori et al. (2004) obtained the clone, which contained *PPR791*, by conducting complementation tests in which this clone resulted in fertile transgenic plants. Komori et al. (2004) identified three additional PPR genes in the region that showed 90.1%, 80.9% and 94.1% homology to the restorer gene *PPR791*, respectively. Akagi et al. (2004) identified nine PPR genes around the *Rf-1* locus. These duplicates may have played diversified roles in RNA processing and/or recombination in mitochondria during the co-evolution of these genes and the mitochondrial genome (Akagi et al. 2004).

For rapeseed, a restorer gene of the PPR type could be isolated for the Ogura-CMS as well as for the Kosena-CMS. The *Rfo* gene from *Raphanus sativus* can restore male fertility in rapeseed carrying the Ogura cytoplasm. Unlike petunia *Rf*, *Rfo* does not affect the transcripts of the corresponding

CMS-associated mitochondrial gene, *orf138*. *Rfo* appears to act at either the translational or post-translational level leading to a reduction of the *orf138* gene product in flowers and leaves (Krishnasamy and Makaroff 1994; Bellaoui et al. 1997). Cloning of the *Rfo* gene was facilitated by using the synteny between *Raphanus* and *Arabidopsis*, although *Arabidopsis* does not have a PPR gene corresponding to the restorer gene *Rfo* (Brown et al. 2003; Desloire et al. 2003). Brown et al. (2003) identified *g26*, encoding a 687-amino acid protein with a predicted mitochondrial targeting sequence, as candidate gene for *Rfo*.

The flanking genes, *g24* and *g27*, also contain multiple PPR motifs, but both lack the third repeat of *g26*. The predicted proteins encoded by the three genes are similar in length: g24p and g27p are 686 and 654 amino acids long, respectively, in comparison to the 687 amino acids of the *Rfo* gene product (*g26p*). Transformation using a clone containing *g26* resulted in fertile regenerated plants (Brown et al. 2003). Desloire et al. (2003) also identified three PPR genes, *PprA*, *PprB* and *PprC*, on the *Raphanus* BAC-clone 64. Sequence analysis of *PprC* suggested that *PprC* is a pseudogene. In conclusion, *Rfo* is likely to correspond to *PprA* or *PprB* or both.

Analyses of the Kosena CMS mtDNA revealed that Kosena CMS carries *orf125* (Iwabuchi et al. 1999), which encodes a 17-kDa protein and has a sequence homologous to that of *orf138*, except for two amino acid substitutions and a 39-bp deletion in the *orf138* coding region. The accumulation of ORF125 and ORF138 is associated with the CMS phenotype in *Brassica napus* (Grelon et al. 1994; Iwabuchi et al. 1999). The *Rf* gene regulates the protein expression at translational level (Koizuka et al. 1998). Two nuclear loci, *Rfk1* and *Rfk2*, with dominant alleles are capable of restoring fertility in the Kosena CMS radish (Koizuka et al. 2000). Although the amount of *orf125* transcript is unchanged in the presence of the *Rfk1* dominant allele, ORF125 protein accumulation is considerably reduced (Iwabuchi et al. 1999; Koizuka et al. 2000). Pursuing a positional cloning strategy, the region of the fertility restorer locus *Rfk1* was delimited to a 43-kb contig in Kosena radish (*Raphanus sativus* L.), which is covered by four lambda clones and one cosmid clone (Imai et al. 2003). To identify *Rfk1*, subclones covering the 43-kb region were introduced into a *B. napus* CMS line via *Agrobacterium*-mediated transformation. The *orf687*, which encodes 687 amino acids with a predicted molecular weight of 76.5 kDa, was identified as *Rfk1*. The recessive allele contains 11 base substitutions. Five of the base substitutions result in four amino acid substitutions, all within the region of PPR repeats (Koizuka et al. 2003). The sequence of the protein ORF687 encoded by *Rfk1* is identical to the protein, g26p. Thus, although there are differences in the mitochondrial CMS determinants between radish Kosena and Ogura cytoplasm, fertility restoration of both systems can be conferred by genes encoding the same polypeptide (Brown et al. 2003).

The fact that a number of restorer genes show PPR motifs will considerably facilitate the isolation of candidate genes for restorer genes in other species (Koizuka et al. 2003).

4 Conclusions

Functional mitochondria are obviously vital for pollen development. CMS systems and the corresponding fertility restorer genes provide interesting possibilities to study the role of mitochondria in pollen development and the interaction of nucleus and mitochondria in this scenario. Whereas the causes of cytoplasmic male sterility have already been analysed in detail on the molecular level, the isolation of the restorer genes has just started. The identification of restorer genes belonging to the PPR gene family might open the possibility of better progress in cloning of further restorer genes by combining map-based cloning with candidate gene approaches. However, only the future will show whether all the restorer genes belong to the same type or whether other types of restorer genes will be identified. For these genes, the cloning procedure will be more time consuming and not for all plants transposon-tagging systems are available, which would be an alternative to the universally applicable map-based-cloning strategy. Although CMS-associated open-reading frames and the encoded proteins have been identified the function and role of these proteins, apart from the maize T-URF13 protein, is still unresolved. Isolation and studies of the function of the restorer gene might allow a better understanding of the processes involved. However, in general it seems that any disturbance of the mitochondrial function is fatal for a process with high energy demands as the development of pollen in higher plants.

The molecular analyses on cytoplasmic male sterility and fertility restoration have also provided molecular markers that allow fingerprinting of hybrids and assessment of purity of hybrid seeds based on PCR-based strategies for commercial applications (Komori and Nitta 2004; Nandakumar et al. 2004).

References

Abad AR, Mehrtens BJ, Mackenzie SA (1995) Specific expression in reproductive tissues and fate of a mitochondrial sterility-associated protein in cytoplasmic male-sterile bean. Plant Cell 7:271–285

Akagi H, Sakamoto M, Shinjyo C, Shimada H, Fujimura T (1994) A unique sequence located downstream from the rice mitochondrial *atp6* may cause male sterility. Curr Genet 25:52–58

Akagi H, Yokozeki Y, Inagaki A, Nakamura A, Fujimura T (1996) A codominant DNA marker closely linked to the rice nuclear restorer, *Rf-1*, identified with inter-SSR fingerprinting. Genome 39:1205–1209

Akagi H, Nakamura A, Yokozeki-MisonoY, Inagaki A, Takahashi H, Mori K, Fujimura T (2004) Positional cloning of the rice *Rf-1* gene, a restorer of BT-type cytoplasmic male sterility that encodes a mitochondria-targeting PPR protein. Theor Appl Genet 108:1449–1457

Aubourg S, Boudet N, Kreis M, Lecharny A (2000) In *Arabidopsis thaliana*, 1% of the genome codes for a novel protein family unique to plants. Plant Mol Biol 42:603–613

Bailey-Serres J, Hanson DK, Fox TD, Leaver CJ (1986a) Mitochondrial genome rearrangement leads to extension and relocation of cytochrome c oxidase subunit I gene in *Sorghum*. Cell 47:567–576

Bailey-Serres J, Dixon LK, Liddell AD, Leaver CJ (1986b) Nuclear-mitochondrial interactions in cytoplasmic male-sterile *Sorghum*. Theor Appl Genet 73:252–260

Balk J, Leaver CJ (2001) The PET1-CMS mitochondrial mutation in sunflower is associated with premature programmed cell death and cytochrome c release. Plant Cell 13:1803–1818

Bellaoui M, Pelletier G, Budar F (1997) The steady-state level of mRNA from the Ogura cytoplasmic male sterility locus in *Brassica* cybrids is determined post-transcriptionally by its 3′ region. EMBO J 16:5057–5068

Bellaoui M, Grelon M, Pelltier G, Budar F (1999) The restorer *Rfo* gene acts post-translationally on the stability of the ORF138 OGURA CMS-associated protein in reproductive tissues of rapeseed cybrids. Plant Mol Biol 40:893–902

Bentolila S, Hanson MR (2001) Identification of a BIBAC clone that co-segregates with the petunia restorer of fertility (*Rf*) gene. Mol Gen Genomics 266:223–230

Bentolila S, Zethof J, Gerats T, Hanson MR (1998) Locating the petunia *Rf* gene on a 650-kb DNA fragment. Theor Appl Genet 96:980–988

Bentolila S, Alfonso AA, Hanson MR (2002) A pentatricopeptide-repeat-containing gene restores fertility to cytoplasmic male-sterile plants. Proc Natl Acad Sci USA 99:10887–10892

Berry ST, Leon AJ, Hanfrey CC, Challis P, Burkholz A, Barnes SR, Rufener GK, Lee M, Caligari PDS (1995) Molecular marker analysis of *Helianthus annuus* L. 2. Construction of a RFLP linkage map for cultivated sunflower. Theor Appl Genet 91:195–199

Börner A, Korzun V, Polley A, Maleyshev S, Melz G (1998) Genetics and molecular mapping of a male fertility restoration locus (*Rfg1*) in rye (*Secale cereale* L). Theor Appl Genet 97:99–102

Bonhomme S, Budar F, Lancelin D, Small I, Defrance MC, Pelltier G (1992) Sequence and transcript analysis of the *Nco*2.5 Ogura-specific fragment correlated with cytoplasmic male-sterility in *Brassica* cybrids. Mol Gen Genet 235:340–348

Boutry M, Faber A, Charbonnier M, Briquet M (1984) Microanalysis of mitochondrial protein synthesis products: detection of variant polypeptides associated with cytoplasmic male sterility. Plant Mol Biol 3:445–452

Brown GG, Formanova N, Jin H, Wargachuk R, Dendy C, Patil P, Laforest M, Zhang J, Cheung WY, Landry BS (2003) The radish *Rfo* restorer gene of Ogura cytoplasmic male sterility encodes a protein with multiple pentatricopeptide repeats. Plant J 35 (2):262–272

Budar F (1998) What can we learn about the interactions between the nuclear and mitochondrial genomes studying cytoplasmic male sterilities? In: Moller IM, Gardeström P, Glimelius K, Glaser E (eds) Plant mitochondria: from gene to function. Backhuys, Leiden, The Netherlands, pp 49–55

Chase CD, Ortega VM (1992) Organization of ATPA coding and 3′ flanking sequences associated with cytoplasmic male sterility in *Phaseolus vulgaris* L. Curr Genet 22:147–153

Cui X, Wise RP, Schnable PS (1996) The *rf2* nuclear restorer gene of male-sterile T-cytoplasm maize. Science 272:1334–1335

De la Canal L, Crouzillat D, Quetier F, Ledoigt G (2001) A transcriptional alteration on the *atp9* gene is associated with a sunflower male-sterile cytoplasm. Theor Appl Genet 102:1185–1189

Delourme R, Foisset N, Horvais R, Barret P, Champagne G, Cheung WY, Landry BS, Renard M (1998) Characterisation of the radish introgression carrying the *Rfo* restorer gene for the Ogu-INRA cytoplasmic male sterility in rapeseed (*Brassica napus* L.). Theor Appl Genet 97:129–134

Desloire S, Gherbi H, Laloui W, Marhadour S, Clouet V, Cattolico L, Falentin C, Giancola S, Renard M, Budar F, Small I, Caboch M, Delourme R, Bendahmane A (2003) Identification of the fertility restoration locus, *Rfo*, in radish, as a member of the pentatricopeptide-repeat protein family. EMBO Rep 4:588–594

Dewey RE, Timothy DH, Levings III CS (1987) A mitochondrial protein associated with cytoplasmic male sterility in the T cytoplasm of maize. Proc Natl Acad Sci 84:5374–5378

Dieterich JH, Braun HP, Schmitz UK (2003) Alloplasmic male sterility in *Brassica napus* (CMS "Tournefortii-Stiewe") is associated with a special gene arrangement around a novel *atp9* gene. Mol Genet Genomics 269 (6):723–731

Dill CL, Wise RP, Schnable PS (1997) *Rf8* and *Rf** mediate unique T-*urf13*-transcript accumulation, revealing a conserved motif associated with RNA processing and restoration of pollen fertility in T-cytoplasm maize. Genetics 147:1367–1379

Dixon LK, Leaver CS (1981) Mitochondrial gene expression and cytoplasmic male sterility in *Sorghum*. Plant Mol Biol 1:89–102

Engelke T, Tatlioglu T (2004) The fertility restorer genes X and T alter the transcripts of a novel mitochondrial gene implicated in CMS1 in chives (*Allium schoenoprasum* L.). Mol Genet Genomics 271:150–160

Flavell RB (1974) A model for the mechanism of cytoplasmic male sterility in plants, with special reference to maize. Plant Sci Lett 3:259–263

Forde BG, Leaver CJ (1980) Nuclear and cytoplasmic genes controlling synthesis of variant mitochondrial polypeptides in male-sterile maize. Proc Natl Acad Sci USA 77:418–422

Gagliardi D, Leaver CJ (1999) Polyadenylation accelerates the degradation of the mitochondrial mRNA associated with cytoplasmic male sterility in sunflower. EMBO J 18:3757–3766

Gallagher LJ, Betz SK, Chase CD (2002) Mitochondrial RNA editing truncates a chimeric open reading frame associated with S male-sterility in maize. Curr Genet 42:179–184

Gentzbittel L, Vear F, Zhang YX, Bervillé A (1995) Development of a consensus linkage RFLP map of cultivated sunflower (*Helianthus annuus* L.). Theor Appl Genet 90:1079–1086

Gentzbittel L, Mestries E, Mouzeyar S, Mazeyrat F, Badaoul S, Vear F, Tourvielle de Labrouhe D, Nicolas P (1999) A composite map of expressed sequences and phenotypic traits of the sunflower (*Helianthus annuus* L.) genome. Theor Appl Genet 99:218–234

Grelon M, Budar F, Bonhomme S, Pelletier G (1994) Ogura cytoplasmic male-sterility (CMS) associated *orf138* is translated into a mitochondrial membrane polypeptide in male-sterile *Brassica* cybrids. Mol Gen Genet 243:540–547

Håkansson G, Van der Mark F, Bonnett HT, Glimelius K (1988) Variant mitochondrial protein and DNA patterns associated with cytoplasmic male-sterile lines of *Nicotiana*. Theor Appl Genet 76:431–437

Handa H, Nakajima K (1992) Different organization and altered transcription of mitochondrial *atp6* gene in the male-sterile cytoplasm of rapeseed, *Brassica napus* L. Curr Genet 21:153–159

Handa H, Gualberto JM, Grienenberger JM (1995) Characterization of the mitochondrial *orfB* and its derivate, *orf224*, a chimeric open reading frame specific to one mitochondrial genome of the "Polima" male-sterile cytoplasm in rapeseed (*Brassica napus* L.). Current Genet 28:546–552

Hanson MR, Bentolila S (2004) Interactions of mitochondrial and nuclear genes that affect male gametophyte development. Plant Cell 16:154–169

He S, Lyznik A, MacKenzie S (1995a) Pollen fertility restoration by nuclear gene *Fr* in CMS bean: nuclear directed alteration of a mitochondrial population. Genetics 139:955–962

He S, Yu ZH, Vallejos CE, Mackenzie SA (1995b) Pollen fertility restoration by nuclear gene *Fr* in CMS common bean: an *Fr* linkage map and the mode of *Fr* action. Theor Appl Genet 90:1056–1062

Heazlewood JL, Whelan J, Millar AH (2003) The products of the mitochondrial *orf25* and *orfB* genes are F_0 components in the plant $F_1 F_0$ ATP synthase. FEBS Lett 540:201–205

Hiesel R, Schobel, Schuster W, Brennicke A (1987) The cytochrome subunit I and subunit III genes in *Oenothera* mitochondria are transcribed from identical promotor sequences. EMBO J 6:29–34

Horn R (2002) Molecular diversity of male sterility inducing and male-fertile cytoplasm in the genus *Helianthus*. Theor Appl Genet 104:562–570

Horn R, Friedt W (1997) Fertility restoration of new CMS sources in sunflower. Plant Breed 116:317–322

Horn R, Friedt W (1999) CMS sources in sunflower: different origin but same mechanism? Theor Appl Genet 98:195–201

Horn R, Köhler RH, Zetsche K (1991) A mitochondrial 16 kDa protein is associated with cytoplasmic male sterility in sunflower. Plant Mol Biol 7:29–36

Horn, R, Hustedt JEG, Horstmeyer A, Hahnen J, Zetsche K, Friedt W (1996) The CMS-associated 16 kDa protein encoded by *orfH522* is also present in other male sterile cytoplasms of sunflower. Plant Mol Biol 30:523–538

Horn R, Kusterer B, Lazarescu E, Prüfe M, Özdemir N, Friedt W (2002) Molecular diversity of CMS sources and fertility restoration in the genus *Helianthus*. Helia 25:29–40

Horn R, Kusterer B, Lazarescu E, Prüfe M, Friedt W (2003) Molecular mapping of the *Rf1* gene restoring pollen fertility in PET1-based F_1 hybrids in sunflower (*Helianthus annuus* L.). Theor Appl Genet 106:599–606

Huang QY, He YQ, Jing RC, Zhu RS, Zhu YG (2000) Mapping of the nuclear fertility restorer for HL CMS in rice using microsatellite markers. Chinese Sci Bull 45:430–432

Ichikawa N, Kishimoto N, Inagaki A, Nakamura A, Koshino Y, Yokozecki Y, Oka HU, Samoto S, Akagi H, Higo K, Shinjyo C, Fujmura T, Shimada H (1997) A rapid PCR-based selection of a rice line containing the *Rf-1* gene which is involved in restoration of the cytoplasmic male sterility. Mol Breed 3:195–202

Imai R, Koizuka N, Fujimoto H, Hayakawa T, Sakai T, Imamura J (2003) Delimitation of the fertility restorer locus *Rfk1* to a 43-kb contig in Kosena radish (*Raphanus sativus* L.). Mol Genet Genomics 269:388–394

Iwabuchi M, Kyozuka J, Shimamoto K (1993) Processing followed by complete editing of an altered mitochondrial *atp6* RNA restores fertility of cytoplasmic male sterile rice. EMBO J 12:1437–1446

Iwabuchi M, Koizuka N, Fujimoto H, Takako S, Imamura J (1999) Identification and expression of the kosena radish (*Raphanus sativus* cv. Kosena) homologue of the ogura radish CMS-associated gene, *orf138*. Plant Mol Biol 39:183–188

Janska H, Mackenzie S (1993) Physical mapping of the mitochondrial genome in CMS common bean and the nature of spontaneous reversion to fertility. In: Brennicke A, Kück U (eds) Plant mitochondria. Springer Verlag, Weinheim, New York, pp 393–401

Janska H, Sarria R, Woloszynska M, Arrietal-Montiel M, Mackenzie S (1998) Stoichiometric shifts in the common bean mitochondrial genome leading to male sterility and spontaneous reversion to fertility. Plant Cell 10:1163–1180

Jean M, Brown GG, Landry BS (1997) Genetic mapping of nuclear fertility restorer genes for the 'Polima' cytoplasmic sterility in canola (*Brassica napus* L.) using DNA markers. Theor Appl Genet 95:321–328

Jia MH, He S, Vanhouten W, Mackenzie S (1997) Nuclear fertility restorer genes map to the same linkage group in cytoplasmic male-sterile bean. Theor Appl Genet 95:205–210

Jing R, Li X, Yi P, Zhu Y (2001) Mapping fertility-restoring genes of rice WA cytoplasmic male sterility using SSLP markers. Bot Bull Acad Sin 42:167–171

Johns C, Lu MQ, Lyznik A, Mackenzie S (1992) A mitochondrial DNA sequence is associated with abnormal pollen development in cytoplasmic male sterile bean plants. Plant Cell 4:435–449

Kamps TL, Chase CD (1997) RFLP mapping of the maize gametophytic restorer-of fertility locus (rf3) and aberrant pollen transmission of the nonrestoring rf3 allele. Theor Appl Genet 95:525–531

Kaspi C, Siedow JM (1993) Cross-linking of the cms-T maize mitochondrial pore-forming protein URF13 by N,N'-dicyclohexylcarboimide and its effect on URF13 sensitivity to fungal toxins. J Biol Chem 268:5828–5833

Kazama T, Toriyama K (2003) A pentatricopeptide repeat-containing gene that promotes the processing of aberrant atp6 RNA of cytoplasmic male-sterile rice. Febs Lett 544:99–102

Kennell JC, Pring DR (1989) Initiation and processing of atp6, T-urf13 and orf221 transcripts from mitochondria of T-cytoplasm maize. Mol Gen Genet 216:16–24

Klein RR, Klein PE, Chhabra AK, Dong J, Pammi S, Childs KL, Mullet JE, Rooney WL, Schertz KF (2001) Molecular mapping of the rf1 gene for pollen fertility restoration in sorghum (Sorghum bicolor L.). Theor Appl Genet 102:1206–1212

Köhler RH, Horn R, Lössl A and Zetsche K (1991) Cytoplasmic male sterility in sunflower is correlated with the co-transcription of a new open reading frame with the atpA gene. Mol Gen Genet 227:369–376

Koizuka N, Fujimoto H, Sakai T, Imamura J (1998) Translational control of ORF125 expression by a radish fertility-restoration gene in Brassica napus. In: Moller IM, Gardeström P, Glimelius K, Glaser E (eds) Plant mitochondria: from gene to function. Backhuys, Leiden, The Netherlands, pp 49–55

Koizuka N, Imai R, Iwabuchi M, Sakai T, Imamura J (2000) Genetic analysis of fertility restoration and accumulation of ORF125 mitochondrial protein in the Kosena radish (Raphanus sativus L. cv. Kosena) and a Brassica napus restorer line. Theor Appl Genet 100:949–955

Koizuka N, Imai R, Fujimoto H, Hayakawa T, Kimura Y, Kohno-Murase J, Sakai T, Kawasaki S, Imamura J (2003) Genetic characterization of a pentatricopeptide repeat protein gene, orf687, that restores fertility in the cytoplasmic male-sterile Kosena radish. Plant J 34:407–415

Komori T, Nitta N (2004) A simple method to control seed purity of japonica hybrid rice varieties using PCR-based markers. Plant Breed 123:549–553

Komori T, Yamamoto T, Takemori N, Kashihara M, Matsushima H, Nitta N (2003) Fine genetic mapping of the nuclear gene Rf-1, that restores the BT-type cytoplasmic male sterility in rice (Oryza sativa L.) by PCR-based markers. Euphytica 129:241–247

Komori T, Ohta S, Murai N, Takakura Y, Kuraya Y, Suzuki S, Hiei Y, Imaseki H, Nitta N (2004) Map-based cloning of a fertility restorer gene, Rf-1, in rice (Oryza sativa L.). Plant J 37:315–325

Korth KL, Kaspi CI, Siedow JN, Levings CSIII (1991) URF13, a maize mitochondrial pore-forming protein, is oligomeric and has a mixed orientation in Escherichia coli plasma membranes. Proc Natl Acad Sci USA 88:10865–10869

Krishnasamy S, Makaroff CA (1993) Characterization of the radish mitochondrial orfB locus: possible relationship with male sterility in Ogura radish. Curr Genet 24:156–163

Krishnasamy S, Makaroff CA (1994) Organ-specific reduction in the abundance of a mitochondrial protein accompanies fertility restoration in cytoplasmic male-sterile radish. Plant Mol Biol 26:935–946

Kurata N, Nagamura Y, Yamamoto K, Haushima Y, Sue N, Wu J, Antonio BA, Somura A, Shimizu T, Lin SY, Inoue T, Fukuda T, Shimano T, Kuboki Y, Toyma T, Miyamoto Y,

Kirihara T, Hayaska K, Miyao A, Monna L, Zhong HS, Tamura Y, Wang ZX, Momma T, Umehara Y, Yano M, Sasaki T, Minobe Y (1994) A 300-kilobase interval genetic map of rice including 883 expressed sequences. Nat Genet 8:365–372

Kusterer B, Horn R, Friedt W (2005) Molecular mapping of the fertility restoration locus *Rf1* in sunflower and development of diagnostic markers for the restorer gene. Euphytica 143:35–43

Landgren M, Zetterstrand M, Sundberg E, Glimelius K (1996) Alloplasmic male-sterile *Brassica* lines containing *B. tournefortii* mitochondria express an ORF 3′ of the *atp6* gene and a 32 kDa protein. Plant Mol Biol 32:879–890

Laporte V, Merdinoglu D, Saumitou-Laprade P, Butterlin G, Vernet P, Cuguen J (1998) Identification and mapping of RAPD and RFLP markers linked to a fertility restorer gene for a new source of cytoplasmic male sterility in *Beta vulgaris* ssp. *maritima*. Theor Appl Genet 96:989–996

Laver HK, Reynolds SJ, Moneger F, Leaver CJ (1991) Mitochondrial genome organization and expression associated with cytoplasmic male sterility in sunflower (*Helianthus annuus*). Plant J 1:185–193

Leclercq P (1984) Identification de gènes de restauration de fertilité sur cytoplasmes stéril-isants chez le tournesol. Agronomie 4:573–576

Levings CS III, Dewey RE (1988) Molecular studies of cytoplasmic male sterility in maize. Philos Trans R Soc Lon B Biol Sci 319:177–186

Liu F, Cui X, Horner HT, Weiner H, Schnable P (2001) Mitochondrial aldehyde dehydroge-nase activity is required for male fertility in maize. Plant Cell 13:1063–1078

Liu L, Guo W, Zhu X, Zhang T (2003) Inheritance and fine mapping of fertility restoration for cytoplasmic male sterility in *Gossypium hirsutum* L. Theor Appl Genet 106:461–469

Liu XQ, Xu X, Tan YP, Li SQ, Hu J, Huang JY, Yang DC, Li YS, Zhu YG (2004) Inheritance and molecular mapping of two fertility-restoring loci for Honglian gametophytic cytoplasmic male sterility in rice (*Oryza sativa* L.). Mol Genet Genomics 271:586–594

Matsui K, Mano Y, Taketa S, Kawada N, Komatsuda T (2001) Molecular mapping of a fertil-ity restoration locus (*Rfm1*) for cytoplasmic male sterility in barley (*Hordeum vulgare* L.). Theor Appl Genet 102:477–482

Michelmore RW, Paran I, Kessli RV (1991) Identification of markers linked to disease resist-ance genes by bulked segregant analysis: a rapid method to detect markers in specific genomic regions using segregating populations. Proc Natl Acad Sci USA 88:9828–9832

Monéger F, Smart CJ, Leaver CJ (1994) Nuclear restoration of cytoplasmic male sterility in sunflower is associated with tissue-specific regulation of a novel mitochondrial gene. EMBO J 13:8–17

Nandakumar N, Singh AK, Sharma RK, Mohapatra T, Prabhu KV, Zaman FU (2004) Molecular fingerprinting of hybrids and assessment of genetic purity of hybrid seeds in rice using microsatellite markers. Euphytica 136:257–264

Nivison HT, Hanson MR (1989) Identification of a mitochondrial protein associated with cytoplasmic male sterility in *Petunia*. Plant Cell 1:1121–1130

Nivison HT, Sutton CA, Wilson RK, Hanson MR (1994) Sequencing, processing, and local-ization of the petunia CMS-associated mitochondrial protein. Plant J 5:613–623

Özdemir N, Horn R, Friedt W (2002) Isolation of HMW DNA from sunflower (*Helianthus annuus* L.) for BAC cloning. Plant Mol Biol Rep 20:239–250

Özdemir N, Horn R, Friedt W (2004) Construction and characterization of a BAC library for sunflower (*Helianthus annuus* L.). Euphytica 138:177–183

Potz H, Tatlioglu T (1993) Molecular analysis of cytoplasmic male sterility in chives (*Allium schoenoprasum* L.). Theor Appl Genet 87:439–445

Pruitt KD, Hanson MR (1991) Transcription of the *Petunia* mitochondrial CMS-associated PCF locus in male sterile and fertility-restored lines. Mol Gen Genet 227:348–355

Quillet MC, Madjidian N, Griveau Y, Serieys H, Tersac M, Lorieux M, Berville A (1995) Mapping genetic factors controlling pollen viability in an interspecific cross in *Helianthus* sect. *Helianthus*. Theor Appl Genet 91:1195–1202

Rhoads DM, Levings CSIII, Siedow JN (1995) URF13, a ligand-gated, pore-forming receptor for T-toxin in the inner membrane of cms-T mitochondria. J Bioenerget Biomemb 27:437–445

Sabar M, Gagliardi D, Balk J, Leaver CJ (2003) ORFB is a subunit of F_1F_0-ATP synthase: insight into the basis of cytoplasmic male sterility in sunflower. EMBO Rep 4:1–6

Scheike R, Gerold E, Brennicke A, Mehring-Lemper M, Wricke G (1992) Unique patterns of mitochondrial genes, transcripts and proteins in different male-sterile cytoplasms of *Daucus carota*. Theor Appl Genet 83:419–427

Schnable PS, Wise RP (1994) Recovery of heritable, transposon-induced mutant alleles of the *rf2* nuclear restorer of T-cytoplasm maize. Genetics 136:1171–1185

Schnable PS, Wise RP (1998) The molecular basis of cytoplasmic male sterility and fertility restoration. Trends in Plant Science 3:175–180

Serieys H (1996) Identification, study and utilisation in breeding programs of new CMS sources. FAO progress report. Helia 19 (special issue):144–158

Shen YW, Guan ZQ, Lu J, Zhuang JY, Zheng KL, Gao MW, Wang XM (1998) Linkage analysis of a fertility restoring mutant generated from CMS rice. Theor Appl Genet 97:261–266

Shinjyo C (1984) Cytoplasmic male sterility and fertility restoration in rice having genome A. In: Tsunoda S, Takahasi N (eds) Biology of rice. Jpn Sci Soc Press, pp 93–95

Shinjyo C, Sato S (1994) Chromosomal location of fertility-restoring gene *Rf2*. Rice Genet Newsl 11:93–95

Small ID, Peeters N (2000) The PPR motif-a TRP-related motif prevalent in plant organellar proteins. Trends Biochem Sci 25:46–47

Stracke S, Schilling AG, Förster J, Weiss C, Glass C, Miedaner T, Geiger HH (2003) Development of PCR-based markers linked to dominant genes for male-fertility restoration in Pampa CMS of rye (*Secale cereale* L.). Theor Appl Genet 106:1184–1190

Tang HV, Chen W, Pring DR (1999) Mitochondrial *orf107* transcription, editing, and nuleolytic cleavage conferred by the gene *Rf3* are expressed in sorghum pollen. Sex Plant Reprod 12:53–59

Tang SX, Kishore VK, Knapp SJ (2003) PCR-multiplexes for a genome-wide framework of simple sequence repeat marker loci in cultivated sunflower. Theor Appl Genet 107:6–19

Vallejos CE, Sakaiyama NS, Chase CD (1992) Molecular marker-based linkage map of *Phaseolus vulgaris* L. Genetics 131:733–740

Wen LY, Chase CD (1999) Pleiotropic effects of a nuclear restorer-of-fertility locus on mitochondrial transcripts in male-fertile and S male-sterile maize. Curr Genet. 35:521–526

Wen L, Tang HV, Chen W, Chang R, Pring DR, Klein PE, Childs KL, Klein RR (2002) Development and mapping of AFLP markers linked to the sorghum fertility restorer gene *rf4*. Theor Appl Genet 104:577–585

Wise RP, Schnable PS (1994) Mapping complementary genes in maize : Positioning the *rf1* and *rf2* nuclear-fertility restorer loci of Texas T-cytoplasm relative to RFLP and visible markers. Theor Appl Genet 88:785–795

Wise RP, Dill CL, Schnable PS (1996) Mutator-induced mutations of the *rf1* nuclear fertility restorer of the T-cytoplasm maize alter the accumulation of *T-urf13* mitochondrial transcripts. Genetics 143:1383–1394

Wise RP, Gobelman WK, Pei D, Dill CL, Schnable PS (1999) Mitochondrial transcript processing and restoration of male fertility in T-cytoplasm maize. J Heredity 90:380–385

Yao FY, Xu CG, Yu SB, Li JX, Gac YJ, Li XH, Zhang Q (1997) Mapping and genetic analysis of two fertility restorer loci in the wild-abortive cytoplasmic male sterility system of rice (*Oryza sativa* L.). Euphytica 98:183–187

Young EG, Hanson MR (1987) A fused mitochondrial gene associated with cytoplasmic male sterility is developmentally regulated. Cell 50:41–49

Zabala G, Gabay-Laughnan S, Laughnan JR (1997) The nuclear *Rf3* affects the expression of the mitochondrial chimeric sequence R implicated in the S-type male sterility in maize. Genetics 147:847–850

Zhang G, Bharaj TS, Lu Y, Virmani SS, Huang N (1997) Mapping of the *Rf-3* nuclear fertility-restoring gene for WA cytoplasmic male sterility in rice using RAPD and RFLP markers. Theor Appl Genet 94:27–33

Zhang J, Stewart JM (2004) Identification of molecular markers linked to the fertility restorer genes for CMS-D8 in cotton. Crop Sci 44:1209–1217

Zhang QY, Liu YG, Mei MT (2002) Molecular mapping of the fertility restorer gene *Rf4* for WA cytoplasmic male sterility. Acta Genetica Sinica 29:1001–1004

Prof. Dr. Renate Horn
Institut für Biowissenschaften
Pflanzengenetik
Universität Rostock
Albert-Einstein-Str3
18051 Rostock
Tel.: +49 381 498 6170
Fax: +49 381 498 6112
e-mail: renate.horn@uni-rostock.de

Function of genetic material:
From genomics to functional markers in maize

Chun Shi, Gerhard Wenzel, Ursula Frei, Thomas Lübberstedt

Abbreviations

cDNA: Copy DNA
QTL: Quantitative trait locus
FM: Functional marker
RDM: Random DNA marker
AFLP: Amplified fragment length polymorphism
RFLP: Restriction fragment length polymorphism
GMO: Genetically modified organism
EST: Expressed sequence tag
BAC: Bacterial artificial chromosome
YAC: Yeast artificial chromosome
PAC: P1-derived artificial chromosome
IRGSP: International rice genome sequencing project
BGI: Beijing genomics institute
RNAi: RNA interference
dsRNA: Double-stranded RNA
SSH: Suppression subtractive hybridization
SAGE: Serial analysis of gene expression
cDNA-AFLP: Complementary DNA - amplified fragment length polymorphism
UV: Ultraviolet
NIL: Near isogenic line
SCMV: Sugarcane mosaic virus
LD: Linkage disequilibrium
SNP: Single nucleotide polymorphism
INDEL: Insertion / deletion polymorphism
DNDF: Neutral detergent fiber digestibility
TILLING: Targeting induced local lesions in genomes
Imap: Integrated map
eEF1A: Elongation factor 1A

Progress in Botany, Vol. 67
© Springer-Verlag Berlin Heidelberg 2006

1 Introduction

Completing the primary genomic sequence of *Arabidopsis thaliana* was a major milestone, being the first plant genome and well established as the premiere model species in plant biology. Since working drafts of rice (*Oryza sativa L.*) genome became available (Yu et al. 2002), it has become the second-best model organism in plants representing monocotyledons. Understanding how the genome sequence comprehensively encodes developmental programs and environmental responses is the next major challenge for all plant genome projects. This requires functional characterization of genes, including identification of regulatory sequences. Several functional genomics approaches were initiated to decode the linear sequence of the model plant *Arabidopsis thaliana*, including full-length cDNA collections, microarrays, natural variation, knockout collections, and comparative sequence analysis (Borevitz and Ecker 2004). Genomics provides the essential tools to speed up the research work of the traditional molecular geneticist, and is now a scientific discipline in its own right (Borevitz and Ecker 2004).

Beside their importance in basic research, markers have entered the field of application in their own right. Frisch (2005) calculated that selection for recombination between a target gene and flanking markers is highly effective even when the marker is rather distant from the target gene. He expects a saving of three backcross generations even with a marker distance of 50 cM. Marker-assisted background selection can be used even for such large distances, since recombinants occur with increasing distances with a higher probability (Frisch et al. 1999). Hoisington and Melchinger (2005) elaborated factors on which a superior selection via markers, and in particular complex QTL in maize breeding, depend, compared with phenotypic selection: the heritability of the trait, the population size of the mapping population employed in QTL mapping, the genetic architecture of the trait, and the total budget of a breeding program. At least some of the early predictions of the usefulness of molecular markers could be verified (Mohler and Singrün 2005).

To convert plant genomics into effective economic and environmental benefits, the knowledge gained must be "translated" into crop varieties with improved characteristics or efficient breeding tools. Functional markers (FMs) are a good "translator" of new knowledge from emerging technologies into improved crop cultivars, or "varieties" (Thro et al. 2004). A concept for definition, development, application, and prospects of FMs in plants has recently been published (Andersen and Lübberstedt 2003). FMs are derived from polymorphic sites within genes causally involved in phenotypic trait variation. Once genetic effects have been assigned to functional sequence motifs, FMs derived from such motifs can be used for fixation of gene alleles (defined by one or several FM alleles) in a number of genetic backgrounds without additional calibration. In contrast, the value of anonymous genetic markers such as random DNA markers (RDMs = microsatellites, AFLPs, RFLPs etc.) depends on the known linkage phase between marker and target locus alleles (Lübberstedt et al. 1998). Thus (quantitative) trait locus mapping is necessary for each cross de novo, as different subsets of QTL are polymorphic in individual populations, and linkage phases between marker and QTL alleles can disagree even in closely related genotypes. FM development requires (1) functionally characterized genes, (2) allele sequences from such genes, (3) identification of polymorphic, functional motifs affecting plant phenotype within these genes, and (4) validation of associations between DNA polymorphisms and trait variation.

Because of its economic importance, and its vigorously active transposable elements, maize has been a focus of interest for plant biologists for many decades. From its early role as a model for the analysis of plant transposons and for gene tagging, maize has, like *Arabidopsis* and rice, recently become the focus of plant genomic research (Varotto and Leister 2002). Here, our major objectives are: (1) with a focus on the major crop maize, to summarize the current status of genomics projects, and (2) to discuss the perspectives of exploitation of this information in terms non-GMO breeding strategies (see Fig. 1).

2 Structural genomics in maize and rice

Projects addressing systematic sequencing of the maize genome contributed large amounts of publicly available sequences during the past few years. More than 417,000 expressed sequence tag (EST) sequences have been released into the public domain (http://www.ncbi.nlm.nih.gov/dbEST/dbEST_summary.html). Additional EST sequences have very recently been made available by private companies (http://www.ncga.com/research/MaizeSeq/). Even more genomic, as compared to the public EST sequences, have been generated within the last 2 years in the context of systematic sequencing of the maize genome (http://pgir.rutgers.edu/). More than 680,000 ends of BAC clones have been sequenced (http://pgir.-rutgers.edu/). Additionally, about about 900,000 genomic sequences have been obtained by methyl filtration and high Cot approaches (http://www.tigr.org/tdb/tgi/maize/release4.0/assembly.shtml). Finally, more than

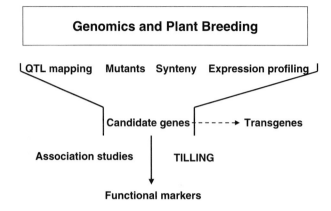

Fig. 1. Diagram indicating how genomics tools can benefit plant breeding

150,000 sequences flanking transposon insertions have been generated (http://www-sequence.stanford.edu/group/maize/maize2.html). Together with systematic genetic and physical mapping (http://www.maizemap. org/), including a genetic high resolution map and more than 400,000 fingerprinted BAC clones arranged in 760 contigs (http://www.genome. arizona.edu/fpc/maize/), these more than 2 million sequence reads provide an excellent basis for gene identification by (i) sequence homology, (ii) synteny to rice or other grasses, and (iii) forward genetic approaches (e.g. map-based gene isolation). Map-based approaches in particular will benefit from the systematic sequencing of BAC contigs planned for the next years (http://www.maizegdb.org/genome/npgi1.pdf). A new project for large scale sequencing of maize has been launched in the USA, building upon the recently established resources (http://www.maizegdb.org/). The resources available for maize genomics research were briefly summarized in Table 1.

Rice has been put forward as a model for crop plants, allowing valuable comparisons to a model dicotyledon (*Arabidopsis thaliana*) and the most important monocotyledons such as maize, wheat and barley (Bennetzen 2002; Schmidt 2000). More than 6000 DNA markers have been mapped in rice, with approximately one marker every 0.25 cM, or every 75–100 kb (Tyagi et al. 2004). SNPs will provide a rich source of DNA based markers since in rice one SNP per 89 bp among various genotypes or one SNP every 232 bp between two randomly selected lines has been reported (Nasu et al. 2002). For physical mapping of the rice genome, various libraries consist-

Table 1. Resources available for maize genomics research

Resource	Amount	Reference
Expressed sequences		
Public ESTs	417,803 (updated 04/02/2005)	http://www.ncbi.nlm.nih.gov/dbEST/dbEST_summary.html
Pioneer/Ceres/Monsanto EST/cDNA	?	http://www.maizeseq.org/
Maps		
Physical map	292, 201 BACs in 760 contigs (updated 25/10/2004)	http://www.genome.arizona.edu/fpc/maize/
Integrated map (imap)	483 BAC contigs anchored onto IBM2 and IBM2 neighbouring maps	http://www.maizemap.org/iMapDB/iMap.html
Comparative map	9 maize maps + 1 rice map + 1 sorghum map	http://www.agron.missouri.edu/cMapDB/cMap.html
Transcript map	2,279 EST mapped (updated 03/12/2004)	http://maize-mapping.plantgenomics.iastate.edu/index.html
Genome-wide sequencing		
Whole-genome (BAC by BAC)	464,544 BACs fingerprinted, 682,116 BAC ends sequenced, 159 BACs sequenced (updated 15/11/2004)	http://pgir.rutgers.edu/
Whole-genome (shotgun)	50,877 (updated 06/04/2004)	http://maize.danforthcenter.org/
Gene-enriched genome (methyl-filtration)	450,166 (updated 23/02/2004)	http://maize.danforthcenter.org/
Gene-enriched genome (high C_0t)	445,565 (updated 23/02/2004)	http://maize.danforthcenter.org/

Table 1. *Continued*

Resource	Amount	Reference
Mutant collection		
Knockout collection (transposons)	43,776 F2 progeny	http://mtm.cshl.edu/
Knockout collection (*RescueMu*)	Over 70,000 *RescueMu* flanking regions sequenced	(Lunde et al. 2003)
TILLING population	3072 lines	http://genome.purdue.edu/maizetilling/
Microarray		
Oligonucleotide array	About 58,000 oligos printed over two slides	http://www.maizearray.org/
Unigene microarray	About 24,000 ESTs spotted over four slides	(Lunde et al. 2003)

ing of large insert clones in vectors such as YACs (Burke et al. 1987), P1-derived artificial chromosomes (PACs) (Ioannou et al. 1994) and BACs (Shizuya et al. 1992), as well as the availability of anchored genetically mapped DNA markers, greatly facilitated the construction of extended contigs. Presently, emphasis is being laid to integrate cytological, genetic and physical maps.

Four versions of the genome sequence from two subspecies of *Oryza sativa* have been released over a short span of time (Buell 2002; Delseny 2003). The vast amount of structural and functional data of the rice genome generated earlier (Sasaki 1998; Sasaki and Burr 2000) provided a strong backbone for the International Rice Genome Sequencing Project (IRGSP), which was started in 1998 (Sasaki and Burr 2000). IRGSP released sequence data up to the phase 2 level (ordered regions of sequences, which can contain gaps) in December 2002 (http://rgp.dna.affrc.go.jp/rgp/Dec18_NEWS.html;http://www.tigr.org/new/press_release_12-18-02.shtml). Furthermore, rice genome projects were carried out by Monsanto (Barry 2001) and Syngenta (Goff et al. 2002). These agribusinesses and the IRGSP focused on the sequencing of the *japonica* cultivar Nipponbare, with its already available molecular and genetic resources. The fourth project conducted at the Beijing Genomics Institute (BGI, http://btn.genomics.org.cn/rice/; Goff et al. 2002) focused on sequencing the tropical *indica* cultivar 93-11. Bioinformatic tools have been used to annotate and analyse the three whole genome draft sequences (two of *japonica* and one of *indica*) of rice. Highly accurate annotated complete sequences of chromosomes 1, 4, and 10 have been made available (Sasaki 2002; Yu et al. 2002, 2003). The final "finished" sequence data, i.e. refined data without gaps, was expected by the end of 2004 (http://www. tigr.org/new/press_release_12-18-02.shtml), setting the stage for future research on rice, similar to the completed sequence of the human genome (Collins et al. 2003). Genome-wide sequencing was preceded by expressed sequence tag (EST) sequencing, as this provides not only an inexpensive sampling method for the expressed fraction of a genome, but also a quantitative profile of expression levels in specific tissues. Currently, there are approximately 266,000 rice ESTs deposited in the GenBank (http://www.ncbi.nlm.nih.gov/dbEST/dbEST_summary.html).

3 Comparative genomics: synteny between maize and rice

One of the cornerstones in the application of molecular markers was the demonstration that large chromosome regions and even whole chromosomes have conserved gene orders across related species within plant families

(Moore et al. 1995). Macrosyntenic relationships between genomes of related species have been displayed in concentric circle models (Moore et al. 1995), with particular emphasis on grass genomes (Devos and Gale 1997). Conservation of gene order has led to the concept of model species in the context of expensive plant genomics projects: once the genome of one species within a plant family or even wider taxonomic unit has been sequenced and the function of genes within this species determined, this information can be exploited for isolating orthologous genes in any related species (reverse genetic approach). Moreover, candidate genes underlying QTL in non-model species would become accessible via synteny relationships to model species (forward genetic approach). The synteny and model species concept has proven successful, such as for isolation of a rust resistance gene in wheat (Huang et al., 2003) and a vernalization response gene in ryegrass (Jensen et al. 2005). Even between the evolutionary distant species *Arabidopsis* and rice, significant microsynteny in small chromosome segments has been discovered (Salse et al. 2004). However, in other studies, rearrangements or deletions impaired the exploitation of synteny for isolation of target genes (Brunner et al. 2003).

The original synteny studies were based on RFLP mapping across species using cDNA probes (Moore et al. 1995). Grass genomes were expected to be organized in a limited number of chromosome blocks. More recent availability of complete genome sequences in *Arabidopsis* and rice as well as sequenced BAC contigs in other species as well as comprehensive (mapped) EST collections led to re-evaluation of the synteny concept (Delseny 2004). These sequencing projects revealed numerous polyploidization events after speciation even in the small genome species *Arabidopsis thaliana* (Blanc et al. 2003), but also in rice and maize (Salse et al. 2004). Furthermore, large genomes such as the maize genome have been invaded by retrotransposons, resulting in reshuffling of original ancestral genomes due to duplications, deletions, illegitimate recombination etc. (Delseny 2004). Since these events occurred after speciation of maize, substantial differences have been reported between larger allelic sequence stretches even within maize among different inbred lines (Fu and Dooner 2002; Brunner et al. 2005). An in-silico alignment of genomic rice sequence with mapped maize EST sequences identified larger collinear chromosome regions between rice and maize in agreement with previous studies (Salse et al. 2004). However, fine-scale analysis revealed, besides several duplicated regions, numerous internal rearrangements within syntenic chromosome blocks. In conclusion, rice can be used to identify candidate genes in a target region identified in maize. However, the order and number of genes might be altered at the microsynteny level between rice and maize.

4 Functional genomics in maize

"Functional genomics" aims at the functional characterization of genes. The main characteristics of genes determined by functional genomics relates to (i) mRNA expression patterns in diverse treatments, tissues, stages, (ii) biochemical classification, and (iii) morphological or phenotypic effects. Several high-throughput approaches have been established to explore the function of genes, as well as to monitor their expression in relation to various other genes of maize. However, systematic assignment of an "agronomic function" to gene variants requires expensive field trials.

Positional cloning of genes underlying QTL will benefit substantially from the availability of physical maps and sequences from whole chromosomes or large chromosome regions, as has been the case for rice (Borevitz and Chory 2004). If knock-out alleles at QTL lack clear phenotypes, map-based gene isolation will be the major route for isolating respective genes.

Maize transposons play an important role as tool in maize gene discovery. The *Ac/Ds* and *MuDR/Mu* maize transposons were widely used in mutagenesis experiments (Walbot 2000). Whereas Mutator transposons are well suited for global mutagenesis and gene discovery, *Ac/Ds* is most convenient for multiple rounds of mutagenesis at a defined target gene (Walbot 2000). New genomics approaches, employing strategies for screening by PCR and for plasmid rescue, are now providing indexed collections of mutations and the sequences flanking transposon insertion sites. Users can identify transposon-generated mutants in specific genes after querying a database rather than searching a cornfield. Comprehensive resources mainly based on the Mutator transposon have been and are being established (http://www.mutransposon.org/project/), both for forward and reverse genetic screening of traits and genes, respectively. For example, the NSF-funded Maize Gene Discovery Project uses *RescueMu* plasmid rescue to create immortalized collections of insertion sites in *E. coli* (Lunde et al. 2003). Over 70,000 RescueMu flanking sequences have been sequenced, while cataloging mutant seed and cob phenotypes of 23,000 maize ears, 6200 families of maize seedlings, and 4000 families of adult maize plants carrying *MuDR/Mu* and *RescueMu* insertion alleles. To obtain seed, users could first search the website database for insertions into genes of interest and then perform PCR or hybridization on column libraries to ascertain which plant has the mutation (Lunde et al. 2003).

Recently, RNAi (RNA interference) has emerged as the method of choice to validate gene function in the context of plant development. The essence of RNAi is the delivery of double-stranded RNA (dsRNA) into an organism, or cell, to induce a sequence-specific RNA degradation mechanism that effectively silences a targeted gene (Waterhouse and Helliwell 2003). An important aspect of using RNAi in plant genomics is the delivery of the silencing-inducing dsRNA. This RNA can be delivered by stably transforming plants with transgenes that encode dsRNA. It can also be transiently delivered by bombarding plants with nucleic-acid-coated beads, by infiltrating plant cells with transgene-carrying *Agrobacterium tumefaciens* or by infecting plants with a virus, either on its own or together with a satellite virus (Waterhouse and Helliwell 2003). However, transformation of maize is still not possible at high throughput. Alternatively, virus-induced gene silencing has been proposed for rapid in-vivo gene function tests in maize based on maize streak virus, wheat streak mosaic virus, or barley stripe mosaic virus (Robertson 2004).

Besides mutagenesis- and genome-based approaches, further high-throughput tools have been established to identify and characterize candidate genes for traits of interest, such as mRNA expression profiling. Transcript profiling methods can be divided into two classes: (1) direct analysis, including procedures involving nucleotide sequencing (EST sequencing, SSH, SAGE) and fragment sizing (e.g. cDNA-AFLP; Baldwin et al. 1999); and (2) indirect analysis (macro- or microarray based expression profiling), involving nucleic acid hybridization of mRNA or cDNA fragments (Donson et al. 2002).

These methods have been extensively implemented in diverse maize research fields, such as, water stress, embryogenesis, UV radiation, plant defences (see Table 2 for a brief summary). Microarray-based expression profiling in particular perfectly matches the ambition in genomics of multi-parallel approaches, studying ideally all genes of an organism in one experiment simultaneously. It has been the predominant method for the parallel analysis of gene expression in functional genomics research. In maize, publicly available microarrays contain currently PCR fragments from more than 10,000 different ESTs (http://www.maizegdb.org/microarray.php), whereas long oligo microarrays include 58,000 different oligonucleotides (http://www.maizearray.org/). We used maize unigene microarrays to identify 497 differentially expressed genes associated with SCMV resistance in the near isogenic line (NIL) pair F7[+] and F7 (Shi et al. 2005). Since current maize microarrays do not include all maize genes, complementary approaches are required to identify, e.g. rare transcripts. Auxiliary techniques include subtraction hybridization (Sargent 1987) and related methods, such as suppression subtractive hybridization (SSH) (Diatchenko et al. 1996). The SSH procedure enriches cDNA libraries for low-abundant and differentially expressed mRNAs by normalization (Diatchenko et al. 1996). Shi et al. (unpublished data) have used SSH combined with macroarray

Table 2. Overview of expression profiling experiments in maize

Experiment	Plant material	Reference
EST sequencing		
Gene discovery	Leaf primordial / 10- to 14-day-old endosperm/1- to 2-cm immature ear/4-day-old root/<2-cm tassel/mixed tassel stages/mixed embryo stages/mixed adult organs/1-mm tassel	(Fernandes et al. 2002)
SSH		
Water stress	Seedlings treated with or without 20% PEG	(Zheng et al. 2004)
Root growth	Elongating zone/fully elongated zones along primary root tips of seedlings	(Bassani et al. 2004)
SCMV resistance	Two-week leafs between NIL pair F7+ and F7	(Shi et al. 2005)
Differential display		
Embryogenesis	Endosperms/embryos	(OpsahlFerstad et al. 1997)
Cytokinin signal transduction	Leaves treated with or without *t*-zeatin	(Sakakibara et al. 1998)
Early androgenesis	Induced microspores isolated from one pretreated tassel/pE5 proembryos	(Magnard et al. 2000)
Germination	Embryo axes germinated for different times with or without osmopriming treatment	(Cruz-Garcia et al. 2003)
Cross-fertilization	Cross-fertilized kernels/self-fertilized kernels at 5, 10, and 15 DAP	(Meng et al. 2005)
CDNA-AFLP		
Fungal resistance	A high disease-resistant line/a susceptible line	(Gao et al. 2001)

Table 2. *Continued*

Experiment	Plant material	Reference
Microarray		
Plant defence	A lesion mimic mutant Les9	(Nadimpalli et al. 2000)
Abiotic stress	Female reproductive tissues	(Zinselmeier et al. 2002)
Embryogenesis	Embryos collected at different times	(Lee et al. 2002)
Water stress	Placenta/endosperm/pedicel in developing kernels at 9 and 12 DAP	(Yu and Setter 2003)
Salinity stress	Unstressed roots/roots after 1, 3, 6, 12, 24 and 72 h treated with 150 mM NaCl	(Wang et al. 2003)
Cell type-specific expression profiling	Epidermal cells/vascular tissues of the coleoptiles using laser-capture microdissection	(Nakazono et al. 2003)
Four UV regimes radiation response	NILs varying in flavonoid content	(Casati and Walbot 2003)
eEF1A content	Endosperms of RIL population crossed between the inbreds with high and low eEF1A content	(Lopez-Valenzuela et al. 2004)
UV-B response	Irradiated tissues (roots, immature ears, and leaves)/shielded tissues with plastic	(Casati and Walbot 2004)
UV-B response	The parental inbred lines, B73 and Mo17, of the IBM mapping population after UV treatment	(Blum et al. 2004)
SCMV resistance	Two-week leaves between NIL pair F7$^+$ and F7	(Shi et al., unpublished data)

hybridization to successfully identify genes differently expressed in near iso-genic lines in the context of a virus resistance in maize.

Continued progress in developing and applying tools for functional genomics approaches will generate data on the function of hundreds or thousands of maize genes. The prospects for identification of genes affecting agronomic characters can be expected to increase substantially within the next decade.

5 Genomics and biodiversity: functional markers in maize

Assignment of an "agronomic function" to short sequence motifs can be achieved by candidate gene based association studies (Risch 2000). This approach is limited by linkage disequilibrium (LD), i.e. haplotype structures in the gene(s) of interest. However, for several genes a generally low LD was detected in maize (Remington et al. 2001; Flint-Garcia et al. 2003), including examples in elite materials (Zein and Wenzel, unpublished data). Thus, candidate gene-based association studies are promising in maize. In heterogeneous genotype collections, associations identified for specific sites might be confounded with effects from other genome regions, especially in the case of population stratification (Pritchard et al. 2000), which needs to be taken into account for interpretation of results from association studies.

In a pioneering study, Thornsberry et al. (2001) demonstrated the feasibility of association studies in maize to identify sequence polymorphisms within genes affecting characters of agronomic significance. While taking population structure into account, nine SNP or INDEL polymorphisms were shown to significantly affect flowering time in a set of 92 diverse maize lines. In part, these results were confirmed in a collection of European elite inbred lines (Andersen et al. 2005). The major reason for non-significance of some of the nine polymorphisms identified by Buckler was probably the much narrower genetic material investigated by Andersen et al. (2005) as compared with Thornsberry et al. (2001).

First reports on association studies for genes involved in cell wall biosynthesis indicate that these pathways are promising targets for identification of polymorphic sites associated with forage quality, and thus FM development (Barriere et al. 2003). Zein and Wenzel (unpublished data) investigated the sequence variation at the Bm3 locus in a collection of 42 European maize inbred lines, contrasting with respect to stover DNDF and relevant for hybrid maize breeding in Central Europe. For association with forage quality, stover digestibility was determined in six environments between 2001 and 2003 in Germany (heritability >0.9). One INDEL polymorphism within the intron revealed significant association with stover digestibility (Lübberstedt et al. 2005). In the study of Guillet-Claude et al. (2004a), polymorphisms both in the AldOMT (=Bm3) and the CCoAOMT2 but not CCoAOMT1 coding genes showed significant association with maize digestibility. Moreover, polymorphisms in the maize peroxidase gene ZmPox3 were also significantly associated with maize digestibility (Guillet-Claude et al. 2004b). Similarly, different genes involved in kernel colour, composition, and starch production have been studied at the level of allelic diversity or association studies (Whitt et al. 2002; Palaisa et al. 2003; Wilson et al. 2004). These studies led to the identification of INDEL or SNP polymorphisms associated with yellow endosperm colour (Y1 gene), kernel composition (genes bt2, sh1, and sh2), starch pasting properties (genes ae1, sh2), and amylose levels (ae1, sh1) (Palaisa et al. 2003; Wilson et al. 2004). In conclusion, availability of qualified candidate genes can effectively converted into informative molecular markers by means of association studies. In maize, comprehensive association studies are ongoing in the group of E. Buckler (http://www.maizegenetics.net/), where 18 genes have been studied in a panel of 102 maize genotypes, and within Genoplante (e.g. Guillet-Claude et al. (2004a)). An overview of alle sequencing or association study conducted in maize is shown in Table 3.

Alternatively, TILLING (McCallum et al. 2000) can be employed to relate sequence polymorphisms with phenotypic variation. Variants for virtually all genes of interest in a fixed genetic background can be produced by TILLING (McCallum et al. 2000). The advantage of TILLING as compared to association studies is that isogenic lines are compared, avoiding statistical artifacts due to population structure effects. The disadvantage of TILLING is that establishing a comprehensive TILLING population covering most genes is quite laborious. Therefore, TILLING populations are usually restricted to one or few genetic backgrounds and the alleles fixed within the respective "background genotypes". Thus, if a knock-out allele is fixed at a locus of interest, it might be not possible to identify revertants. Two TILLING populations for maize have been produced at Purdue (http://genome. purdue.edu/maizetilling/) in B73 and W22 background, available for the maize research community. Within the next few years, 150 maize genes will undergo systematic studies using this resource. (http://genome.purdue.edu/ maizetilling/). In the longer run, establishment of homologous recombination as established in moss (*Physcomitrella patens*) (Schaefer and Zryd 1997) would be desirable to generate isogenic genotypes with defined polymorphic differences and, if possible, in any genetic background.

One major recent finding in maize is that of "non-shared sequences" when comparing allelic genome regions (Brunner et al. 2005). In contrast to

Table 3. Overview over allele sequencing / association studies conducted in maize

Gene	Plant material	Trait	Association	Reference
Kernel properties				
Phytoene synthase (Y1)	75 inbred lines (worldwide origin)	Endosperm color (yellow, orange, white)	+	(Palaisa et al. 2003)
		Pasting characteristics	+	
Amylose extender1 (ae1)		Amylose content	+	
		Kernel composition	+	
Brittle endosperm2 (bt2)		Pasting characteristics	+	
	102 inbred lines (worldwide origin)	Kernel composition	+	
Shrunken1 (sh1)		Pasting characteristics	+	(Wilson et al. 2004)
		Amylose content	+	
		Kernel composition	+	
Shrunken2 (sh2)		Amylose content	+	
Sugary1 (su1)		–	–	
Waxy1 (wx1)		–	–	
Amylose extender1 (ae1)		?	–	
Brittle endosperm2 (bt2)	30 inbred lines, 10 ssp	?	–	(Whitt et al. 2002)
		?	–	
Shrunken1 (sh1)	parviglumis	?	–	
Shrunken2 (sh2)	Tricapsum	?	–	
Sugary1 (su1)	dactyloides (Mexico)		–	
Waxy1 (wx1)		?		
Adh1	36 maize			
Stearoyl-ACP-desasturase	genotypes (mainly US origin)	–	–	(Ching et al. 2002)
Acetyl-CoAC-acyltransferase		–	–	
		–	–	

Table 3. *Continued*

Gene	Plant material	Trait	Association	Reference
Whole plant digestibility				
Peroxidase ZmPox3	37 inbred lines temperate regions germplasm)	*Cell wall digestibility*	+	(Guillet-Claude et al. 2004b)
Caffeoyl-CoA 3-O-methyltransferase (COMT)	42 inbred lines (European origin)	DNDF (digestible neutral detergent fiber)	+	(Lübberstedt et al. 2005)
Caffeoyl-CoA 3-O-methyltransferase (CCoAOMT2)			+	
Caffeoyl-CoA 3-O-methyltransferase (CCoAOMT1)	34 inbred lines	Cell wall digestibility	−	(Guillet-Claude et al. 2004a)
Aldehyde O-methyltransferase (AldOMT)			−	
Morphology				
Dwarf 8 (d8)	92 inbred lines (US origin)	Plant height	+	(Thornsberry et al. 2001)
	71 inbred lines	Flowering date	+	(Andersen et al. 2005)
		Plant height	+	
Dwarf 8 (d8)	(European origin) 12 Zea and Tripsacum lines	Flowering date	+	
Terminal ear1 (te1)	(US, Middle and South American)	Morphology	−	(White and Doebley 1999)
Indeterminate (id1)		?	−	
Teosinte branched1 (tb1)	102 inbred lines (worldwide)	?	−	(Remington et al. 2001)
Dwarf8 (d8)		?	−	
Dwarf3 (d3)	24 ssp. mays, 22 ssp.	?	−	
Teosinte branched1 (tb1)	parviglumis, 13 ssp. mexicana, 1 ssp. diplo-perennis	?	−	(Clark et al. 2004)

Table 3. *Continued*

Gene	Plant material	Trait	Association	Reference
Teosinte branched1 (tb1)	11 archeological cobs (Mexico)	?	–	(Jaenicke-Despres et al. 2003)
Prolamine box binding factor (pbf)		?	–	
Sugary1 (su1)		?	–	

the original assumption of sequence variation among genes in identical order explaining for phenotypic variation between genotypes, a surprisingly high level of genomic rearrangements has been found when comparing large stretches of allelic DNA in maize. Therefore, more often then expected, allelic regions will not just be sequence variants but a "functional" allele will pair with a deletion for the respective locus. Moreover, different composition of neighbouring regions might affect expression of the gene of interest by regulatory elements acting over long-distance or through altered chromatin structure (Brunner et al. 2005). For these reasons association studies might have to take the composition of neighboring regions into consideration.

6 Conclusions and outlook

In plant breeding, FMs would be superior as compared with anonymous markers for selection of, e.g. parent materials to build segregating populations, as well as subsequent development of inbred lines. FMs would also be useful for variety registration based on presence/absence of specific alleles at morphological trait loci currently used to discriminate varieties. Rapid progress in maize genomics will shift the current bottleneck for FM development from the availability of candidate genes and the availability of allele sequence information to the assignment of "agronomic function" with sequence polymorphisms, which is currently not systematically considered in functional genomics projects. First reports on association studies in maize are promising. The number of sequence polymorphisms useful for FM development can be expected to increase substantially within the next 5–10 years with the availability of further association studies and the completion of TILLING experiments.

References

Andersen JR, Lübberstedt T (2003) Functional markers in plants. Trends Plant Sci 8:554–560

Andersen JR, Schrag T, Lübberstedt T (2005) Validation of Dwarf8 polymorphisms associated with flowering time in elite European inbred lines of maize (*Zea mays* L.) Theor Appl Genet 111:206–207

Baldwin D, Crane V, Rice D (1999) A comparison of gel-based, nylon filter and microarray techniques to detect differential RNA expression in plants. Curr Opin Plant Biol 2:96–103

Barriere Y, Guillet C, Goffner D, Pichon M (2003) Genetic variation and breeding strategies for improved cell wall digestibility in annual forage crops. A review. Anim Res 52:193–228

Barry GF (2001) The use of the Monsanto draft rice genome sequence in research. Plant Physiol 125:1164–1165

Bassani M, Neumann PM, Gepstein S (2004) Differential expression profiles of growth-related genes in the elongation zone of maize primary roots. Plant Mol Biol 56:367–380

Bennetzen J (2002) The rice genome. Opening the door to comparative plant biology. Science 296:60–63

Blanc G, Hokamp K, Wolfe KH (2003) A recent polyploidy superimposed on older large-scale duplications in the *Arabidopsis* genome. Genome Res 13:137–144

Blum JE, Casati P, Walbot V, Stapleton AE (2004) Split-plot microarray design allows sensitive detection of expression differences after ultraviolet radiation in the inbred parental lines of a key maize mapping population. Plant Cell Environ 27:1374–1386

Borevitz JO, Chory J (2004) Genomics tools for QTL analysis and gene discovery. Curr Opin Plant Biol 7:132–136

Borevitz JO, Ecker JR (2004) Plant genomics: the third wave. Annu Rev Genom Hum Genet 5:443–477

Brunner S, Keller B, Feuillet C (2003) A large rearrangement involving genes and low-copy DNA interrupts the microcollinearity between rice and barley at the Rph7 locus. Genetics 164:673–683

Brunner S, Fengler K, Morgante M, Tingey S, Rafalski A (2005) Evolution of DNA sequence nonhomologies among maize inbreds. Plant Cell 17:343–60

Buell CR (2002) Obtaining the sequence of the rice genome and lessons learned along the way. Trends Plant Sci 7:538–542

Burke DT, Carle GF, Olson MV (1987) Cloning of large segments of exogenous DNA into yeast by means of artificial chromosome vectors. Science 236:806–812

Casati P, Walbot V (2003) Gene expression profiling in response to ultraviolet radiation in maize genotypes with varying flavonoid content. Plant Physiol 132:1739–1754

Casati P, Walbot V (2004) Rapid transcriptome responses of maize (*Zea mays*) to UV-B in irradiated and shielded tissues. Genome Biology 5:R16

Ching A, Caldwell KS, Jung M, Dolan M, Smith OS, Tingey S, Morgante M, Rafalski AJ (2002) SNP frequency, haplotype structure and linkage disequilibrium in elite maize inbred lines. BMC Genet 3:19

Clark RM, Linton E, Messing J, Doebley JF (2004) Pattern of diversity in the genomic region near the maize domestication gene tb1. Proc Natl Acad Sci USA 101:700–707

Collins FS, Green ED, Guttmacher AE, Guyer MS (2003) A vision for the future of genomics research. Nature 422:835–847

Cruz-Garcia F, Gomez A, Zuniga JJ, Plasencia J, Vazquez-Ramos JM (2003) Cloning and characterization of a COBRA-like gene expressed de novo during maize germination. Seed Sci Re 13:209–217

Delseny M (2003) Towards an accurate sequence of the rice genome. Curr Opin Plant Biol 6: 101–105

Delseny M (2004) Re-evaluating the relevance of ancestral shared synteny as a tool for crop improvement. Curr Opin Plant Biol 7:126–131

Devos KM, Gale MD (1997) Comparative genetics in the grasses. Plant Mol Biol 35:3–15

Diatchenko L, Lau YF, Campbell AP, Chenchik A, Moqadam F, Huang B, Lukyanov S, Lukyanov K, Gurskaya N, Sverdlov ED, Siebert PD (1996) Suppression subtractive hybridization: a method for generating differentially regulated or tissue-specific cDNA probes and libraries. Proc Natl Acad Sci USA 93:6025–6030

Donson J, Fang Y, Espiritu–Santo G, Xing W, Salazar A, Miyamoto S, Armendarez V, Volkmuth W (2002) Comprehensive gene expression analysis by transcript profiling. Plant Mol Biol 48:75–97

Fernandes J, Brendel V, Gai X, Lal S, Chandler VL, Elumalai RP, Galbraith DW, Pierson EA, Walbot V (2002) Comparison of RNA expression profiles based on maize expressed sequence tag frequency analysis and micro-array hybridization. Plant Physiol 128:896–910

Flint-Garcia SA, Thornsberry JM, Buckler ESt (2003) Structure of linkage disequilibrium in plants. Annu Rev Plant Biol 54:357–374

Frisch M (2005) Breeding strategies: optimum design of markers—assisted backcross programs. In: Loerz H, Wenzel G (eds) Biotechnology in agriculture and forestry vol 55. Molecular marker systems in plant breeding and crop improvement. Springer-Verlag, Berlin, pp 319–334

Frisch M, Bohn M, Melchinger AE (1999) Minimum sample size and optimal positioning of flanking markers in marker-assisted backcrossing for transfer of a target gene. Crop Sci 39: 967–975

Fu H, Dooner HK (2002) Intraspecific violation of genetic colinearity and its implications in maize. Proc Natl Acad Sci USA 99:9573–9578

Gao ZH, Xue YB, Dai JR (2001) cDNA-AFLP analysis reveals that maize resistance to *Bipolaris maydis* is associated with the induction of multiple defense-related genes. Chinese Sci Bull 46:1454–1458

Goff SA, Ricke D, Lan TH, Presting G, Wang RL, Dunn M, Glazebrook J, Sessions A, Oeller P, Varma H, Hadley D, Hutchinson D, Martin C, Katagiri F, Lange BM, Moughamer T, Xia Y, Budworth P, Zhong JP, Miguel T, Paszkowski U, Zhang SP, Colbert M, Sun WL, Chen LL, Cooper B, Park S, Wood TC, Mao L, Quail P, Wing R, Dean R, Yu YS, Zharkikh A, Shen R, Sahasrabudhe S, Thomas A, Cannings R, Gutin A, Pruss D, Reid J, Tavtigian S, Mitchell J, Eldredge G, Scholl T, Miller RM, Bhatnagar S, Adey N, Rubano T, Tusneem N, Robinson R, Feldhaus J, Macalma T, Oliphant A, Briggs S (2002) A draft sequence of the rice genome (*Oryza sativa* L. ssp japonica). Science 296:92–100

Guillet-Claude C, Birolleau-Touchard C, Manicacci D, Fourmann M, Barraud S, Carret V, Martinant JP, Barriere Y (2004a) Genetic diversity associated with variation in silage corn digestibility for three O-methyltransferase genes involved in lignin biosynthesis. Theor Appl Genet 110:126–135

Guillet-Claude C, Birolleau-Touchard C, Manicacci D, Rogowsky PM, Rigau J, Murigneux A, Martinant JP, Barriere Y (2004b) Nucleotide diversity of the ZmPox3 maize peroxidase gene:relationships between a MITE insertion in exon 2 and variation in forage maize digestibility. BMC Genet 5:19

Hoisington DA, Melchinger AE (2005) From theory to practice: marker-assisted selection in maize. In: Loerz H, Wenzel G (eds) Biotechnology in agriculture and forestry vol 55. Molecular marker systems in plant breeding and crop improvement. Springer-Verlag, Berlin, pp 335–352

Huang L, Brooks SA, Li WL, Fellers JP, Trick HN, Gill BS (2003) Map-based cloning of leaf rust resistance gene Lr21 from the large and polyploid genome of bread wheat. Genetics 164:655–664

Ioannou PA, Amemiya CT, Garnes J, Kroisel PM, Shizuya H, Chen C, Batzer MA, de Jong PJ (1994) A new bacteriophage P1-derived vector for the propagation of large human DNA fragments. Nat Genet 6:84–89

Jaenicke-Despres V, Buckler ES, Smith BD, Gilbert MT, Cooper A, Doebley J, Paabo S (2003) Early allelic selection in maize as revealed by ancient DNA. Science 302:1206–1208

Jensen LB, Andersen JR, Lübberstedt T (2005) QTL mapping of vernalization response in perennial ryegras reveals cosegregation with an orthologue of wheat VRN1. Theor Appl Genet (in press)

Lee JM, Williams ME, Tingey SV, Rafalski JA (2002) DNA array profiling of gene expression changes during maize embryo development. Funct Integr Genom 2:13–27

Lopez-Valenzuela JA, Gibbon BC, Holding DR, Larkins BA (2004) Cytoskeletal proteins are coordinately increased in maize genotypes with high levels of eEF1A. Plant Physiol 135:1784–1797

Lübberstedt T, Melchinger AE, Fahr S, Klein D, Dally A, Westhoff P (1998) QTL mapping in testcrosses of flint lines of maize: III. Comparison across populations for forage traits. Crop Sci 38:1278–1289

Lübberstedt T, Zein I, Andersen JR, Wenzel G, Krüzfeldt B, Eder J, Ouzunova M, Shi C (2005) Development and application of functional markers in maize. Euphytica (in press)

Lunde CF, Morrow DJ, Roy LM, Walbot V (2003) Progress in maize gene discovery: a project update. Funct Integr Genomics 3:25–32

Magnard JL, Le Deunff E, Domenech J, Rogowsky PM, Testillano PS, Rougier M, Risueno MC, Vergne P, Dumas C (2000) Genes normally expressed in the endosperm are expressed at early stages of microspore embryogenesis in maize. Plant Mol Biol 44:559–574

McCallum CM, Comai L, Greene EA Henikoff S (2000) Targeted screening for induced mutations. Nat Biotechnol 18:455–457

Meng FR, Ni ZF, Wu LM, Sun QX (2005) Differential gene expression between cross-fertilized and self-fertilized kernels during the early stages of seed development in maize. Plant Sci 168:23–28

Mohler V, Singrün C (2005) General considerations: marker–assisted selection. In: Loerz H, Wenzel G (eds) Biotechnology in agriculture and forestry, vol 55. Molecular marker systems in plant breeding and crop improvement. Springer-Verlag, Berlin, pp 305–318

Moore G, Devos KM, Wang Z, Gale MD (1995) Cereal genome evolution—grasses, line up and form a circle. Curr Biol 5:737–739

Nadimpalli R, Yalpani N, Johal GS, Simmons CR (2000) Prohibitins, stomatins, and plant disease response genes compose a protein superfamily that controls cell proliferation, ion channel regulation, and death. J Biol Chem 275:29579–29586

Nakazono M, Qiu F, Borsuk LA, Schnable PS (2003) Laser-capture microdissection, a tool for the global analysis of gene expression in specific plant cell types: identification of genes expressed differentially in epidermal cells or vascular tissues of maize. Plant Cell 15:583–596

Nasu S, Suzuki J, Ohta R, Hasegawa K, Yui R, Kitazawa N, Monna L, Minobe Y (2002) Search for and analysis of single nucleotide polymorphisms (SNPs) in rice (Oryza sativa, Oryza rufipogon) and establishment of SNP markers. DNA Res 9:163–171

OpsahlFerstad HG, LeDeunff E, Dumas C, Rogowsky PM (1997) ZmEsr, a novel endosperm-specific gene expressed in a restricted region around the maize embryo. Plant J 12:235–246

Palaisa KA, Morgante M, Williams M, Rafalski A (2003) Contrasting effects of selection on sequence diversity and linkage disequilibrium at two phytoene synthase loci. Plant Cell 15:1795–1806

Pritchard JK, Stephens M, Rosenberg NA, Donnelly P (2000) Association mapping in structured populations. Am J Hum Genet 67:170–181

Remington DL, Thornsberry JM, Matsuoka Y, Wilson LM, Whitt SR, Doeblay J, Kresovich S, Goodman MM, Buckler ES (2001) Structure of linkage disequilibrium and phenotypic associations in the maize genome. Proc Natl Acad Sci USA 98:11479–11484

Risch NJ (2000) Searching for genetic determinants in the new millennium. Nature 405:847–856

Robertson D (2004) VIGS vectors for gene silencing: many targets, many tools. Annu Rev Plant Biol 55:495–519

Sakakibara H, Suzuki M, Takei K, Deji A, Taniguchi M, Sugiyama T (1998) A response-regulator homologue possibly involved in nitrogen signal transduction mediated by cytokinin in maize. Plant J 14:337–344

Salse J, Piegu B, Cooke R, Delseny M (2004) New in silico insight into the synteny between rice (*Oryza sativa* L.) and maize (*Zea mays* L.) highlights reshuffling and identifies new duplications in the rice genome. Plant J 38:396–409

Sargent TD (1987) Isolation of differentially expressed genes. Meth Enzymol 152:423–432

Sasaki T (1998) The rice genome project in Japan. Proc Natl Acad Sci USA 95:2027–2028

Sasaki T (2002) Rice genomics to understand rice plant as an assembly of genetic codes. Curr Sci 83:834–839

Sasaki T, Burr B (2000) International Rice Genome Sequencing Project: the effort to completely sequence the rice genome. Curr Opin Plant Biol 3:138–141

Schaefer DG, Zryd JP (1997) Efficient gene targeting in the moss *Physcomitrella patens*. Plant J 11:1195–1206

Schmidt R (2000) Synteny:recent advances and future prospects. Curr Opin Plant Biol 3:97–102

Shi C, Ingvardsen C, Thuemmler F, Melchinger AE, Wenzel G, Lübberstedt T (2005) Identification of differentially expressed genes between maize near-isogenic lines in association with SCMV resistance using suppression subtractive hybridization. Mol Genet Genom DoI 10.1007/S00438-004-1103-8

Shizuya H, Birren B, Kim UJ, Mancino V, Slepak T, Tachiiri Y, Simon M (1992) Cloning and stable maintenance of 300-kilobase-pair fragments of human DNA in *Escherichia coli* using an F-factor-based vector. Proc Natl Acad Sci USA 89:8794–8797

Thornsberry JM, Goodman MM, Doebley J, Kresovich S, Nielsen D, Buckler ES (2001) Dwarf8 polymorphisms associate with variation in flowering time. Nat Genet 28:286–289

Thro AM, Parrott W, Udall JA, Beavis WD (2004) This issue in crop science—symposium on genomics and plant breeding: the experience of the initiative for future agricultural and food systems—introduction. Crop Sci 44:1893–1893

Tyagi AK, Khurana JP, Khurana P, Raghuvanshi S, Gaur A, Kapur A, Gupta V, Kumar D, Ravi V, Vij S, Sharma S (2004) Structural and functional analysis of rice genome. J Genet 83:79–99

Varotto C, Leister D (2002) Maize in the genomics era. Maydica 47:203–211

Walbot V (2000) Saturation mutagenesis using maize transposons. Curr Opin Plant Biol 3:103–107

Wang H, Miyazaki S, Kawai K, Deyholos M, Galbraith DW, Bohnert HJ (2003) Temporal progression of gene expression responses to salt shock in maize roots. Plant Mol Biol 52:873–891

Waterhouse PM, Helliwell CA (2003) Exploring plant genomes by RNA-induced gene silencing. Nat Rev Genet 4:29–38

White SE, Doebley JF (1999) The molecular evolution of terminal earl, a regulatory gene in the genus Zea. Genetics 153:1455–1462

Whitt SR, Wilson LM, Tenaillon MI, Gaut BS, Buckler ESt (2002) Genetic diversity and selection in the maize starch pathway. Proc Natl Acad Sci USA 99:12959–12962

Wilson LM, Whitt SR, Ibanez AM, Rocheford TR, Goodman MM, Buckler ESt (2004) Dissection of maize kernel composition and starch production by candidate gene association. Plant Cell 16:2719–2733

Yu J, Hu SN, Wang J, Wong GKS, Li SG, Liu B, Deng YJ, Dai L, Zhou Y, Zhang XQ, Cao ML, Liu J, Sun JD, Tang JB, Chen YJ, Huang XB, Lin W, Ye C, Tong W, Cong LJ, Geng JN, Han YJ, Li L, Li W, Hu GQ, Huang XG, Li WJ, Li J, Liu ZW, Li L, Liu JP, Qi QH, Liu JS, Li L, Li T, Wang XG, Lu H, Wu TT, Zhu M, Ni PX, Han H, Dong W, Ren XY, Feng XL, Cui P, Li XR, Wang H, Xu X, Zhai WX, Xu Z, Zhang JS, He SJ, Zhang JG, Xu JC, Zhang KL, Zheng

XW, Dong JH, Zeng WY, Tao L, Ye J, Tan J, Ren XD, Chen XW, He J, Liu DF, Tian W, Tian
    CG, Xia HG, Bao QY, Li G, Gao H, Cao T, Wang J, Zhao WM, Li P, Chen W, Wang XD,
    Zhang Y, Hu JF, Wang J, Liu S, Yang J, Zhang GY, Xiong YQ, Li ZJ, Mao L, Zhou CS, Zhu
    Z, Chen RS, Hao BL, Zheng WM, Chen SY, Guo W, Li GJ, Liu SQ, Tao M, Wang J, Zhu LH,
    Yuan LP, Yang HM (2002) A draft sequence of the rice genome (*Oryza sativa* L. ssp indica)
    Science 296:79–92
Yu LX, Setter TL (2003) Comparative transcriptional profiling of placenta and endosperm in
    developing maize kernels in response to water deficit. Plant Physiol 131:568–582
Yu YS, Rambo T, Currie J, Saski C, Kim HR, Collura K, Thompson S, Simmons J, Yang TJ,
    Nah G, Patel AJ, Thurmond S, Henry D, Oates R, Palmer M, Pries G, Gibson J, Anderson
    H, Paradkar M, Crane L, Dale J, Carver MB, Wood T, Frisch D, Engler F, Soderlund C,
    Palmer LE, Tetylman L, Nascimento L, de la Bastide M, Spiegel L, Ware D, O'Shaughnessy
    A, Dike S, Dedhia N, Preston R, Huang E, Ferraro K, Kuit K, Miller B, Zutavern T,
    Katzenberger F, Muller S, Balija V, Martienssen RA, Stein L, Minx P, Johnson D, Cordum
    H, Mardis E, Cheng ZK, Jiang JM, Wilson R, McCombie WR, Wing RA, Yuan QP, Shu OY,
    Liu J, Jones KM, Gansberger K, Moffat K, Hill J, Tsitrin T, Overton L, Bera J, Kim M, Jin
    SH, Tallon L, Ciecko A, Pai G, Van Aken S, Utterback T, Reidmuller S, Bormann J,
    Feldblyum T, Hsiao J, Zismann V, Blunt S, de Vazeilles A, Shaffer T, Koo H, Suh B, Yang
    Q, Haas B, Peterson J, Pertea M, Volfovsky N, Wortman J, White O, Salzberg SL, Fraser
    CM, Buell CR, Messing J, Song RT, Fuks G, Llaca V, Kovchak S, Young S, Bowers JE,
    Paterson AH, Johns MA, Mao L, Pan HQ, Dean RA and Cons RCS (2003) In-depth view
    of structure, activity, and evolution of rice chromosome 10. Science 300:1566–1569
Zheng J, Zhaol JF, Tao YZ, Wang JH, Liu YJ, Fu JJ, Jin Y, Gao P, Zhang JP, Bai YF and Wang GY
    (2004) Isolation and analysis of water stress induced genes in maize seedlings by subtrac-
    tive PCR and cDNA macroarray. Plant Mol Biol 55:807–823
Zinselmeier C, Sun YJ, Helentjaris T, Beatty M, Yang S, Smith H, Habben J (2002) The use of
    gene expression profiling to dissect the stress sensitivity of reproductive development in
    maize. Field Crops Res 75:111–121

Cun Shi
Gerhard Wenzel
Technical University of Munich, Chair of Plant Breeding
Am Hochanger 2, D-85350 Freising, Germany
e-mail: gwenzel@wzw.tum.de

Ursula Frei
Thomas Lübberstedt
Danish Institute of Agricultural Sciences,
Research Centre Flakkebjerg,
Department of Genetics and Biotechnology,
DK-4200 Slagelse, Denmark
Thomas.luebberstedt@agrsci.dk

Extranuclear inheritance:
Gene transfer out of plastids

Ralph Bock

1 Introduction: the evolutionary significance of gene transfer

The emergence of eukaryotes from prokaryotes is considered one of the major evolutionary transitions, and has triggered a dramatic increase in complexity of life on earth (Szathmary and Smith 1995). Szathmáry and Smith define three features shared by all major transitions in evolution:

(i) Loss of independent replication by formerly independent entities becoming parts of larger units.
(ii) Appearance of division of labor.
(iii) Significant changes in the way in which (genetic) information is stored and transmitted (Szathmary and Smith 1995).

Clearly, the evolution of the eukaryotic cell fulfills all three criteria: it involved the association of smaller independently replicating entities to form a larger whole by combining altogether three organisms and their genomes in one and the same cell (Fig. 1). In an endosymbiosis-like process, the pre-eukaryotic host successively engulfed two symbionts: an α-proteobacterium that gave rise to mitochondria and a cyanobacterium that gave rise to plastids. The endosymbiotic uptake of the two eubacterial cells was followed by the gradual integration of the endosymbionts into the metabolism of the host cell by establishing a division of labor and inventing sophisticated regulatory networks to co-ordinate the host's gene expression with that of the endosymbionts. This process was accompanied by a drastic restructuring of all three genomes (Fig. 1) and involved (i) the loss of dispensable genetic information (such as genes for bacterial cell wall biosynthesis), (ii) the elimination of redundant genetic information (for example, genes for amino acid biosyntheses present in all three genomes), (iii) the acquisition of new gene functions to co-ordinate gene expression and metabolism in the three genetic compartments (for example, by establishing new signal transduction chains), and (iv) the massive translocation of genetic information between the three genomes (Fig. 1; Martin and Herrmann 1998). The main direction of this

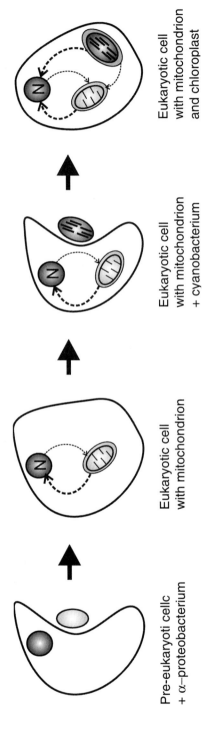

Pre-eukaryoti cellc
+ α–proteobacterium

Eukaryotic cell
with mitochondrion

Eukaryotic cell
with mitochondrion
+ cyanobacterium

Eukaryotic cell
with mitochondrion
and chloroplast

Fig. 1. The endosymbiotic origin of cell organelles and the directions of gene transfer between the three genetic compartments of the plant cell. Massive gene transfer has occurred from the organellar genomes to the nuclear genome (thick arrows) whereas gene transfer from the nuclear genome to the mitochondrion is much rarer (thin arrow), and no formerly nuclear sequences have been detected in plastid genomes. Gene transfer between the two DNA-containing cell organelles has also been documented but seems to be a one-way street from the plastid to the mitochondrion (thin arrow).

gene transfer has been from the endosymbionts' genomes to the host genome (Fig. 1), leading to a dramatic reduction in genome size and gene content of both plastid and mitochondrial genomes. As a consequence, contemporary organellar genomes are greatly reduced and contain only a small proportion of the genes that their free-living ancestors had possessed. Whereas eubacterial genomes usually carry a few thousand genes (the cyanobacterium *Synechocystis*, for example, has approximately 3200 genes; Kaneko et al. 1996; Kaneko and Tabata 1997; Kotani and Tabata 1998), higher plant plastid genomes (plastomes) harbor only about 130 genes (Sugiura 1989; Wakasugi et al. 2001) and plant mitochondrial genomes have an even lower coding capacity of approximately 60 genes (Unseld et al. 1997; Knoop, 2004).

The massive translocation of genetic information between the nuclear, plastid and mitochondrial genomes represents a hallmark of the evolutionary shaping of the plant cell following the endosymbiotic uptake of the eubacterial progenitors of the two types of cell organelles. This review discusses the transfer of genetic information from the plastid genome to the two other genomes of the plant cell in the nucleus and the mitochondrion.

1. The currently available information regarding functional and non-functional gene transfers from the plastid genome to the nuclear and mitochondrial genomes is summarized.
2. Novel experimental approaches aiming at the reconstruction in the laboratory of gene transfer processes from the plastid to the nuclear genome are reviewed.
3. Possible mechanisms of intercompartmental DNA translocation are highlighted and the implications of frequent gene transfer out of plastids for the biosafety of plants with transgenic plastids are discussed.

2 Gene transfer from the plastid to the nuclear genome

Two types of plastid sequences are found in the nuclear genome of higher plants: (i) functional nuclear genes of plastid origin and (ii) presumably non-functional plastid sequences referred to as "promiscuous DNA". Functional gene transfer events involve the translocation of plastid-encoded genes into the nuclear genome followed by acquisition of functionality of the formerly plastid genes in the nuclear environment (Fig. 2). Once the transferred gene has become functional in the nucleus, the plastid-encoded original is no longer needed and can degenerate through accumulation of deleterious mutations. Presence in the plastid genome of pseudogene-like remnants of transferred genes is generally taken as an

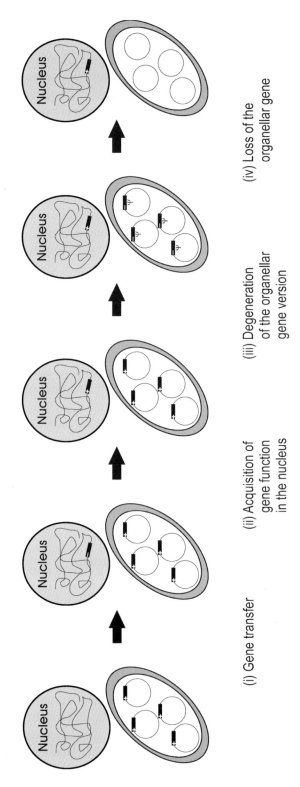

Fig. 2. Model for functional gene transfer from the organelle to the nucleus. (+) indicates functional genes that are actively expressed, (−) indicates non-functional genes and Ψ denotes pseudogenes. Initially, transferred genes are probably inactive or only poorly expressed in the nucleus and may have to acquire mutations facilitating their efficient expression from the nuclear genome. Such mutations may involve both point mutations (e.g. to adjust the codon usage) and more complex rearrangements that create (or capture from existing nuclear genes) a eukaryotic promoter and terminator structure, a polyadenylation signal and, if needed, a transit peptide for re-directing the gene product into the organellar compartment. Once the transferred genes have acquired functionality in the nucleus, their organellar homologues are no longer required and thus can accumulate deleterious mutations without negative phenotypic consequences. In this way, mutational inactivation of the organellar gene leads to irreversible fixation of the nuclear gene version.

(i) Gene transfer   (ii) Acquisition of gene function in the nucleus   (iii) Degeneration of the organellar gene version   (iv) Loss of the organellar gene

indication of the transfer event having taken place relatively recently (Shimada and Sugiura, 1991; Millen et al., 2001). However, these are exceptional cases and the vast majority of genes from the cyanobacterial endosymbiont that was transferred to the nuclear genome have disappeared completely from present-day plastid genomes (Fig. 2). By contrast, non-functional transfer events involve plastid sequences that (i) are still present and fully functional in the plastid genome and (ii) are highly unlikely to function in the nucleus because they are not expressed there. To what extent such 'promiscuous' DNA sequences are intermediates of functional gene transfers is currently unknown.

2.1 Functional gene transfer from the plastid to the nuclear genome

Sequencing of the entire genome of the model plant *Arabidopsis thaliana* (The Arabidopsis Genome Initiative 2000) has made it possible to determine the extent of functional gene transfer from the plastid to the nuclear genome. Bioinformatics analysis of 25,000 protein-coding genes revealed the presence of approximately 4500 cyanobacterial genes in the nucleus of *Arabidopsis* (Martin et al. 2002) indicating that as much as 18% of all plant genes were acquired from the cyanobacterial ancestor of plastids. It is important to note that not all gene products of transferred plastid genes are re-targeted to the plastid compartment, but instead can function in the cytosol or even in the mitochondrion. In fact, the majority of *Arabidopsis* genes of cyanobacterial origin are targeted to subcellular compartments other than the plastid (Martin et al. 2002).

In some cases, it has been possible to reconstruct the molecular events involved in functional gene transfer processes from the plastome to the nuclear genome. The *infA* gene encoding the plastid translation initiation factor 1 provides a particularly interesting example of evolutionary recent gene transfer events (Millen et al. 2001). It had long been known that *infA*, while being a functional gene in the plastid genome of the liverwort *Marchantia polymorpha* and rice, exists as a pseudogene in the tobacco plastome (Shinozaki et al. 1986; Ohyama et al., 1986; Hiratsuka et al. 1989; Shimada and Sugiura 1991). Presence of an intact *infA* in some higher plant plastomes and retention of a highly homologous pseudogene in tobacco strongly suggest that the *infA* pseudogene in tobacco represents an evolutionary intermediate on the way to complete elimination of the gene from the plastid genome (Fig. 2).

A systematic phylogenetic study of *infA* structure in angiosperm plastid genomes revealed that the gene has repeatedly become non-functional in approximately 24 separate lineages of angiosperm evolution. A search for nuclear *infA* copies in four of these lineages resulted in identification of expressed nuclear *infA* genes whose gene products are targeted to plastids. Molecular analysis of the nuclear loci (exon-intron structure, transit peptide sequence) provided strong evidence for four independent gene transfer events (Millen et al. 2001).

Multiple recent gene transfer events as demonstrated for the *infA* gene indicate that the transfer of plastid genes to the nucleus is a still ongoing process, and that the gene content of present-day plastomes represents by no means an evolutionary static state of affairs.

2.2 Non-functional gene transfer from the plastid to the nuclear genome

Pieces of plastid and mitochondrial DNA are often found in nuclear genomes and, assuming accidental escape from the organelle, these sequences are commonly referred to as promiscuous DNA. Promiscuous DNA sequences in the nucleus lack any apparent function, but, on an evolutionary timescale, may provide the raw material for converting organellar genes into functional nuclear genes (Fig. 2). Furthermore, promiscuous DNA of mitochondrial origin has recently been implicated in DNA repair in yeast by patching broken chromosomes (Ricchetti et al., 1999; Yu and Gabriel, 1999).

In most higher plants, the nuclear genome harbors a significant amount of promiscuous DNA of plastid origin (Timmis and Scott 1983; Scott and Timmis 1984; Ayliffe and Timmis 1992; Ayliffe et al. 1998; reviewed in Thorsness and Weber 1996; Timmis et al. 2004). Data obtained for the *rbcL* region of the tobacco plastid genome suggest that there is a minimum of 15 nuclear plastid DNA insertions that contain at least a proportion of the *rbcL* gene (Ayliffe and Timmis 1992). Many of the integrants are rather large, spanning several kb of plastid DNA, and it has been suggested that, in an extreme case, the size of the integrated tract may even exceed the size of the 156 kb plastid genome monomer (Ayliffe and Timmis 1992), indicating that more than one plastome copy contributed to the insertion. Interestingly, nuclear genome size and amount of promiscuous plastid DNA in the nucleus appear to be negatively correlated, as comparably little promiscuous plastid DNA is found in the small genome of the genetic model plant *Arabidopsis* (Ayliffe et al., 1998).

3 Gene transfer from the plastid to the mitochondrial genome

The mitochondrial genome (chondriome) of land plants has a circular map (Oda et al. 1992; Brennicke et al. 1996; Unseld et al. 1997), although linear genome molecules may be predominantly present *in vivo* (Bendich and Smith 1990; Oldenburg and Bendich 1998, 2001). Plant mitochondrial genomes are generally far bigger than animal chondriomes and display considerable size variation, ranging from 186.8 kb in the liverwort *Marchantia polymorpha* to more than 2000 kb in some Cucurbitaceae species (reviewed, for example, in Knoop 2004). However, the gene numbers in land plant chondriomes display only little variation and there is no correlation between genome size and coding capacity. Almost any size increase can be attributed to the accumulation of additional non-coding DNA, which is predominantly found in large intergenic spacers, but to some extent also in introns (Kao et al. 1984; Albertazzi et al. 1998). Much of this seemingly non-functional DNA lacks any homology to known sequences, making it currently impossible to trace back its evolutionary origin. However, in some instances, striking homology with chloroplast genes revealed that DNA transfer from the plastid to the mitochondrion has contributed significantly to the genome size of present-day plant mitochondria. In many cases, this mitochondrial DNA of plastid origin seems to be non-functional and hence truly "promiscuous". There are, however, several examples where transferred plastid genes have been utilized to make functional mitochondrial genes.

3.1 Non-functional gene transfer from the plastid to the mitochondrial genome

The first example of plastid genes residing in a plant mitochondrial genome was described in 1982, when Stern and Lonsdale found that the plastid and mitochondrial genomes of maize (*Zea mays*) share a 12 kb DNA fragment (Stern and Lonsdale, 1982).

This fragment appeared to stem from the inverted repeat region of the chloroplast genome and contained the 16S rRNA gene, most of the spacer between the 16S and the downstream 23S rRNA gene [this spacer harbors two intron-containing tRNA genes, tRNA-Ile(GAU) and tRNA-Ala(UGC) and a large region upstream of the 16S rRNA gene (Stern and Lonsdale 1982)]. Analysis of maize lines carrying different cytoplasms revealed that, although the transferred plastid fragment was present in all of them, there was significant length variation that could be accounted for by internal deletions (Stern and Lonsdale 1982). This may indicate an evolutionary tendency to lose transferred promiscuous DNA due to the lack of functional constraints.

It soon became clear that the presence of promiscuous DNA of chloroplast origin is not a curiosity but rather a typical feature of higher plant mitochondrial genomes (Schuster and Brennicke 1987, 1988; Nugent and Palmer 1988; Jubier et al. 1990; Nakazono and Hirai 1993). An analysis of the completely sequenced rice (*Oryza sativa*) chondriome revealed that as much as 6.3% of this mitochondrial genome is of plastid origin (Notsu et al. 2002). There is significant variation between species regarding which regions of the plastome are found incorporated into the chondriome. The picture is further complicated by the recent finding that during evolution, multiple independent transfer events may have occurred. For example, the plastid-encoded gene for the large subunit of Rubisco, *rbcL*, has been repeatedly transferred to the mitochondrial genome and a phylogenetic study has identified at least five independent transfer events in angiosperms (Cummings et al. 2003).

It seems unlikely that these transferred (photosynthesis-related) gene sequences are functional inside the mitochondrion, and most of the promiscuous plastid DNA sequences in mitochondrial genomes can be regarded as pseudogenes. Even in those cases where complete genes were transferred, the coding sequences often have accumulated mutations and the transcripts, although sometimes detectable, are not properly processed (Zeltz et al. 1996). Also, to date no evidence for active translation of transcripts from promiscuous plastid DNA fragments has been provided.

A 6.8 kb plastid DNA insertion in the rice mitochondrial genome comprises almost the complete *rpoB* operon (Nakazono and Hirai 1993), which consists of three genes for subunits of the *E. coli*-like plastid RNA polymerase: *rpoB*, *rpoC1* and *rpoC2*. However, several lines of evidence make it highly unlikely that there is a functional *E. coli*-like RNA polymerase in rice mitochondria:

(i) The gene for the essential α-subunit of the RNA polymerase (*rpoA*) was not transferred from the plastome to the chondriome;
(ii) Most of the *rpoC2* coding region is missing from the promiscuous fragment in the chondriome (Nakazono and Hirai 1993; Zeltz et al. 1996);
(iii) The *rpoB* coding region has accumulated several point mutations (Zeltz et al. 1996); and
(iv) The three RNA editing sites undergoing post-transcriptional C-to-U conversion in the plastid *rpoB* transcript are not edited in mitochondria (Zeltz et al. 1996; Bock 2000, 2001a) leading to additional amino acid sequence deviations between the plastid and mitochondrial *rpoB* versions.

Remarkably, no promiscuous plastid sequences are found in the mitochondrial genomes of the liverwort *Marchantia polymorpha* (Oda et al. 1992a,b) and the green algae *Chlamydomonas reinhardtii* and *Chlamydomonas eugametos* (Denovan-Wright et al. 1998). This could indicate that either the competence to take up plastid DNA or the capability to tolerate promiscuous plastid sequences in the chondriome has appeared relatively late in plant evolution. The recent spectacular finding that higher plant mitochondria are capable of actively importing DNA via the permeability transition pore complex

(Koulintchenko et al. 2003) may provide us with a first clue about the mechanism of DNA transfer from the plastome to the chondriome. Whether or not this DNA import capacity is restricted to higher plant mitochondria and thus could explain the absence of promiscuous plastid DNA from lower plant chondriomes remains to be investigated.

3.2 Functional gene transfer from the plastid to the mitochondrial genome

Most plastid genes are likely to be useless when transferred to the mitochondrial genome. This is because the majority of protein-coding genes on the plastid genome are involved in photosynthesis. However, what plastids and mitochondria have in common is a prokaryotic-type genetic apparatus and, in theory, at least some components of the gene expression machinery should be exchangeable between the two organelles. Indeed, this seems to be the case for tRNA genes. The mitochondrial genomes of angiosperms contain four tRNA genes, *tRNA-Met(CAU)*, *tRNA-His(GUG)*, *tRNA-Asn(GUU)* and *tRNA-Trp(CCA)*, that clearly come from the plastid genome (Binder et al. 1990; Ambrosini et al. 1992; Weber-Lotfi et al. 1993; Veronico et al. 1996; Unseld et al. 1997) and appear to have replaced the homologous mitochondrial tRNA genes. Besides tRNAs, no other example of transferred chloroplast genes that have become functional in the mitochondrion is known, possibly suggesting that a transferred small RNA gene, like a tRNA, can become functional more easily than a protein-coding gene or a large RNA gene (such as an rRNA gene) that has co-evolved with dozens of interacting proteins.

Interestingly, plastid-derived tRNA genes are not found in the completely sequenced mitochondrial genome of the bryophyte *Marchantia polymorpha* (Oda et al. 1992a,b), indicating that plant mitochondria had a full set of tRNA genes derived from the α-proteobacterial endosymbiont before these tRNA gene transfer events have occurred. Thus, it seems reasonable to assume that the transfer of a functional tRNA from the plastid genome allowed for the subsequent loss of the corresponding mitochondrial tRNA gene. Whether or not such replacements of mitochondrial by chloroplast tRNA genes occurred by chance is currently unknown. It seems difficult to envisage that there could have been any selective advantage of recruiting plastid tRNAs for mitochondrial translation. In theory, evolutionary optimization of the resident mitochondrial tRNAs and their expression by playing with mutations should be a simpler and more efficient strategy for shaping the translational apparatus of plant mitochondria. However, making functional mitochondrial tRNAs out of transferred plastid DNA fragments may be easier to accomplish than we currently anticipate: Chloroplasts and mitochondria both have a bacteriophage-like RNA polymerase that is nuclear-encoded. If the mitochondrial enzyme would recognize the promoters of plastid tRNAs, then transferred plastid tRNA genes would immediately be functional in the mitochondrion. Some higher plants like *Arabidopsis thaliana* and *Nicotiana sylvestris* (Hedtke et al. 2000; Kobayashi et al. 2001, 2002) even have three nuclear genes for organellar RNA polymerases with one of them being targeted to mitochondria, a second to plastids and the third one being even dually targeted to both plastids and mitochondria (Hedtke et al. 2000; Kobayashi et al. 2001). Whether or not this dually targeted RNA polymerase is capable of transcribing plastid and mitochondrial tRNA genes is currently unknown, but it seems conceivable that similar (if not identical) RNA polymerase activities and similar promoter structures in both organelles have facilitated the replacement of mitochondrial by plastid tRNA genes during evolution.

An interesting case of reutilization of a promiscuous plastid DNA sequence in the mitochondrial genome has been described in rice. Here, a 4 kb plastid fragment inserts 355 bp upstream of the mitochondrial *nad9* gene. The *nad9* gene is transcribed from multiple promoters. Surprisingly, all of the altogether seven mapped transcription initiation sites are located within the plastid-derived sequence (Nakazono et al. 1997). Some of the putative promoter sequences harbor the canonical CRTA motif described for mitochondrial promoters in higher plants. It is noteworthy in this respect that a highly similar promoter consensus sequence is recognized by the bacteriophage-like RNA polymerase in plastids (Liere and Maliga, 1999). However, whether or not any of the promoters utilized in mitochondria is also active in rice plastids, is not yet known.

4 Gene transfer from the mitochondrial to the nuclear genome

Following their endosymbiotic uptake, α-proteobacteria underwent a similarly drastic genome reduction as the cyanobacterial endosymbiont, indicating that massive translocation of genetic information has taken place also during mitochondrial evolution (Fig. 1). There are many reasons to believe that the mechanisms of functional gene transfer from the mitochondrion to the nucleus are similar to if not largely identical with gene transfer events from the plastid to the nuclear genome. Although the scope of this review are gene transfer processes involving genetic information from plastids, a few interesting examples for transfer of mitochondrial sequences exemplarily shall be mentioned.

A gigantic mitochondrial insertion of 620 kb was identified in chromosome 2 of *Arabidopsis thaliana* (Stupar et al. 2001). Remarkably, the size of this insertion is far bigger than the entire mitochondrial genome of *Arabidopsis* (367 kb; Unseld et al. 1997). Fine mapping revealed that, although the insertion is not co-linear with the chondriome due to structural rearrangements, it probably encompasses the entire sequence of the *Arabidopsis* mitochondrial genome. Duplications of two regions of the chondriome account for the size increase of the integrated mitochondrial genome sequence (Stupar et al. 2001).

A 3.9 kb mitochondrial DNA insertion into the *Arabidopsis* polyubiquitin gene appears to be one of the most recent integration events identified to date (Sun and Callis 1993). The insertion is present in the polyubiquitin locus of the ecotype Columbia but absent from all other ecotypes, indicating that it occurred after diversification of the species *Arabidopsis* into distinct ecotypes. The lack of sequence homology between the nuclear integration site and the borders of the mitochondrial sequence suggests that integration occurred via non-homologous recombination (Sun and Callis 1993).

Another interesting transfer event of mitochondrial DNA to the nucleus involves part of a group II intron within the mitochondrial *cox2* gene of the monocotyledonous species *Ruscus aculeatus* (Kudla et al. 2002). A 154 bp piece of this mitochondrial intron was found to be part of an intron in a nuclear gene (*adhB*), suggesting that *cox2* intron sequences were transferred to the nuclear genome and reused there to build a spliceosomal intron (Kudla et al. 2002).

While the above examples are readily explained by DNA-mediated transfer, there is also evidence for an RNA/cDNA-mediated transfer mechanism (Nugent and Palmer 1991; Covello and Gray 1992; reviewed in Henze and Martin 2001). Absence of introns and lack of mitochondrial RNA editing sites (in that Ts are present where Cs undergo C-to-U editing in mitochondria) are generally taken as indications for reverse transcription of fully processed mitochondrial mRNAs followed by integration of the cDNA into the nuclear genome. Interestingly, mitochondrial pseudogenes are sometimes found as putative intermediates in the evolutionary replacement of mitochondrial genes by a transferred nuclear copy (Covello and Gray 1992; Kadowaki et al. 1996; Fig. 2). In several cases, it has also been possible to reconstruct how transferred mitochondrial genes became functional in the nucleus by capturing promoters and transit peptide sequences for re-targeting of the cytosolically made protein products into the mitochondrial compartment (Figueroa et al. 1999; Kubo et al. 1999; Kadowaki et al. 1996)

5 Gene transfer from the nuclear to organellar genomes

While plastid genomes seem to be largely resistant to promiscuous DNA, a significant proportion of the mitochondrial genomes in higher plants is made up of foreign sequences. These sequences not only come from the plastome, as discussed above, but can also be derived from the plant's nuclear genome (Marienfeld et al. 1999). The nuclear contribution to the rice chondriome is even larger than that of the plastome: as much as 13.4% of the sequence of the mitochondrial genome has been identified to be of nuclear origin (Notsu et al. 2002). Analysis of the *Arabidopsis* chondriome sequence revealed that most of this promiscuous nuclear DNA is derived from retrotransposons (Brennicke et al. 1996; Knoop et al. 1996; Unseld et al. 1997). Nuclear retrotransposons are subdivided into three classes termed the Ty1/*copia*, the Ty3/*gypsy* and the non-LTR/LINE families. Interestingly, fragments from all three classes are represented in the mitochondrial genome of *Arabidopsis*. The predominant insertion of retrotransposons might be mechanistically linked to the DNA transfer process in that the relatively small and freely mobile transposition intermediates may be taken up more readily by the mitochondrion than chromosomal DNA. Large deletions, accumulated frame-shift mutations and in-frame stop codons clearly render all retrotransposon sequences identified to date in plant mitochondrial genomes non-functional (Knoop et al. 1996).

Like promiscuous DNA of chloroplast origin, nuclear DNA is not present in the mitochondrial genome of the liverwort *Marchantia polymorpha* (Oda et al. 1992a,b) and the chondriomes of the two sequenced *Chlamydomonas* species (Denovan-Wright et al. 1998), lending further support to the idea that lower plants may lack the capacity to take up (or maintain) promiscuous DNA.

Yet another source of foreign DNA in the mitochondrial genome appears to be horizontal gene transfer from other organisms. Horizontal gene transfer has been suggested to occur both between mitochondrial genomes of higher plant species (Bergthorsson et al. 2003; Won and Renner 2003; Davis and Wurdack 2004) and between non-plant organisms and higher plant mitochondria (Vaughn et al. 1995; Marienfeld et al. 1997). Virus-mediated gene transfer (Marienfeld et al. 1997) and host-parasite interactions (Davis and Wurdack 2004; Nitz et al. 2004) have been postulated to promote such horizontal transfers of genetic information.

6 Experimental approaches to investigate gene transfer to the nucleus

Until recently, gene transfer from organellar genomes to the nucleus was mainly indirectly inferred from hybridization experiments, DNA sequencing projects or molecular phylogenetic studies. Gene transfer was believed to occur only on an evolutionary timescale and thought to be far too infrequent to be caught in the act by the experimentalist.

Things changed when transgenic tools for cell organelles became available and stringent selection schemes for gene transfer events were worked out which allowed the process to be monitored in real time. An organelle transformation technology first became available for yeast mitochondria (Johnston et al. 1988; Butow and Fox 1990) and subsequently, was also developed for chloroplasts (Boynton et al. 1988; Svab et al. 1990; Svab and Maliga 1993). How can the technology be employed to pick up gene transfer events? The general experimental setup is illustrated in Fig. 3. Along with the selectable marker gene usually required to generate transgenic organellar genomes, a second selection marker is integrated that is driven by nucleus-specific expression signals (Fig. 3). As they are of eukaryotic type, these expression signals (promoter and transcription terminator) are not efficiently recognized by the organellar gene expression machinery, which is prokaryotically organized. Consequently, the resistance trait encoded by the second selection marker is not expressed. Marker gene translocation into the nuclear genome, however, would activate the marker gene resulting in expression of the resistance and facilitating growth in the presence of the selecting agent (Fig. 3). Using this experimental strategy, it has indeed been possible to detect gene transfer events from organellar genomes to the nucleus in yeast mitochondria (Thorsness and Fox 1990) and, more recently, also in tobacco chloroplasts (Huang et al. 2003a; Stegemann et al. 2003; Fig. 3).

In two parallel studies, a kanamycin resistance gene embedded in a nucleus-specific expression cassette was used to determine the gene transfer frequency from the plastid to the nuclear genome (Fig. 3). Cells from tobacco plants with transgenic chloroplasts ("transplastomic plants") were subjected to selection for kanamycin resistance. In one of the studies, this was done by placing leaf explants on a plant regeneration medium and subjecting them to stringent selection for kanamycin resistance (Stegemann et al. 2003), whereas in the other study, wild-type plants were fertilized with pollen from a transplastomic plant (Huang et al. 2003a) and the resulting seeds

Plastid genome

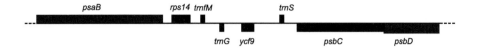

Plastid transformation vector

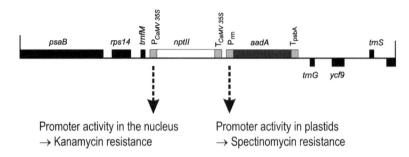

Promoter activity in the nucleus
→ Kanamycin resistance

Promoter activity in plastids
→ Spectinomycin resistance

Fig. 3. Experimental strategy to detect gene transfer events from the plastid to the nuclear genome. The upper panel shows the physical map of a region of the tobacco chloroplast genome (modified after Stegemann et al. 2003). Genes above the line are transcribed from the left to the right, genes below the line are transcribed in the opposite direction. The lower panel shows the restriction fragment cloned from this region and used for construction of a plastid transformation vector. The intergenic spacer between the tRNA-glycine and tRNA-N-formyl-methionine genes was used to insert two selectable marker genes: a chimeric spectin-omycin resistance gene (*aadA*) carrying plastid-specific expression signals (a promoter derived from the ribosomal RNA operon and a 3′ untranslated region taken from the plastid *psbA* gene) and an *nptII* gene driven by nuclear expression signals (35S promoter and termi-nator from the cauliflower mosaic virus CaMV). While the *aadA* gene is required to success-fully select chloroplast transformants, the kanamycin resistance gene is not efficiently expressed in transgenic chloroplasts. Gene transfer of the *nptII* out of the plastid into the nuclear genome activates *nptII* gene expression resulting in high level kanamycin resistance (Stegemann et al. 2003). Therefore, gene transfer events can be identified by large-scale selec-tion experiments for kanamycin-resistant cell lines (Stegemann et al. 2003) or seedlings (Huang et al. 2003a).

were germinated in the presence of kanamycin. Both studies provided strong experimental evidence for an ongoing gene transfer from the plastid to the nuclear genome and measured a surprisingly high transfer frequency. The frequency in vegetative cells was estimated to be approximately one transfer event per 5 million cells (Stegemann et al. 2003), whereas in pollen grains, the frequency was even higher (one transfer event in 16,000 pollen grains; Huang et al. 2003a; for a plausible mechanistic explanation of the different frequencies see below). Roughly, these frequencies are in a similar range to the rate of mitochondrial gene transfer to the nuclear genome in yeast (determined to be approximately 2×10^{-5} per cell per generation; Thorsness and Fox, 1990). It should be noted that these estimates, despite being shockingly high, are conservative ones: for technical reasons and limitations of the selection schemes, part of the gene transfer events that have occurred go undetected in these experiments suggesting that the real transfer frequency might be even higher than what has been calculated from the experimental data.

A molecular analysis of the nuclear loci created by newly transferred plastid DNA sequences revealed several interesting features (Huang et al., 2004):

(i) Long tracts of plastid DNA (between 6 and >22 kb) are integrated into the nuclear genome.
(ii) The nuclear loci harboring transferred plastid sequences can have a complex structure and can be composed of multiple ptDNA fragments.
(iii) Microhomologies of 2–5 bp are found at the integration sites.
(iv) ptDNA integration is likely to proceed via non-homologous recombination (Huang et al. 2004).

Perhaps the most dramatic conclusions from these gene transfer experiments concern the genetic heterogeneity of cell populations in an individual, a given tissue or organ: Clearly, a single tobacco leaf has more than 5 million cells. Finding a plastid DNA transfer frequency of at least one event per 5 million leaf cells suggests that the cells in one and the same leaf are not genetically uniform, but instead may differ with respect to the pattern of organellar DNA insertions in their nuclear genomes. Consequently, similar to transposon mobilization, the frequent integration of organellar DNA into the nuclear genome may contribute substantially to intraspecific and intraorganismic genetic variation. Moreover, if one assumes that with a certain probability, transferred plastid DNA sequences will become integrated into coding regions, promiscuous organellar DNA must also be considered as a possible cause of somatic mutations.

The successful development of systems suitable to study gene transfer in the laboratory and in real time now facilitates in-depth investigation of the evolutionary mechanisms underlying the transfer of genetic information from organellar genomes to the nucleus. This in turn will allow experimental reconstruction of an important aspect of genome evolution in eukaryotes.

7 Mechanisms of gene transfer from the plastid to the nucleus

Although the relative contributions of DNA-mediated versus cDNA-mediated gene transfer processes to the evolutionary translocation of organellar genes to the nucleus are debatable (Timmis et al. 2004), it is now well established that organellar nucleic acids can escape to the nucleus and integrate there into chromosomal DNA at a significant frequency. Irrespective of the transferred genetic information being DNA or cDNA, any attempt to explain the mechanisms of gene transfer from organellar genomes to the nucleus must address one crucial question: How do nucleic acids escape from the plastid and how can they get into the nucleus? It should be borne in mind that, at least in dividing cells, the nuclear envelope does not pose a serious obstacle to gene transfer into the nuclear genome because it dissolves during mitosis and meiosis and thus, released plastid DNA (or cDNA) floating in the cytoplasm could readily get in physical contact with chromosomes. In contrast, the escape of DNA from the plastid seems more difficult to envisage, since the double membrane of the plastid is believed to be impermeable to nucleic acids. Theoretically, at least three possible mechanisms of DNA release from the plastid must be considered: (i) active export, (ii) occasional passive release, for example by DNA molecules slipping out of the plastid during organelle division, or (iii) release during organelle degradation. At present, there is no definitive support for either mechanism. However, several lines of circumstantial evidence may lend support to the idea that it is chloroplast destruction that releases the DNA from the organelle, which then can integrate into the nuclear genome. First, the frequency of gene transfer to the nuclear genome was found to be significantly higher in pollen (Huang et al. 2003a) than in vegetative leaf cells (Stegemann et al. 2003). Although it cannot entirely be ruled out that the different frequencies measured in the two studies can be attributed to differences in the experimental setups (Huang et al. 2003a; Stegemann et al. 2003), a higher frequency in pollen would be compatible with the idea that plastid disintegration promotes gene transfer to the nucleus. This is because plastids are excluded from pollen transmission and uniparentally maternally inherited in most angiosperm species (Birky Jr 1995; Hagemann 2002). At least in a number of plant species, the elimination of plastids from sperm cells during pollen grain maturation has been demonstrated to involve plastid degradation (Hagemann 2002), a process that can be expected to result in the massive release of plastid DNA into the cytosol which in turn may lead to a drastically increased rate of ptDNA integration into the nuclear genome.

Another line of evidence has come from a study aiming at detecting gene transfer events in *Chlamydomonas reinhardtii*, a unicellular green alga that possesses only one single big chloroplast. Using a similar experimental approach as taken to detect plastid-to-nuclear gene transfer in tobacco cells, a bleomycin resistance gene driven by nucleus-specific expression signals was integrated into the chloroplast genome of *Chlamydomonas* followed by selection for antibiotic resistance (Lister et al. 2003). Although large numbers of algal cells ($>10^9$) were subjected to this screening for bleomycin resistance, no evidence of gene translocation to the nucleus could be detected suggesting that plastid-to-nuclear gene transfer in *Chlamydomonas*, if it occurs at all, is far more infrequent than in higher plants. Again, this provides indirect evidence for gene transfer being promoted by organelle degradation, because the *Chlamydomonas* cell most probably cannot afford losing its one and only chloroplast. The chloroplast provides an essential cellular compartment in that it harbors a large number of biochemical pathways (including fatty acid, heme and amino acid biosyntheses). Moreover, the expression of chloroplast genome-encoded genes appears to be essential in both tobacco (Drescher et al. 2000; Ahlert et al. 2003) and *Chlamydomonas* cells (Boudreau et al. 1997; Rochaix 1997). Chloroplast degradation is, therefore, most probably lethal in *Chlamydomonas*, which could provide a simple explanation for the failure to detect any gene transfer event in the laboratory. Interestingly, this explanation is also supported by a preliminary bioinformatics analysis of the draft version of the fully sequenced *Chlamydomonas* nuclear genome which detected no promiscuous DNA of chloroplast origin, confirming that gene translocation from the plastid to the nuclear genome is much rarer in *Chlamydomonas reinhardtii* than in higher plants (Lister et al. 2003) and again suggesting that the transfer of nucleic acids out of organelles into the nucleus may be dependent upon organelle destruction.

Many other open questions remain to be addressed. Is the integration of organellar DNA into the plant's nuclear genome linked to the repair of DNA double-strand breaks as has been proposed for the integrations of mitochondrial DNA into the yeast genome (Ricchetti et al. 1999; Yu and Gabriel 1999)? If so, will DNA damage further enhance the frequency of gene transfer? How can a transferred organellar gene acquire functionality in its new environment that requires eukaryotic expression signals? How does the nucleus cope with all this immigrant genetic information? Has the nucleus evolved elimination mechanisms that prevent it from getting swamped with organellar DNA? How does gene transfer from plastids to mitochondria occur? Certainly, the new experimental systems based on transplastomic technologies provide a rich platform for future research on the mechanisms of intercompartmental gene transfer.

8 Implications for plastid biotechnology

Recent technological progress has made it possible to integrate foreign genes
into the plastid genome (Svab and Maliga 1993; Sidorov et al. 1999; Ruf et al.
2001; reviewed, e.g. in Bock 2001b; Bock and Khan 2004; Maliga 2004). One
of the great attractions of placing transgenes into the plastid genome rather
than in the nucleus is that this so-called transplastomic technology provides
an effective strategy for transgene containment (Daniell 2002). This is
because plastids are uniparentally maternally inherited in many higher
plants and most agriculturally important crops, so that plastid transgenes
are usually not transmitted via pollen. In this way, the technology may help
to allay concerns over uncontrolled spreading of transgenes from fields with
transgenic plants to neighboring fields with non-transgenic plants. The
experimental findings that plastid genes and transgenes can escape to the
nucleus at high frequency (Huang et al. 2003a; Stegemann et al. 2003) have
evoked controversial discussions about the significance of this finding and
the level of transgene containment achievable by chloroplast transformation. Initially, raised technical criticisms concerning a possible co-transformation of the plastid and nuclear genomes (Daniell and Parkinson 2003)
proved unsubstantiated (Huang et al. 2003b; Stegemann et al. 2003) and
were unequivocally disproven by experimental data (Stegemann et al. 2003).
However, the most critical question remains: does transgene escape to the
nucleus pose a serious threat to the biological safety of transplastomic plants
and if so, is there anything we can do about it?

The frequency of gene transfer measured experimentally is certainly in
the range that is highly significant for plant growth at the field level.
However, the experimental setup used to demonstrate gene transfer to the
nucleus is radically different from a normal plastid transformation experiment in that a plastid transgene (*nptII*) was equipped with nuclear expression signals (Fig. 3). Normally, transgenes intended to be expressed from the
plastome are tethered to plastid-specific, prokaryotic-type expression signals: a plastid promoter, a Shine–Dalgarno sequence serving as binding site
for prokaryotic 70S ribosomes and a 3′ untranslated region folding into a
stemloop-type RNA secondary structure and conferring transcript stability.
Thus a typical plastid transgene is not expected to be expressed when transferred to the nucleus. As in most gene transfer lines selected from the transgenic experiments (Huang et al. 2003a; Stegemann et al. 2003; Fig. 3), the
plastid-specific antibiotic resistance gene (*aadA*) was co-transferred to the
nucleus together with the *nptII* (Fig. 3), this scenario could be tested directly
by assaying the gene transfer lines for their resistance to spectinomycin and
streptomycin (after having crossed out the transgenic chloroplasts;

Stegemann et al. 2003). All lines turned out to be sensitive to both drugs, indicating that there is no significant gene expression from the plastid expression signals in the nucleus (Huang et al. 2003a; Stegemann et al. 2003). Of course, it is conceivable that the plastid transgenes could be expressed if it lands in a favorable nuclear location, for example downstream of a strong nuclear promoter. However, this probability is likely to be orders of magnitude lower than the measured gene transfer frequency. Nonetheless, the possibility of transgene escape to the nucleus should be taken into consideration when designing containment strategies based on plastid transformation. A number of straightforward tools are available that can be used to prevent the escape of plastid transgenes to the nuclear genome, or make sure that a transferred plastid transgene, even when landing downstream of a nuclear promoter, cannot be expressed. Below, three such strategies, which could be employed alone or in combination, are briefly discussed.

A strategy suitable to suppress transgene escape would be the incorporation of nucleus-specific suicide cassettes close to or within the plastid transgene. Such suicide genes can be, for example, genes for ribosome-inactivating proteins, RNases or proteases that either are not expressed in the plastid (because they carry nuclear expression signals or their gene products require nucleus-specific processing or modification steps, such as spliceosomal intron removal or protein glycosylation) and/or have no targets or substrates inside the plastid compartment.

A suitable strategy to prevent the expression of escaped plastid transgenes in the nucleus would be, for example, to make transgene expression dependent on plastid-specific RNA maturation processes. Group II intron splicing and RNA editing are good candidate processes here and their suitability for driving RNA processing-dependent plastid transgene expression has been demonstrated already experimentally (Bock and Maliga 1995; Chaudhuri et al. 1995; Chaudhuri and Maliga 1996).

Introns are present in a number of plastid-encoded genes. In higher plants, most plastid introns belong to the so-called group II introns with the single exception of the tRNA-Leu(UAA) gene that harbors a group I intron (Kuhsel et al. 1990). Both group I and group II introns are radically different from the nuclear spliceosomal introns in that they fold into a complex RNA secondary structure (Michel et al. 1989; Sharp 1994; Michel and Ferat 1995) and need organelle-specific *trans*-acting factors (splicing factors) for the splicing reaction to occur. The splicing of all plastid group II introns requires the assistance of proteinaceous splicing factors some of which appear to be intron-specific whereas others are somewhat more general and are involved in the splicing of a subset of plastid introns (Jenkins et al. 1997; Vogel et al. 1999; Jenkins and Barkan 2001; Till et al. 2001; Ostheimer et al. 2003).

Incorporation of a plastid group II intron into a transgene makes faithful transgene expression dependent on splicing (Bock and Maliga 1995) and, moreover, restricts expression to the plastid compartment where the essential splicing factors are present. In this way, insertion of a group II intron into a plastid transgene (Bock and Maliga 1995) is a suitable strategy to prevent nuclear expression of escaped plastid transgenes.

RNA editing in plastids proceeds by cytidine-to-uridine conversion at highly specific sites (for review see, e.g. Bock 2000, 2001a). As several RNA editing sites have been found that create a start codon for translation by converting a genomically encoded ACG codon into a functional AUG initiator codon (Hoch et al. 1991; Kudla et al. 1992; Bock et al. 1993; Wakasugi et al. 1996, fusing an ACG codon in an editable sequence context with the coding region of the plastid transgene of interest will make transgene expression dependent on RNA editing (Chaudhuri et al. 1995; Chaudhuri and Maliga 1996) and also will restrict transgene expression to the plastid compartment, since C-to-U RNA editing is absent from the nucleocytoplasmic compartment in higher plants. It should be noted, however, that this strategy may not reduce the risk of unwanted nuclear expression of escaped plastid transgenes to zero since in-frame insertion of the transgene into an active nuclear gene, in theory, could give rise to a functional fusion protein.

In summary, nuclear escape of plastid transgenes is unlikely to pose a serious threat to the containment level provided by plastid transformation. Moreover, appropriate construction of transformation vectors can help to install additional safeguarding mechanisms that (i) reduce the frequency of transgene escape, and (ii) prevent nuclear expression of escaped plastid transgenes.

Acknowledgements. I thank Stephanie Ruf for critical reading of this manuscript. Work on gene transfer in my laboratory is supported by grants from the Deutsche Forschungsgemeinschaft and the Max-Planck Society.

References

Ahlert D, Ruf S, Bock R (2003) Plastid protein synthesis is required for plant development in tobacco. Proc Natl Acad Sci USA 100:15730–15735

Albertazzi FJ, Kudla J, Bock R (1998) The cox2 locus of the primitive angiosperm plant *Acorus calamus*: molecular structure, transcript processing and RNA editing. Mol Gen Genet 259: 591–600

Ambrosini M, Ceci LR, Fiorella S, Gallerani R (1992) Comparison of regions coding for tRNAHis genes of mitochondrial and chloroplast DNA in sunflower: a proposal concerning the classification of "CP-like" tRNA genes. Plant Mol Biol 20:1–4

Arabidopsis Genome Initiative (2000) Analysis of the genome sequence of the flowering plant *Arabidopsis thaliana.* Nature 408:796–815

Ayliffe MA, Timmis JN (1992) Tobacco nuclear DNA contains long tracts of homology to chloroplast DNA. Theor Appl Genet 85:229–238

Ayliffe MA, Scott NS, Timmis JN (1998) Analysis of plastid DNA-like sequences within the nuclear genomes of higher plants. Mol Biol Evol 15:738–745

Bendich AJ, Smith SB (1990) Moving pictures and pulsed-field gel electrophoresis show linear DNA molecules from chloroplasts and mitochondria. Curr Genet 17:421–425

Bergthorsson U, Adams KL, Thomason B, Palmer JD (2003) Widespread horizontal transfer of mitochondrial genes in flowering plants. Nature 424:197–201

Binder S, Schuster W, Grienenberger J-M, Weil JH, Brennicke A (1990) Genes for tRNAGly, tRNAHis, tRNAPhe, tRNASer and tRNATyr are encoded in *Oenothera* mitochondrial DNA. Curr Genet 17:353–358

Birky Jr CW (1995) Uniparental inheritance of mitochondrial and chloroplast genes: mechanisms and evolution. Proc Natl Acad Sci USA 92:11331–11338

Bock R (2000) Sense from nonsense: how the genetic information of chloroplasts is altered by RNA editing. Biochimie 82:549–557

Bock R (2001a) RNA editing in plant mitochondria and chloroplasts. In: Bass B (ed): Frontiers in molecular biology: RNA editing. Oxford University Press, New York, pp 38–60

Bock R (2001b) Transgenic chloroplasts in basic research and plant biotechnology. J Mol Biol 312:425–438

Bock R, Khan MS (2004) Taming plastids for a green future. Trends Biotechnol 22:311–318

Bock R, Maliga P (1995) Correct splicing of a group II intron from a chimeric reporter gene transcript in tobacco plastids. Nucleic Acids Res 23:2544–2547

Bock R, Hagemann R, Kössel H, Kudla J (1993) Tissue- and stage-specific modulation of RNA editing of the psbF and psbL transcript from spinach plastids—a new regulatory mechanism? Mol Gen Genet 240:238–244

Boudreau E, Turmel M, Goldschmidt-Clermont M, Rochaix J-D, Sivan S, Michaels A, Leu S (1997) A large open reading frame (orf1995) in the chloroplast DNA of *Chlamydomonas reinhardtii* encodes an essential protein. Mol Gen Genet 253:649–653

Boynton JE, Gillham NW, Harris EH, Hosler JP, Johnson AM, Jones AR, Randolph-Anderson BL, Robertson D, Klein TM, Shark KB, Sanford JC (1988) Chloroplast transformation in *Chlamydomonas* with high velocity microprojectiles. Science 240:1534–1538

Brennicke A, Klein M, Binder S, Knoop V, Grohmann L, Malek O, Marchfelder A, Marienfeld J, Unseld M (1996) Molecular biology of plant mitochondria. Naturwissenschaften 83:339–346

Butow RA, Fox TD (1990) Organelle transformation: shoot first, ask questions later. Trends Biochem Sci 15:465–468

Chaudhuri S, Maliga P (1996) Sequences directing C to U editing of the plastid psbL mRNA are located within a 22 nucleotide segment spanning the editing site. EMBO J 15:5958–5964

Chaudhuri S, Carrer H, Maliga P (1995) Site-specific factor involved in the editing of the psbL mRNA in tobacco plastids. EMBO J 14:2951–2957

Covello PS, Gray MW (1992) Silent mitochondrial and active nuclear genes for subunit 2 of cytochrome c oxidase (cox2) in soybean: evidence for RNA-mediated gene transfer. EMBO J 11:3815–3820

Cummings MP, Nugent JM, Olmstead RG, Palmer JD (2003) Phylogenetic analysis reveals five independent transfers of the chloroplast gene rbcL to the mitochondrial genome in angiosperms. Curr Genet 43:131–138

Daniell H (2002) Molecular strategies for gene containment in transgenic crops. Nature Biotechnol 20:581–586

Daniell H, Parkinson CL (2003) Jumping genes and containment. Nature Biotechnol 21:374–375

Davis CC, Wurdack KJ (2004) Host-to-parasite gene transfer in flowering plants: phylogenetic evidence from Malpighiales. Science 305:676–678

Denovan-Wright EM, Nedelcu AM, Lee RW (1998) Complete sequence of the mitochondrial DNA of *Chlamydomonas eugametos*. Plant Mol Biol 36:285–295

Drescher A, Ruf S, Calsa Jr. T, Carrer H, Bock R (2000) The two largest chloroplast genome-encoded open reading frames of higher plants are essential genes. Plant J 22:97–104

Figueroa P, Gómez I, Holuigue L, Araya A, Jordana X (1999) Transfer of rps14 from the mitochondrion to the nucleus in maize implied integration within a gene encoding the iron-sulphur subunit of succinate dehydrogenase and expression by alternative splicing. Plant J 18:601–609

Hagemann R (2002) Milestones in plastid genetics of higher plants. Prog Bot 63:1–51

Hedtke B, Börner T, Weihe A (2000) One RNA polymerase serving two genomes. EMBO Rep 1:435–440

Henze K, Martin W (2001) How do mitochondrial genes get into the nucleus. Trends Genet 17:383–387

Hiratsuka J, Shimada H, Whittier R, Ishibashi T, Sakamoto M, Mori M, Kondo C, Honji Y, Sun C-R, Meng B-Y, Li Y-Q, Kanno A, Nishizawa Y, Hirai A, Shinozaki K, Sugiura M (1989) The complete sequence of the rice (*Oryza sativa*) chloroplast genome: intermolecular recombination between distinct tRNA genes accounts for a major plastid DNA inversion during the evolution of cereals. Mol Gen Genet 217:185–194

Hoch B, Maier RM, Appel K, Igloi GL, Kössel H (1991) Editing of a chloroplast mRNA by creation of an initiation codon. Nature 353:178–180

Huang CY, Ayliffe MA, Timmis JN (2003) Direct measurement of the transfer rate of chloroplast DNA into the nucleus. Nature 422:72–76

Huang CY, Ayliffe MA, Timmis JN (2003) Organelle evolution meets biotechnology. Nature Biotechnol 21:489–490

Huang CY, Ayliffe MA, Timmis JN (2004) Simple and complex nuclear loci created by newly transferred chloroplast DNA in tobacco. Proc Natl Acad Sci USA 101:9710–9715

Jenkins BD, Barkan A (2001) Recruitment of a peptidyl-tRNA hydrolase as a facilitator of group II intron splicing in chloroplasts. EMBO J 20:872–879

Jenkins BD, Kulhanek DJ, Barkan A (1997) Nuclear mutations that block group II RNA splicing in maize chloroplasts reveal several intron classes with distinct requirements for splicing factors. Plant Cell 9:283–296

Johnston SA, Anziano PQ, Shark K, Sanford JC, Butow RA (1988) Mitochondrial transformation in yeast by bombardment with microprojectiles. Science 240:1538–1541

Jubier M-F, Lucas H, Delcher E, Hartmann C, Quétier F, Lejeune B (1990) An internal part of the chloroplast atpA gene sequence is present in the mitochondrial genome of *Triticum aestivum*: molecular organization and evolutionary aspects. Curr Genet 17:523–528

Kadowaki K-I, Kubo N, Ozawa K, Hirai A (1996) Targeting presequence acquisition after mitochondrial gene transfer to the nucleus occurs by duplication of existing targeting signals. EMBO J 15:6652–6661

Kaneko T, Tabata S (1997) Complete genome structure of the unicellular cyanobacterium *Synechocystis* sp. PCC6803. Plant Cell Physiol 38:1171–1176

Kaneko T, Sato S, Kotani H, Tanaka A, Asamizu E, Nakamura Y, Miyajima N, Hirosawa M, Sugiura M, Sasamoto S, Kimura T, Hosouchi T, Matsuno A, Muraki A, Nakazaki N, Naruo K, Okumura S, Shimpo S, Takeuchi C, Wada T, Watanabe A, Yamada M, Yasuda M, Tabata S (1996) Sequence analysis of the genome of the unicellular cyanobacterium *Synechocystis* sp. strain PCC6803. II. Sequence determination of the entire genome and assignment of potential protein-coding regions. DNA Res 3:109–136

Kao T, Moon E, Wu R (1984) Cytochrome oxidase subunit II gene of rice has an insertion sequence within the intron. Nucleic Acids Res 12:7305–7315

Knoop V (2004) The mitochondrial DNA of land plants: peculiarities in phylogenetic perspective. Curr Genet 46:123–139

Knoop V, Unseld M, Marienfeld J, Brandt P, Sünkel S, Ullrich H, Brennicke A (1996) copia-, gypsy- and LINE-like retrotransposon fragments in the mitochondrial genome of *Arabidopsis thaliana*. Genetics 142:597–585

Kobayashi Y, Dokiya Y, Sugita M (2001) Dual targeting of phage-type RNA polymerase to both mitochondria and plastids is due to alternative translation initiation in single transcripts. Biochem Biophys Res Commun 289:1106–1113

Kobayashi Y, Dokiya Y, Kumazawa Y, Sugita M (2002) Non-AUG translation initiation of mRNA encoding plastid-targeted phage-type RNA polymerase in *Nicotiana sylvestris*. Biochem Biophys Res Commun 299:57–61

Kotani H, Tabata S (1998) Lessons from sequencing of the genome of a unicellular cyanobacterium, Synechocystis SP. PCC6803. Annu Rev Plant Physiol Plant Mol Biol 49: 151–171

Koulintchenko M, Konstantinov Y, Dietrich A (2003) Plant mitochondria actively import DNA via the permeability transition pore complex. EMBO 22:1245–1254

Kubo N, Harada K, Hirai A, Kadowaki K-i (1999) A single nuclear transcript encoding mitochondrial RPS14 and SDHB of rice is processed by alternative splicing: common use of the same mitochondrial targeting signal for different proteins. Proc Natl Acad Sci USA 96:9207–9211

Kudla J, Igloi GL, Metzlaff M, Hagemann R, Kössel H (1992) RNA editing in tobacco chloroplasts leads to the formation of a translatable psbL mRNA by a C to U substitution within the initiation codon. EMBO J 11:1099–1103

Kudla J, Albertazzi FJ, Blazevic Hermann, M, Bock R (2002) Loss of the mitochondrial cox2 intron 1 in a family of monocotyledonous plants and utilization of mitochondrial intron sequences for the construction of a nuclear intron. Mol Genet Genomics 267:223–240

Kuhsel MG, Stickland R, Palmer JD (1990) An ancient group I intron shared by eubacteria and chloroplasts. Science 250:1570–1573

Liere K, Maliga P (1999) *In vitro* characterization of the tobacco rpoB promoter reveals a core sequence motif conserved between phage-type plastid and plant mitochondrial promoters. EMBO J 18:249–257

Lister DL, Bateman JM:,Purton,S, Howe CJ (2003) DNA transfer from chloroplast to nucleus is much rarer in *Chlamydomonas* than in tobacco. Gene 316:33–38

Maliga P (2004) Plastid transformation in higher plants. Annu Rev Plant Biol 55:289–313

Marienfeld JR, Unseld M, Brandt P, Brennicke A (1997) Viral nucleic acid sequence transfer between fungi and plants. Trends Genet 13:260–261

Marienfeld J, Unseld M, Brennicke A (1999) The mitochondrial genome of *Arabidopsis* is composed of both native and immigrant information. Trends Plant Sci 4:495–502

Martin W, Herrmann RG (1998) Gene transfer from organelles to the nucleus: how much, what happens, and why? Plant Physiol 118:9–17

Martin W, Rujan T, Richly E, Hansen A, Cornelsen S, Lins T, Leister D, Stoebe B, Hasegawa M, Penny D (2002) Evolutionary analysis of *Arabidopsis*, cyanobacterial, and chloroplast genomes reveals plastid phylogeny and thousands of cyanobacterial genes in the nucleus. Proc Natl Acad Sci USA 99:12246–12251

Michel F, Ferat J-L (1995) Structure and activities of group II introns. Annu Rev Biochem 64:435–461

Michel F, Umesono K, Ozeki H (1989) Comparative and functional anatomy of group II catalytic introns—a review. Gene 82:5–30

Millen RS, Olmstead RG, Adams KL, Palmer JD, Lao NT, Heggie L, Kavanagh TA, Hibberd JM, Gray JC, Morden CW, Calie PJ, Jermiin LS, Wolfe KH (2001) Many parallel losses of infA from chloroplast DNA during angiosperm evolution with multiple independent transfers to the nucleus. Plant Cell 13:645–658

Nakazono M, Hirai A (1993) Identification of the entire set of transferred chloroplast DNA sequences in the mitochondrial genome of rice. Mol Gen Genet 236:341–346

Nakazono M, Nishiwaki S, Tsutsumi N, Hirai A (1997) A chloroplast-derived sequence is utilized as a source of promoter sequences for the gene for subunit 9 of NADH dehydrogenase (nad9) in rice mitochondria. Mol Gen Genet 252:371–378

Nitz N, Gomes C, De Cássia Rosa A, D'Souza-Ault MR, Moreno F, Lauria-Pires L, Nascimento RJ, Teixeira ARL (2004) Heritalbe integration of kDNA minicircle sequences from *Trypanosoma cruzi* into the avian genome: insights into human Chagas disease. Cell 118:175–186

Notsu Y, Masood S, Nishikawa T, Kubo N, Akiduki G, Nakazono M, Hirai A, Kadowaki K (2002) The complete sequence of the rice (*Oryza sativa* L.) mitochondrial genome: frequent DNA sequence acquisition and loss during the evolution of flowering plants. Mol Genet Genom 268:434–445

Nugent JM, Palmer JD (1988) Location, identity, amount and serial entry of chloroplast DNA sequences in crucifer mitochondrial DNAs. Curr Genet 14:501–509

Nugent JM, Palmer JD (1991) RNA mediated transfer of the gene coxII from the mitochondrion to the nucleus during flowering plant evolution. Cell 66:473–481

Oda K, Yamato K, Ohta E, Nakamura Y, Takemura M, Nozato N, Akashi K, Kanegae T, Ogura Y, Kohchi T, Ohyama K (1992a) Gene orgaization deduced from the complete sequence of liverwort *Marchantia polymorpha* mitochondrial DNA—a primitive form of plant mitochondrial genome. J Mol Biol 223:1–7

Oda K, Kohchi T, Ohyama K (1992b) Mitochondrial DNA of *Marchantia polymorpha* as a single circular form with no incorporation of foreign DNA. Biosci Biotech Biochem 56: 132–135

Ohyama K, Fukuzawa H, Kohchi T, Shirai H, Sano T, Sano S, Umesono K, Shiki Y, Takeuchi M, Chang Z, Aota S-i, Inokuchi H, Ozeki H (1986) Chloroplast gene organization deduced from complete sequence of liverwort *Marchantia polymorpha* chloroplast DNA. Nature 322:572–574

Oldenburg DJ, Bendich AJ (1998) The structure of mitochondrial DNA from the liverwort, Marchantia polymorpha. J Mol Biol 276:745–758

Oldenburg DJ, Bendich AJ (2001) Mitochondrial DNA from the liverwort *Marchantia polymorpha*: circularly permuted linear molecules, head-to-tail concatemers, and a 5'protein. J Mol Biol 310:549–562

Ostheimer GJ, Williams-Carrier R, Belcher S, Osborne E, Gierke J, Barkan A (2003) Group II intron spicing factors derived by diversification of an ancient RNA-binding domain. EMBO J 22:3919–3929

Ricchetti M, Fairhead C, Dujon B (1999) Mitochondrial DNA repairs double-strand breaks in yeast chromosomes. Nature 402:96–100

Rochaix J-D (1997) Chloroplast reverse genetics: new insights into the function of plastid genes. Trends Plant Sci 2:419–425

Ruf S, Hermann M, Berger IJ, Carrer H, Bock R (2001) Stable genetic transformation of tomato plastids and expression of a foreign protein in fruit. Nature Biotechnol 19:870–875

Schuster W, Brennicke A (1987) Plastid, nuclear and reverse transcriptase sequences in the mitochondrial genome of *Oenothera*: is genetic information transferred between organelles via RNA? EMBO J 6:2857–2863

Schuster W, Brennicke A (1988) A plastid fragment from the psbE-psbF coding region in the mitochondrial genome of *Oenothera berteriana*. Nucleic Acids Res 16:7728

Scott NS, Timmis JN (1984) Homologies between nuclear and plastid DNA in spinach. Theor Appl Genet 67:279–288

Sharp PA (1994) Split genes and RNA splicing. Cell 77:805–815

Shimada H, Sugiura M (1991) Fine structural features of the chloroplast genome: comparison of the sequenced chloroplast genomes. Nucleic Acids Res 19:983–995

Shinozaki K, Ohme M, Tanaka M, Wakasugi T, Hayashida N, Matsubayashi T, Zaita N, Chunwongse J, Obokata J, Yamaguchi-Shinozaki K, Ohto C, Torazawa K, Meng BY, Sugita M, Deno H, Kamogashira T, Yamada K, Kusuda J, Takaiwa F, Kato A, Tohdoh N, Shimada

H, Sugiura M (1986) The complete nucleotide sequence of the tobacco chloroplast genome: its gene organization and expression. EMBO J 5:2043–2049

Sidorov VA, Kasten D, Pang S-Z, Hajdukiewicz PTJ, Staub JM, Nehra NS (1999) Stable chloroplast transformation in potato: use of green fluorescent protein as a plastid marker. Plant J 19:209-216

Stegemann S, Hartmann S, Ruf S, Bock R (2003) High-frequency gene transfer from the chloroplast genome to the nucleus. Proc Natl Acad Sci USA 100:8828–8833

Stern DB, Lonsdale DM (1982) Mitochondrial and chloroplast genomes of maize have a 12-kilobase DNA sequence in common. Nature 299:698–702

Stupar RM, Lilly JW, Town CD, Cheng Z, Kaul S, Buell CR, Jiang J (2001) Complex mtDNA constitutes an approximate 620-kb insertion on *Arabidopsis thaliana* chromosome 2: implication of potential sequencing errors caused by large-unit repeats. Proc Natl Acad Sci USA 98:5099–5103

Sugiura M (1989) The chloroplast chromosomes in land plants. Annu Rev Cell Biol 5:51–70

Sun C-W, Callis J (1993) Recent stable insertion of mitochondrial DNA into an *Arabidopsis* polyubiquitin gene by nonhomologous recombination. Plant Cell 5:97–107

Svab Z, Maliga P (1993) High-frequency plastid transformation in tobacco by selection for a chimeric aadA gene. Proc Natl Acad Sci USA 90:913–917

Svab Z, Hajdukiewicz P, Maliga P (1990) Stable transformation of plastids in higher plants. Proc Natl Acad Sci USA 87:8526–8530

Szathmary E, Smith JM (1995) The major evolutionary transitions. Nature 374:227–232

Thorsness PE, Fox TD (1990) Escape of DNA from mitochondria to the nucleus in *Saccharomyces cerevisiae*. Nature 346:376–379

Thorsness PE, Weber ER (1996) Escape and migration of nucleic acids between chloroplasts, mitochondria and the nucleus. Int Rev Cytol 165:207–234

Till B, Schmitz-Linneweber C, Williams-Carrier R, Barkan A (2001) Crs1 is a novel group II intron splicing factor that was derived from a domain of ancient origin. RNA 7:1227–1238

Timmis JN, Scott NS (1983) Spinach nuclear and chloroplast DNAs have homologous sequences. Nature 305:65–67

Timmis JN, Ayliffe MA, Huang CY, Martin W (2004) Endosymbiotic gene transfer: organelle genomes forge eukaryotic chromosomes. Nature Rev Genet 5:123–136

Unseld M, Marienfeld JR, Brandt P, Brennicke A (1997) The mitochondrial genome of *Arabidopsis thaliana* contains 57 genes in 366,924 nucleotides. Nature Genetics 15:57–61

Vaughn JC, Mason MT, Sper-Whitis GL, Kuhlman P, Palmer JD (1995) Fungal origin by horizontal gene transfer of a plant mitochondrial group I intron in the chimeric coxI gene of *Peperomia*. J Mol Evol 41:563–572

Veronico P, Gallerani R, Ceci LR (1996) Compilation and classification of higher plant mitochondrial tRNA genes. Nucleic Acids Res 24:2199–2203

Vogel J, Börner T, Hess WR (1999) Comparative analysis of splicing of the complete set of chloroplast group II introns in three higher plant mutants. Nucleic Acids Res 27:3866–3874

Wakasugi T, Hirose T, Horihata M, Tsudzuki T, Kössel H, Sugiura M (1996) Creation of a novel protein-coding region at the RNA level in black pine chloroplasts: the pattern of RNA editing in the gymnosperm chloroplast is different from that in angiosperms. Proc Natl Acad Sci USA 93:8766–8770

Wakasugi T, Tsudzuki T, Sugiura M (2001) The genomics of land plant chloroplasts: gene content and alteration of genomic information by RNA editing. Photosynth Res 70:107–118

Weber-Lotfi F, Marechal-Drouard L, Folkerts O, Hanson M, Grienenberger JM (1993) Localization of tRNA genes on the Petunia hybrida 3704 mitochondrial genome. Plant Mol Biol 21:403–407

Won H, Renner SS (2003) Horizontal gene transfer from flowering plants to *Gnetum*. Proc Natl Acad Sci USA 100:10824–10829

Yu X, Gabriel A (1999) Patching broken chromosomes with extranuclear cellular DNA. Mol Cell 4:873–881

Zeltz P, Kadowaki K-i, Kubo N, Maier RM, Hirai A, Kössel H (1996) A promiscuous chloroplast DNA fragment is transcribed in plant mitochondria but the encoded RNA is not edited. Plant Mol Biol 31:647–656

Professor Dr. Ralph Bock
Max-Planck-Institut für Molekulare Pflanzenphysiologie
Am Mühlenberg 1
D-14476 Potsdam-Golm
Tel.: +49(0)331 567-8700
Fax: +49 (0)331 567-8701
e-mail: rbock@mpimp-golm.mpg.de

Molecular cell biology:
Epigenetic gene silencing in plants

Roman A. Volkov, Nataliya Y. Komarova, Ulrike Zentgraf,
Vera Hemleben

1 Introduction

The highly organized arrangement of the DNA in the cell nucleus and
structurally defined compartments within the nucleus are important pre-
requisites for a timely and spatially correct gene expression and for the
functioning of the eukaryotic cell. Since it is known that nuclear DNA is
wrapped up into a nucleoprotein chromatin complex forming the period-
ically arranged nucleosomes, the factors involved in chromatin stabilization and remodeling during different stages of cell activity have attracted a
lot of interest (Reyes et al. 2002). During the 1980s, much emphasis was
laid on the unraveling of gene structure and regulation of transcription
catalyzed by RNA polymerases mediated and stimulated by transcription
factors. In the last decade, the structural organization of the chromatin
within the cell nucleus during different stages of the cell cycle underwent
a revival in research activity, and gene silencing was investigated in more
detail.

The central component of chromatin, the nucleosome, contains the
highly conserved histone proteins, which are now known to be subject to a
wide range of post-translational modifications and which act as recognition
sites for the binding of chromatin-associated factors. Earlier it was already
known that, in addition to these histone modifications (Loidl 2004), DNA
methylation could also have a dramatic influence on gene expression
(Bowler et al. 2004). Detailed cytological investigations with improved multicolor fluorescent in-situ hybridization (FISH) techniques (Lichter 1997)
elucidated the higher order structure of the chromosomes; recently more
and more information has been gained regarding this structural organization, not only at the chromosome level of a mitotic or meiotic cell, but
mainly at the eu- and heterochromatic interphase stage of the cell nucleus
(Tariq and Paszkowski 2004).

Recent studies have demonstrated that many important aspects of plant
development are accompanied by heritable changes in gene expression

Progress in Botany, Vol. 67
© Springer-Verlag Berlin Heidelberg 2006

that do not involve changes at the nucleotide sequence level. Rather, these regulatory mechanisms involve modifications of the chromatin structure giving access of target genes to regulatory factors that can control their expression. Epigenetic mutation, heritable developmental variation not based on a change in nucleotide sequence, is widely reported in plants. Heritable patterns of gene activity and gene silencing arise through the formation and the propagation of specific chromatin states that restrict or permit gene expression. The developmental and evolutionary significance of such epigenetically controlled mutations remains an interesting field in plant biology.

In this review article, we would like to concentrate on the role of DNA modification and chromatin packaging and factors involved in structural rearrangements resulting in epigenetically controlled gene silencing and/or modification and modulation of gene activity, such as DNA methylation at cytosines catalyzed by DNA methyltransferases, histone methyltransferases, histone acetyltransferases/deacetylases and chromatin remodeling factors. The role of small RNAs interacting and interfering with transcription products and the subsequent influence on developmentally processes will be discussed, as well as epigenetically regulated de- and reactivation of ribosomal RNA genes in nucleolar dominance and of transposable elements, since various recent studies indicate a clear correlation between DNA methylation, histone modification, chromatin remodeling and RNA interference.

2 Molecular mechanisms of gene silencing

2.1 Methylation of cytosine and DNA methyltransferases

2.1.1 Cytosine methylation

Methylation of cytosine at the fifth position of the pyrimidine ring (5mC; Jost and Saluz 1993; Colot and Rossignol 1999) is the most common modification of DNA in higher eukaryotes. The presence of modified nucleotides increase the amount of genetic information encoded by DNA sequence. The patterns of 5mC can be stably inherited providing a molecular basis for an "epigenetic memory". It has become apparent that cytosine methylation can fulfill pleitropic functions depending on the organism and the genomic location. In prokaryotes, methylation of cytosine and adenosine is used for cell protection via the restriction-modification system, which maintains specific methylation patterns of the host cellular DNA and destroys foreign unmethylated DNA. In addition, an important role in DNA repair and replication was demonstrated (for review, see Messer and Noyer-Weidner 1988;

Noyer-Weidner and Trautner 1993). In contrast, methylation of DNA in eukaryotes appears to have diverse functions, including gene silencing, genomic imprinting, transposon inactivation, regulation of tissue specific expression and epigenetic inactivation of chromosomal domains.

In addition to the symmetric CG motifs, which are the main target of cytosine methylation in vertebrates, plant DNA is also methylated at symmetric CNG as well as asymmetric CHH sequences (where N is any nucleotide and H is A, C or T). In addition, density of methylation versus methylation at specific sites should be distinguished (Gruenbaum et al. 1981; Bender 2004; Tariq and Paszkowski 2004).

5mC is non-randomly distributed through the genome, but mainly concentrated in repeated sequence elements including 5S and 35S rRNA genes (rDNA), satellite DNA elements (Hemleben et al. 1982), transposons and endogenous viruses, which are present in an inactive form in heterochromatin. In contrast, protein coding sequences are usually unmethylated (Guseinov et al. 1975; Bennetzen et al. 1988; Bennetzen 1996; Kovarik et al. 2000; Mathieu et al. 2002a,b; Kato et al. 2003; Palmer et al. 2003; Rabinowicz et al. 2003; Lippman et al. 2004). Accordingly, methylation of heterochromatic B chromosomes in *Crepis capillaries* is higher than that of A chromosomes (Luchniak et al. 2002). In plants, compared with animals and fungi, the proportion of 5mC is higher, usually ranging from 5% to 25% of total cytosine (Matassi et al. 1992), and one reason for that appears to be increased fractions of repeated sequences in plants, which is especially high in large genomes. Accordingly, a positive correlation between genome size and 5mC content was found (Bender 2004; Rangwala and Richards 2004). However, in several species with a small genome, a high content of 5mC was also found, e.g. in rice (33%; Thomas and Sherratt 1956) or in nettle (19%; R.A. Volkov, G.P. Miroshnichenko, A.S. Antonov, unpublished results). Methylation of cytosine at many loci can be stably transmitted through mitosis and meiosis, indicating that 5mC patterns are repeatedly reproduced during the cell cycle. Changes in cytosine methylations have also been demonstrated, i.e. during plant development (see below). Respectively, two aspects (i) maintenance of already existing patterns and (ii) de novo establishment of cytosine methylation could be distinguished.

2.1.2 DNA methyltransferases

In all eukaryotes, methylation of cytosine occurs post-replicatively, directly at the replication fork and also later during cell cycle progression (e.g. Vanyushin and Kirnos 1988). In plants, at least three classes of methyltransferases (MT) provide transfer of a methyl group from S-adenosyl-L-methionine to carbon 5 of cytosine within DNA helix (for review, see Finnegan and Kovac 2000).

The first class of MT represents the *METI* family, which comprises genes homologous to the DNMT1, a maintenance MT in the mouse. The observation that some motifs are conserved between DNMT1 and prokaryotic MT allowed the isolation of homologous proteins from plants (Bestor et al. 1988; Finnegan and Dennis 1993). In spite of the sequence similarity, eukaryotic MT of the first class are larger than prokaryotic enzymes because they contain a large relatively variable N-terminal domain in addition to the C-terminal conservative catalytic domain. In mouse, the N-terminal domain directs the enzyme to the nucleus, targets it to the replication fork and provides discrimination between unmethylated and hemi-methylated DNA (Bestor et al. 1992; Leonhardt et al. 1992; Bestor and Verdine 1994; Liu et al. 1998). Similar functions were also suggested for the homologous plant MTs (Finnegan and Kovac 2000). The *METI* gene of *Arabidopsis* belongs to a small multi-gene family, which includes up to five members (Finnegan and Dennis 1993; Genger et al. 1999). The *METI* gene appears to be predominantly transcribed in meristematic, vegetative and floral tissues, whereas other members of the multigene family seem to be only weakly or not expressed (Genger et al. 1999). Several *METI* homologs were identified in carrot, pea, tomato, tobacco, maize and rice (Finnegan and Kovac 2000; Nakano et al. 2000; Teerawanichpan et al. 2004).

In transgenic *Arabidopsis* and tobacco plants expressing *METI* homologs in antisense orientation, methylation of cytosine was reduced preferentially in CG dinucleotides, indicating that METI is a functionally active enzyme (Finnegan and Dennis 1993; Finnegan et al. 1996; Nakano et al. 2000). Regarding the sequence similarity to mouse DNMT1, a maintenance function of METI in plants, i.e. methylation of the newly synthesized DNA strand during cell cycle, was suggested (Finnegan and Dennis 1993; Finnegan et al. 1996). This was confirmed by observations that transcripts of tobacco, *NtMET1* (Nakano et al. 2000), as well as of rice, *OsMET1-1* and *OsMET1-2* (Teerawanichpan et al. 2004), are localized in actively proliferating tissues. Recent studies of *Arabidopsis METI* loss-of-function mutants containing a structurally disrupted MET1 catalytic domain confirm the requirement of MET1 to maintain CG methylation, and show in addition that the enzyme also contributes to CG de novo methylation (Aufsatz et al. 2004).

The second class of MT is represented by the plant specific chromomethylases (CMT), which contain a chromo-domain imbedded into a C-terminal methyltransferase domain (Henikoff and Comai 1998). The chromo-domain, which was identified early in several groups of proteins in animals and plants, is sufficient for protein targeting to heterochromatin (Paro and Harte 1996; Ingram et al. 1999). Similar to MT of class I, CMT also possess a variable N-terminal domain, which, however, lack a sequence similarity to the METI, suggesting distinct functions of CMT. Three CMT genes are present in the genome of *Arabidopsis*. Among them, CMT2 is highly transcribed in comparison with the two other members of the gene family (Genger et al. 1999; Finnegan and Kovac 2000). In *Arabidopsis*, CMT3 is involved in methylation at CNG motifs (Bartee et al. 2001). In a loss-of-function mutant of maize, carrying a Mu insertion in the *ZMET2* gene coding for a CMT, methylation of cytosine was remarkably decreased in CNG trinucleotides, especially in heterochromatic DNA such as centromeric, ribosomal and knob repeats, whereas methylation of CG dinucleotides was not altered (Finnegan and Kovac 2000).

Domains rearranged methyltransferases (DRM), which demonstrate similarity to mammalian de novo DNMT3, belong to the third class of plant MT. *DRM* genes were first described for *Arabidopsis* and maize (Cao et al. 2000). These proteins demonstrate a novel arrangement of the motifs required for DNA methyltransferase catalytic activity. Interestingly, the N termini of DRM contain a series of ubiquitin-associated (UBA) domains, which may be involved in ubiquitin binding, providing a possible link between DNA methylation and ubiquitin/proteasome pathways. In *Arabidopsis, DRM* genes are required for the initial establishment of methylation of cytosines in all known sequence contexts (Cao and Jacobsen 2002a), and for the maintenance of asymmetric DNA methylation (Cao and Jacobsen 2002b). At some loci, *DRM* act redundantly with *CMT3*, so that only in *drm1 drm2 cmt3* triple mutants all asymmetric methylation is lost. Similarly, in tobacco, NtDRM1 is able to methylate cytosine effectively at non-CG motives in non-methylated substrates both *in vivo* and *in vitro*. Remarkably, methylation activity for hemimethylated substrate was lower, indicating that the NtDRM1 is a de novo MT, which actively excludes CG substrates. Transcripts of *NtDRM1* were found in all tissues and during the cell cycle in cultured tobacco cells (Wada et al. 2003).

In addition, sequences encoding predicted proteins with similarity to other eukaryotic MT, such as MASC1 of *Ascobolus* and DNMT2 of mouse, were identified in the *Arabidopsis* database (Finnegan and Kovac 2000), suggesting that respective classes of MT may be present in plants.

2.2 Histone modifications

2.2.1 Histones as targets for post-translational modifications

In the nuclei of all eukaryotic cells, genomic DNA is highly folded, constrained, and compacted by histone and non-histone proteins in a dynamic structure called chromatin. The activity of chromatin is regulated by epigenetic events often mediated by covalent modifications of nuclear proteins. Chromatin is composed of nucleosomes, each of which contains approximately 145 bp of core DNA wrapped around an octamer of basic histone proteins (two sets of histones H2A, H2B, H3 and H4). Histone H1 interacts with linker DNA sequence between adjacent nucleosomes (Fischer et al. 1994; Luger 2003; Loidl 2004).

Histone proteins represent not only a structural component of chromatin, but are also actively involved in regulation of gene expression, which is connected with differential post-transcriptional modification of histone molecules at different places in the genome. Several modifications have been reported, including acetylation, methylation, phosphorylation, carbonylation, biotinylation, glycosylation, ADP ribosilation, ubiquitinilation and sumoylation. All core histones can be targets of modifications at distinct amino acid residues, preferentially within terminal extensions, and several modifications can be simultaneously present in the same histone molecule. Modifications in some positions were found in different groups of eukaryotes, whereas other modifications are plant-specific. Interestingly, two individual molecules of the same histone class within a nucleosome can be differentially modified. Therefore, a huge number of combinations of different modifications within a single nucleosome is possible, allowing establishment of a very specific histone modification landscape through the genome (Jenuwein and Allis 2001; Turner 2002; Wagner 2003; Loidl 2004). All modifications change the chemical properties of histone molecules, which results in rearrangement of chromatin organization. Some of these modifications may be necessary for histone turnover, e.g. ubiquitinilation as a target for subsequent proteolytic degradation, whereas other appears to play a role in modulation of transcription. In particular, the importance of two modifications of core histones, acetylation and methylation discovered 4 decades ago (Allfrey et al. 1964), was intensively studied recently.

2.2.2 Acetylation of histones

Acetylation is the best characterized modification of histones. Through this modification, histone acetyltransferase (HAT) catalyzes transfer of acetyl group from acetyl-CoA to the free amino groups of lysine (Lys) in the N-terminal extension of core histones. In plants, acetylation of histone H3 was found at Lys 9, 14, 18, and 23, and of histone H4 at Lys 5, 8,

12, 16 and 20. Histones H2A and H2B are acetylated to a lower extend (Waterborg 1990, 1992; Reyes et al. 2002). These modifications can be reverted by histone deacetylase (HDAC). Both enzymes, HAT and HDAC, are encoded by multigene families in *Arabidopsis* (Pandey et al. 2002).

Various HAT demonstrating obvious sequence similarity in their catalytic domains interact with similar proteins in different species and appear to have similar functions in all eukaryotes. According to subcellular localization, HAT can be classified into nuclear (type A) and cytoplasmic (type B) varieties (Roth et al. 2001). Type A comprises four distinct classes of HAT: (i) the MYST family; (ii) the p300/CBP co-activator family; (iii) the $TAF_{II}250$-related family; and (iv) GCN5 family (Marmorstein 2001; Pandey et al. 2002). Open reading frames homologous to members of all these groups were identified in the *Arabidopsis* genome, and HAT activity was demonstrated for representatives of p300 and GCN5 families (Bordoli et al. 2001; Stockinger et al. 2001). Plant type B cytoplasmic HAT appears to be heterodimeric complexes involved in acetylation of histone H4 at positions 5 and 12 before its incorporation into nucleosomes (Lusser et al. 1999).

In plants, at least three classes of HDAC can be distinguished: (i) the RPD3/HDA1 family; (ii) the sirtuin family related to yeast SIR2; and (iii) the HD2 family (Pandey et al. 2002). The last group comprises plant specific deacetylases, which are unrelated to the other HDAC gene families but distantly related to *cis-trans* isomerases, a group of enzymes with distinct function (Lusser et al. 1997; Dangl et al. 2001; Khochbin et al. 2001; Pandey et al. 2002). In *Arabidopsis*, at least five HDAC of class I are present. HDA19 is expressed at high levels in leaves, stems, flowers and young siliques and appears to be involved into transcriptional repression via acetylation of histone H4 affecting different developmental processes (Wu et al. 2000; Tian and Chen 2001). HDA6 seems to have a more specific function in gene silencing (Murfett et al. 2001). A plant-specific HDAC of class III is represented by the maize HD2. This protein is localized in the nucleolus, indicating a possible role in the regulation of rRNA genes. Four genes (*HDA3, HDA4, HDA11,* and *HDA13*) demonstrating high sequence similarity with the maize *HD2* were found in *Arabidopsis* (Wu et al. 2000). One of these genes, *HDA3*, appears to be involved in embryo development.

It was reported that at least in animals, HAT and HDAC could modify not only histones, but also other proteins (Gu and Roeder 1997; Chan et al. 2001; Vervoorts et al. 2003; Zhang et al. 2003). Hence, at the moment it is unclear if HAT and HDAC are predominantly histone modifiers or rather protein acetyltransferases/deacetylases with broad specificity (Loidl 2004).

2.2.3 Methylation of histones

Methylation of histones is catalyzed by histone methyltransferase (HMT), which transfer a methyl group to arginine or lysine residues, mostly within the N-terminal extensions of histones H3 and H4 (Loidl 2004). Arginine residues can be mono- or dimethylated, symmetrically or asymmetrically, by protein arginine methyltransferases (PMMT), whereas lysines are modified by members of the SET-domain HMT family to mono-, di- or trimethylated level (Kouzarides 2002; Santos-Rosa et al. 2002; Lachner et al. 2003). Histone methylation occurring at the same site can correlate either with repression or activation of transcription, depending on the target gene, individual HMT involved or the level of methylation (Santos-Rosa et al. 2002; Loidl 2004). Combinations of active (H3-Lys4, H3-Lys36, H3-Lys79) or repressive (H3-Lys9, H3-Lys27, and H4-Lys20) modifications are specific for different chromosomal subdomains. In mammals, H3-Lys9 trimethylation, H3-Lys27 monomethylation (Peters et al. 2003; Rice et al. 2003), and H4-Lys20 trimethylation (Schotta et al. 2004) are features of constitutive heterochro-matin. Also in *Drosophila* and *Arabidopsis*, heterochromatin is characterized by high levels of dimethylated H3-Lys9, whereas an increased level of dimethylated H3-Lys4 was found for euchromatin of *Arabidopsis* (Schotta et al. 2002; Jasencakova et al. 2003).

Interestingly, a high level of dimethylated H3-Lys9 in constitutive heterochromatin was found in species with small genomes. In contrast, dimethylated H3-Lys9 appears to be randomly distributed in plants with large genomes, suggesting that in these plants dispersed repetitive sequences are silenced also within euchromatic regions with a high level of methylated H3-Lys4 (Houben et al. 2003).

Originally, HMT Su(var)3-9 was identified in *Drosophila* (Rea et al. 2000; Schotta et al. 2002). The gene belongs to a large group of *suppressor of position-effect variegation, Su(var)* loci, which comprise more than 50 genes (Reuter and Spierer 1992; Ebert et al. 2004). Molecular characterization of about 15 *Su(var)* genes revealed that this group of genes encode not only HMT, but also other components of heterochromatin, such as the zinc finger protein *Su(var)3-7* (Delattre et al. 2000; Jaquet et al. 2002), the chromo-domain protein HP1 (Eissenberg et al. 1990; Eissenberg and Elgin 2000) and HDAC1 (DeRubertis et al. 1996; Mottus et al. 2000). HMT Su(var)3-9 and its yeast and human homologs contain a SET domain, which is necessary for specific methylation of Lys9 in histone H3, a modification required for heterochromatin formation (Nakayama et al. 2001). In the genome of *Arabidopsis* 37 SET domain proteins were identified (Baumbusch et al. 2001), ten of which demonstrate homology to Su(var)3-9, i.e. represent putative HMT. For one of them, SUVH4 (KRYPTONITE, KYP), histone methyltransferase activity was shown: the SUVH4 (KYP) methylates histone H3 at Lys 9 (Jackson et al. 2002). Loss of SUVH4 as well as of SUVH1 causes only a weak reduction of heterochromatic histone H3-Lys9 dimethylation, whereas in SUVH2 null plants levels of mono- and dimethyl H3-Lys9, mono- and dimethyl H3-Lys27 and monomethyl H4-Lys20 are significantly reduced (Naumann et al. 2004). Several SET-domain genes were also detected and partially characterized in tobacco (Shen 2001), spinach and pea (Trievel et al. 2002), maize (Springer et al. 2002), and rice (Liang et al. 2003).

2.3 Cross-talk between DNA methylation and modifications of histones

Available data show that certain modifications of histone proteins as well as methylation of cytosines in DNA appear to be non-randomly distributed between euchromatin and heterochromatin. But are they rather causally connected? Formation of transcriptionally inactive heterochromatin is associated with hypermethylation of DNA at CG sites and with histone H3-Lys9 methylation, both in mammals and plants (Jenuwein and Allis 2001). In *Neurospora* and *Arabidopsis*, the methylation status of histone H3 at Lys9 determines patterns of DNA methylation (Tamaru and Selker 2001; Tamaru et al. 2003). However, a feedback loop from DNA methylation to histone methylation is less well understood. To answer the question whether CG methylation of DNA can affect H3-Lys9 methylation, *Arabidopsis* mutants with a partial loss of function of MET1, which is necessary for maintenance of CG methylation, were investigated. However, the results obtained were partially in conflict (Johnson et al. 2002; Soppe et al. 2002), and the most probable reason for the discrepancy seemed to be an incomplete loss (about 50%) of CG methylation in the mutant studied (Vongs et al. 1993; Bartee and Bender 2001; Kankel et al. 2003). Recently, other *MET1* gene mutants of *Arabidopsis*, which demonstrate complete elimination of CG methylation, were characterized (Saze et al. 2003). In the *met1* strain, a clear loss of histone H3-Lys9 methylation in heterochromatin was found. Interestingly, the demethylation of histone H3 occurred without any detectable activation of transcription at heterochromatic loci and without alterations in heterochromatin structure (Tariq et al. 2003).

Interplay between the DNA and histone H3 methylation was also studied using a *SUPERMAN(SUP)/clark kent (clk)* model in *Arabidopsis*. The *SUP* and *clk* represent two epigenetically distinct alleles of a floral developmental gene, which significantly differ in expression although both alleles are identical in the nucleotide sequence. The switch from active SUP to inactive *clk* correlates with cytosine methylation at CG, CNG and CHH sites (Jacobsen and Meyerowitz 1997). In order to find genes controlling the *SUP/clk* switch, gene-silencing suppressor screens were performed, resulting in isolation of 16 mutants exhibiting wild type phenotype (Lindroth et al. 2001; Jackson et al. 2002). These mutants can be classified into three classes. The first class combines nine mutations in a gene coding for plant specific DNA-MT of class II, *CMT3*. The *cmt3* mutants show a strong decrease of cytosine methylation at plant-specific symmetric CNG or asymmetric CNN, but not at CG sites (Lindroth et al. 2001). The second class includes three mutants with a defective SET-domain histone methyltransferase SUVH4(KYP). Similar to the *cmt3* mutants the loss-of-function *kyp* mutants demonstrate a reduced cytosine methylation at non-CG sites (Jackson et al. 2002), indicating that KYP-dependent methylation of histone H3 can direct CNG methylation by CMT3.

2.4 Chromatin remodeling

To regulate gene activation/silencing via chromatin remodeling, other proteins distinct from the described above appear to be necessary. One of these proteins is a heterochromatin protein 1 (HP1), a conserved heterochromatin-associated protein found in animals, fungi and plants. The HP1 contains a chromo-shadow domain and a chromo-domain, which specifically recognize Lys9-methylated H3, resulting in association of HP1 with heterochromatin (Bannister et al. 2001; Lachner et al. 2003). Oligomerization of HP1 via the chromo-shadow domain seems to be necessary for maintenance of heterochromatin structures and gene silencing (Jenuwein 2001). The *Arabidopsis* HP1 homolog, LHP1 (LIKE HP1) can interact with histone H3 methylated at Lys 9 (Jackson et al. 2002) and with the CMT3 (Lindroth et al. 2001).

A model of epigenetic gene silencing that summarizes available data about the interplay between DNA and histone modifications during *SUP/clk* transcriptional switch (see above) was proposed (Fig. 1; see Jackson et al. 2002; Lachner 2002). Accordingly, the active state of *SUP* locus is connected with histone acetylation. As a first step, HDAC removes the acetyl groups, and then *SUVH4(KYP)* mediates methylation of histone H3 at Lys9, which is required for binding of LHP1. Subsequent recruitment of CMT3 by LHP1 results in methylation at CNG sites and shut down expression of the gene. In this silencing pathway, histone H3 methylation at Lys9 is placed upstream of DNA methylation.

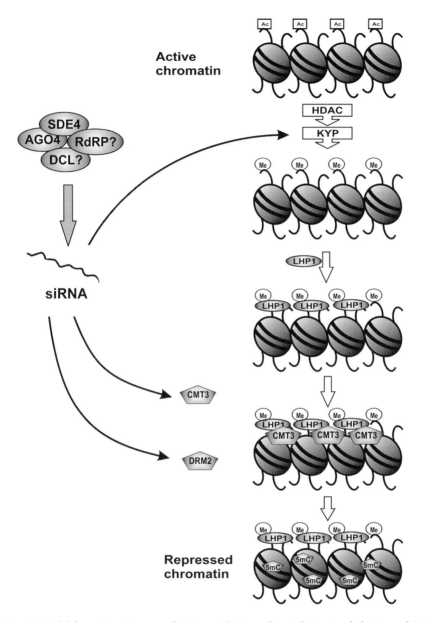

Fig. 1. Model for epigenetic gene silencing via histone deacetylation/methylation and DNA cytosine methylation at non-CG sites under the control of the AGO4/siRNA pathway. For details see chapter 2.4 and 2.5. *Ac* histone acetylation; *Me* histone methylation; *5mC* 5-methylcytosine; *HDAC* histone deacetylase; *KYP* histone methyltransferase KRYPTONITE; *LPH1* like heterochromatin-associated protein 1; *CMT3* chromomethylase 3; *DRM2* domains rearranged methyltransferase 2; *SDE4* SILENCING DEFECTIVE 4, a protein of unknown function; *AGO4* ARGONAUTE 4 protein; *RdRP* RNA-dependent RNA polymerase; *DCL* DICER-like enzyme; *siRNA* small interfering RNA. Modified from Jackson et al. (2002), Lachner (2002) and Zilberman et al. (2004)

A further link between DNA methylation, chromatin remodeling and gene silencing is provided by the methyl-CG-binding domain (MBD) proteins, which were first identified in animals (Hendrich and Bird 1998; Ballestar and Wolffe 2001). The MBD proteins function in transcriptional repression binding to methylated CG sites and recruiting HDAC complexes (Ahringer 2000; Ng et al. 2000) and HMT activity (Fuks et al. 2003). Recent studies demonstrated that during DNA replication in mammals, MBD1 recruits a histone H3-Lys9 methylase SETDB1 to the large subunit of the chromatin assembly factor CAF-1 to form an S phase-specific CAF-1/MBD1/SETDB1 complex that facilitates methylation of H3 at Lys9 during replication-coupled chromatin assembly. The data suggest a model in which H3-Lys9 methylation by SETDB1 is dependent on MBD1, which is heritably maintained through DNA replication to support the formation of stable heterochromatin at methylated DNA (Sarraf and Stancheva 2004). The *Arabidopsis* genome contains 12 putative genes encoding proteins with domains similar to MBD, of which at least three bind symmetrically methylated DNA. The *AtMBD* genes are active and differentially expressed in diverse tissues (Berg et al. 2003).

Thus, several groups of proteins such as HAT, HDAC, HMT, LHP, DNA-MT and MBD proteins cooperate to establish specific patterns of modifications of histones and DNA, which appear to modulate the local chromatin organization and control expression/silencing of respective genes. Are these modifications per se sufficient to change the chromatin structure or, alternatively, are some other proteins required for recognition, interpretation and maintenance of the modification patterns? The first direct evidence confirming the second possibility was obtained by isolation of *Arabidopsis* mutants suffering on decreased DNA methylation (*ddm*; Vongs et al. 1993). In the *ddm1-2* mutant plants about 70% of methylation was lost, although MT activity was the same as in wild type plants. Hypomethylation was initially observed only for repeated sequences, and after several generations of selfing also for unique sequences (Vongs et al. 1993; Kakutani et al. 1996, 1999).

The *DDM1* gene encodes a member of the SNF2/SWI2 family of the ATP-dependent chromatin remodeling factors (CRM; Peterson and Herskowitz 1992; Varga-Weisz 2001), which represent multiprotein complexes able to modify DNA-histone interactions by introducing superhelical torsion into DNA (Havas et al. 2000). Activity of CRMs can change nucleosome position or spacing (sliding) or accessibility to nucleosomal DNA, or provoke histone eviction, resulting in all of the cases in an increase of chromatin fluidity (Reyes et al. 2002).

Demethylation of DNA in *ddm1* mutants is accompanied by a shift toward euchromatin and transcriptional derepression, which are associated with histone H3 methylation at Lys4 and depletion of methylation at Lys9 (Grendel et al. 2002). It was proposed that the loss of *DDM1* function negatively affects chromatin remodeling, resulting in the alteration of histone H3 methylation. Respectively, the loss of 5mC in heterochromatic sequences occurred because no H3-Lys9 methylation is present to guide the cytosine methylation machinery. Remarkably, in the *ddm1* mutants there is no significant change in the overall abundance of Lys9-methylated histon H3 in contrast to the significant reduction in 5mC. Apparently DDM1 is not required for methylation of histone H3 at Lys9, but it seems to be necessary to restrict the histone methylation to specific regions in the genome, which can explain ectopic cytosine methylation observed in euchromatic regions in *ddm1* mutants (Jacobsen et al. 2000).

A chromatin remodeling machinery associated with transcription activation was described in *Arabidopsis:* FAcilitates Chromatin Transcription (FACT) complex, consisting of the Spt16 and SSRP1 proteins. The FACT subunits co-localize to cytologically defined euchromatin of the majority of cell types in embryos, shoots and roots, but never in terminally differentiated cells such as mature trichoblasts and root cap. FACT localizes to inducible genes only after induction of transcription, e.g. HSP70 and salicylic acid-inducible PR-1, and the association of the complex with the genes correlates with the level of transcription, indicating that FACT assists transcription elongation through plant chromatin (Duroux et al. 2004).

2.5 RNA silencing

Generally, RNA silencing describes a series of events that leads to the targeted degradation of cellular mRNA and thereby to the silencing of corresponding gene expression (Fig. 2). One of the first papers describing RNA silencing was published as early as 1928, although the phenomenon could not be explained at that time. Virus infected tobacco plants showed symptoms only on the initially infected leaves, whereas the upper leaves had somehow become immune to this virus and resistant to secondary infection (Wingard 1928). Now it has become clear that RNA silencing was involved. The different mechanisms are summarized in Fig. 2.

Introduction of a gene into the host genome can initiate RNA silencing of a gene that is homologous to the integrated gene. After transcription of the introduced gene, double-stranded RNA (dsRNA) is formed, which is cut into smaller dsRNA species called small interfering RNAs (siRNAs). This degradation is accomplished by an RNaseIII-like enzyme termed DICER. The siRNAs are then incorporated into a multiprotein complex termed RNA-induced silencing complex (RISC). RISC gets activated and a helicase function separates the two siRNA strands so that the remaining

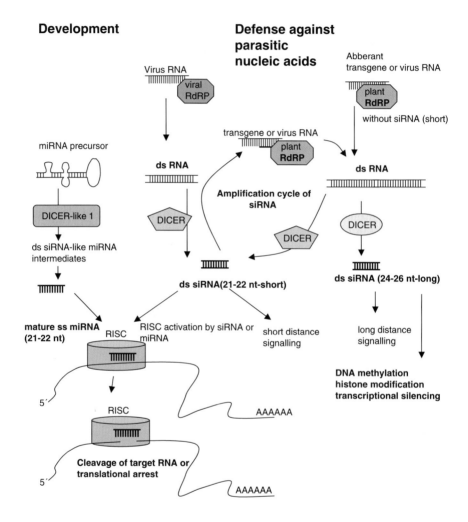

ds RNA: double-stranded RNA
siRNA: short interferring RNA
miRNA: micro RNA
DICER: RNAse-III like enzyme
RISC: RNA-induced silencing complex
RdRP: RNA-dependent RNA polymerase

strand guides the sequence-specific cleavage of the single-stranded complementary target RNA, most likely by proteins of the ARGONAUTE (AGO) family (for review, see Baulcombe 2004; Susi et al. 2004). At least some of the ten AGO homologues in *Arabidopsis* may be associated with effector complexes of RNA silencing that might be adapted to silence genes in specialized cells or at particular developmental stages as indicated by the *ago7* mutant, which shows alterations in the timing of the phase change between juvenile and adult leaves (Hunter et al. 2003; Vaucheret et al. 2004). Recently, nuclear RISC complexes have been described (Cerutti, 2003), and two of the four DICER enzymes that have been identified in the *Arabidopsis* genome, DICER-like 1 (DCL1) and DICER-like 4 (DCL4), contain nuclear localization signals (Schauer et al. 2002). Therefore, the processing reaction of RNA by DICER enzymes might proceed not only in the cytoplasm but also in the nucleus.

Another pathway is the silencing of endogenous messenger RNA by micro RNA (miRNA; Fig. 2). MiRNAs are generated from endogenous hairpin precursors, which are encoded by their own genes. These hairpin structures are diced by DCL1 into short ds siRNA-like miRNA intermediates. Mature miRNAs are 21-22 nt single stranded RNAs that negatively regulate gene expression by base pairing to specific mRNA, resulting in either RNA cleavage or arrest of protein translation (Baulcombe 2004; Susi et al. 2004).

Fig. 2. RNA-dependent gene silencing. In plants, RNA silencing can be divided into two main pathways: developmental regulation and defense against parasitic nucleic acids. In the initiation step of developmental regulation, short, imperfect double-stranded (ds) RNA precursors are cleaved by DICER-LIKE1 (DCL1) into 21–22 nucleotide (nt) single stranded RNAs. Subsequently, miRNAs are recruited into the RNA-induced silencing complex (RISC), which regulates the endonucleolytic cleavage or translational arrest of target mRNAs by the perfect or near-perfect base pairing between siRNAs or miRNAs and the targeted sequences. The defense pathways are also initiated by cleavage of ds or highly structured viral, transgene and transgenic aberrant RNAs into 21–22 nt (short) and 24–26 nt (long) siRNAs. Viral RNA-dependent RNA polymerase (RdRP) produced large amount of ds viral RNAs that are cleaved into short ds siRNAs. By contrast, dsRNAs (which are generated from aberrant transgenic or viral transcript by a plant RdRP without short siRNA guides) are processed to short and long ds siRNAs. These siRNAs are distinct in size and function, and probably arise from separate DICER activities. Short siRNAs activate RISC for target cleavage or translational arrest and also guide plant RdRP to amplify dsRNAs, which are cleaved again to short and long siRNAs. Moreover, these short siRNAs are also responsible for short-distance signaling, whereas the long siRNAs are probably involved in long-distance silencing and transcriptional silencing. (Modified from Silhavy and Burgyan 2004; Susi et al. 2004; Baulcombe 2004)

RNA silencing is linked to chromatin remodeling. In fission yeast, heterochromatin formation at centromere boundaries is associated with siRNA (Volpe et al. 2002). Also in *Arabidopsis*, the AGO4 protein appears to be required for silencing of the *SUP* locus, because the *ago4-1* mutant reactivated silent *SUP* alleles and decreased CNG and asymmetric DNA methylation as well as histone H3 methylation at Lys9. In addition, accumulation of 25-nucleotide siRNA, which corresponds to the retroelement *AtSN1* as well as histone and DNA methylation, was blocked in the *ago4-1* mutant, suggesting that AGO4 and long siRNAs direct chromatin modifications, including histone methylation and non-CG DNA methylation. The AGO4 of *Arabidopsis* probably encodes a component of a silencing system that generates long siRNA. The system also includes a DICER-like enzyme, an RNA-dependent RNA polymerase and an unidentified product of the *SDE4* locus. The long siRNA guide KYP-dependent histone methylation and CMT3- and DRM-dependent DNA methylation to specific regions of chromatin (Fig. 1; Zilberman et al. 2003).

In plants, RNA silencing appears to serve as a defense mechanism against viral pathogens and also to suppress the activity of virus-like mobile genetic elements and transgenes. However, not all transgenes are silenced. No case of silencing was observed in 132 independent transgenic lines with various sites of T-DNA integration. Below a certain number of identical transgenes in the genome, gene copy number and expression were positively correlated. Expression was high and stable over several generations, and expression levels were comparable among independent lines carrying the same copy number of a particular transgene. However, if the transcript level of a transgene surpassed a gene-specific threshold, RNA silencing was induced. It is proposed that the RNA sensing mechanism described is a genome surveillance system that eliminates RNA corresponding to excessively transcribed genes, including transgenes, and so plays an important role in genome defense (Schubert et al. 2004).

During infection by a conventional RNA virus, the entire process of RNA silencing is probably cytoplasmic with the general trigger of the double-stranded replicative intermediate, which is cleaved into siRNA (see Fig. 2). Specific degradation of single-stranded (+) or (−) viral RNA occurs by RISC-mediated cleavage. In addition, further double-stranded RNA might be produced by an RNA-directed RNA polymerase (RdRP) resulting in the generation of secondary siRNAs. This might also result in a phenomenon called transitive silencing signifying the spread of silencing to neighboring genes (Hutvagner and Zamore 2002).

In some plant systems, there are additional silencing processes such as systemic spread of silencing and RNA-directed methylation of homologous genomic DNA (reviewed in Wassenegger 2000; Bender 2001; Rangwala and Richards 2004). Methylation of genomic DNA also occurred when the silencing was induced by an RNA virus that replicates exclusively in the cytoplasm (Jones et al. 1998), suggesting communication between the cytoplasm and the nucleus (Mlotshwa et al. 2002). Therefore, dsRNA generated in the cytoplasm can potentially play a dual role by initiating post-transcription gene silencing (PTGS) and by entering the nucleus to direct methylation of homologous DNA sequences. It is conceivable that the small RNAs also guide DNA methyltransferase to homologous sequences in the genome (Mette et al. 2000). Interestingly, a characteristic for RNA-directed DNA methylation is a very dense methylation of cytosine sites at the target locus also at non-CG sites, and the removal of the RNA trigger leads to the eventual loss of cytosine methylation at these non-CG sites, although CG methylation can still be maintained by the methylase MET1 (Jones et al. 2001; Aufsatz et al. 2002; Chan et al. 2004).

Furthermore, the finding that silencing is not cell autonomous but can spread from cell to cell, also over long distances, suggests the existence of an as yet unidentified mobile silencing signal as a component of the RNA silencing pathway (Mlotshwa et al. 2002). In contrast to animals, plant RNA silencing is more complicated, since different classes of siRNAs, short (21–22 nt) and long (24–26 nt) siRNAs, can be distinguished (Fig. 2). Whereas the long siRNAs are dispensable for sequence-specific mRNA degradation but correlate with systemic silencing and methylation of homologous DNA, the short siRNA class correlates with mRNA degradation but not with systemic signaling or methylation (Hamilton et al. 2002).

3 Cellular processes regulated via gene silencing/chromatin remodeling

3.1 Control of developmental processes

Originally, chromatin remodeling factors were identified as genetic modifiers of developmental mutations in plants. These mutations result in lethality in metazoans, whereas mutant plants are viable and a wide range of developmental and physiological processes is affected.

In *Drosophila*, chromatin remodeling provides one way in which on/off states of gene expression can be fixed and inherited through mitotic division (Lawrence and Struhl 1996). Genetic and biochemical analyses have elucidated that fixation of the chromatin state during development is conferred by Polycomb group proteins (PcG). Two main classes of PcG were originally described as repressors of homeotic genes in *Drosophila*: (i) PRC1 and (ii) the extra sex combs (ESC)-E(z) complex (for review see Francis and Kingston 2001). In plants, representatives of the second group were identified (Reyes et al. 2002), suggesting that the PcG-mediated cellular memory system is an important mechanism of transcription control in plants as well.

Genetic screens for mutations causing parent-of-origin effects on seed development in *Arabidopsis* revealed lesions in *FERTILIZATION INDEPEN-DENT ENDOSPERM (FIE,* Ohad et al. 1996, 1999), *FERTILIZATION INDE-PENDENT SEED 2 (FIS2,* Chaudhury et al. 2001*)* or *MEDEA (MEA,* Grossniklaus et al. 1998; Kiyosue et al. 1999) loci. Molecular cloning and investigation of expression patterns of *MEA, FIE* and *FIS2* revealed that they are homologs of the animal PcG complex and are necessary to prevent transcription of target genes involved in regulation of endosperm development (Luo et al. 2000; Spillane et al. 2000; Yadegari et al. 2000). MEA is coding for a SET-domain protein closely related to *Drosophila* ENHANCER OF ZESTE, FIE encodes a WD-40 protein with high similarity to EXTRA SEX COMB, and FIS2 is a C2H2 zinc finger homolog of *Drosophila* SUPPRESSOR OF ZESTE. It has been shown that the MEA-FIE complex can bind PHERES1 (PHE1), a MADS-domain protein that is expressed transiently after fertilization and controls seed viability. PHE1 gene expression dramatically increases in *mea, fie* or *fis2* mutants, and its expression continues until the mutant seeds abort, suggesting that the MEA-FIE complex is required for repressed state of *PHE1* during seed development (Köhler et al. 2002, 2003).

In *Arabidopsis, FIE* is expressed ubiquitously throughout development (Ohad et al. 1999) and probably participates in the formation of distinct PcG complexes at multiple developmental stages. It has been shown that *FIE,* in addition to its function in the central cell and developing seed, mediates the PcG complex formation that suppresses flowering during early plant development, controlling several floral meristem and floral homeotic genes such as *LEAFY, AGAMOUS* and *PISTILLATA* (Kinoshita et al. 2001). FIE can interact with other PcG proteins to repress floral gene expression during embryo and seedling development. One example is EMBRYONIC FLOWER2 (ENF2), a zinc finger PcG protein similar to FIS2. Genetic analysis showed that *EMF2* is epistatic to flowering time genes such as *CONSTANT (CO), APETALA1 (AP1)* and *FLOWERING LOCUS T (FT),* suggesting that EMF2 plays an early role in repressing the transition to reproduction (Chen et al. 1997; Haung and Yang 1998; Yoshida et al. 2001). Another PcG proteins involved in developmental regulation are CURLY LEAF (CLF), which prevents inappropriate expression of the MADS box homeotic gene *AG* in leaves (Kinoshita et al. 2001; Yoshida et al. 2001) and *VERNALIZATION2 (VRN2),* coding for a FIS2-like zinc finger PcG protein, which control the transition from vegetative growth to reproduction by repression of *FLOWERING LOCUS C (FLC)* in temperate biennial plants (Sheldon et al. 1999; Gendall et al. 2001).

Recent studies on *Arabidopsis* developmental mutants revealed a number of loci being under control of the chromatin remodeling machinery. Representatives of jumonji/zinc-finger-class of transcription factors EARLY FLOWERING 6 (ELF6) and RELATIVE EARLY FLOWERING 6 (REF6) are involved in control of photoperiod-independent and FLC flowering pathways, respectively, whereas repression of FLC expression is accompanied with histone modifications (Noh et al. 2004). The PHOTOPERIOD-INDEPENDENT EARLY FLOWERING 1 (PIE1) is a further chromatin remodeling protein similar to the SWI2/SNF2 ATP-dependent protein family, which is required for appropriate expression of FLC and plays a role in petal development (Noh and Amasino 2003). Several other homologs of SWI/SNF complex proteins have been characterized to control developmental processes in *Arabidopsis*. SPLAYED (SYD) was identified to act as LEAFY (LFY)-dependent repressor of the meristem identity switch in floral transition in response to environmental stimuli (Wagner and Meyerowitz 2002).

AtBRM, a homolog of BRAHMA, an ATPase identified from *Drosophila* as a chromatin remodeling protein, has been described to interact with CHB4 (AtSWI3C) protein to regulate reproductive development in the photoperiod-dependent flowering pathway (Farrona et al. 2004). Another member of the SWI3 family, CHB2 (AtSWI3B), seems to play a global role in the regulation of expression of genes involved in plant growth and development (Sarnowski et al. 2002; Zhou et al. 2003)

Silencing of endogenous messenger RNAs by miRNAs is an important regulatory mechanism during plant development. These miRNAs negatively regulate gene expression by base pairing to specific mRNAs, resulting in either RNA cleavage or arrest of protein translation (Papp et al. 2003). It has been estimated that the *Arabidopsis* genome has about 100 miRNA loci. However, the range of targets is not restricted to "developmental" genes because there are also miRNAs that increase or decrease in abundance following cold or drought stress or sulphur starvation (Jones-Rhoades et al. 2004).

Interestingly, unrelated suppressors from multiple viruses were shown also to inhibit miRNA activities and trigger an overlapping series of severe developmental defects in transgenic *Arabidopsis*. This suggests that interference with miRNA-directed processes may contribute to the pathogenicity of many viruses (Chapman et al. 2004). These developmental defects were remarkably similar, and their penetrance correlated with inhibition of miRNA-guided cleavage of endogenous transcripts and not with altered miRNA accumulation per se (Dunoyer et al. 2004).

3.2 Ribosomal DNA transcriptional regulation and nucleolar dominance

Higher plants often contain more repeated sequence elements in the genome than other eukaryotes. The relative content of repeated sequences correlates with genome size, and it amounts up to 90% of the total nuclear DNA in plants with large genomes (e.g. *Allium*). The majority of the repeated sequence elements represent retrotransposons and/or satellite DNA elements (Hemleben et al. 2000). They are mostly highly methylated and transcriptionally silenced. In contrast, genes coding for rRNA, which also belong to the repeated sequences, are transcriptionally active and ultimately necessary for surviving of the cell. The nuclear loci coding for the 18S, 5.8S and 25/28S rRNA (35S rDNA) are arranged in head-to-tail tandem arrays, often comprising between 500 and more than 30,000 members depending on the plant species (Hemleben et al. 1988). The transcribed genes can be cytologically distinguished as nucleolus in interphase nuclei or as secondary constrictions (SC) or nucleolus organizer region (NOR) in mitosis/meiosis (Heitz 1931; McClintock 1934; for review, see Leweke and Hemleben 1982; Hemleben and Zentgraf 1994; Moss and Stefanovsky 1995). In the transition area from fibrillar centers to the dense fibrillar component (FC/DFC) of the nucleolus, the genes for the 18S, 5.8S and 25/28S rRNA are commonly transcribed by RNA polymerase I (pol I) as a large precursor, which in plants is approximately 35S in size (35S pre-rRNA); this 35S pre-rRNA is then step-wisely processed into the respective mature rRNAs (Volkov et al. 2004). The 5S rDNA is also arranged in head-to-tail tandem arrays. Usually, 5S rDNA is located at other regions of the genome as 35S rDNA and is transcribed by RNA polymerase III (pol III) into 5S rRNA (Paule and White 2000; Mathieu et al. 2003).

Similar to other eukaryotes, plants often contain much more 35S rDNA repeats than are used for the production of cellular rRNA. Respectively, only a fraction of rDNA is transcriptionally active at any time. Functional activity of 35S rDNA is regulated at two levels, (i) by controlling the number of active 35S rDNA repeats existing in open chromatin state (dosage control) and (ii) by modulating pol I initiation frequency (for reviews, see Hemleben et al. 1988, 1998; Grummt and Pikaard 2003; Volkov et al. 2004). Accordingly, transcriptional silencing of rDNA often correlates with methylation of cytosine (Flavell et al. 1988; Sardana et al. 1993; Torres-Ruiz and Hemleben 1994; Chen and Pikaard 1997a,b; Houchins et al. 1997; Santoro et al. 2002), although several exceptions were also described (Macleod and Bird 1982; Chen and Pikaard, 1997; Papazova et al. 2001). A special case of regulation of functional activity of 35S rDNA represents the phenomenon of nucleolar dominance, often occurring in interspecific hybrids. By nucleolar dominance, rDNA inherited from one of the parental species appears transcriptionally active, whereas rDNA of the other crossing partner is silenced (Pikaard 2000; Volkov et al. 2004).

In *Arabidopsis,* presumably inactive rRNA repeats exhibit hypomethylated promoters and are associated with H3-trimethyl-Lys4 and pol I. In contrast, promoters that are hypermethylated associate with H3-dimethyl-Lys9 and are presumably silenced. Hence, concerted changes in the density of cytosine methylation at the rDNA promoter and specific histone modifications dictate the on and off states of the rRNA genes. A key component of the transcriptional switch is HDT1, a plant-specific histone deacetylase that localizes to the nucleolus and is required for H3-Lys9 deacetylation and subsequent

H3-Lys9 methylation. The data support a model in which cytosine methylation and histone deacetylation are each upstream of one another in a self-reinforcing repression cycle. A similar regulation mechanism is used for gene dosage control and nucleolar dominance (Lawrence et al. 2004).

Recently, it was shown that the *sil1* mutant of *Arabidopsis* represents a defect allele of *HDA6*. The *sil1* mutation results in reactivation of certain transcriptionally silent transgenes and endogenous repeats and also influences histone acetylation levels. Remarkably, significant hyperacetylation of histone H4 restricted to NOR was found in the mutant plants, whereas total level of H4 acetylation was only slightly increased. These alterations correlated with an increase of histone H3 methylation at Lys4, modification of rDNA methylation pattern, and a concomitant decondensation of rDNA chromatin. Nevertheless, the changes at rDNA loci seem to occur without major changes in transcription rates. Together, the data indicate that HDA6 might play a role in regulating activity of rRNA genes, and this control might be functionally linked to silencing of other repetitive templates and to the presumptive role of HDA6 in RNA-directed DNA methylation (Probst et al. 2004).

Thus, regulation of the number of potentially active rDNA repeats (dosage control) is accomplished via chromatin remodeling. Also in hybrids, modulation of transcription/silencing of parental rDNA represents a manifestation of the dosage control (Lawrence et al. 2004). However, it remains illusive which molecular mechanism provides differential inactivation of one of the parental rDNA. In hybrids of *Solanum* differential expression of parental rDNA correlates with the number of conserved repeated sequence elements (CE) downstream of the transcription start and with cytosine methylation (Volkov et al. 1999a; Komarova et al. 2004). Interestingly, computer simulation shows that each of CE can form a double-stranded stem-loop RNA structure (Volkov et al. 1999b, and unpublished results), which could be a signature of RNA-directed DNA silencing (Wassenegger et al. 2000; Matzke et al. 2004). It is tempting to postulate that these sequence elements downstream of the transcription start could represent targets for RNA-directed DNA methylation/silencing involved in regulation of rDNA transcription.

Studies on *Nicotiana*, *Secale* and *Arabidopsis* showed that 5S rDNA represents a highly methylated multigene family (Fulnecek et al. 1998, 2002). However, in contrast to 35S rDNA, transcription of these genes by RNA polymerase III appears not to be modulated by DNA methylation (Mathieu et al. 2002a, b), but chromatin remodeling events are involved (Mathieu et al. 2003).

3.3 Silencing of transposons

An important role of RNA silencing at the chromatin level is probably protecting the genome against damage caused by active transposons or by extreme amplification of repetitive elements (for review, see: Lippman and Martienssen 2004) leading to heterochromatin formation. Heterochromatin is found near centromeres and telomeres, but interstitial sites of heterochromatin (knobs) are common in plant genomes and were first described in maize as regions that contain repetitive and late-replicating DNA components.

As shown in *Drosophila*, heterochromatin influences gene expression, a heterochromatin phenomenon called position effect variegation (see above). Similarities between position effect variegation in *Drosophila* and gene silencing in maize mediated by "controlling elements" (that is, transposable elements) led in part to the proposal that heterochromatin is composed of transposable elements, and that such elements scattered throughout the genome might regulate development.

Recent progress in understanding the silencing of transposable elements in the model plant *Arabidopsis* has revealed an interplay between DNA methylation, histone methylation and siRNAs. However, DNA and histone methylation are not always sufficient to maintain silencing, and RNA-based reinforcement can be needed to maintain as well as initiate it (Zilberman and Henikoff 2004).

In the genus *Ipomoea*, especially in three species, *I. nil* (the Japanese morning glory), *I. purpurea* (the common morning glory), and *I. tricolor*, numerous mutations affecting flower pigmentation were found to be caused by the insertion of DNA transposable elements in the genes for the anthocyanin pigmentation (Iida et al. 2004). The flecked, speckled, r-1, and purple mutations in *I. nil* were caused by insertions of Tpnl and its relatives in the En/Spm superfamily, Tpn2, Tpn3, and Tpn4, into the genes for anthocyanin coloration in flowers, i.e. *DFR* (dihydoflavonol reductase)-B, *CHI* (chalcone isomerase) and *CHS* (chalcone synthase)-D, respectively. Similarly, the flaked and pink mutants of *I. purpurea* have distantly related elements, Tip100 and Tip201, in the Ac/Ds superfamily inserted into the CHS-D and F3'H (flavonol 3' hydroxylase) genes, respectively. The flower variegation patterns can be determined by the frequency and timing of the excision of these transposons, and their stable insertions produce plain colored flowers without generating pigmented spots or sectors. Both genetic and epigenetic regulation appeared to play important roles in determining the frequency and timing of the excision of the transposons.

From studies of the endogenous *Arabidopsis* transposon CACTA (Kato et al. 2004), it was proposed that the inheritance of epigenetic gene silencing over generations could function as a transgenerational genome defense mechanism against deleterious movement of transposons. Previously, it was reported that silent CACTA1 is mobilized by the DNA hypomethylation mutation *ddm1* (see above). CACTA activated by the *ddm1* mutation remains mobile in the presence of the wild-type *DDM1* gene, suggesting that de novo silencing is not efficient for the defense of the genome against CACTA movement. The defense depends on maintenance of transposon silencing over generations. In addition, the activated CACTA1 element transposes throughout the genome in DDM1 wild type plants. Furthermore, the CACTA1 element integrated into both the *ddm1*-derived and the *DDM1*-derived chromosomal regions in the *DDM1* wild-type plants, demonstrating that this transposable element does not exhibit targeted integration into heterochromatin, despite its accumulation in the pericentromeric regions in natural populations.

It was shown that heterochromatin in *Arabidopsis* is determined by transposable elements and related tandem repeats, under the control of the chromatin remodeling ATPase DDM1. Small interfering RNAs correspond to these sequences, suggesting a role in guiding DDM1 (Zilberman et al. 2003, 2004). Transposable elements can regulate genes epigenetically, but only when inserted within or very close to them.

In a number of organisms, transgenes containing transcribed inverted repeats (IRs) that produce hairpin RNA can trigger RNA-mediated silencing, which is associated with 21-24 nucleotide siRNAs. In plants, IR-driven RNA silencing also causes extensive cytosine methylation of homologous DNA in both the transgene "trigger" and any other homologous DNA sequences ("targets"). Endogenous genomic sequences, including transposable elements and repeated elements, are also subject to RNA-mediated silencing. The RNA silencing gene *AGO4* is required for maintenance of DNA methylation at several endogenous loci, e.g. for the establishment of methylation at the endosperm development associated gene *FWA* (Lippman et al. 2004). Mutation of *AGO4* substantially reduces the maintenance of DNA methylation triggered by IR transgenes, but *AGO4* loss-of-function does not block the initiation of DNA methylation by IRs (Zilberman et al. 2003, 2004). AGO4 primarily affects non-CG methylation of the target sequences, while the IR trigger sequences lose methylation in all sequence contexts. Finally, *AGO4* and the *DRM* methyltransferase genes are required for maintenance of siRNAs at a subset of endogenous sequences, but *AGO4* is not required for the accumulation of IR-induced siRNAs or a number of endogenous siRNAs, suggesting that AGO4 may function downstream of siRNA production.

4 Conclusions and perspectives

In a eukaryotic cell of a differentiated organism, not all genes are actively transcribed; mechanisms of gene silencing acting also on single-copy genes have been extensively analyzed in recent years. Therefore, it became obvious that epigenetically controlled gene silencing and modulation of gene activity play a pivotal role, and basic mechanisms such as DNA methylation, histone modifications and a complex pattern of chromatin remodeling factors are involved in combination with RNA interference mechanisms not only at the constitutive heterochromatin, but also at various loci resulting in facultatively regulated gene silencing.

The task for the future remains to unravel the processes where internal and external factors meet to obtain a co-integrated picture of the regulation of cell activity and cooperation of this activity within a multicellular organism.

References

Ahringer J (2000) NuRD and SIN3; histone deacetylase complexes in development. Trends Genet 16:351–356

Allfrey VG (1964) Acetylation and methylation of histons and their possible role in the regulation of RNA synthesis. Proc Natl Acad Sci USA 51:786–784

Aufsatz W, Mette MF, van der Winden J, Matzke AJ, Matzke M (2002) Proc Natl Acad Sci USA 99:16499–16506

Aufsatz W, Mette MF, Matzke AJM, Matzke M (2004) The role of MET1 in RNA-directed de novo and maintenance methylation of CG dinucleotides. Plant Mol Biol 54:793–804

Ballestar E, Wolffe AP (2001) Methyl-CpG-binding proteins. Targeting specific gene repression. Eur J Biochem 268:1–6

Bannister AJ, Zegerman P, Partridge JF, Miska EA, Thomas JO, Allshire RC, Kouzarides T (2001) Selective recognition of methylated lysine 9 on histone H3 by the HP1 chromo domain. Nature 410:120–124

Bartee L, Bender J (2001) Two *Arabidopsis* methylation-deficiency mutations confer only partial effects on a methylated endogenous gene family. Nucleic Acids Res 29:2127–2134

Bartee L, Malagnac F, Bender J (2001) *Arabidopsis* cmt3 chromomethylase mutations block non-CG methylation and silencing of an endogenous gene. Genes Dev 15:1753–1758

Baulcombe DC (2004) RNA silencing in plants. Nature 431:356–363

Baumbusch LO, Thorstensen T, Krauss V, Fischer A, Naumann K, Assalkhou R, Schulz I, Reuter G, Aalen RB (2001) The *Arabidopsis thaliana* genome contains at least 29 active genes encoding SET domain proteins that can be assigned to four evolutionarily conserved classes. Nucleic Acids Res 29:4319–4333

Bender J (2001) A vicious cycle: RNA silencing and DNA methylation in plants. Cell 106:129–132

Bender J (2004) DNA methylation and epigenetics. Annu Rev Plant Biol 55:41–68

Bennetzen JL (1996) The contributions of retroelements to plant genome organization, function and evolution. Trends Microbiol 4:347–353

Bennetzen JL, Brown WE Springer PS (1988) The state of DNA modification within and flanking maize transposable elements. In: Nelson O (ed) Plant transposable elements. Plenum, New York, pp 237–250

Berg A, Meza TJ, Mahic M, Thorstensen T, Kristiansen K, Aalen RB (2003) Ten members of the *Arabidopsis* gene family encoding methyl-CpG-binding domain proteins are transcriptionally active and at least one, AtMBD11, is crucial for normal development. Nucleic Acids Res 31:18 5291–5304

Bestor TH, Verdine GL (1994) DNA methyltransferases. Curr Opin Cell Biol 6:380–389

Bestor T, Laudano A, Mattaliano R, Ingram V (1988) Cloning and sequencing of a cDNA encoding DNA methyltransferase of mouse cells. J Mol Biol 203:971–983

Bestor TH, Gundersen G, Kolsto AB, Prydz H (1992) CpG islands in mammalian gene promoters are inherently resistant to de novo methylation. Genet Anal Tech Appl 9:48–53

Bordoli L, Netsch M, Luthi U, Lutz W, Eckner R (2001) Plant orthologs of p300/CBP: conservation of a core domain in metazoan p300/CBP acetyltransferase-related proteins. Nucleic Acids Res 29:589–597

Bowler C, Benvenuto G, Laflamme P, Molino D, Probst AV, Tariq M, Paszkowski J (2004) Chromatin techniques for plant cells. Plant J 39:776–789

Cao X, Jacobsen SE (2002a) Role of the *Arabidopsis DRM* methyltransferases in de novo DNA methylation and gene silencing. Curr Biol 12:1138–1144

Cao X, Jacobsen SE (2002b) Locus-specific control of asymmetric and CpNpG methylation by the DRM and CMT3 methyltransferase genes. Proc Natl Acad Sci USA 99 (Suppl 4):16491–16498

Cao X, Springer NM, Muszynski MG, Phillips RL, Kaeppler S, Jacobsen SE (2000) Conserved plant genes with similarity to mammalian de novo DNA methyltransferases. Proc Natl Acad Sci USA 97:4979–4984

Cerutti H (2003) RNA interference: traveling in the cell and gaining functions? Trends Genet 19:39–46

Chan HM, Krstic-Demonacos M, Smith L, Demonacos C, La Thangue NB (2001) Acetylation control of the retinoblastoma tumour-suppressor protein. Nat Cell Biol 3:667–674

Chan SW, Zilbermann D, Xie Z, Johannsen LK, Carrington JC, Jacobsen SE (2004) RNA silencing genes control de novo DNA methylation. Science 303:1336

Chapman EJ, Prokhnevsky AI, Gopinath K, Dolja VV, Carrington JC (2004) Viral RNA silencing suppressors inhibit the microRNA pathway at an intermediate step. Genes Dev 18:1179–1186

Chaudhury AM, Koltunov A, Payne T, Luo M, Tucker MR and Dennis Peacock WJ (2001) Control of early seed development. Annu Rev Cell Dev Biol 17:677–699

Chen ZJ, Pikaard CS (1997a) Epigenetic silencing of RNA polymerase I transcription: a role for DNA methylation and histone modification in nucleolar dominance. Genes Dev 11:2124–2136

Chen ZJ, Pikaard CS (1997b) Transcriptional analysis of nucleolar dominance in polyploid plants: biased expression/silencing of progenitor rRNA genes is developmentally regulated in *Brassica*. Proc Natl Acad Sci USA 94:3442–3447

Chen L, Cheng JC, Castle L, Sung ZR (1997) EMF genes regulate *Arabidopsis* inflorescence development. Plant Cell 9:2011–2024

Colot V, Rossignol JL (1999) Eukaryotic DNA methylation as an evolutionary device. Bioessays 21:402–411

Dangl M, Brosch G, Haas H, Loidl P, Lusser A (2001) Comparative analysis of HD2 type histone deacetylases in higher plants. Planta 213:280–285

Delattre M, Spierer A, Tonka CH, Spierer P (2000) The genomic silencing of position-effect variegation in *Drosophila melanogaster*: interaction between the heterochromatin-associated proteins Su(var)3–7 and HP1. J Cell Sci 13:4253–4261

DeRubertis F, Kadosh D, Henchoz S, Pauli D, Reuter G, Struhl K, Spierer P (1996) The histone deacetylase RPD3 counteracts genomic silencing in *Drosophila* and yeast. Nature 384:589–591

Dunoyer P, Lecellier CH, Parizotto EA, Himber C, Voinnet O (2004) Probing the microRNA and small interfering RNA pathways with virus-encoded suppressors of RNA silencing. Plant Cell 16:1235–1250

Duroux M, Houben A, Ruzicka K, Friml J, Grasser KD (2004) The chromatin remodelling complex FACT associates with actively transcribed regions of the *Arabidopsis* genome. Plant J 40:660–671

Ebert A, Schotta G, Lein S, Kubicek S, Krauss V, Jenuwein T, Reuter G (2004) Su(var) genes regulate the balance between euchromatin and heterochromatin in *Drosophila*. Genes Dev 18:2973–2983

Eissenberg JC, Elgin SC (2000) The HP1 protein family: getting a grip on chromatin. Curr Opin Genet Dev 10:204–210

Eissenberg JC, James TC, Foster-Hartnett DM, Hartnett T, Ngan V, Elgin SC (1990) Mutation in a heterochromatin-specific chromosomal protein is associated with suppression of position-effect variegation in *Drosophila melanogaster*. Proc Natl Acad Sci USA 87: 9923–9927

Farrona S, Hurtado L, Bowman JL, Reyes JC (2004) The *Arabidopsis thaliana* SNF2 homolog AtBRM controls shoot development and flowering. Development 131:4965–4975

Finnegan EJ, Dennis ES (1993) Isolation and identification by sequence homology of a putative cytosine methyltransferase from *Arabidopsis thaliana*. Nucl Acids Res 21:2383–2388

Finnegan EJ, Kovac KA (2000) Plant DNA methyltransferases. Plant Mol Biol 43:189–201

Finnegan EJ, Peacock WJ, Dennis ES (1996) Reduced DNA methylation in *Arabidopsis thaliana* results in abnormal plant development. Proc Natl Acad Sci USA 93:8449–8454

Fischer TC, Groner S, Zentgraf U, Hemleben V (1994) Evidence for nucleosomal phasing and a novel protein specifically binding to cucumber satellite DNA. Z Naturforsch 149:79–86

Flavell RB, O'Dell M, Thompson WF (1988) Regulation of cytosine methylation in ribosomal DNA and nucleolus organizer expression in wheat. J Mol Biol 204:523–534

Francis NJ, Kingston RE (2001) Mechanisms of transcriptional memory. Nat Rev Mol Cell Biol 2:409–421

Fuks F, Hurd PJ, Wolf D, Nan X, Bird AP, Kouzarides T (2003) The methyl-CpG-binding protein MeCP2 links DNA methylation to histone methylation. J Biol Chem 278: 4035–4040

Fulnecek J, Matyasek R, Kovarik A, Bezdek M (1998) Mapping of 5-methylcytosine residues in *Nicotiana tabacum* 5S rRNA genes by genomic sequencing. Mol Gen Genet 259:133–141

Fulnecek J, Matyasek R, Kovarik A (2002) Distribution of 5-methylcytosine residues in 5S rRNA genes in *Arabidopsis thaliana* and *Secale cereale*. Mol Genet Genomics 268:510–517

Gendall AR, Levy YY, Wilson A, Dean C (2001) The VERNALIZATION 2 gene mediates the epigenetic regulation of vernalization in *Arabidopsis*. Cell 107:525–355

Genger RK, Kovac KA, Dennis ES, Peacock WJ, Finnegan EJ (1999) Multiple DNA methyl-transferase genes in *Arabidopsis thaliana*. Plant Mol Biol 41:269–278

Grendel AV, Lippman Z, Yordan C, Colot V, Martienssen R (2002). Dependence of heterochro-matic histone H3 methylation patterns on the *Arabidopsis* gene *DDM1*. Science 20:20

Grossniklaus U, Vielle-Calzada JP, Hoeppner MA, Gagliano WB (1998) Maternal control of embryogenesis by MEDEA, a polycomb group gene in *Arabidopsis*. Science 280:446–450

Gruenbaum Y, Naveh-Many T, Cedar H, Razin A (1981) Sequence specificity of methylation in higher plant DNA. Nature 292:860–862

Grummt I, Pikaard CS (2003) Epigenetic mechanisms controlling RNA polymerase I tran-scription. Nat Rev Mol Cell Biol 4 641–64

Gu W, Roeder RG (1997) Activation of p53 sequence-specific DNA binding by acetylation of the p53 C-terminal domain. Cell 90:595–606

Guseinov VA, Kiryanov GI, Vanyushin BF (1975) Intragenome distribution of 5-methylcyto-sine in DNA of healthy and wilt-infected cotton plants (*Gossypium hirsutum* L.). Plant Mol Biol Rep 2:59–63

Hamilton A, Voinnet O, Chappell L, Baulcombe D (2002) Two classes of short interfering RNA in RNA silencing. EMBO J 21:4671–4679

Haung MD, Yang CH (1998) EMF genes interact with late-flowering genes to regulate *Arabidopsis* shoot development. Plant Cell Physiol 39:382–393

Havas K, Flaus A, Phelan M, Kingston R, Wade PA, Lilley DM, Owen-Hughes T (2000) Generation of superhelical torsion by ATP-dependent chromatin remodeling activities. Cell 103:1133–1142

Heitz E (1931) Nucleolen und Chromosomen in der Gattung *Vicia*. Planta 15:495–505

Hemleben V, Zentgraf U (1994) Structural organization and regulation of transcription by RNA polymerase I of plant nuclear ribosomal RNA genes. In: Nover L (ed) Plant promot-ers and transcription factors. Results and problems in cell differentiation. Springer, Berlin Heidelberg, 20, pp 3–24

Hemleben V, Leweke B, Roth A, Stadler J (1982) Organization of highly repetitive satellite DNA of two Cucurbitaceae species (*Cucumis melo* and *Cucumis sativus*). Nucleic Acids Res 10:631–644

Hemleben V, Ganal M, Gerstner J, Schiebel K, Torres RA (1988) Organization and length het-erogeneity of plant ribosomal RNA genes. In: Kahl G (ed) The architecture of eukaryotic gene. VHC, Weinheim, pp 371–384

Hemleben V, Zanke C, Panchuk II, Volkov RA (1998) Repetitive elements as molecular mark-ers in potato breeding. Beiträge zur Züchtungsforschung 4:61–66

Hemleben V, Schmidt T, Torres-Ruiz RA, Zentgraf U (2000) Molecular cell biology: role of repetitive DNA in nuclear architecture and chromosome structure. In: Esser et al. (eds) Progress in botany. Springer Verlag, Berlin, Heidelberg, New York, 61, pp 91–117

Hendrich B, Bird A (1998) Identification and characterization of a family of mammalian methyl-CpG binding proteins. Mol Cell Biol 18:6538–6547

Henikoff S, Comai L (1998) A DNA methyltransferase homolog with a chromodomain exists in multiple polymorphic forms in *Arabidopsis*. Genetics 149:307–318

Houben A, Demidov D, Gernand D, Meister A, Leach CR, Schubert I (2003) Methylation of histone H3 in euchromatin of plant chromosomes depends on basic nuclear DNA content. Plant J 33:967–973

Houchins K, O'Dell M, Flavell RB, Gustafson JP (1997) Cytosine methylation and nucleolar dominance in cereal hybrids. Mol Gen Genet 55:294–301

Hunter C, Sun H, Poethig RS (2003) The *Arabidopsis* heterochronic gene *ZIPPY* is an ARG-ONAUTE family member. Curr Biol 13:1734–1739

Hutvagner G, Zamore PD (2002) RNAi: nature abhors a double-strand. Curr Opin Genet Dev 12:225–232

Iida S, Morita Y, Choi JD, Park KI, Hoshino A (2004) Genetics and epigenetics in flower pigmentation associated with transposable elements in morning glories. Adv Biophys 38: 141–159

Ingram R, Charrier B, Scollan C, Meyer P (1999) Transgenic tobacco plants expressing the *Drosophila* Polycomb (Pc) chromodomain show developmental alterations: possible role of Pc chromodomain proteins in chromatin-mediated gene regulation in plants. Plant Cell 11:1047–1060

Jackson JP, Lindroth AM, Cao X, Jacobsen SE (2002) Control of CpNpG DNA methylation by the KRYPTONITE histone H3 methyltransferase. Nature 416:556–560

Jacobsen SE, Meyerowitz EM (1997) Hypermethylated SUPERMAN epigenetic alleles in *Arabidopsis*. Science 277:1100–1103

Jacobsen SE, Sakai H, Finnegan EJ, Cao X, Meyerowitz EM (2000) Ectopic hypermethylation of flower-specific genes in *Arabidopsis*. Curr Biol 10:179–186

Jaquet Y, Delattre M, Spierer A, Spierer P (2002) Functional dissection of the *Drosophila* modifier of variegation Su(var)3–7. Development 129:3975–3982

Jasencakova Z, Soppe WJJ, Meister A, Gernand D, Turner BM, Schubert I (2003) Histone modifications in *Arabidopsis*—high methylation of H3 lysine 9 is dispensable for constitutive heterochromatin. Plant J 33:471–480

Jenuwein T, Allis CD (2001) Translating the histone code. Science 293:1074–1080

Jones AL, Thomas CL, Maule AJ (1998) De novo methylation and co-suppression induced by a cytoplasmically replicating plant RNA virus. EMBO J 17:6385–6393

Jones L, Ratcliff F, Baulcombe DC (2001) RNA-directed transcriptional gene silencing in plants can be inherited independently of the RNA trigger and requires Met1 for maintenance. Curr Biol 11:747–757

Jones-Rhoades MW, Bartel DP (2004) Computational identification of plant microRNAs and their targets, including a stress-induced miRNA. Mol Cell 14:787–799

Johnson L, Cao X, Jacobsen SE (2002) Interplay between two epigenetic marks: DNA methylation and histone H3 lysine 9 methylation. Curr Biol 12:1360–1367

Jost JP, Saluz HP (1993) Steroid hormone dependent changes in DNA methylation and its significance for the activation or silencing of specific genes. EXS 64:425–451

Kakutani T, Jeddeloh JA, Flowers SK, Munakata K, Richards EJ (1996) Developmental abnormalities and epimutations associated with DNA hypomethylation mutations. Proc Natl Acad Sci USA 93:12406–12411

Kakutani T, Munakata K, Richards EJ, Hirochika H (1999) Meiotically and mitotically stable inheritance of DNA hypomethylation induced by ddm1 mutation of *Arabidopsis thaliana*. Genetics 151:831–838

Kankel MW, Ramsey DE, Stokes TL, Flowers SK., Haag JR, Jeddeloh JA, Riddle NC, Verbsky ML, Richards EJ (2003) *Arabidopsis* MET1 cytosine methyltransferase mutants. Genetics 163:1109–1122

Kato M, Miura A, Bender J, Jacobsen SE, Kakutani T (2003) Role of CG and non-CG methylation in immobilization of transposons in *Arabidopsis*. Curr Biol 13:421–642

Kato M, Takashima K, Kakutani T (2004) Epigenetic control of CACTA transposon mobility in *Arabidopsis thaliana*. Genetics 168:961–969

Kinoshita T, Harada JJ, Goldberg RB, Fischer RL (2001) Polycomb repression of flowering during early plant development. Proc Natl Acad Sci USA 98:14156–14161

Kiyosue T, Ohad N, Yadegari R, Hannon M, Dinneny J, Wells D, Katz A, Margossian L, Harada JJ, Goldberg RB, Fischer RL (1999) Control of fertilization-independent endosperm development by the MEDEA polycomb gene in *Arabidopsis*. Proc Natl Acad Sci USA 96: 4186–4191

Khochbin S, Verdel A, Lemercier C, Seigneurin-Berny D (2001) Functional significance of histone deacetylase diversity. Curr Opin Genet Dev 11:162–166

Köhler C, Grossniklaus U (2002) Epigenetic inheritance of expression states in plant development: the role of Polycomb group proteins. Curr Opin Cell Bio 14:773–779

Köhler C, Hennig L, Spillane C, Pien S, Gruissem W, Grossniklaus U (2003) The Polycomb-group protein MEDEA regulates seed development by controlling expression of the MADS-box gene PHERES1. Genes Dev 17:1540–1553

Komarova NY, Grabe T, Huigen DJ, Hemleben V, Volkov RA (2004) Organization, differential expression and methylation of rDNA in artificial *Solanum* allopolyploids. Plant Mol Biol 56:439–463

Kouzarides T (2002) Histone methylation in transcriptional control. Curr Opin Genet Dev 12:198–209

Kovarik A, Koukalova B, Lim KY, Matyasek R, Lichtenstein CP, Leitch AR, Bezdek M (2000) Comparative analysis of DNA methylation in tobacco heterochromatic sequences. Chromosome Res 8:527–541

Lachner M (2002) Epigenetics: SUPERMAN dresses up. Curr Biol 12:434–436

Lachner M, O'Sullivan RJ, Jenuwein T (2003) An epigenetic road map for histone lysine methylation. J Cell Sci 116:2117–2124

Lawrence PA, Struhl G (1996) Morphogenes, compartment, and pattern: lessons from *Drosophila*? Cell 85:951–961

Lawrence RJ, Earley K, Pontes O, Silva M, Chen ZJ, Neves N, Viegas W, Pikaard CS (2004) A concerted DNA methylation/histone methylation switch regulates rRNA gene dosage control and nucleolar dominance. Mol Cell 13:599–609

Leonhardt H, Page AW, Weier HU, Bestor TH (1992) A targeting sequence directs DNA methyltransferase to sites of DNA replication in mammalian nuclei. Cell 71:865–873

Leweke B, Hemleben V (1982) Organization of rDNA in chromatin: plants. In: Busch H, Rothblum L (eds) The cell nucleus (part B, vol XI). Academic Press, New York, pp 225–253

Liang YK, Wang Y, Zhang Y, Li SG, Lu XC, Li H, Zou C, Xu ZH, Bai SN (2003) OsSET1, a novel SET-domain-containing gene from rice. J Exp Bot 54:1995–1996

Lichter P (1997) Multicolor FISHing: what's the catch? Trends Genet 13:475–479

Lindroth AM, Cao X, Jackson JP, Zilberman D, McCallum CM, Henikoff S, Jacobsen SE (2001) Requirement of CHROMOMETHYLASE3 for maintenance of CpXpG methylation. Science 292:2077–2080

Lippman Z, Martienssen R (2004) The role of RNA interference in heterochromatic silencing. Nature 431:364–370

Lippman Z, Gendrel AV, Black M, Vaughn MW, Dedhia N, McCombie WR, Lavine K, Mittal V, May B, Kasschau KD, Carrington JC, Doerge RW, Colot V, Martienssen R (2004) Role of transposable elements in heterochromatin and epigenetic control. Nature 430: 471–476

Liu YL, Oakeley EJ, Sun LJ, Jost JP (1998) Multiple domains are involved in the targeting of the mouse DNA methyltransferase to the DNA replication foci. Nucleic Acids Res 26:1038–1045

Loidl P (2004) A plant dialect of the histone language. Trends Plant Sci 9:84–90

Luchniak P, Maluszynska J, Olszewska MJ (2002) Different DNA methylation pattern in A and B chromosomes of *Crepis capillaris* detected by in situ nick-translation. Comparison with molecular methods. Folia Histochem Cytobiol 40:325–330

Luo M, Bilodeau P, Dennis ES, Peacock WJ, Chaudhury A (2000) Expression and parent-of-origin effects for FIS2, MEA, and FIE in the endosperm and embryo of developing *Arabidopsis* seeds. Proc Natl Acad Sci USA 97:10637–10642

Luger K (2003) Structure and dynamic behavior of nucleosomes. Curr Opin Genet Dev 13: 127–135

Lusser A, Brosch G, Loidl A, Haas H, Loidl P (1997) Identification of maize histone deacetylase HD2 as an acidic nucleolar phosphoprotein. Science 277:88–91

Lusser A, Eberharter A, Loidl A, Goralik-Schramel M, Horngacher M, Haas H, Loidl P (1999) Analysis of the histone acetyltransferase B complex of maize embryos. Nucleic Acids Res 27:4427–4435

Macleod D, Bird A (1982) DNAase I sensitivity and methylation of active versus inactive rRNA genes in *Xenopus* species hybrids. Cell 29:211–218

Marmorstein R (2001) Structure and function of histone acetyltransferases. Cell Mol Life Sci 58:693–703

Matassi G, Melis R, Kuo KC, Macaya G, Gehrke CW, Bernardi G (1992) Large-scale methylation patterns in the nuclear genomes of plants. Gene 122:239–245

Mathieu O, Picard G, Tourmente S (2002a) Methylation of a euchromatin-heterochromatin transition region in *Arabidopsis thaliana* chromosome 5 left arm. Chromosome Res 10: 455–466

Mathieu O, Yukawa Y, Sugiura M, Picard G, Tourmente S (2002b) 5S rRNA genes expression is not inhibited by DNA methylation in *Arabidopsis*. Plant J 29:313–323

Mathieu O, Jasencakova Z, Vaillant I, Gendrel AV, Colot V, Schubert I, Tourmente S (2003) Changes in 5S rDNA chromatin organization and transcription during heterochromatin establishment in *Arabidopsis*. Plant Cell 15:2929–2939

Matzke M, Aufsatz W, Kanno T, Daxinger L, Papp I, Mette MF, Matzke AJ (2004) Genetic analysis of RNA-mediated transcriptional gene silencing. Biochim Biophys Acta 1677: 129–141

McClintock B (1934) The relationship of a particular chromosomal element to the development of the nucleoli in *Zea mays*. Z Zellforsch Mikrosk Anat 21:294–328

Messer W, Noyer-Weidner M (1988) Timing and targeting: the biological functions of dam methylation in *E. coli*. Cell 54:735–737

Mette MF, Aufsatz W, van der Winden J, Matzke MA, Matzke AJ (2000) Transcriptional silencing and promoter methylation triggered by double-stranded RNA. EMBO J 19: 5194–5201

Mlotshwa S, Voinnet O, Mette MF, Matzke M, Vaucheret H, Ding SW, Pruss G, Vance VB (2002) RNA silencing and the mobile silencing signal. Plant Cell 14: S289–S301

Moss T, Stefanowsky VY (1995) Promotion and regulation of ribosomal transcription in eukaryotes by RNA polymerase I. Prog Nucl Acid Res Mol Biol 50:25–66

Mottus R, Sobel RE, Grigliatti TA (2000) Mutational analysis of a histone deacetylase in *Drosophila melanogaster*: missense mutations suppress gene silencing associated with position effect variegation. Genetics 154:657–668

Murfett J, Wang XJ, Hagen G, Guilfoyle TJ (2001) Identification of *Arabidopsis* histone deacetylase hda6 mutants that affect transgene expression. Plant Cell 13:1047–1061

Nakano Y, Steward N, Sekine M, Kusano T, Sano H (2000) A tobacco NtMET1 cDNA encoding a DNA methyltransferase: molecular characterization and abnormal phenotypes of transgenic tobacco plants. Plant Cell Physiol 41:448–457

Nakayama J, Rice JC, Strahl BD, Allis CD, Grewal SI (2001) Role of histone H3 lysine 9 methylation in epigenetic control of heterochromatin assembly. Science 292:110–113

Naumann F, Remus R, Schmitz B, Doerfler W (2004) Gene structure and expression of the 5'-
(CGG)(n)-3'-binding protein (CGGBP1). Genomics 83:106–118

Ng HH, Jeppesen P, Bird A (2000) Active repression of methylated genes by the chromosomal
protein MBD1. Mol Cell Biol 20:1394–1406

Noh YS, Amasino RM (2003) PIE1, an ISWI family gene, is required for FLC activation and
floral repression in *Arabidopsis*. Plant Cell 15:1671–1682

Noh B, Lee SH, Kim HJ, Yi G, Shin EA, Lee M, Jung KJ, Doyle MR, Amasino RM, Noh
YS (2004) Divergent roles of a pair of homologous jumonji/zinc-finger-class transcription factor proteins in the regulation of *Arabidopsis* flowering time. Plant Cell 16:
2601–2613

Noyer-Weidner M, Trautner TA (1993) Methylation of DNA in prokaryotes. In: Jost JP, Saluz
HP (eds) DNA methylation: molecular biology and biological significance. Birkhauser
Verlag, Basel, pp 39–108

Ohad N, Margossian L, Hsu YC, Williams C, Repetti P, Fischer RL (1996) A mutation that allows
endosperm development without fertilization. Proc Natl Acad Sci USA 93:5319–5324

Ohad N, Yadegari R, Margossian L, Hannon M, Michaeli D, Harada JJ, Goldberg RB, Fischer
RL (1999) Mutations in FIE, a WD polycomb group gene, allow endosperm development
without fertilization. Plant Cell 11:407–416

Palmer LE, Rabinowicz PD, O'Shaughnessy AL, Balija VS, Nascimento LU, Dike S, de la
Bastide M, Martienssen RA, McCombie WR (2003) Maize genome sequencing by methylation filtrations. Science 302:2115–2117

Pandey R, Muller A, Napoli CA, Selinger DA, Pikaard CS, Richards EJ, Bender J, Mount DW,
Jorgensen RA (2002) Analysis of histone acetyltransferase and histone deacetylase families
of *Arabidopsis thaliana* suggests functional diversification of chromatin modification
among multicellular eukaryotes. Nucleic Acids Res 30:5036–5055

Papazova N, Hvarleva T, Atanassov A, Gecheff K (2001) The role of cytosine methylation
for rRNA gene expression in reconstructed karyotypes of barley. Biotechnol Equipment 15:
35–44

Papp I, Mette MF, Aufsatz W, Daxinger L, Schauer SE, Ray A, van der Winden J, Matzke M,
Matzke AJM (2003) Evidence for nuclear processing of plant microRNA and short interfering RNA precursors. Plant Physiol 132:1382–139

Paro R, Harte PJ (1996) The role of Polycomb group and trithorax group chromatin com-
plexes in the maintenance of determined cell states. In: Russo VEA, Martienssen RA, Riggs
AD (eds) Epigenetic mechanisms of gene regulation. Cold Spring Harbor Laboratory
Press, Cold Spring Harbor, N.Y., pp 507–528

Paule MR, White RJ (2000) Transcription by RNA polymerases I and III. Nucl Acids Res 28:
1283–1298

Peters AH, Kubicek S, Mechtler K, O'Sullivan RJ, Derijck AA, Perez-Burgos L, Kohlmaier A,
Opravil S, Tachibana M, Shinkai Y, Martens JH, Jenuwein T (2003) Partitioning and
plasticity of repressive histone methylation states in mammalian chromatin. Mol Cell 12:
1577–1589

Peterson CL, Herskowitz I (1992) Characterization of the yeast SWI1, SWI2, and SWI3 genes,
which encode a global activator of transcription. Cell 68:573–583

Pikaard CS (2000) Nucleolar dominance: uniparental gene silencing on a multi-megabase
scale in genetic hybrids. Plant Mol Biol 43:163–177

Probst AV, Fagard M, Proux F, Mourrain P, Boutet S, Earley K, Lawrence RJ, Pikaard CS,
Murfett J, Vaucheret IFH, Scheida OM (2004) *Arabidopsis* histone deacetylase HDA6 is
required for maintenance of transcriptional gene silencing and determines nuclear organization of rDNA repeats. Plant Cell 16:1021–1034

Rabinowicz PD, Palmer LE, May BP, Hemann MT, Lowe SW, McCombie WR, Martienssen RA (2003) Genes and transposons are differentially methylated in plants, but not in mammals. Genome Res 13:2658–2664

Rangwala SH, Richards EJ (2004) The value-added genome: building and maintaining genomic cytosine methylation landscapes. Curr Opin Genet Dev 14:686–691

Rea S, Eisenhaber F, O'Carroll D, Strahl BD, Sun ZW, Schmid M, Opravil S, Mechtler K, Ponting CP, Allis CD, Jenuwein T (2000) Regulation of chromatin structure by site-specific histone H3 methyltransferases. Nature 406:593–599

Reuter G, Spierer P (1992) Position effect variegation and chromatin proteins. Bioessays 14: 605–612

Reyes JC, Hennig L, Gruissem W (2002) Chromatin-remodeling and memory factors. New regulators of plant development. Plant Physiol 130:1090–1101

Rice JC, Briggs SD, Ueberheide B, Barber CM, Shabanowitz J, Hunt DF, Shinkai Y, Allis CD (2003) Histone methyltransferases direct different degrees of methylation to define distinct chromatin domains. Mol Cell 12:1591–1598

Roth SY, Denu JM, Allis CD (2001) Histone acetyltransferases. Annu Rev Biochem 70:81–120

Santoro R, Li J, Grummt I (2002) The nucleolar remodeling complex NoRC mediates heterochromatin formation and silencing of ribosomal gene transcription. Nature Genet 32: 393–396

Santos-Rosa H, Schneider R, Bannister AJ, Sherriff J, Bernstein BE, Emre NC, Schreiber SL, Mellor J, Kouzarides T (2002) Active genes are tri-methylated at K4 of histone H3. Nature 419:407–411

Sardana R, O'Dell A, Flavell R (1993) Correlation between the size of the intergenic regulatory region, the status of cytosine methylation of rRNA genes and nucleolar expression in wheat. Mol Gen Genet 236:155–162

Sarnowski TJ, Swiezewski S, Pawlikowska K, Kaczanowski S (2002) AtSWI3B, an *Arabidopsis* homolog of SWI3, a core subunit of yeast Swi/Snf chromatin remodeling complex, interacts with FCA, a regulator of flowering time. Nucleic Acids Res 30:3412–3421

Sarraf SA, Stancheva I (2004) Methyl-CpG binding protein MBD1 couples histone H3 methylation at lysine 9 by SETDB1 to DNA replication and chromatin assembly. Mol Cell 15:595–605

Saze H, Mittelsten Scheid O, Paszkowski J (2003) Maintenance of CpG methylation is essential for epigenetic inheritance during plant gametogenesis. Nat Genet 34:65–69

Schauer SE, Jacobsen SE, Meinke DW, Ray A (2002) DICER-LIKE1: blind men and elephants in *Arabidopsis* development. Trends Plant Sci 7:487–491

Schotta G, Ebert A, Krauss V, Fischer A, Hoffmann J, Rea S, Jenuwein T, Dorn R, Reuter G (2002) Central role of Drosophila SU(VAR)3–9 in histone H3-K9 methylation and heterochromatic gene silencing. EMBO J 21:1121–1131

Schotta G, Lachner M, Sarma K, Ebert A, Sengupta R, Reuter G, Reinberg D, Jenuwein T (2004) A silencing pathway to induce H3-K9 and H4-K20 trimethylation at constitutive heterochromatin. Genes Dev 18:1251–1262

Schubert D, Lechtenberg B, Forsbach A, Gils M, Bahadur S, Schmidt R (2004) Silencing in *Arabidopsis* T-DNA transformants: the predominant role of a gene-specific RNA sensing mechanism versus position effects. Plant Cell 16:2561–2572

Sheldon CC, Burn JE, Perez PP, Metzger J, Edwards JA, Peacock WJ, Dennis ES (1999) The FLF MADS box gene: a repressor of flowering in *Arabidopsis* regulated by vernalization and methylation. Plant Cell 11:445–458

Shen WH (2001) NtSET1, a member of a newly identified subgroup of plant SET-domain-containing proteins, is chromatin-associated and its ectopic overexpression inhibits tobacco plant growth. Plant J 28:371–383

Silhavy D, Burgyán J (2004) Effects and side-effects of viral RNA silencing suppressors on short RNAs. Trends Plant Sci 9:76–83

Soppe WJJ, Jasencakova Z, Houben A, Kakutani T, Meister A, Huang MS, Jacobsen SE, Schubert I, Fransz PF (2002) DNA methylation controls histone H3 lysine 9 methylation and heterochromatin assembly in *Arabidopsis*. EMBO J 21:6549–6559

Spillane C, McDougall C, Stock C, Kohler C, Vielle-Calzada JP (2000) Interaction of the *Arabidopsis* polycomb group proteins FIE and MEA mediates their common phenotypes. Curr Biol 10:1535–1538

Springer NM, Danilevskaya ON, Hermon P, Helentjaris TG, Phillips RL, Kaeppler HF, Kaeppler SM (2002) Sequence relationships, conserved domains, and expression patterns for maize homologs of the polycomb group genes E(z), esc, and E(Pc). Plant Physiol 128:1332–1345

Stockinger EJ, Mao Y, Regier MK, Triezenberg SJ, Thomashow MF (2001) Transcriptional adaptor and histone acetyltransferase proteins in *Arabidopsis* and their interactions with CBF1, a transcriptional activator involved in cold-regulated gene expression. Nucleic Acids Res 29:1524–1533

Susi P, Hohkuri M, Wahlroos T, Kilby NJ (2004) Characteristics of RNA silencing in plants: similarities and differences across kingdoms. Plant Mol Biol 54:157–174

Tamaru H, Selker EU (2001) A histone H3 methyltransferase controls DNA methylation in *Neurospora crassa*. Nature 414:277–283

Tamaru H, Zhang X, McMillen D, Singh PB, Nakayama J, Grewal SI, Allis CD, Cheng X, Selker EU (2003) Trimethylated lysine 9 of histone H3 is a mark for DNA methylation in *Neurospora crassa*. Nat Genet 34:75–79

Tariq M, Paszkowski J (2004) DNA and histone methylation in plants. Trends Genet 20:244–251

Tariq M, Saze H, Probst AV, Lichota J, Habu Y, Paszkowski J (2003). Erasure of CpG methylation in *Arabidopsis* alters patterns of histone H3 methylation in heterochromatin. Proc Natl Acad Sci USA 100:8823–8827

Teerawanichpan P, Chandrasekharan MB, Jiang Y, Narangajavana J, Hall TC (2004) Characterization of two rice DNA methyltransferase genes and RNAi-mediated reactivation of a silenced transgene in rice callus. Planta 218:337–3496

Thomas AJ, Sherratt HS (1956) The isolation of nucleic acid fractions from plant leaves and their purine and pyrimidine composition. Biochem J 62:1–4

Tian L, Chen ZJ (2001) Blocking histone deacetylation in *Arabidopsis* induces pleiotropic effects on plant gene regulation and development. Proc Natl Acad Sci USA 98:200–205

Torres-Ruiz RA, Hemleben V (1994) Pattern and degree of methylation in ribosomal RNA genes of *Cucurbita pepo* L. Plant Mol Biol 26 1167–1179

Trievel RC, Beach BM, Dirk LM, Houtz RL, Hurley JH (2002) Structure and catalytic mechanism of a SET domain protein methyltransferase. Cell 111:91–103

Turner BM (2002) Cellular memory and the histone code. Cell 111:285–291

Vanyushin BF, Kirnos MD (1988) DNA methylation in plants. Gene 4:117–121

Varga-Weisz P (2001) ATP-dependent chromatin remodeling factors: nucleosome shufflers with many missions. Oncogene 20:3076–3085

Vaucheret H, Vazquez F, Crete P, Bartel DP (2004) The action of *ARGONAUTE1* in the miRNA pathway and its regulation by the miRNA pathway are crucial for plant development. Genes Dev 18:1187–1197

Vervoorts J, Luscher-Firzlaff JM, Rottmann S, Lilischkis R, Walsemann G, Dohmann K, Austen M, Luscher B (2003) Stimulation of c-Myc transcriptional activity and acetylation by recruitment of the cofactor CBP. EMBO Rep 4:484–490

Volkov RA, Borisjuk NV, Panchuk II, Schweizer D, Hemleben V (1999a) Elimination and rearrangement of parental rDNA in allotetraploid *Nicotiana tabacum*. Mol Biol Evol 16: 311–320

Volkov RA, Bachmair A, Panchuk II, Kostyshyn SS, Schweizer D (1999b) 25S-18S rDNA intergenic spacer of *Nicotiana sylvestris* (Solanaceae): primary and secondary structure analysis. Plant Syst Evol 218:89–97

Volkov RA, Medina FJ, Zentgraf U, Hemleben V (2004) Molecular cell biology: organization and molecular evolution of rDNA, nucleolar dominance, and nucleolus structure. Prog Bot 65:106–146

Volpe TA, Kidner C, Hall IM, Teng G, Grewal SIS, Martienssen RA (2002) Regulation of heterochromatic silencing and histone H3 lysine-9 methylation by RNAi. Science 297:1833–1837

Vongs A, Kakutani T, Martienssen RA, Richards EJ (1993) *Arabidopsis thaliana* DNA methylation mutants. Science 260:1926–1928

Wada Y, Ohya H, Yamaguchi Y, Koizumi N, Sano H (2003) Preferential de novo methylation of cytosine residues in non-CpG sequences by a domains rearranged DNA methyltransferase from tobacco plants. J Biol Chem 278:42386–42393

Wagner D (2003) Chromatin regulation of plant developpment. Curr Opin Plant Biol 6:20–28

Wagner D, Meyerowitz EM (2002) SPLAYED, a novel SWI/SNF ATPase homolog, controls reproductive development in *Arabidopsis*. Curr Biol 12:85–94

Wassenegger M (2000) RNA-directed DNA methylation. Plant Mol Biol 43:203–220

Waterborg JH (1990) Sequence analysis of acetylation and methylation in two histone H3 variants of alfalfa. J Biol Chem 265:17157–17161

Waterborg JH (1992) Identification of five sites of acetylation in alfalfa histone H4. Biochemistry 31:6211–6219

Wingard SA (1928) Hosts and symptoms of ring spot, a virus disease of plants. J Agric Res 37:127–153

Wu K, Malik K, Tian L, Brown D, Miki B (2000) Functional analysis of a RPD3 histone deacetylase homologue in *Arabidopsis thaliana*. Plant Mol Biol 44:167–176

Yadegari R, Kinoshita T, Lotan O, Cohen G, Katz A, Choi Y, Katz A, Nakashima K, Harada JJ, Goldberg RB, Fischer RL, Ohad N (2000) Mutations in the FIE and MEA genes that encode interacting polycomb proteins cause parent-of-origin effects on seed development by distinct mechanisms. Plant Cell 12:2367–2382

Yoshida N, Yanai Y, Chen L, Kato Y, Hiratsuka J, Miwa T, Sung ZR, Takahashi S (2001) EMBRYONIC FLOWER2, a novel polycomb group protein homolog, mediates shoot development and flowering in *Arabidopsis*. Plant Cell 13:2471–2481

Zhang Y, Li N, Caron C, Matthias G, Hess D, Khochbin S, Matthias P (2003) HDAC-6 interacts with and deacetylates tubulin and microtubules *in vivo*. EMBO J 22:1168–1179

Zhou C, Miki B, Wu K (2003) CHB2, a member of the SWI3 gene family, is a global regulator in *Arabidopsis*. Plant Mol Biol 52:1125–1134

Zilberman D, Henikoff S (2004) Silencing of transposons in plant genomes: kick them when they're down. Genome Biol 5:249

Zilberman D, Cao X, Jacobsen SE (2003) ARGONAUTE4 control of locus specific siRNA accumulation and DNA and histone methylation. Science 299:716–719

Zilberman D, Cao X, Johansen LK, Xie Z, Carrington JC, Jacobsen SE (2004) Role of *Arabidopsis* ARGONAUTE4 in RNA-directed DNA methylation triggered by inverted repeats. Curr Biol.14:1214–1220

Prof. Dr. Vera Hemleben
Dr. Nataliya Y. Komarova
PD Dr. Roman A. Volkov
PD Dr. Ulrike Zentgraf
Lehrstuhl für Allgemeine Genetik
Center of Plant Molecular Biology (ZMBP)
University of Tübingen
Auf der Morgenstelle 28
72076, Tübingen, Germany

Genetics of phytopathology:
Secondary metabolites as virulence determinants of fungal plant pathogens

Eckhard Thines, Jesús Aguirre, Andrew J. Foster, Holger B. Deising

1 Introduction

The kingdom fungi comprises a highly diverse array of species. Where species diversification has been studied in detail, fungal species can be 6 times as numerous as those of flowering plants. On this basis, since approximately 270,000 flowering plants are known today, it is estimated that more than 1.5 million fungal species may exist (Carlile and Watkinson 1994; Hawksworth 2001)). Fungi are able to colonize a broad range of substrata; they may live as saprophytes, associated with plants as mycorrhiza and as pathogens of plants, animals and microorganisms (Dix and Webster 1995). As saprophytes, fungi contribute to the ecological balance of the environment by degrading organic materials, such as decaying plant material, dung and organic pollutants. In mycorrhiza, fungi provide phosphate for the associated plant while they gain carbohydrates in exchange (Smith and Read 1997). However, many plant-associated fungi do not provide any mutual benefit for the colonized hosts. Whereas biotrophic fungi rely on the living host for nutrient provision, others (necrotrophs) often kill and exploit the surrounding tissue. The high bio-diversity of fungal species is also reflected by exceedingly rich diversity of metabolism and the corresponding metabolites that affect the attacked plant.

A large group of pathogens known as necrotrophs secrete phytotoxic secondary metabolites to kill the host tissue and to avoid initiation of defense responses. Secondary metabolites have been defined as metabolic products not essential for growth and without obvious function for the producing cell during its life cycle (Aharonowitz and Demain 1980). Secondary metabolite production depends both on the genotype of the organism and the environmental conditions under which growth takes place, for example the composition of the medium used to culture the organism (Weinberg 1974; Bu'Lock 1975). So far, several thousand secondary metabolites have been identified from natural sources, with sponges and corals being the most prolific producers of secondary metabolites (Berlepsch 1980; Faulkner 1993). The majority of secondary metabolites identified to date, however, derive from bacteria, fungi and plants.

The existence of such a large number of chemically highly diverse metabolites has led to various speculations as to why secondary metabolites are produced. A general hypothesis was provided by Zähner et al. (1983), who suggested that secondary metabolites evolved continuously in a "biochemical playground" of the cell. Beneath the supply of precursors and energy from the primary metabolism, the closely related intermediary metabolism, differentiation and morphogenesis determine the evolution of secondary metabolism. In fact, fungi exhibit high mutation rates, due to exposure of air-borne spores to UV light, which, in combination with short generation times may result in extremely efficient and well-regulated intermediary and secondary metabolism, allowing them to colonize a broad range of living or dead substrata. It should be emphasized that secondary metabolites not only function as toxins required for plant colonization, but also provide an advantage for the colonization of the substratum for fungi sharing the environment and competing for nutrients with many other microorganisms. Plants are very frequently colonized by endophytic fungi that do not cause disease symptoms. In contrast, plant pathogenic fungi cause necrosis, rotting, chlorosis or bleaching of different tissues of their host plants. Toxic secondary metabolites synthesised by the fungus *in vivo* have been identified as essential determinants of pathogenicity. They are secreted into and spread in plants by diffusion and either act as virulence factors, i.e. they can intensify disease symptoms, or as pathogenicity factors, i.e. they are exclusively responsible for the development of disease symptoms. For decades, the only known agents conferring specificity in interactions between microbes and plants were host-selective toxins (Walton and Panaccione 1993; Walton 1996). Host-selective toxins (HS-toxins) are active only on host plants carrying genetically determined sensitivity for the particular toxin (see below).

The majority of the phytotoxic secondary metabolites produced by plant pathogenic fungi belong to the non-specific or non-host-specific toxins (NHS-toxins). NHS toxins are poisonous to all plants. As many plants are susceptible to NHS-toxins the most destructive pathogens in agriculture produce these compounds. This review provides insight into biosynthesis and evaluation of fungal secondary metabolites as disease determinants in plants. For each of the diverse groups of chemicals we will focus on one or a few toxins and give examples of their mode(s) of action.

2 Secondary metabolism and its biochemical precursors

Even though several thousand secondary metabolites from fungal sources have been identified to date, the pathways involved in secondary metabolism are still not very well understood compared with those of intermediary metabolism (Zähner et al. 1983). Primary metabolism is relatively well understood and involves a limited number of well-known intermediates and enzymes. Certain enzymes and genes involved in secondary metabolism are currently the focus of intensive investigations, with polyketide synthases and non-ribosomal polypeptide synthases representing excellent examples. The major source for energy in most heterotrophic organisms is glucose, usually derived from complex carbohydrates in the environment. The breakdown of glucose however, does not just provide energy, but also provides the precursors necessary for secondary metabolism. Figure 1 illustrates the carbon fluxes through primary and intermediary to secondary metabolism. Glucose degradation via the pentose phosphate cycle, as well as triose and pentose formation, results in a tetrose (erythrose-4-phosphate), which can react with phosphoenolpyruvate to yield shikimic acid. Triose generated via glycolysis is converted to pyruvate, and subsequently to acetyl-coenzyme A, which is possibly the most predominant building block in fungal secondary metabolism. Condensation of three acetyl-coenzyme A-units gives rise to mevalonic acid, the key intermediate in terpene biosynthesis. Additionally, acetyl-coenzyme A can condense with oxalacetate, by which carbon enters the tricarboxylic acid cycle. This cycle also functions as a source for carbon skeletons for several amino acids (Turner 1971).

Certain enzymes of primary metabolism catalyse the formation of products that can be channeled into the pathways of secondary metabolites. Despite the fact that only few relatively simple biochemical precursors originate from intermediary metabolism, secondary metabolites are of enormous chemical diversity. Secondary metabolites can be categorized according to the key intermediates they derive from. Major building blocks for the assembly of fungal secondary metabolites are acetyl-coenzyme A, shikimate, amino acids and glucose. Acetyl-coenzyme A is used as a building block for polyketides, polyenes terpenoids, steroids and carotenoids, while shikimate precursors lead to the assembly of aromatic secondary metabolites. Peptides and alkaloids derive from amino acids, and simple sugars are used to form glycosides. An overview of the biosynthetic pathways leading to the formation of fungal secondary metabolites is given in Fig. 1.

The following section gives four important pathways as examples of how secondary metabolites act as virulence determinants of fungal plant pathogens. Phytotoxic and non-toxic polyketides, terpenoids, glycosids and aromatic compounds including molecular evidence for their role in pathogenesis will be discussed in detail.

3 Fungal secondary metabolites as phytotoxins and virulence determinants

Secondary fungal products can either act as directly phytotoxic compounds or as compounds that mediate the generation of toxic molecules. Furthermore, secondary metabolites may be non-toxic but add structural features obligately

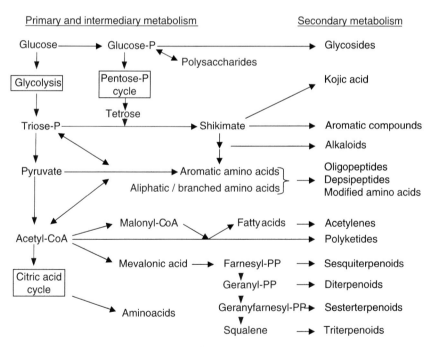

Fig. 1. Metabolic pathways leading to the major groups of secondary metabolites; after Zähner et al. (1983).

required for fungal virulence or pathogenicity. Examples for all of these modes of action can be found among the polyketides (Fig. 2), and it is therefore not surprising that this group of secondary metabolites has received significant attention during the last decades.

3.1 Polyketides

Polyketides are a group of secondary metabolites possessing remarkable diversity in their structure as well as in their function. Turner (1971) classified these molecules according to the number of C_2 units in the molecule chain in tetraketides, pentaketides, hexaketides, etc. These metabolites are ubiquitous in distribution and have been reported in fungi, bacteria, plants, sponges, molluscs and insects (O'Hagan 1992). In addition to several virulence or pathogenicity factors (see below), fungal polyketides include pigments, signaling molecules, carcinogenic mycotoxins (e.g. aflatoxin) and the anticholesterol compound lovastatin (Kroken et al. 2003), and references therein). Polyketide biosynthesis is carried out by polyketide synthases (PKSs), which are highly specialized, large multifunctional enzyme complexes catalyzing different sequential enzymatic reactions. Based on their molecular structure and function, three types of fungal PKSs can be distinguished (Gokhale and Tuteja 2001). Fungal type I PKSs consist of large multifunctional proteins involved in the biosynthesis of toxins, such as aflatoxin. Type II PKSs are functionally homologous to bacterial fatty acid synthases. They typically catalyze the synthesis of compounds that require aromatization and cyclisation, but not extensive reduction. Type III PKSs are mainly involved in the biosynthesis of secondary metabolites, such as stilbenes, precursors for flavonoids and chalcones in plants and bacteria.

Type I fungal PKSs are closely related to fatty acid synthetases (FASs). Both are multifunctional enzymes with the same enzymatic domain structure, i.e. ketoacyl synthase (KS), acyl transferase (AT), ketoreductase (KR), dehydratase (DH), enoyl reductase (ER), and acyl carrier protein domain with a phosphopantetheine attachment site (PP domain). While KS, AT, and PP domains are essential for all PKSs, some or all of the other domains may be absent. In a sequential reaction, KR, DH, and ER domains catalyze reduction of a keto to a hydroxyl group, dehydration of the hydroxyl to an enoyl group, and reduction of the enoyl to an alkyl group (Khosla et al. 1999; Kroken et al. 2003).

Most fungal type I PKSs are iterative monomodular enzymes that use their active sites repeatedly to synthesize a polyketide, adding an activated

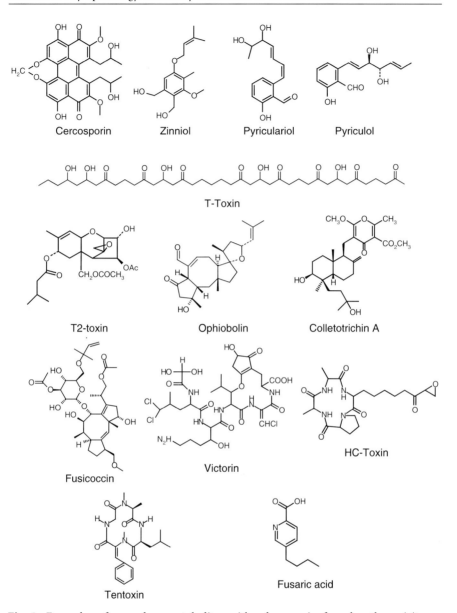

Fig. 2. Examples of secondary metabolites with relevance in fungal pathogenicity or virulence

two-carbon molecule (i.e. a CoA ester) to the growing chain with each condensation. However, as PKS can use 2–, 3– or 4-carbon starter blocks (acetyl-, propionyl- and butyryl-CoA) and their activated derivates malonyl-, methylmalonyl- and ethylmalonyl-CoA, highly variable and complex polyketides are formed (Khosla et al. 1999). The non-iterative fungal PKSs perform only one condensation cycle, synthesizing a diketide. Genes encoding non-iterative PKSs seem to occur in a cluster along with a *PKS* gene encoding an iterative PKS (Abe et al. 2002; Kroken et al. 2003). The products of an iterative and a non-iterative PKS can be joined to form a branched polyketide. Formation of branches adds to the diversity of polyketides generated through the contribution of the three optional PKS reducing domains and the use of different starter blocks.

Selective protein–protein interactions direct channelling of intermediates between individual polyketide synthase modules. The C- and N-terminal ends of adjacent PKS polypeptides are capped by short peptides of 20–40 residues. While matched sequences can facilitate the channelling of intermediates between PKS modules, mismatched sequences abolish chain interdomain transfer, without affecting the activity of individual modules. Thus, in addition to substrate-PKS interactions and domain-domain interactions, short interpolypeptide sequences represent a third determinant of selective chain transfer (Tsuji et al. 2001).

Modern genetic methods and in particular the availability of complete genome information of microorganisms have led to the identification of several PKSs. In fact, a thorough phylogenomic analysis of type I *PKS* genes indicated that ascomycetes of the subphylum Pezizomycotina, but not early diverging ascomycetes like *Saccharomyces cerevisiae* or *Schizosaccharomyces pombe*, had large numbers (7–25) of *PKS* genes (Kroken et al. 2003). Based on these data, one may speculate that *PKS* genes play specific roles in filamentous growth, morphogenesis, and pathogenicity or virulence. Two clades of fungal PKSs (i.e. reducing and non-reducing PKSs), each falling into four sub-clades, based on their different molecular organization, have been identified (Kroken et al. 2003). By means of genetic manipulation, genes involved in the biosynthesis of polyketides have been functionally characterized.

In the following paragraphs, we will discuss biosynthetic aspects of the synthesis of individual compounds (Fig. 2) or groups of compounds. If molecular analyses of the role(s) of such compounds are available, we will outline these as well.

A decarboxylative condensation is the key step in the chain elongation in biosynthesis polyketide-biosynthesis. This reaction is analogous to the chain elongation step of fatty acid biosynthesis, whereby PKSs and fatty acid synthases show remarkable genetic, mechanistic and protein-structural similarities (Khosla et al. 1999). In contrast to fatty acid biosynthesis, in which the chain elongation is followed by a conserved mechanism of ketore-duction, dehydration and enoyl reduction, biosynthesis of polyketides has no fixed scheme of reactions after the condensation. Variation in the chemical reactions following the chain elongation contributes to the chemical diversity of the emerging compounds.

3.1.1 Cercosporin

Different species of the genus *Cercospora* cause foliar diseases widespread among cereals, grasses, field crops, vegetables and trees. About 5 decades ago, the non-host-specific phytotoxin (NHST) cercosporin (Fig. 2) was identified as a virulence factor in diseases caused by *Cercospora* in several plant species (Kuyama and Tamura 1957). Mutants, obtained by spontaneous and UV-induced mutations deficient in cercosporin production, produce few small lesions on infected leaves in contrast to large necrotic lesions caused by the wild-type isolates (Upchurch et al. 1991).

The biosynthetic origin of cercosporin from acetate has been demon-strated by feeding experiments using ^{14}C- and ^{13}C-labelled precursors (Yamazaki and Ogawa 1972) (Fig. 1). The observed labeling pattern of cer-cosporin from $[1-^{13}C]$-, $[2-^{13}C]$ and $[1,2-^{13}C2]$actetate was found to be con-sistent with its formation from a heptaketide chain with decarboxylation and oxidative dimerization, while the O-alkyl groups derive from formate (Okubo et al. 1975; Turner and Aldridge 1983). The existence of *PKS* genes in species of the genus *Cercospora* would fit with the view that that biosyn-thesis of the molecule may proceed via the polyketide pathway (Chung et al. 2003).

The photodynamic pigment cercosporin is in fact not a host-selective phy-totoxin, since the compound is also lethal to bacteria, most fungi and animals (Daub and Ehrenshaft 2000). Following cercosporin production the com-pound is excreted by the fungus and activated (Fig. 3). Cercosporin belongs to a group of compounds activated by visible wavelength of light and generating reactive oxygen intermediates (ROIs), which are toxic to living cells (Spikes 1989; Heiser et al. 1998). In plants, pathogen attack causes the rapid produc-tion of ROIs (superoxide, hydrogen peroxide and hydroxyl radicals). In par-ticular, H_2O_2 has been shown to play a key role in the orchestration of a localized hypersensitive response during the expression of disease resistance (Levine et al. 1994). While programmed cell death efficiently protects plants against biotrophic pathogens such as powdery mildews or rusts (Kogel et al. 1994; Deising et al. 2002), some necrotrophic pathogens may thrive in an

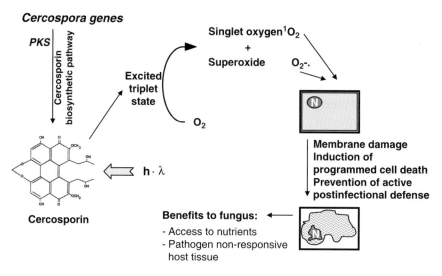

Fig. 3. Role of cercosporin in fungal pathogenesis. Light activation of cercosporin leads to formation of reactive oxygen species, which cause damage to plant cells. The toxin thus prevents activation of defense responses, and gains access to plant metabolites; after Daub and Ehrenshaft (2000).

ROI-rich environment, and can therefore exploit this host defense mechanism for their pathogenicity. The grey mold fungus *Botrytis cinerea*, for example, may actively generate ROIs to trigger hypersensitive cell death, which facilitates colonization of the host plants. In *Cercospora*, the secreted and activated (oxidized) photosensitizer generates superoxide (O_2^-) and singlet oxygen ($^1O_2$) *in vivo*, and in particular production of the latter is essential for the toxicity. ROIs derived from the toxin may cause membrane peroxidation and leakage of nutrients from plant cells (Daub and Ehrenshaft 2000), or may present a signal triggering programmed cell death. Interestingly, sugarbeets carrying superoxide dismutase transgenes exhibited increased tolerance to pure cercosporin, as well as to leaf infection with the fungus *C. beticola* (Tertivanidis et al. 2004). The fact that plant pathogenic fungi belonging to at least eight different genera (Daub and Ehrenshaft 2000; Heiser et al. 2003) synthesize photosensitizing prenylenequinone toxins may be taken as an indication that formation of ROIs during pathogenesis is a successful strategy during host colonization.

Some *Cercospora* species can tolerate up to millimolar concentrations of cercosporin in the light, without observable toxic effects (Rollins et al. 1993). Targeted disruption of the single genomic copy a gene encoding the cercosporin facilitator protein CFP resulted in a drastic reduction in cercosporin production, greatly reduced virulence of the fungus to soybean, and increased fungal sensitivity to exogenous cercosporin (Callahan et al. 1999). In contrast, over-expression of the *CFP* gene in *Cercospora kikuchii* up-regulated production and secretion of cercosporin (Upchurch et al. 2001). These data indicate that efficient secretion of the toxin confers cercosporin tolerance to *Cercospora* species. One may speculate that expression of the cercosporin transporter gene could result in transgenic plants with increased resistance *Cercospora* and to the toxin. The *C. kikuchii* cercosporin export gene, *CFP*, has been introduced into *Beta vulgaris* by conjugation with *Rhizobium radiobacter*, and was stably maintained during vegetative propagation (Kuykendall and Upchurch 2004). Analysis of the relative susceptibility of CFP-transgenic sugar beet plants will allow the evaluation of this new strategy in *Cercospora* disease management.

Using *Saccharomyces cerevisiae* as a model system, Ververidis et al. (2001) identified two genes expressed in high-copy number vectors conferring cercosporin resistance to an otherwise sensitive strain. One gene codes for the well-characterized multidrug efflux transporter Snq2p. The other, *CPD1* (Cercosporin Photosensitizer Detoxification), encodes a protein with similarity to the FAD-dependent pyridine nucleotide reductases. Over-expression of either of these proteins can also mediate resistance to cercosporin and other singlet oxygen-generating compounds. The involvement of Snq2p and Cpd1p in photosensitizer detoxification reinforces previous observations which suggested that resistance to cercosporin is mediated by a mechanism involving toxin efflux and/or toxin reduction (Daub and Ehrenshaft 2000; Ververidis et al. 2001). In previous studies, the cellular resistance of *Cercospora* species to cercosporin has been correlated with the ability to maintain cercosporin in a chemically reduced state (Daub et al. 2000). Localization of reduced cercosporin in fungal cells has been studied using a fluorescence assay and laser scanning confocal microscopy. This assay showed a uniform green fluorescence, indicative of reduced cercosporin, in the cytoplasm of hyphal cells treated with cercosporin (Daub et al. 2000). The *C. nicotianae pdx1* and *crg1* genes have previously been identified as genes required for resistance to the singlet oxygen-generating toxin cercosporin (Chung et al. 2002). The *pdx1* gene has been shown to be involved in pyridoxine biosynthesis, and pyridoxine (vitamin B_6) and its derivatives are efficient singlet oxygen quenchers and potential fungal antioxidants (Bilski et al. 2000). Thus, in addition to efficient secretion of the toxin, maintenance of reduced conditions in cercosporin-producing hyphae is essential for self-protection against ROIs.

3.1.2 Zinniol

The genus *Alternaria* is responsible for some of the world's most devastating plant diseases, and at the same time is considered one of the most important producers of fungal allergens (Rotem 1994; Cramer and Lawrence 2003). Among the species of the genus *Phoma*, the causal agent of blackleg or stem canker of oilseed rape, *Phoma lingam* (teleomorph: *Leptosphaeria maculans*), a soil-borne fungus, is one of the economically most important oilseed rape diseases around the world. Two distinct groups of the fungus are described as toxin-producing aggressive and toxin non-producing non-aggressive. Both fungi are true necrotrophs during certain stages of their life cycle.

Zinniol (Fig. 2) is a phytotoxic secondary metabolite isolated from *Alternaria* and *Phoma* species (Cotty and Misaghi 1984; Sugawara and Strobel 1986). Feeding experiments with $[1,2\text{-}^{13}C_2]$acetate led to the isolation of labelled zinniol from *Alternaria solani* and the observed pattern of intact acetate units is consistent with a normal tetraketide origin with *O*-prenylation (Stoessl et al. 1979). Zinniol production is believed to be one determinant of pathogenicity in the field, since the symptoms caused by the compound on detached leaves match the disease symptoms in the field (Cotty and Misaghi 1984). It has been reported that the presence of two hydroxymethyl-functionalities are essential for the phytotoxic activity of Zinniol (Barash et al. 1981).

Zinniol belongs to the group of NHS-toxins and is therefore toxic to several other plants. Many other NHS-toxins have been reported to have binding sites in the plasma membrane of plant cells. It was found that zinniol binds to isolated membranes of carrot protoplasts in a saturable and reversible manner. Furthermore, zinniol stimulates the entry of calcium into chloroplasts and thereby leading to a perturbation which may have a lethal effect for the cell (Thuleau et al. 1988).

3.1.3 Pyriculol and pyriculariol

The heptaketides pyriculol and pyriculariol (Fig. 2) have been isolated from several species of the rice-blast fungus *Magnaporthe grisea* (anamorph *Pyricularia oryzae*) and their impact in the disease has been discussed elsewhere in detail (Nukina et al. 1981; Talbot 2003). Several other toxicity-related toxins have been reported in *M. grisea*, most of which appear to be non host-specific.

Interestingly, recently a putative PKS/peptide synthase has been identified and characterized by molecular genetic methods. It was shown that strains carrying the gene *ACE1* encoding avirulence conferring enzyme 1 are recognized by rice (*Oryza sativa*) cultivars carrying the resistance gene *Pi33*. Analysis of the gene showed that it encodes a putative hybrid enzyme between a PKS and a non-ribosomal peptide synthases. The gene is expressed exclusively during penetration of plant leaves by the fungus. As a single amino acid exchange in the putative catalytic site of the β-ketoacyl synthase domain of Ace1 abolishes recognition of the fungus and defense response in resistant rice varieties, it was suggested that the secondary metabolite triggers a defense response in the plant (Böhnert et al. 2004). Attempts to identify the Ace1 metabolite by liquid chromatography tandem mass spectrometry analyses in barley leaves inoculated with virulent or avirulent isolates of *M. grisea* were as yet unsuccessful (Böhnert et al. 2004).

3.1.4 T-toxin

The linear long-chain polyketide (C35 to C41) T-toxin (Fig. 2) is produced by race T of *Cochliobolus heterostrophus*, the causal agent for southern corn leaf blight (Kono and Daly 1979). T-toxin is a well-studied host-selective toxin that is well characterized regarding its mode of action. As extensive reviews exist (Walton 1996; Wirsel et al. 2004), this toxin is only briefly discussed. *C. heterosporohus* race T attacks only corn carrying the cytoplasmically inherited gene *tms*, for pollen sterility (male sterile cytoplasma). T-toxin binds to the specific protein URF13 unique to the inner cell wall of T-cytoplasma rice. The protein is encoded by the gene *T-urf13* located on the mitochondrial chromosome of the T-cytoplasma. Upon binding of the T-toxin to the URF13-protein, pores are formed in the inner membrane and subsequently NAD^+ and small molecules, e.g. calcium, leak out of the mitochondrion (Siedow et al. 1995).

Following targeted disruption of the gene *PKS1*, encoding a polyketide synthase occurring in race T but not in race O strains, T-toxin production was eliminated and virulence was reduced. In the toxin-deficient mutant, T-toxin is not necessary for pathogenicity of *C. heterostrophus* race T, since the strain was still able to cause disease symptoms (Yang et al. 1996).

3.1.5 Non-toxic polyketides essential for pathogenicity

Apart from toxic secondary metabolites, fungi have been known for more than 40 years to produce pigments known as melanins, predominantly dihydroxy phenylalanine (DOPA)-melanin and 1,8-dihydroxynaphthalene

(DHN)-melanin (Langfelder et al. 2003). Melanin is incorporated into fungal cell walls, and, depending on the fungal lifestyle, this polymer may act as a reactive oxygen scavenger, add to the rigidity of the cell wall, or it may restrict the pore size of appressorial cell walls. In mammalian pathogens such as *Cryptococcus neoformans* or *Aspergillus fumigatus*, melanin has been shown to quench free radicals and is thought to be an important virulence factor (Langfelder et al. 2003). Results obtained from agar penetration assays with melanin deficient isolates of the human pathogenic black yeast *Wangiella dermatitidis*, and with a wild type isolate treated with the melanin biosynthesis inhibitor tricyclazole indicated that melanin improves biomechanical characteristics and may be important for virulence and disease progression in human and animal mycoses (Brush and Money 1999). Apart from acting as a scavenger molecule for reactive oxygen species, melanin creates a semi-permeable membrane in the inner cell wall of specialized infection cells (appressoria) of phytopathogenic fungi such as *Magnaporthe grisea*, different *Colletotrichum* species or *Phyllosticta ampelicida* (Deising et al. 2000; Wirsel et al. 2004). In melanized appressoria, an osmolyte is retained inside the cell, whereas water is able to permeate the semi-permeable layer. As a consequence of water uptake, an enormous turgor of up to 8 MPa is generated inside the appressoria which drives the infection peg though a pore at the basis of the infection structure and through the plant cuticle (Howard et al. 1991). The resulting force, which is exerted onto the host surface at the appressorial base, is thought to be sufficient to allow forceful penetration of the plant epidermis (Bechinger et al. 1999; Deising et al. 2000). Thus, mutants deficient in melanin biosynthesis fail to penetrate the plant surface and to colonize the host tissue. Inhibitors of DHN-melanin biosynthesis have successfully been used as protectants against blast disease in rice fields for almost 30 years. As fungi, but not animals or plants, use DHN as the melanin precursor, key enzymes in the biosynthetic pathway of DHN melanin are excellent targets for specific fungicides. These fungicides interfere with the infection-related differentiation process in phytopathogenic fungi, but they do not affect vegetative fungal growth.

3.2 Isoprenoids and terpenoids

The terpenes and steroids constitute a large group of secondary metabolites, many of which exhibit significant biological activities (Turner 1971). Some well-known examples of this group of chemicals synthesized by fungi include the trichothecene mycotoxins, gibberellic acid, ergosterol and β-carotene. They are biosynthetically derived from the C_5 "isoprene" unit (isopentenyl pyrophosphate, IPP), which itself is formed from acetate via mevalonate (Fig. 1) (Bennett 1983). The synthesis of mevalonate includes the condensation of three units of acetyl-CoA to 3-hydroxy-3-methylglutaryl-CoA (HMG-CoA), followed by a reduction to mevalonate. Two successive phosphorylation steps at C-5 of mevalonate catalyzed by the mevalonate and the phosphomevalonate kinase and a decarboxylation step lead to the basic C5 isoprene unit. IPP is added to prenyl pyrophosphate cosubstrates to form longer chains, such as geranyl pyrophosphate (C_{10}), farnesyl pyrophosphate (C_{15}) and geranylgeranyl pyrophosphate (C_{20}). These intermediates may then selfcondense, be utilized in alkyation reactions or undergo internal cyclization to create the basic skeletons of various terpenoid families (McGarvey and Croteau 1995). Terpenes are classified according their number of IPP-units, e.g. C10: monoterpenes, C15: sesquiterpenes, C20: diterpenes, C25 sesterterpenes (Fig. 1) (Turner 1971).

3.2.1 Trichothecenes

Trichothecenes constitute a large family of epoxide-containing sesquiterpenes, well known as mycotoxins commonly found as food and feed contaminants. Deoxynivalenon (or vomitoxin), nivalenol and T2-toxin (Fig. 2) are the most prominent examples of this mycotoxin family. They are produced by many fungal genra, including *Fusarium*, *Trichoderma*, *Myrothecium*, and cause listlessness or inactivity, diarrhea, vomiting, dermatitis (upon skin contact) and degeneration of the cells of the bone marrow, the lymph nodes and intestines (Marasas et al. 1984; Joffe 1986). Since trichothecenes are of enormous economic significance in agriculture, the biosynthesis and the mode of action of the compounds has been well studied. Members of this toxin family are known as potent inhibitors of eukaryotic protein biosynthesis, with different members interfering with initiation, elongation or termination stages (Bennett and Klich 2003). Within the molecule, the 12,13-epoxide-moiety is essential for the inhibition of protein biosynthesis (McLaughlin et al. 1977).

Apart from their importance as mycotoxins, many trichothecenes show potent phytotoxic activity. Phytotoxic properties of individual trichothecenes were discovered more than 4 decades ago [Brian 1961]. These early studies indicated that diacetoxyscirpenol reduced root elongation of cress. At very low concentrations trichothecenes, such as T2-toxin (Fig. 2) cause wilting, necrosis, and inhibition of germination and elongation of pea, barley, tomato and wheat seedlings (Cutler 1988; Wakulinski 1989).

Toxic sesquiterpenes have, however, not been detected in muskmelon seedlings infected with the trichothecene-producing strain *Myrothecium roridum*. This finding led to the question as to whether the fungal toxins are correlated with virulence of the pathogen. In order to investigate the role of trichothecenes in plant pathogenesis, the biosynthesis of the terpenes was blocked in *Fusarium* and *Giberella* strains by UV-induced or site-directed mutagenesis. The UV-mutants of *Fusarium sporotrichoides* failed to produce T2-toxin and they were found to be blocked at different steps of toxin biosynthesis (Beremand 1987; Plattner et al. 1989). In a plant assay on parsnip roots, only the wild type and a diacetoxyscirpenol-producing mutant were highly virulent, whereas a trichodiene- and a calonectrin-analogue producing mutant were significantly reduced in virulence. Further assays revealed, that the mutants were able to complement each other in restoring T2-toxin production *in vitro* and to partially restore virulence on parsnip roots. It was therefore suggested that T2-toxin contributes to the virulence of *F. sporotrichoides* on parsnip roots. In a candidate-gene approach, the *tox5* gene, encoding a trichodiene synthase putatively involved in diacetoxyscirpenol biosynthesis in *G. pulicaris*, was disrupted. Mutants unable to produce trichothecenes in culture were found to be significantly reduced in virulence compared to the wild-type strain (Desjardins et al. 1992, 1993). However, further experiments on potato tubers showed that toxin production had no influence on virulence on this host.

Trichothecene-non-producing mutants were generated via transformation-mediated disruption of a gene (*Tri5*) putatively encoding a trichodiene synthase. This enzyme catalyzes cyclization of farnesylpyrophosphate to trichodiene and is the first enzyme in the trichothecene biosynthetic pathway. *Tri5* mutants showed strongly reduced virulence. Complementation of the mutation in a *Tri5*-disrupted mutant restored the ability to produce trichothecenes and wild-type or near wild-type levels of virulence on wheat seedlings. These results provide further evidence that trichothecenes contribute to the virulence of plant-pathogenic fungi (Proctor et al. 1997).

3.2.2 Ophiobolin A

The maize pathogenic fungus *Helminthosporium maydis* and other members of the genus were found to produce the secondary metabolite ophiobolin A (Fig. 2). The sesquiterpene causes leakage of electrolytes and metabolites from cells in roots of maize seedlings (Tipton et al. 1977) and is believed to cause the symptoms of brown spot disease in rice (Narain and Biswal 1992). It was found that a possible inhibition of calmodulin *in vivo* was responsible for the phytotoxicity of the compound in maize roots (Leung et al. 1985). Mode of action studies showed that ophiobolin A binds directly to calmodulin (Leung et al. 1988). In bovine brain calmodulin-binding of the toxin to a single lysine residue (Lys-75) was responsible for the inhibitory effect of the compound. Since the structure of calmodulin is well conserved it is believed that this proposed mode of action is also valid in plant cells (Au and Leung 1998).

3.2.3 Colletotrichin

Several secondary metabolites have been identified from *Colletotrichum* species, but only a few of them were shown to possess phytotoxic activity. Colletotrichins (Fig. 2), secondary metabolites from *C. nicotina,* were identified as non-host specific phytotoxins. When the colletotrichins A, B or C were applied to tobacco leaves, they induced symptoms similar to those of tobacco anthracnose caused by *C. nicotina* (Gohbara et al. 1978). The chemical structure of the compound consists of a unique norditerpene and a polysubstituted γ-pyrone. According to feeding experiments with labeled precursors, the pyrone derives from a polyketide origin, while the terpene part of the molecule is synthesized via geranylgernyl phosphate (Turner and Aldridge 1983).

The colletotrichin A causes rapid loss of membrane integrity. Utrastructural observations indicated that the plasmalemma was damaged by an unknown mechanism (Gohbara et al. 1978). Other investigations showed that colletotrichins bind to the oxidized form of cytochrome and thereby inhibit the electron flow in the mitochondrial respiration chain (Halestrap 1982). These compounds may thus act via two independent mechanisms.

3.2.4 Fusicoccin

Fusicoccin A (Fig. 2) is a terpene phytotoxin produced by the fungus *Fusicoccum amygdali,* a parasite of peach and almond trees (Agrios 1997). It consists of a sugar and a terpene moiety, originally thought of as a degraded sesterterpene. Incorporation studies revealed, however, that fusicoccin is a diterpene rather than a sesterterpene (Turner and Aldridge 1983). Within host leaves, fusicoccin activates the plasma membrane H^+-ATPase by binding to 14-3-3 proteins. In order to bind fusicoccin, the 14-3-3 protein requires the presence of H^+-ATPase (Baunsgaard et al. 1998), resulting in hyperpolarization of the plasma membrane, accompanied by an acidification of the cell wall. An irreversible opening of the stomata is thereby triggered, which causes wilting of leaves and death of the plant.

3.3 Aromatic compounds and peptides

A large number of nitrogen-containing secondary metabolites have been identified from fungal sources. Many of these compounds are biosynthesized using common amino acids as building blocks. In this chapter we will discuss phytotoxins whose carbon skeletons derive mainly from amino acids.

3.3.1 Victorin

The host-specific HV-toxin (victorin) is an acyclic combination of glyoxylic acid with five unusual amino acids (Fig. 2) produced by the pathogenic fungus *Cochliobolus (Helminthosporium) victoriae*, which causes victoria blight of oats. Only oat cultivars carrying the chromosomal marker *Vb* are susceptible to the toxin (Wolpert et al. 2002). Victorin represents a pathogenicity factor, since strains of the fungus deficient in toxin biosynthesis fail to cause disease symptoms in host plants. Furthermore, the purified toxin provokes all disease symptoms when applied to a susceptible host leaf. Resistance and susceptibility of the plant are conferred by a dominant allele at the *Vb* locus. This *Vb* locus appears to be either identical to the gene *Pc-2* of oat plants or closely linked. The gene *Pc-2* is essential for the race-specific resistance against the biotrophic rust fungus *Puccinia coronata* (Wolpert et al. 2002).

In vivo labelling experiments have shown that victorin C binds to a 100 kDa protein called victorin-binding protein. This protein is present only in susceptible cultivars. The interaction of victorin and the *Vb* gene product induced responses in *Avena sativa* characteristic of programmed cell death (PCD) (Yao et al. 2002; Coffeen and Wolpert 2004), i.e. apoptotic DNA laddering, heterochromatin activation and cell shrinkage (Wolpert et al. 2002). Apart from the induction of PCD, victorin has been shown to contribute to mitochondrial dysfunction (Curtis and Wolpert 2002).

In addition to the 100 kDa protein, the toxin binds to a 15 kDa protein from both, susceptible and resistant oats. While both proteins seem to be components of the glycine decarboxylase complex, it is unclear whether the victorin binding protein is a product of the *Vb* locus. However, biochemical studies have shown that victorin inhibits the activity of glycine decarboxylase (Navarre and Wolpert 1995) and induces specific proteolytic cleavage of ribulose-1,5-bisphosphate carboxylase/oxygenase (Navarre and Wolpert 1999).

3.3.2 HC-toxin

In contrast to victorin, HC-toxin (Fig. 2), produced by the maize-pathogenic fungus *Cochliobolus* (*Helminthosporium*) *carbonum* has been classified as a virulence factor (for review, see Walton 1996). The cyclic tetrapeptide is produced by only one out of three races of the pathogen. This host-selective toxin is not required for the pathogenicity of *C. carbonum* on maize, but lesion size is increased upon infection with toxin-producing strains. Resistance against toxin-producing races is conferred by a gene called *Hm1*, responsible for reductive detoxification in resistant maize cultivars (Meeley et al. 1992). The gene *Hm1* encodes a carbonyl reductase, which reduces the ketone function on the side chain of the 2-amino-8-oxo-9,10-epoxyoctde-canoic acid. Deleting the *Hm1* gene resulted in varieties highly susceptible to both HC-toxin and HC toxin-producing races of *C. carbonum* (Walton 1996).

HC toxin is a cytostatic compound, as visualized by the inhibition of root growth in susceptible maize in the presence of the toxin. The mode of action of HC toxin is the inhibition of histone deacetylase (Brosch et al. 1995, 2001). The compound, as well as structurally related tetrapeptides, inhibits histone deacetylases from a variety of organisms, but the enzyme of the producing organism is relatively insensitive to HC-toxin (Baidyaroy et al. 2002).

3.3.3 Tentoxin

The cyclotetrapeptide tentoxin (Fig. 2) is produced by *Alternaria* species and acts in a non-host-selective manner. It has been shown to inhibit energy transfer in chloroplasts during light dependent phosphorylation and causes chlorosis in germinating seedlings of sensitive species (Prell and Day 2001). In isolated thylakoids, tentoxin inhibits ATP synthesis at micromolar concentrations (Arntzen 1972). It acts as inhibitor of chloroplast F1-ATPase (CF_1) in susceptible plant species, but not on homologous enzymes from chloroplasts of non-susceptible plants, bacteria or mitochondria (Pavlova et al. 2004). Crystal structure studies revealed that tentoxin binds with high affinity to the $\alpha\beta$-subunit interface (Groth 2002). At higher concentrations, the toxin binds to further low affinity binding sites, hereby re-activating ATP hydrolysis (Pavlova et al. 2004).

3.3.4 Fusaric acid

Fusaric acid (5-butylpicolinic acid) (Fig. 2) was first identified as a plant-growth inhibitor from *Fusarium moniliforme* (*Giberella fujikoroi*) (Yabuta et al. 1934). Natural compounds from fungal origin bearing a pyridine ring are not very common. Radioactive labelling experiments have shown that radioactive aspartate and acetate are incorporated in the pyridine ring (Turner 1971). Fusaric acid was isolated from a number of *Fusarium* species (Bacon et al. 1996), and is one of the first fungal metabolites linked with the pathogenesis of *Fusarium oxysporum* which causes tomato wilt symptoms. Whether fusaric acid is essential for pathogenesis *in planta* has not been proven, however, it has been reported that the toxin was not responsible for the formation of disease symptoms (Turner 1971). In addition to its suggested role in plant pathogenesis, fusaric acid is a mycotoxin potentially toxic to animals.

4 Regulation of secondary metabolism during pathogenic development

Surprisingly few data exist directly documenting the regulation of secondary metabolism during fungal pathogenesis. It is known that secondary metabolite production is often closely linked to fungal morphogenesis, i.e. differentiation of infection structures (Bennett 1983). The production of alkaloids, for example, coincides with the formation of conidiospores in *Penicillium cyclopium* and clamydospores in *Claviceps purpurea*. In *Cephalosporium acremonium*, cephalosporin synthesis is concomitant with arthrospore development (Zähner et al. 1983). These developmental processes are generally characteristic responses to nutrient deprivation and it is well documented that secondary metabolite production often occurs primarily within the stationary phase (Bennett 1983). In *C. carbonum*, synthesis of the HC-toxin is regulated during infection-related morphogenesis. Analysis of spore germination fluids by plasma desorption mass spectrometry revealed that spores induced to form appressoria *in vitro* synthesised and released the toxin at a time coincident with maturation of appressoria. Spores incubated under conditions that did not induce appressorium formation failed to produce toxin (Weiergang et al. 1996). Given these observations, one might anticipate that pathways regulating secondary metabolite production may, at least to some degree, share common components with pathways regulating responses to nutrient deprivation and/or cellular differentiation. Evidence for signalling components common to secondary metabolite formation and sporulation has come with the finding that both processes in *A. nidulans* are negatively regulated by G-protein mediated signalling (Hicks et al. 1997). The G-protein mediated signalling pathway also plays a critical role in disease related development in several fungi which cause disease of plants (Bölker 1998), and there is now evidence that G-protein signalling also may influence trichothecene production in *Fusarium sporotrichioides* (Tag et al. 2000). Targeted disruption of *CZK3,* a gene encoding a MAP kinase homolog in *Cercospora zeae-maydis,* suppressed expression of genes predicted to participate in cercosporin biosynthesis and abolished cercosporin production. The mutants grew faster on agar media than the wild type but were deficient in conidiation and elicited only small chlorotic spots on inoculated maize leaves compared with rectangular necrotic lesions incited by the wild type, indicating that cercosporin is a virulence factor in *C. zeae-maydis* (Shim and Dunkle 2003). The above data raise the possibility that conserved signalling pathways might also control the timely production of other disease relevant secondary metabolites.

Melanin biosynthetic genes of *Magnaporthe grisea* and *Colletotrichum lagenarium* appear to be transcriptionally activated by the putative transcription factors Pig1 and Cmr1, respectively (Sweigard et al. 1998; Tsuji et al. 2000). Although structurally very similar, Cmr1 appears to regulate mycelial melanin production while Pig1 seems to regulate the production of the melanin required for appressorium function in *M. grisea.* The control of the transcription of the genes required for the biosynthesis of disease-related toxins is only known in any detail for the HC-toxin biosynthetic genes of *Cochliobolus carbonum.* Transcription of these genes is controlled via *TOXE,* which encodes a protein containing a bZIP basic region and four C-terminal ankyrin repeats but lacking a true leucine zipper (Pedley and Walton 2001). TOXE has been proposed to represent a member of a novel class of transcription factors termed bANK proteins (Bussink et al. 2001). Whether this class of transcription factors plays a conserved role in regulating toxin production is presently unknown. TOXE has been shown to bind *in vitro* to the consensus sequence ATCTCNCGNA present in the promoters of *TOX2* genes, encoding HC-toxin synthetase. Basic residues at the N-terminus and the C-terminal ankyrin repeats of the transcription factor were shown to be essential for DNA binding (Pedley and Walton 2001). TOXE is not known to regulate genes other than those of the *TOX2* cluster within which it resides. In this manner, TOXE resembles the pathway-specific transcription factors TRI6 of *Fusarium sporotrichiodes,* and AflR of species of *Aspergillus,* governing trichothecene and aflatoxin biosynthesis, respectively (Woloshuk et al. 1994; Proctor et al. 1995; Hohn et al. 1999). A structurally similar putative transcription factor, ORFR, occurs within the cluster of genes responsible for the biosynthesis of the

AK-toxin of *Alternaria alternata* (Tanaka and Tsuge 2000), although the role of this gene in regulating expression of the genes within the cluster is not known. In *Fusarium verticillioides*, the production of fumonisin is controlled by the putative transcription factor Zfr1, whose activity in turn requires the presence of the cyclin C like protein FCC1 (Shim and Woloshuk 2001; Flaherty and Woloshuk 2004). In *Aspergillus nidulans*, a candidate global regulator of secondary metabolism, the nuclear protein LaeA, which positively regulates the AlfR transcription factor, has been discovered recently (Bok and Keller 2004). This factor is required for the expression of several secondary metabolite producing gene clusters and, interestingly, putative LaeA homologs are present in a number of other filamentous fungi, including important phytopathogenic species such as *M. grisea* and *F. sporotrichioides* (Bok and Keller 2004). If these LaeA homologues have a conserved regulatory function, not only would they shed light on the control of secondary metabolite production in general, but additionally their manipulation may present a future route towards the identification of novel fungal secondary metabolites with a role in plant disease.

Despite the great chemical diversity and effects of fungal secondary metabolites, many of these share a common point of regulation. As indicated by Fig. 4B, non-ribosomal peptide synthetases (NRPSs), polyketide synthases (PKSs), hybrid NRPS/PKSs (Kroken et al. 2003; Lee et al. 2005) and fatty acid synthases are involved in pathogenesis-related secondary metabolism and therefore play a role in fungal pathogenicity or virulence. For these groups of enzymes, activation by covalent attachment of the 4'-phosphopantetheine (P-pant) moiety of coenzyme A, and thus 4'-phosphopantetheinyl transferase (PPTase) activity, is essential (Fig. 4A) (Fichtlscherer et al. 2000; Mootz et al. 2002; Keszenman-Pereyra et al. 2003; Oberegger et al. 2003) (Fig. 4A). PPTases, due to their central role in fungal secondary metabolism, may be regarded as key-elements in pathogenicity or virulence, and could represent excellent fungicide targets.

5 Concluding remarks

Fungal secondary metabolism is characterized by an enormous diversity of products and, consequently, of metabolic pathways and enzymes involved in their synthesis. As several of these metabolites have been proven to be essential for pathogenicity, either biosynthetic or regulatory enzymes may be excellent targets in chemical plant protection. The wealth of secondary metabolites could also be used as a source of inhibitors of infection-related morphogenesis. For example, different glisoprenins isolated from submerged cultures of the deuteromycete *Gliocladium roseum* inhibited appres-

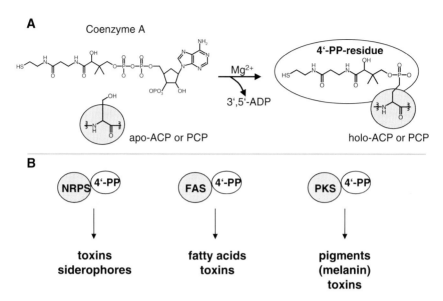

Fig. 4. Mechanism and role of 4′-phosphopantetheinyl transferases (4′-PPTases). **A** 4′-PPTases catalyze the posttranslational transfer of the 4′-phosphopantetheine moiety of CoA onto a conserved serine residue within acyl- or peptidyl carriers. Thereby, the carrier proteins are converted from their inactive apo form into the active holo form. The reaction is dependent on Mg_2 and yields 3,5-ADP as a second product. After Mootz et al. (2001). **B** In order to synthesize secondary metabolites with relevance in pathogenesis (e.g. toxins, siderophores, melanin) different enzymes such as non-ribosomal peptide synthases (NRPS), fatty acid synthases (FAS), and polyketide synthases (PKS) require activation by 4′-phosphopantetheinylation.

sorium formation of *M. grisea* on inductive surfaces. As these compounds did not exhibit phytotoxic activities, they may well serve as lead structures for efficient rice blast fungicides (Thines et al. 1997). It is known that several other fungal secondary metabolites exist that interfere with infection structure differentiation (Thines et al. 2004). These compounds may, in the future, serve as highly specific fungicides directed against discreet stages of the infection process. There is no doubt that new fungicides are urgently required, as frequent occurrence of fungicide resistance drastically reduces the number of useful fungicides that are available in modern agriculture (Deising et al. 2002; Reimann and Deising 2005).

References

Abe Y, Suzuki T, Ono C, Iwamoto K, Hosobuchi M, Yoshikawa H (2002) Molecular cloning and characterization of an ML-236B (compactin) biosynthetic gene cluster in *Penicillium citrinum*. Mol Genet Genomics 267:636–646

Agrios GN (1997) Plant pathology, 4th edn. Academic Press., San Diego

Aharonowitz Y, Demain AL (1980) Thoughts on secondary metabolism. Biotechnol Bioeng 22:5–9

Arntzen CJ (1972) Inhibition of photophosphorylation by tentoxin, a cyclic tetrapeptide. Biochim Biophys Acta 283:539–542

Au TK, Leung PC (1998) Identification of the binding and inhibition sites in the calmodulin molecule for ophiobolin A by site-directed mutagenesis. Plant Physiol 118:965–997

Bacon CW, Porter JK, Norred WP, Leslie JF (1996) Production of fusaric acid by *Fusarium* species. Appl Environ Microbiol 62:4039–4043

Baidyaroy D, Brosch G, Graessle S, T, Rojer P, Walton JD (2002) Characterization of inhibitor-resistant histone deacetylase activity in plant-pathogenic fungi. Eukaryot Cell 1:538–547

Barash I, Mor H, Netzer D, Kashman Y (1981) Production of zinniol by *Alternaria dauci* and its phytotoxic effect on carrot. Physiol Plant Pathol 19:7–15

Baunsgaard L, Fuglsang AT, Jahn T, Korthout HAAJ, de Boer AH, Palmgren MG (1998) The 14–3-3 proteins associate with the plant plasma membrane H+-ATPase to generate a fusic-occin binding complex and a fusicoccin responsive system. Plant J 13:661–671

Bechinger C, Giebel K-F, Schnell M, Leiderer P, Deising HB, Bastmeyer M (1999) Optical measurements of invasive forces exerted by appressoria of a plant pathogenic fungus. Science 285:1896–1899

Bennett JW (1983) In: Bennett JW, Ciegler A (eds) Secondary metabolism and differentiation in fungi. Dekker, New York

Bennett JW, Klich M (2003) Mycotoxins. Clin Microbiol Rev 16:497–516

Beremand MN (1987) Isolation and characterization of mutants blocked in T-2 toxin biosynthesis. Appl Environ Microbiol 53:1855–1859

Berlepsch KV (1980) Drugs from marine organisms. Naturwiss 67:338–242

Bilski P, Li MY, Ehrenshaft M, Daub ME, Chignell CF (2000) Vitamin B$_6$ (pyridoxine) and its derivatives are efficient singlet oxygen quenchers and potential fungal antioxidants. Photochem Photobiol 71:129–134

Böhnert HU, Fudal I, Dioh W, Tharreau D, Notteghem JL, Lebrun MH (2004) A putative polyketide synthase/peptide synthetase from *Magnaporthe grisea* signals pathogen attack to resistant rice. Plant Cell 16:2499–2513

Bok JW, Keller NP (2004) LaeA, a regulator of secondary metabolism in *Aspergillus* spp. Eukaryotic Cell 3:527–535

Bölker M (1998) Sex and crime: heterotrimeric G proteins in fungal mating and pathogenesis. Fung Genet Biol 25:143–156

Brosch G, Ransom R, Lechner T, Walton JD, Loidl P (1995) Inhibition of maize histone deacetylases by HC toxin, the host-selective toxin of *Cochliobolus carbonum*. Plant Cell 7:1941–1950

Brosch G, Dangl M, Graessle S, Loidl A, Trojer P, Brandtner EM, Mair K, Walton JD, Baidyaroy D, Loidl P (2001) An inhibitor-resistant histone deacetylase in the plant pathogenic fungus *Cochliobolus carbonum*. Biochem 40:12855–12863

Brush L, Money NP (1999) Invasive hyphal growth in *Wangiella dermatitidis* is induced by stab inoculation and shows dependence upon melanin biosynthesis. Fung Genet Biol 28:190–200

Bu'Lock JD (1975) Secondary metabolism in fungi and its relationship to growth and development. In: Smith JE, Berry DR (eds) The filamentous fungi, vol 1, industrial mycology. Edward Arnold Press, London, and Halsted Press, N.Y., USA

Bussink HJ, Clark A, Oliver R (2001) The *Cladosporium fulvum Bap1* gene: evidence for a novel class of Yap-related transcription factors with ankyrin repeats in phytopathogenic fungi. Eur J Plant Pathol 107:655–659

Callahan TM, Rose MS, Meade MJ, Ehrenshaft M, Upchurch RG (1999) CFP, the putative cercosporin transporter of *Cercospora kikuchii*, is required for wild type cercosporin production, resistance, and virulence on soybean. Mol Plant Microbe Interact 12:901–910

Carlile MJ, Watkinson SC (1994) The fungi. Academic Press, London, Boston, San Diego

Chung KR, Ehrenshaft M, Daub ME (2002) Functional expression and cellular localization of cercosporin-resistance proteins fused with the GFP in *Cercospora nicotianae*. Curr Genet 41:159–167

Chung KR, Ehrenshaft M, Wetzel DK, Daub ME (2003) Cercosporin-deficient mutants by plasmid tagging in the asexual fungus *Cercospora nicotianae*. Mol Genet Genom 270:103–113

Coffeen WC, Wolpert TJ (2004) Purification and characterization of serine proteases that exhibit caspase-like activity and are associated with programmed cell death in *Avena sativa*. Plant Cell 16: 857–873.

Cotty PJ, Misaghi IJ (1984) Zinniol production by *Alternaria* species. *Phytopathology* 74:785–788

Cramer RA, Lawrence CB (2003) Cloning of a gene encoding an Alt a 1 isoallergen differentially expressed by the necrotrophic fungus *Alternaria brassicicola* during *Arabidopsis* infection. Appl Environ Microbiol 69:2361–2364

Curtis MJ, Wolpert TJ (2002) The oat mitochondrial permeability transition and its implication in victorin binding and induced cell death. Plant J 29:295–312

Cutler H (1988) Trichothecenes and their role in the expression of plant disease. Biotechnol Crop Protect. ACS Symposium Series 379:50–72

Daub ME, Ehrenshaft M (2000) The photoactivated *Cercospora* toxin cercosporin: contributions to plant disease and fundamental biology. Annu Rev Phytopathol 38:461–490

Daub ME, Li M, Bilski P, Chignell CF (2000) Dihydrocercosporin singlet oxygen production and subcellular localization: a possible defense against cercosporin phototoxicity in *Cercospora*. Photochem Photobiol 71:135–140

Deising HB, Werner S, Wernitz M (2000) The role of fungal appressoria in plant infection. Microbes Infect 2:1631–1641

Deising HB, Reimann S, Peil A, Weber WE (2002) Disease management of rusts and powdery mildews. In: Kempken F (ed) The mycota XI. Application in agriculture. Springer, Berlin, pp 243–269

Desjardins AE, Hohn TM, McCormick SP (1992) Effect of gene disruption of trichodiene synthase on the virulence of *Gibberella pulicaris*. Mol Plant Microbe Interact 5:214–222

Desjardins AE, Hohn TM, McCormick SP (1993) Trichothecene biosynthesis in *Fusarium* species: chemistry, genetics, and significance. Microbiol Rev 57:595–604

Dix NJ, Webster J (1995) Fungal ecology. Chapman & Hall, London

Faulkner DJ (1993) Academic chemistry and the discovery of bioactive marine natural products. In: Attaway DH, Zaborsky OR (eds) Marine biotechnology, vol 1: pharmaceuticals and bioactive natural products. Plenum Press, New York

Fichtlscherer F, Wellein C, Mittag M, Schweizer E (2000) A novel function of yeast fatty acid synthase. Subunit alpha is capable of self-pantetheinylation. Eur J Biochem 267:2666–2671

Flaherty JE, Woloshuk CP (2004) Regulation of fumonisin biosynthesis in *Fusarium verticillioides* by a zinc binuclear cluster-type gene, *ZFR1*. Appl Environ Microbiol 70:2653–2659

Gohbara M, Kosuge Y, Yamasaki S, Kimura Y, Suzuki A, Tamura S (1978) Isolation, structures and biological-activities of colletotrichins, phytotoxic substances from *Colletotrichum nicotianae*. Agric Biol Chem 42:1037–1043

Gokhale RS, Tuteja D (2001) Biochemistry of polyketide synthases. In: Rehm H-J, Reed G (eds) Special processes, vol 10. Wiley-VCH, Weinheim, pp 342–372.

Groth G (2002) Structure of spinach chloroplast F1-ATPase complexed with the phytopathogenic inhibitor tentoxin. Proc Natl Acad Sci USA 99:3464–3468

Halestrap A (1982) The pathway of electron flow through ubiquinol:cytochrome c oxidoreductase in the respiratory chain. Evidence from inhibition studies for a modified "Q cycle". Biochem J 204:49–59

Hawksworth DL (2001) The magnitude of fungal diversity: the 1.5 million species estimate revisited. Mycol Res 105:1422–1432

Heiser I, Oßwald W, Elstner EF (1998) The formation of reactive oxygen species by fungal and bacterial phytotoxins. Plant Physiol Biochem 36:703–713

Heiser I, Sachs E, Liebermann B (2003) Photodynamic oxygen activation by rubellin D, a phytotoxin produced by *Ramularia collo-cygni* (Sutton et Waller). Physiol Mol Plant Pathol 62:29–36

Hicks JK, Yu JH, Keller NP, Adams TH (1997) *Aspergillus* sporulation and mycotoxin production both require inactivation of the FadA G alpha protein-dependent signaling pathway. EMBO J 16:4916–4923

Hohn TM, Krishna R, Proctor RH (1999) Characterization of a transcriptional activator controlling trichothecene toxin biosynthesis. Fungal Genet Biol 26:224–235

Howard RJ, Ferrari MA, Roach DH, Money NP (1991) Penetration of hard substances by a fungus employing enormous turgor pressures. Proc Natl Acad Sci USA 88:11281–11284

Joffe AZ (1986) *Fusarium* Species: their biology and toxicology. Wiley, New York

Keszenman-Pereyra D, Lawrence S, Twfieg ME, Price J, Turner G (2003) The npgA/cfwA gene encodes a putative 4′-phosphopantetheinyl transferase which is essential for penicillin biosynthesis in *Aspergillus nidulans*. Curr Genet 43:186–90

Khosla C, Gokhale RS, Jacobsen JR, Cane DE (1999) Tolerance and specificity of polyketide synthases. Annu Rev Biochem 68:219–253

Kogel K-H, Beckhove U, Drescher J, Münch S, Rommé Y (1994) Acquired resistance in barley. The mechanism induced by 2,6-dichloroisonicotinic acid is a phenocopy of a genetically based mechanism governing race-specific powdery mildew resistance. Plant Physiol 106:1269–1277

Kono Y, Daly J (1979) Characterization of the host-specific pathotoxin produced by *Helminthosporium maydis,* race T, affecting corn with Texas male sterile cytoplasm. Bioorg Chem 8:391–397

Kroken S, Glass NL, Taylor JW, Yoder OC, Turgeon BG (2003) Phylogenomic analysis of type I polyketide synthase genes in pathogenic and saprobic ascomycetes. Proc Natl Acad Sci USA 100:15670–15675

Kuyama S, Tamura T (1957) Cercosporin. A pigment of *Cercospora kikuchii* Matsumoto et Tomoyasu. I. Cultivation of fungus, isolation and purification of pigment. J Am Chem Soc 79:5725–5726

Kuykendall LD, Upchurch RG (2004) Expression in sugar beet of the introduced cercosporin toxin export (CFP) gene from *Cercospora kikuchii*, the causative organism of purple seed stain in soybean. Biotechnol Lett 26:723–727

Langfelder K, Streibel M, Jahn B, Haase G, Brakhage AA (2003) Biosynthesis of fungal melanins and their importance for human pathogenic fungi. Fungal Genet Biol 38:143–158

Lee BN, Kroken S, Chou DY, Robbertse B, Yoder OC, Turgeon BG (2005) Functional analysis of all nonribosomal peptide synthetases in *Cochliobolus heterostrophus* reveals a factor, NPS6, involved in virulence and resistance to oxidative stress. Eukaryotic Cell 4:545–555

Leung PC, Taylor WA, Wang JH, Tipton CL (1985) Role of cal-modulin inhibition in the mode of action of ophiobolin A. Plant Physiol 177:303–308

Leung PC, Graves LM, Tipton CL (1988) Characterization of the interaction of ophiobolin A and calmodulin. Int J Biochem 20:1351–1359

Levine A, Tenhaken R, Dixon R, Lamb C (1994) H_2O_2 from the oxidative burst orchestrates the plant hypersensitive disease resistance response. Cell 79:583–593

Marasas WFO, Nelson PE, Tousson TA (1984) Toxigenic *Fusarium* species: identity and mycotoxicology. Pennsylvania State University Press, University Park, Pennsylvania

McGarvey DJ, Croteau R (1995) Terpenoid metabolism. Plant Cell 7:1015–1026

McLaughlin CS, Vaughn MH, Campbell JM, Wei CM, Stafford ME (1977) Inhibition of protein synthesis by trichothecenes. In: Rodricks JV, Hesseltine CW, Mehlman MA (eds) Mycotoxins in human and animal health. Pathotoxin Publishers, Park Forest, Ill., pp 263–273

Meeley RB, Johal GS, Briggs SP, Walton JD (1992) A biochemical phenotype for a disease resistance gene of maize. Plant Cell 4:71–77

Mootz HD, Finking R, Marahiel MA (2001) 4′-Phosphopantetheine transfer in primary and secondary metabolism of *Bacillus subtilis*. J Biol Chem 276:37289–37298

Mootz HD, Schorgendorfer K, Marahiel MA (2002) Functional characterization of 4′-phosphopantetheinyl transferase genes of bacterial and fungal origin by complementation of *Saccharomyces cerevisiae lys5*. FEMS Microbiol Lett 213:51–57

Narain A, Biswal G (1992) *Helminthosporium oryzae* toxin (ophiobolin) and its involvement with pathogenesis on rice. Int J Trop Plant Dis 10:1–8

Navarre DA, Wolpert TJ (1995) Inhibition of the glycine decarboxylase multienzyme complex by the host-selective toxin victorin. Plant Cell 7:463–471

Navarre DA, Wolpert TJ (1999) Victorin induction of an apoptotic/senescence-like response in oats. Plant Cell 11:237–249

Nukina M, Sassa T, Ikeda M, Umezawa T, Tasaki H (1981) Pyriculariol, a new phytotoxic metabolite of *Pyricularia oryzae*. Agric Biol Chem 45:2161–2162

Oberegger H, Eisendle M, Schrettl M, Graessle S, Haas H (2003) 4′-Phosphopantetheinyl transferase-encoding npgA is essential for siderophore biosynthesis in *Aspergillus nidulans*. Curr Genet 44:211–215

O'Hagan D (1992) Biosynthesis of polyketide metabolites. Nat Prod Rep 9:447–479

Okubo A, Yamazak S, K. Fuwa K (1975) Biosynthesis of cercosporin. Agric Biol Chem 39: 1173–1175

Pavlova P, Shimabukuro K, Hisabori T, Groth G, Lill H, Bald D (2004) Complete inhibition and partial re-activation of single F1-ATPase molecules by tentoxin: new properties of the re-activated enzyme. J Biol Chem 279:9685–9688

Pedley KF, Walton JD (2001) Regulation of cyclic peptide biosynthesis in a plant pathogenic fungus by a novel transcription factor. Proc Natl Acad Sci USA 98:14174–14179

Plattner RD, Tjarks LW, Beremand MN (1989) Trichothecenes accumulated in liquid culture of a mutant of Fusarinum sporotrichioides NRRL 3299. Appl Environ Microbiol 55:2190–2194

Prell HH, Day P (2001) Plant–fungal pathogen interaction—a classical and molecular view. Springer Verlag, Berlin

Proctor RH, Hohn TM, McCormick SP (1995) Reduced virulence of *Gibberelle zeae* caused by disruption of a trichothecene toxin biosynthetic gene. Molec Plant Microbe Interact 8:593–601

Proctor RH, Hohn TM, McCormick SP (1997) Restoration of wild-type virulence to *Tri5* disruption mutants of *Gibberella zeae* via gene reversion and mutant complementation. Microbiology 143:2583–2591

Reimann S, Deising HB (2005) Inhibition of efflux transporter-mediated fungicide resistance in *Pyrenophora tritici-repentis* by a derivative of natural 4′-hydroxyflavone and potentiation of fungicide activity. Appl Environ Microbiol 71:3269–3275

Rollins JA, Ehrenshaft M, Upchurch RG (1993) Effects of light and altered cercosporin phenotypes on gene expression in *Cercospora kikuchii*. Can J Microbiol 39:118–124

Rotem J (ed) (1994) The genus *Alternaria*. Biology, epidemiology, and pathology. APS Press, St Paul, Minnesota

Shim WB, Dunkle LD (2003) CZK3, a MAP kinase kinase kinase homolog in *Cercospora zeae-maydis*, regulates cercosporin biosynthesis, fungal development, and pathogenesis. Mol Plant Microbe Interact 16:760–768

Shim WB, Woloshuk CP (2001) Regulation of fumonisin B(1) biosynthesis and conidiation in *Fusarium verticillioides* by a cyclin-like (C-type) gene, *FCC1*. Appl Environ Microbiol 67:1607–1612

Siedow JN, Rhoads DM, Ward GC, Levings CS (1995) The relationship between the mitochondrial gene *T-urf13* and fungal pathotoxin sensitivity in maize. Biochim Biophys Acta 1271:235–240

Smith SE, Read DJ (1997) Mycorrhizal symbiosis. Academic Press, London.

Spikes JD (1989) Photosensitization. In: Smith KC (ed) The science of photobiology, 2nd edn. Plenum Press, New York

Stoessl A, Unwin CH, Stothers JB (1979) Metabolites of *Alternaria-solani*. 5. Biosynthesis of altersolanol-A and incorporation of altersolanol-A. A-C-13(X) into altersolanol-B and macrosporin. Tetrahed Lett 27:2481–2484

Sugawara F, Strobel G (1986) Zinniol, a phytotoxin from *Phoma macdonaldi*. Plant Sci 43:19–23

Sweigard JA, Carroll AM, Farrall L, Chumley FG, Valent B (1998) *Magnaporthe grisea* pathogenicity genes obtained through insertional mutagenesis. Mol Plant Microbe Interact 11:404–412

Tag A, Hicks J, Garifullina G, Ake CJ, Phillips TD, Beremand M, Keller N (2000) G-protein signalling mediates differential production of toxic secondary metabolites. Mol Microbiol 38:658–665

Talbot NJ (2003) On the trail of a cereal killer: exploring the biology of *Magnaporthe grisea*. Annu Rev Microbiol 57:177–202

Tanaka A, Tsuge T (2000) Structural and functional complexity of the genomic region controlling AK-toxin biosynthesis and pathogenicity in the Japanese pear pathotype of *Alternaria alternata*. Mol Plant Microbe Interact 13:975–986

Tertivanidis K, Goudoula C, Vasilikiotis C, Hassiotou E, Perl-Treves R, Tsaftaris A (2004) Superoxide dismutase transgenes in sugarbeets confer resistance to oxidative agents and the fungus *C. beticola*. Transgen Res 13:225–333

Thines E, Eilbert F, Sterner O, Anke H (1997) Glisoprenin A, an inhibitor of the signal transduction pathway leading to appressorium formation in germinating conidia of *Magnaporthe grisea* on hydrophobic surfaces. FEMS Microbiol Lett 151:219–224

Thines E, Anke H, Weber RW (2004) Fungal secondary metabolites as inhibitors of infection-related morphogenesis in phytopathogenic fungi. Mycol Res 108:14–25

Thuleau P, Graziana A, Rossignol M, Kauss H, Auriol P, Ranjeva R (1988) Binding of the phytotoxin zinniol stimulates the entry of calcium into plant protoplasts. Proc Natl Acad Sci USA 85:5932–5935

Tipton CL, Paulsen PV, Betts RE (1977) Effects of ophiobolin A on ion leakage and hexose uptake by maize roots. Plant Physiol 59:907–910

Tsuji G, Kenmochi Y, Takano Y, Sweigard J, Farrall L, Furusawa I, Horino O, Kubo Y (2000) Novel fungal transcriptional activators, Cmr1p of *Colletotrichum lagenarium* and pig1p of *Magnaporthe grisea*, contain Cys2His2 zinc finger and Zn(II)2Cys6 binuclear cluster DNA-binding motifs and regulate transcription of melanin biosynthesis genes in a developmentally specific manner. Mol Microbiol 38:940–954

Tsuji SY, Cane DE, Khosla C (2001) Selective protein-protein interactions direct channeling of intermediates between polyketide synthase modules. Biochem 40:2326–2331

Turner WB (1971) Fungal metabolites. Academic Press, London

Turner WB, Aldridge DC (1983) Fungal metabolites, 1st edn. Academic Press, London

Upchurch RG, Rose MS, Eweida M (2001) Over-expression of the cercosporin facilitator protein, CFP, in *Cercospora kikuchii* up-regulates production and secretion of cercosporin. FEMS Microbiol Lett 204:89–93

Upchurch RG, Walker DC, Rollins JA, Ehrenshaft M, Daub ME (1991) Mutants of *Cercospora kikuchii* altered in cercosporin synthesis and pathogenicity. Appl Environ Microbiol 57: 2940–2945

Ververidis P, Davrazou F, Diallinas G, Georgakopoulos D, Kanellis AK, Panopoulos N (2001) A novel putative reductase (Cpd1p) and the multidrug exporter Snq2p are involved in resistance to cercosporin and other singlet oxygen-generating photosensitizers in *Saccharomyces cerevisiae*. Curr Genet 39:127–136

Wakulinski W (1989) Phytotoxicity of Fusarium metabolites in relation to pathogenicity. In: Chelkowski J (ed) Fusarium: mycotoxins, taxonomy and pathogenicity. Elsevier, Amsterdam

Walton JD (1996) Host-selective toxins: agents of compatibility. Plant Cell 8:1723–1733

Walton JD, Panaccione DG (1993) Host-selective toxins and disease specificity: Perspectives and progress. Annu Rev Phytopathol 31:275–303

Weiergang I, Dunkle LD, Wood KV, Nicholson RL (1996) Morphogenic regulation of pathotoxin synthesis in *Cochliobolus carbonum*. Fungal Genet Biol 20:74–78

Weinberg ED (1974) Secondary metabolism: control by temperature and inorganic phosphate. Dev Ind Microbiol 15:70–81

Wirsel SGR, Reimann S, Deising HB (2004) Genetics of phytopathology: fungal morphogenesis and plant infection. In: Esser K, Lüttge U, Beyschlag W, Murata J (eds) Progress in botany. Springer-Verlag, Berlin Heidelberg, pp 147–178

Woloshuk CP, Foutz KR, Brewer JF, Bhatnagar D, Cleveland TE, Payne GA (1994) Molecular characterization of *aflR*, a regulatory locus for aflatoxin biosynthesis. Appl Environ Microbiol 60:2408–2414

Wolpert TJ, Dunkle LD, Ciuffetti LM (2002) Host-selective toxins and avirulence determinants: what's in a name? Annu Rev Phytopathol 40:251–285

Yabuta T, Kambe K, Hayashi T (1934) Biochemistry of the bakanae-fungus. I. Fusarinic acid, a new product of the bakanae-fungus. J Agric Chem Soc Jpn 10:1059–1068

Yamazaki S, Ogawa. T (1972) The chemistry and stereochemistry of cercosporin. Agric Biol Chem 36:1707–1718

Yang G, Rose MS, Turgeon BG, Yoder OC (1996) A polyketide synthase is required for fungal virulence and production of the polyketide T-toxin. Plant Cell 8:2139–2150

Yao N, Tada Y, Sakamoto M, Nakayashiki H, Park P, Tosa Y, Mayama S (2002) Mitochondrial oxidative burst involved in apoptotic response in oats. Plant J 30:567–579

Zähner H, Anke H, Anke T (1983) Evolution of secondary pathways. In: Bennett JW, Ciegler E (eds) Differentiation and secondary metabolism in fungi. Marcel Dekker, New York, pp 153–171

Eckhard Thines
Universität Kaiserslautern, Fachbereich Biologie, Abteilung Biotechnologie,
Postfach 3049,
67663 Kaiserslautern, Germany.
e-mail: thines@rhrk.uni-kl.de

Jesús Aguirre
Instituto de Fisiología Celular-Universidad Nacional Autónoma de México,
Apartado Postal 70–242,
04510 México, D.F., México.
e-mail: jaguirre@ifc.unam.mx

Andrew J. Foster
Institute of Biotechnology and Drug Research,
Erwin-Schrödinger-Str. 56, 67663 Kaiserslautern, Germany.
e-mail: foster@ibwf.de

Holger B. Deising
Martin-Luther-University Halle-Wittenberg, Faculty of Agriculture,
Phytopathology and Plant Protection,
Ludwig-Wucherer-Str. 2,
D-06099 Halle (Saale), Germany.
e-mail: Deising@landw.uni-halle.de

Plant Breeding:
MADS ways of memorizing winter: vernalization in weed and wheat

Günter Theißen

Abbreviations:

AGLn: *AGAMOUS-LIKE GENE No. n*
AP1: *APETALA1*
CAL: *CAULIFLOWER*
Col: *Arabidopsis* ecotype Columbia
CO: *CONSTANS*
ELFn: *EARLY FLOWERINGn*
FLC: *FLOWERING LOCUS C*
FLD: *FLOWERING LOCUS D*
FRI: *FRIGIDA*
FUL: *FRUITFULL*
Ler: *Arabidopsis* ecotype Landsberg *erecta*
LFY: *LEAFY*
MADS: acronym for the genes *MCM1, AG, DEF, SRF* (founding members of the MADS-box gene family)
MAFn: *MADS AFFECTING FLOWERINGn*
PIE1: *PHOTOPERIOD-INDEPENDENT EARLY FLOWERING1*
SOC1: *SUPPRESSOR OF OVEREXPRESSION OF CONSTANS1*
VIN3: *VERNALIZATION INSENSITIVE3*
VIPn: *VERNALIZATION INDEPENDENCEn*
VRNn: *VERNALIZATIONn*

1 Introduction

Flowering at the wrong time may seriously hamper reproductive success. Therefore, flowering plants (angiosperms) have evolved multiple genetic pathways to regulate the timing of the transition from vegetative development to flowering in response to environmental stimuli and developmental cues. Since plants live under very different environmental conditions and follow diverse life strategies, the mechanisms controlling the floral transition vary a lot, often even within single species.

Flower development can be subdivided into several major steps, such as floral induction, floral meristem formation, and floral organ development. Precise genetic control of the different steps of flower development is achieved by a hierarchy of interacting regulatory genes, many of which encode transcription factors (for reviews, see Simpson et al. 1999; Theißen 2001a; Mouradov et al. 2002). Close to the top of that hierarchy are "flowering time genes", which are triggered by developmental cues and environmental factors such as plant age, day length and temperature. Flowering time genes control the switch from vegetative to reproductive development by activating "meristem identity genes". Meristem identity genes mediate the transition from vegetative to inflorescence and floral meristems and work as upstream regulators of "floral organ identity genes". Combinatorial interactions of these genes specify the identity of the different floral organs by activating organ-specific "realizator genes" (Theißen 2001b). Most of the genes controlling flower development belong to highly conserved gene families, most prominently MADS-box genes encoding MADS-domain transcription factors (Theißen 2001b; Becker and Theißen 2003; Kaufmann et al. 2005).

The analysis of natural variants ("ecotypes") and of mutants that flower later or earlier than wild type has revealed more than 80 gene loci that affect flowering time in the model plant *Arabidopsis thaliana* (henceforth termed *Arabidopsis*). These flowering time genes may contribute to two different components of the floral transition: the production of flowering signals and the competence of the shoot apical meristem to respond to these signals. The flowering time mutants can be grouped into different classes defining different pathways of floral induction. *Arabidopsis* is a facultative long-day plant that responds to long days (indicating spring and summer) by flowering earlier than when grown in short days. One class of mutants displays a reduced response to changes in photoperiod (day length) when compared with wild type. The corresponding genes may therefore participate in a "photoperiod promotion pathway". A second class of late-flowering mutants are unaffected by changes in photoperiod. The corresponding genes thus may be involved in an "autonomous promotion pathway". This pathway monitors the signals of an internal developmental clock that measures plant age. A third pathway that mediates floral induction, the "gibberellic acid promotion pathway", depends on the plant hormone gibberellic acid. Here we are concerned with a fourth pathway, which confers susceptibility to an extensive exposure to cold.

Many varieties of *Arabidopsis* and a wide range of other plants require prolonged exposure to low temperatures to flower. In northern latitudes, this requirement, often combined with responsiveness to long-day photoperiods, makes sure that the winter has passed and thus spring has actually arrived before these plants invest their resources into flower formation (Henderson et al. 2003; Amasino 2004). The Latin word for spring is *vernum*, and hence the whole process has been termed vernalization. Vernalization requiring plants are often winter annual or biennial plants, but also many perennials require a promotion of flowering by cold.

Besides being an ecologically important trait, the requirement of vernalization is also of great agronomical importance. In crop plants such as common wheat (*Triticum aestivum*), a vernalization requirement distinguishes winter varieties from spring varieties. A vernalization requirement may prevent some otherwise beneficial varieties from being cultivated in areas without a pronounced winter season, while winter varieties may exploit the growing season more comprehensively (a potential advantage in cold climates).

Obviously, varieties that require vernalization and others that do not are often found within the same species; even single gene changes can convert plants without a vernalization requirement into plants that require vernalization, or vice versa, suggesting that the genetic basis of vernalization is not very complex. However, vernalization has some remarkable features, which raises intriguing questions about its molecular mechanism. For example, vernalization establishes a cellular memory that is stable through mitotic cell divisions, but which is reset after meiosis, otherwise biennials would only be biennial for one generation (Amasino 2004).

In recent years, researchers have begun to characterize the genes involved, providing breakthroughs in our understanding of the molecular mechanisms of vernalization in both the weed *Arabidopsis* and wheat, a plant of prime agronomic importance. This demonstrated also both the power and limitations of *Arabidopsis* as a flowering plant model system. So let us see what has recently been learned about vernalization in weed and wheat.

2 Vernalization in *Arabidopsis*

2.1 The major genes

Natural accessions of *Arabidopsis* have different requirements for vernalization. The majority of ecotypes, especially those from higher latitudes, are extremely late flowering if not exposed to longer periods of cold. The requirement for vernalization in these ecotypes, combined with a long-day promotion of flowering, ensures that flowering occurs in spring to provide the optimal conditions for seed set before the next winter season (Henderson et al. 2003). In contrast, ecotypes found at low latitudes, such as the Cape Verde Island ecotype, and the laboratory 'working horses' Landsberg *erecta* (L*er*) and Columbia (Col), are early flowering even without cold exposure.

Genetic analyses that included pioneering work by Napp-Zinn (1955, 1957, 1987) more than half a century ago demonstrated that the differences in flowering time and vernalization requirement between the different ecotypes of *Arabidopsis* are mostly due to just two genetic loci, termed *FLOWERING LOCUS C* (*FLC*) and *FRIGIDA* (*FRI*); both loci act synergistically to repress flowering (Clarke and Dean 1994; Koornneef et al. 1994, 1998; Henderson et al. 2003).

Molecular cloning of both *FRI* and *FLC* provided the starting point for detailed studies on the molecular mechanism of vernalization, which is currently fueling an almost explosive increase in our knowledge about this process. *FRI* encodes a novel protein with two potential coiled-coil domains (Johanson et al. 2000). The role of *FRI* is to elevate the expression of *FLC* (Fig. 1), but by which exact mechanism has remained unknown so far (Amasino 2004). The FRI protein has a nuclear localization and is strongly expressed in meristematic regions.

2.2 The central role of *FLC*

Molecular cloning of *FLC* (also known as *FLF*) revealed that it encodes a MADS-domain protein that acts as a potent repressor of flowering (Michaels and Amasino 1999; Sheldon et al. 1999, 2000). Expression of *FLC*, e.g. in transgenic plants, is sufficient to block flowering, and the role of *FRI* is to elevate the expression of *FLC* to levels that block flowering. *FLC* blocks flowering by inhibiting the expression of *SOC1* (also a MADS-box gene; formerly known as *AGL20*) and *FT*, which are both promoters of flowering that upregulate the floral meristem identity genes *LEAFY* (*LFY*) and *APETALA1* (*AP1*), respectively (Fig. 1). At least in the case of *SOC1*, repression by *FLC* appears to be direct, because the FLC protein binds, possibly together with other proteins, to a *cis*-regulatory DNA sequence element termed "CArG-box" in the promoter of the *SOC1* gene (Hepworth et al. 2002). In this way, FLC may prevent CONSTANS (CO) from binding to a nearby promoter element. *CO* is a key gene in the *Arabidopsis* photoperiod promotion pathway (Mouradov et al. 2002).

Vernalization promotes flowering by repressing *FLC* expression (Fig. 1). The crucial role of *FLC* in the vernalization process is indicated by the observation that there is a quantitative relationship between the duration of cold treatment and the extent of down-regulation of *FLC* mRNA and protein (Sheldon et al. 2000). The vernalization-induced reduction in FLC

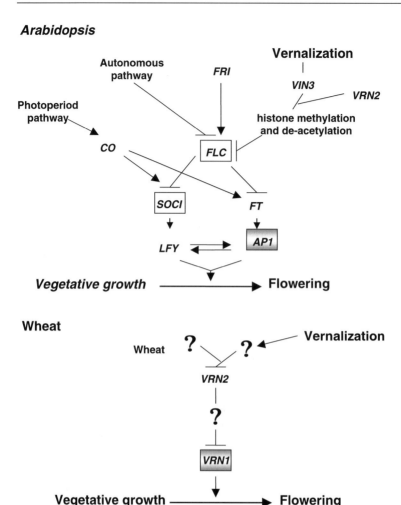

Fig. 1. The role of vernalization in flowering pathways in weed (*Arabidopsis*; upper part) and wheat (*Triticum*; lower part). Lines with arrows denote regulatory interactions resulting in up-regulation of gene expression; lines with bars denote repression of gene expression. MADS-box genes are boxed; boxes of the putatively orthologous *SQUA*-like MADS-box genes are shaded. For *Arabidopsis*, only some of the many recently identified genes regulating *FLC* expression are shown. For details, see text.

expression is mitotically stable and occurs in all plant tissues; *FLC* expression is restored in each generation, however, as is the characteristic resetting of the requirement for low-temperature induction (Sheldon et al. 2000). *FLC* thus shows all the characteristic features expected from a central player in the vernalization process. Other vernalization-responsive late-flowering mutants, which are disrupted in genes that encode regulators of *FLC*, are late-flowering due to elevated levels of *FLC* expression (Sheldon et al. 2000). This raises the question as to how expression of the *FLC* gene is regulated, especially with respect to exposure to low temperatures.

2.3 Complex regulation of *FLC* expression

Mutant analyses demonstrated that *FLC* is positively or negatively regulated by a surprising number of genes, including *FCA*, *FY*, *FPA*, *FVE* (Rouse et al. 2002), *VIP3* (Zhang et al. 2003), *VIP4* (Zhang and van Nocker 2002), *VIP5*, *VIP6* (Oh et al. 2004), *PIE1* (Noh and Amasino 2003), *ELF5* (Noh et al. 2004), *ELF7*, *ELF8* (He et al. 2004), *FLD* (He et al. 2003), *FLK* (Lim et al. 2004), *VER-NALIZATION1* (*VRN1*) (Levy et al. 2002), *VRN2* (Gendall et al. 2001), *VERNALIZATION INSENSITIVE3* (*VIN3*) (Sung and Amasino 2004) and *HUA2* (Doyle et al. 2005). The still rapidly growing list of known regulators of *FLC* not only includes genes involved in vernalization, but also genes of the autonomous promotion pathway. Beyond its central role in vernalization, *FLC* is thus a more general convergence point for flowering signals (Fig. 1).

Intriguingly, many of the genes that control *FLC* expression appear to either alter chromatin structure or to be involved in RNA processing (Henderson and Dean 2004). For example, VRN2 is related to the polycomb-group protein SUPPRESSOR OF ZESTE-12, VIN3 contains a protein domain termed PHD, and VIP5, VIP6, ELF7 and ELF8 are homologous to components of the PAF1 complex of baker's yeast. In fungi and animals relatives of these proteins are involved in chromatin-remodelling complexes, which often catalyze the modification of specific histone residues. These observations suggest that vernalization functions via changes in histone modification and its effects on gene expression, i.e. a "histone code" (Bastow et al. 2004). Work along these lines is still in its infancy. However, examination of *FLC* chromatin has already revealed vernalization-mediated changes, such as reduction in acetylation or increase in methylation of specific lysine residues in histone 3 (Bastow et al. 2004; He et al. 2004; Sung and Amasino 2004). Elevated methylation of the respective lysine residues in histone 3 is associated with the formation of stable heterochromatin in human and fruit fly cells, suggesting that vernalization-mediated formation of heterochromatin at the *FLC* locus might account for the vernalized state, at least in part (Amasino 2004).

However, vernalization might not be the only process that can lead to heterochromatin formation at the *FLC* locus. Some alleles of *FLC* present in certain rapid flowering varieties of *Arabidopsis* such as L*er* are "resistant" to upregulation by *FRI*. An example is the allele present in L*er*, which has a transposable element inserted in one of its introns (Michaels et al. 2003). The presence of this element renders *FLC* susceptible to silencing RNAs (siRNAs) generated by homologous transposons elsewhere in the genome (Liu et al. 2004). Targeting of the transposon by the siRNAs creates an island of a chromatin modification (methylation of a specific lysine residue of histone 3) typical for vernalization-induced heterochromatin at *FLC* (Liu et al. 2004). In this way, transcription of *FLC* is shut down and flowering is promoted as during vernalization, except that the long cold kiss of winter is substituted by a hot cross-talk between transposable elements.

2.4 Role of other *FLC*-like genes

Null mutants of *FLC* are still responsive to vernalization, indicating that cold can promote flowering by mechanisms other than the repression of *FLC* (Michaels and Amasino 2001). There are five closely related paralogues of *FLC* (*FLC*-like genes) in the *Arabidopsis* genome, termed *MADS AFFECTING FLOWERING1* (*MAF1*)–*MAF5* (some of them, however, have also been published under other names) (Becker and Theißen 2003; Ratcliffe et al. 2003); these genes are good candidates for providing functions similar to those of *FLC*. The available experimental evidence suggests that *MAF1–MAF5* are indeed involved in the floral transition, where all except one (*MAF5*) may act as floral repressors (Ratcliffe et al. 2003). Both *MAF1* and *MAF5* appear to contribute to the vernalization response, at least in some genetic backgrounds, explaining how vernalization can promote flowering independent of the repression of *FLC* (Ratcliffe et al. 2003).

An interesting twist in the tale of *Arabidopsis* vernalization became apparent through detailed analysis of the *MAF2* gene. *maf2* mutant plants show a pronounced vernalization response when subjected to relatively short periods of cold. These brief cold periods are insufficient to elicit a strong flowering response in wild type plants even though they result in a strong reduction in *FLC* transcript and protein accumulation. *MAF2* expression is less sensitive to vernalization than that of *FLC*, and the gene exerts its activity as a floral repressor independently or downstream of *FLC* transcription (Ratcliffe et al. 2003). *MAF2* prevents vernalization by short periods of cold, thus possibly compensating for a decrease in *FLC* levels that occurs already after short cold exposures. *MAF2* might hence be part of a mechanism that exists to ensure that vernalization does not occur in response to periods of cold that last only for a few days (Ratcliffe et al. 2003). Having such a mechanism operating might be advantageous for the plant, because a few cold days do not reliably indicate the passage of winter; if these occur in the autumn and are followed by a number of warm days, as is often the case in many habitats, without the *MAF2*-based mechanism flower development may set in before the 'real winter' has even arrived and reproductive success will almost certainly be undermined.

3 Vernalization in winter varieties of wheat

Due to the small size of its completely sequenced genome (Arabidopsis Genome Initiative 2000), short life cycle, low requirements for growth space, and easy transformability, *Arabidopsis* is generally considered THE "model plant". The detailed insights into the vernalization process in *Arabidopsis* sketched above thus raise the question as to which extent the same mechanisms apply to other flowering plants, such as winter varieties of cereals. The fact that neither *FRI*-like nor *FLC*-like genes have been found so far in the draft sequence of the complete rice genome (Feng et al. 2002; Goff et al. 2002; Sasaki et al. 2002; Yu et al. 2002) may represent an "autapomorphy" of a subtropical grass that does not require adaptation to cold winters and hence has lost the corresponding genes; alternatively, the system present in *Arabidopsis* might be a relatively recent evolutionary achievement, and vernalization might work quite different in grasses. So let's move 'from weed to wheat' to see how cereals "feel the chill before the bloom" (Henderson and Dean 2004).

Common wheat, or bread wheat (*Triticum aestivum*), belongs to the most important crop plants for human consumption on a global scale. Since varieties with different growth habits are available, wheat can be grown in very different environments. Winter wheats are sown in the fall and require vernalization in order to flower, thus avoiding damage of the temperature-sensitive floral meristems and organs by the cold. In contrast, spring wheats can be sown in the spring, since they do not require vernalization, and hence can be cultivated in climates without cold winters.

In diploid wheat (*Triticum monococcum*), two major genes are involved in the vernalization response, termed *VRN1* and *VRN2* (note that they are unrelated to the genes with similar names in *Arabidopsis*) (Yan et al. 2003, 2004). Most of the variation in vernalization requirement in the agronomically important polyploidy wheats, such as common wheat, is controlled by the *VRN1* locus, which divides wheats into winter and spring varieties (Yan et al. 2003). In both diploid wheat and barley (*Hordeum vulgare*), *VRN1* is dominant for spring growth habit, whereas *VRN2* is dominant for winter growth habit. Similar epistatic interactions and chromosomal locations indicate that wheat and barley vernalization genes are orthologous (Yan et al. 2003, 2004, and references cited therein).

Both *VRN1* and *VRN2* have recently been molecularly cloned by positional approaches in *Triticum monococcum*, which is a remarkable achievement in a huge genome rich in repetitive DNA such as that of wheat (Yan et al. 2003, 2004). *VRN1* represents a MADS-box gene with high similarity to the floral meristem identity genes *AP1* from *Arabidopsis* and *SQUAMOSA* (*SQUA*) from *Antirrhinum*; it is hence a member of the MADS-box gene subfamily termed *SQUA*-like genes (for a recent classification of MADS-box genes, see Becker and Theißen 2003). In *Arabidopsis*, expression of *AP1* is sufficient to trigger the transition to flowering (Mandel and Yanofsky 1995), suggesting that *VRN1* is an activator of flowering and that its upregulation is part of the vernalization response of wheat. In line with this idea, prolonged cold exposure upregulates *VRN1* expression in winter wheat, but not in spring varieties (Yan et al. 2003).

   VRN2 is a repressor of *VRN1* whose expression is repressed by vernalization. *VRN2* encodes a putative zinc finger transcription factor; a short region of the protein probably involved in nuclear localization has similarity to the flowering time gene *CONSTANS* (*CO*) and other *CO*-like genes from *Arabidopsis*. *CO* is a key gene in the *Arabidopsis* photoperiod promotion pathway (Mouradov et al. 2002). Expression of *VRN2* is downregulated during vernalization, in line with the gene being a repressor of flowering. Accordingly, loss-of-function mutations in *VRN2* inhibiting the activity of the gene in winter wheat produce spring varieties (Yan et al. 2004). Downregulation of the *VRN2* gene is concomitant with an increase in *VRN1* expression, consistent with the epistatic interaction between the two genes. *VRN1* alleles from spring varieties have a deletion of 20 nucleotides in the putative promoter region of the gene and hence are not repressed by *VRN2*. Whether the respective DNA sequence constitutes a binding site for a transcriptional repressor and whether that repressor is VRN2 is not known so far—repression of *VRN1* by VRN2 could thus also be indirect (Fig. 1).

4 Evolution of the vernalization requirement

The increasing insights into the phylogeny of flowering plants and the molecular mechanisms of vernalization also significantly further the understanding of the evolution of the vernalization requirement. The lineages that led to extant *Arabidopsis* (a higher eudicotyledonous angiosperm) and wheat (a monocotyledonous angiosperm) separated relatively early during flowering plant evolution, according to recent molecular data about 140–160 million years ago (Chaw et al. 2004). Major groups of angiosperms may have evolved in warm climates in which a vernalization response was not required, and the ability to respond to vernalization may have originated independently as different plant groups colonized habitats with a winter season (Amasino 2004). This makes it appear likely that the absence of a vernalization response is the ancestral state in the angiosperms, and that vernalization requirements originated independently in the lineages that led to *Arabidopsis* and wheat. The clear differences in the molecular components of vernalization between *Arabidopsis* and wheat strongly corroborate that view (Fig. 1). For example, the central floral repressors in both systems have been recruited from different gene families, MADS-box genes (*FLC* in *Arabidopsis*) and *CO*-like genes *sensu lato* (*VRN2* in wheat). The *VRN1* gene from wheat is a distantly related homologue of *FLC* from *Arabidopsis* within the family of MIKC-type MADS-box genes; however, both genes are clearly not orthologues, but members of different subfamilies of MIKC-type genes, termed *SQUA*- and *FLC*-like genes, respectively (Becker and Theißen 2003). Neither have *FLC*-like genes ever been found so far in any monocot genome, including the completely sequenced rice genome, nor has a *SQUA*-like gene ever been found being involved in vernalization outside of the grasses. For example, there are three functional *SQUA*-like genes in the *Arabidopsis* genome, *AP1*, *CAULIFLOWER* (*CAL*), and *FRUITFULL* (*FUL*, also known as *AGL8*) (Becker and Theißen 2003). Phylogeny reconstructions suggest that the ancestral function of *SQUA*-like genes was in specifying inflorescence or floral meristem identity, which is maintained in many extant genes, and that additional functions in specifying organ identity of sepals and petals (*AP1*), or fruit valve identity (*FUL*) are probably derived (Theißen et al. 2000; Becker and Theißen 2003). Importantly, none of these genes is involved in vernalization. So the involvement of the wheat gene *VRN1* in vernalization is probably a recent addition to the growing list of *SQUA*-like genes that underwent neofunctionalization events during evolution. The gene may represent an ancestral reproductive meristem identity gene that might have been recruited for vernalization when it got under the control of a floral repressor (*VRN2*) responding to cold treatment.

On the other hand, comparison of vernalization in *Arabidopsis* and wheat shows also striking similarities (Fig. 1). For example, vernalization in both systems works via the repression of floral repressors probably encoding transcription factors (FLC and VRN2, respectively), even though these are encoded by members of different gene families. In both *Arabidopsis* and wheat, MADS-box genes have obtained crucial functions during the vernalization-response (Fig. 1). Even though these are from different subfamilies and the similarities in both systems are quite probably the result of convergent evolution, this observation adds to the remarkable versatility of MIKC-type MADS-box genes in controlling diverse steps of flower development, from very early to very late ones (Theißen et al. 2000; Theißen 2001a). The predominance of MIKC-type genes in the control of flower development might be based on the special domain structure of MIKC-type proteins that facilitates the formation of multimeric protein complexes. These might be especially well suited for the accurate recognition of target genes as required during the evolution of increasingly complicated developmental processes (such as flower development) generating increasingly complex and diverse structures (such as inflorescences and flowers) (Kaufmann et al. 2005). When and at what taxonomic level the *Arabidopsis*-type vernalization system originated is not clear.

All extant grasses (Poaceae) probably evolved from subtropical ancestors that, like rice, had no vernalization requirement (Yan et al. 2004). Functional conservation of the *VRN2* gene of wheat was corroborated by comparison with other systems that are, however, evolutionary quite closely related. Downregulation of the *VRN2* gene by RNA interference (RNAi) in winter hexaploid wheat (variety Jagger) resulted in plants with an upregulation of a *VRN1* gene orthologue and earlier flowering (Yan et al. 2004). Moreover, screening of a collection of barley varieties from different parts of the world showed presence of the *VRN2* gene in all winter varieties and complete deletion of the gene as well as a similar gene in close vicinity in all but one spring variety (Yan et al. 2004). Thus, the wheat-type vernalization system may have originated in the lineage that led to extant Pooideae (including wheat and barley) after the lineage that led to extant Ehrhartoideae (including rice) had branched off.

In addition to these insights into long-term processes the recent molecular data also revealed microevolutionary (intraspecific) events in both *Arabidopsis* and wheat. Allelic variation at the *FRI* locus is an important determinant of flowering time in wild *Arabidopsis* populations (reviewed by Henderson et al. 2003). Many non-vernalization responsive, early flowering varieties carry recessive null alleles of *FRI*. Molecular analyses of *FRI* alleles in wild accessions indicated that the early flowering lifestyle originated several times independently from winter-annual, vernalization-responsive ancestors in *Arabidopsis*, possibly, e.g. to enable more than one generation of plants per year (Henderson et al. 2003). An alternative route to the evolution of summer-annual flowering behavior in *Arabidopsis* was provided by the origin of weak (but not null) alleles of *FLC* (Michaels et al. 2003). Molecular analyses suggest that weak *FLC* alleles have arisen independently at least twice during the course of the evolution of *Arabidopsis* (Michaels et al. 2003), e.g. by the insertion of a transposable element into an intron (see above). Caicedo et al. (2004) have recently shown that a latitudinal cline in flowering time under over-winter conditions in the field is generated by epistatic interactions between *FRI* and *FLC*.

Spring varieties evolved in parallel in wheat, barley and rye (*Secale cereale*), possibly by loss of VRN2-binding sites in the promoter of *VRN1*

(Yan et al. 2003), and also by loss-of-function mutations at the *VRN2* locus (Yan et al. 2004). Thus the wide adaptability of temperate cereals was facilitated at least to considerable extents by mutational events at just two genetic loci.

5 Future prospects

Vernalization appears to be a suitable system to study as to how the evolution of genes, gene regulatory mechanisms and phenotypic traits brings about plant adaptation at different time scales. Nevertheless, many basic questions about the molecular mechanisms of vernalization still remain to be answered even within the prime model system *Arabidopsis*, such as: what is the primary cold sensor, and how is the duration of cold measured at the molecular level? In a strict sense, all components that have been identified so far in both *Arabidopsis* and wheat are just targets of the vernalization pathway (Amasino 2004).

However, studies on vernalization are not only of interest for basic research; also its agronomic implications are profound. Quite a number of major field crops, including onion (*Allium cepa*), beet (*Beta vulgaris*), cauliflower (*Brassica oleracea*), carrot (*Daucus carota*), pea (*Pisum sativum*) and spinach (*Spinacia oleracea*) require vernalization. In addition to wheat, barley and rye, the list of vernalization requiring crops includes even more major cereal grasses, e.g. oat (*Avena sativa*) (Henderson et al. 2003). The increasing knowledge about the control of flowering by environmental cues, both in eudicots and cereal grasses, might provide the tools to match crops in a better way to their environment and hence to increase productivity, e.g. by transgenic technology or marker assisted breeding (Theißen 2002).

Even though there appear to be no *FLC*-like genes in the rice genome, ectopic expression of the *Arabidopsis* gene *FLC* delays flowering in rice, and the up-regulation of the *SOC1* orthologue of rice, *OsSOC1*, at the onset of flowering initiation is delayed in the transgenic lines expressing the *Arabidopsis FLC* gene. This suggests that some components of the flowering pathways are shared between rice and *Arabidopsis*, and that it might be possible to manipulate flowering time in cereals employing *Arabidopsis* genes (Tadege et al. 2003). However, the increasing knowledge about the molecular details of vernalization in cereals itself will also facilitate approaches employing homologous rather than heterologous genetic tools.

6 Concluding remarks

There are two distantly related angiosperm model systems with completely sequenced genomes and powerful tools for functional genomics available now, *Arabidopsis* and rice (Theißen and Becker 2004). However, for investigations on vernalization in wheat, they were only of limited help, because *Arabidopsis* has a system that originated independently of and is hence quite different from that of wheat, and rice has no vernalization-requirement at all. The investigators of wheat vernalization thus had to go through the laborious and time consuming efforts of positional approaches in a large genome (rather than a candidate gene approach) to molecularly clone the major genes involved (Yan et al. 2003, 2004). This is not to say that the rice genome sequence, which is largely collinear with those of all other grasses and hence is considered the reference genome for the world's most important crop plants (Shimamoto and Kyozuka 2002), was not helpful. On the contrary, genomic resources of rice facilitated considerably cloning of the *VRN1* gene (Yan et al. 2003). However, when it comes to the specific mechanistic details of a physiological process, vernalization shows not only the power, but also the severe limitations of the model system attitude in current research.

No one knows how many different systems of vernalization work in our crops, or in plants in general. Knowledge about *Arabidopsis*, rice and wheat may provide inspiration concerning general mechanisms involved (such as chromatin remodelling), but will not necessarily provide us with good candidate genes in other systems, especially outside of the grasses (Poaceae) and Brassicaceae (the plant families where wheat and *Arabidopsis*, respectively, belong to). To understand even better the evolution and agronomic potential of vernalization there thus will hardly be an alternative to detailed, comparative studies in a number of well-chosen systems. Let us prepare for the fact that a number of them will be no 'model plants' in a strict sense.

References

Amasino R (2004) Vernalization, competence, and the epigenetic memory of winter. Plant Cell 16:2553–2559

Arabidopsis Genome Initiative (2000) Analysis of the genome sequence of the flowering plant *Arabidopsis thaliana*. Nature 408:796–815

Bastow R, Mylne JS, Lister C, Lippman Z, Martienssen RA, Dean C (2004) Vernalization requires epigenetic silencing of *FLC* by histone methylation. Nature 427:164–167

Becker A, Theißen G (2003) The major clades of MADS-box genes and their role in the development and evolution of flowering plants. Mol Phyl Evol 29:464–489

Caicedo AL, Stinchcombe JR, Olsen KM, Schmitt J, Purugganan MD (2004) Epistatic interaction between *Arabidopsis FRI* and *FLC* flowering time genes generates a latitudinal cline
in a life history trait. Proc Natl Acad Sci USA 101:15670–15675

Chaw S-M, Chang C-C, Chen H-L, Li W-H (2004) Dating the monocot-dicot divergence and
the origin of core eudicots using whole chloroplast genomes. J Mol Evol 58:424–441

Clarke JH, Dean C (1994) Mapping *FRI*, a locus controlling flowering time and vernalization
response. Mol Gen Genet 242:81–89

Doyle MR, Bizzell CM, Keller MR, Michaels SD, Song J, Noh Y-S, Amasino RM (2005) *HUA2* is
required for the expression of floral repressors in *Arabidopsis thaliana*. Plant J 41:376–385

Feng Q, Zhang Y, Hao P et al. (2002) Sequence and analysis of rice chromosome 4. Nature
420:316–320

Gendall AR, Levy YY, Wilson A, Dean C (2001) The *VERNALIZATION2* gene mediates the
epigenetic regulation of vernalization in *Arabidopsis*. Cell 107:525–535

Goff SA, Ricke D, Lan TH et al. (2002). A draft sequence of the rice genome (*Oryza sativa* L.
ssp. *japonica*). Science 296:92–100

He Y, Michaels SD, Amasino RM (2003) Regulation of flowering time by histone acetylation
in *Arabidopsis*. Science 302:1751–1754

He Y, Doyle MR, Amasino RM (2004) PAF1-complex-mediated histone methylation of
FLOWERING LOCUS C chromatin is required for the vernalization-responsive, winter-
annual habit in *Arabidopsis*. Genes Dev 18:2774–2784

Henderson IR, Dean C (2004) Control of *Arabidopsis* flowering: the chill before the bloom.
Development 131:3829–3838

Henderson IR, Shindo C, Dean C (2003) The need for winter in the switch to flowering. Annu
Rev Genet 37:371–392

Hepworth SR, Valverde F, Ravenscroft D, Mouradov A, Coupland G (2002) Antagonistic regulation of flowering-time gene *SOC1* by CONSTANS and FLC via separate promoter
motifs. EMBO J 21:4327–4337

Johanson U, West J, Lister C, Michaels S, Amasino R, Dean C (2000) Molecular analysis of
FRIGIDA, a major determinant of natural variation in *Arabidopsis* flowering time. Science
290:344–347

Kaufmann K, Melzer R, Theißen G (2005) MIKC-type MADS-domain proteins: structural
modularity, protein interactions and network evolution in land plants. Gene 347:183–198

Koornneef M, Blankestijn-De Vries H, Hanhart C, Soppe W, Peeters T (1994) The phenotype
of some late-flowering mutants is enhanced by a locus on chromosome 5 that is not effective in the Landsberg *erecta* wild-type. Plant J 6:911–919

Koornneef M, Alonso-Blanco C, Blankestijn-De Vries H, Hanhart CJ, Peeters AJM (1998)
Genetic interactions among late flowering mutants of *Arabidopsis*. Genetics
148:885–892

Levy YY, Mesnage S, Mylne JS, Gendall AR, Dean C (2002) Multiple roles of *Arabidopsis VRN1*
in vernalization and flowering time control. Science 297:243–246

Lim MH, Kim J, Kim YS, Chung KS, Seo YH, Lee I, Kim J, Hong CB, Kim HJ, Park CM (2004)
A new *Arabidopsis* gene, *FLK*, encodes an RNA binding protein with K homology motifs
and regulates flowering time via *FLOWERING LOCUS C*. Plant Cell 16:731–740

Liu J, He Y, Amasino R, Chen X (2004) siRNAs targeting an intronic transposon in the regulation of natural flowering behaviour in *Arabidopsis*. Genes Dev 18:2873–2878

Mandel MA, Yanofsky MF (1995) A gene triggering flower formation in *Arabidopsis*. Nature
377:522–524

Michaels SD, Amasino RM (1999) *FLOWERING LOCUS C* encodes a novel MADS domain
protein that acts as a repressor of flowering. Plant Cell 11:949–956

Michaels SD, Amasino RM (2001) Loss of *FLOWERING LOCUS C* activity eliminates the
late-flowering phenotype of *FRIGIDA* and autonomous pathway mutations but not
responsiveness to vernalization. Plant Cell 13:935–941

Michaels SD, He Y, Scortecci KC, Amasino RM (2003) Attenuation of FLOWERING LOCUS C activity as a mechanism for the evolution of summer-annual flowering behavior in *Arabidopsis*. Proc Natl Acad Sci USA 100:10102–10107

Mouradov A, Cremer F, Coupland G (2002) Control of flowering time: interacting pathways as a basis for diversity. Plant Cell 14:S111–130

Napp-Zinn K (1955) Genetische Grundlagen des Kältebedürfnisses bei *Arabidopsis thaliana* (L.) Heynh. Naturwissenschaften 42:650

Napp-Zinn K (1957) Untersuchungen über das Vernalisationsverhalten einer winterannuellen Rasse von *Arabidopsis thaliana*. Planta 50:177

Napp-Zinn K (1987) Vernalization: environmental and genetic regulation. In: Atherton JG (ed) Manipulation of flowering. Butterworths, London pp 123–132

Noh Y-S, Amasino RM (2003) *PIE1*, and ISWI family gene, is required for *FLC* activation and floral repression in *Arabidopsis*. Plant Cell 15:1671–1682

Noh YS, Bizzell CM, Noh B, Schomburg FM, Amasino RM (2004) EARLY FLOWERING 5 acts as a floral repressor in *Arabidopsis*. Plant J 38:664–672

Oh S, Zhang H, Ludwig P, van Nocker S (2004) A mechanism related to the yeast transcriptional regulator Paf1c is required for expression of the *Arabidopsis FLC/MAF* MADS box gene family. Plant Cell 16:2940–2953

Ratcliffe OJ, Kumimoto RW, Wong BJ, Riechmann JL (2003) Analysis of the Arabidopsis *MADS AFFECTING FLOWERING* gene family: *MAF2* prevents vernalization by short periods of cold. Plant Cell 15:1159–1169

Rouse DT, Sheldon CC, Bagnall DJ, Peacock WJ, Dennis ES (2002) FLC, a repressor of flowering, is regulated by genes in different inductive pathways. Plant J 29:183–191

Sasaki T, Matsumoto T, Yamamoto K et al. (2002) The genome sequence and structure of rice chromosome 1. Nature 420:312–316

Sheldon CC, Burn JE, Perez PP, Metzger J, Edwards JA, Peacock WJ, Dennis ES (1999) The *FLF* MADS box gene: a repressor of flowering in *Arabidopsis* regulated by vernalization and methylation. Plant Cell 11:445–458

Sheldon CC, Rouse DT, Finnegan EJ, Peacock WJ, Dennis ES (2000) The molecular basis of vernalization: the central role of *FLOWERING LOCUS C (FLC)*. Proc Natl Acad Sci USA 97:3753–3758

Shimamoto K, Kyozuka J (2002) Rice as a model for comparative genomics of plants. Annu Rev Plant Biol 53:399–419

Simpson GG, Gendall AR, Dean C (1999) When to switch to flowering. Annu Rev Cell Dev Biol 99:519–550

Sung SB, Amasino RM (2004) Vernalization in *Arabidopsis thaliana* is mediated by the PHD finger protein VIN3. Nature 427:159–164

Tadege M, Sheldon CC, Helliwell CA, Upadhyaya NM, Dennis ES, Peacock WJ (2003) Reciprocal control of flowering time by OsSOC1 in transgenic *Arabidopsis* and by FLC in transgenic rice. Plant Biotech J 1:361–369

Theißen G (2001a) Flower development, genetics of. In: Brenner S, Miller JH (eds) Encyclopedia of genetics. Academic Press, London pp 713–717

Theißen G (2001b) Development of floral organ identity: stories from the MADS house. Curr Opin Plant Biol 4:75–85

Theißen G (2002) Key genes of crop domestication and breeding: molecular analyses. Prog Bot 63:189–203

Theißen G, Becker A (2004) The ABCs of flower development in *Arabidopsis* and rice. Prog Bot 65:193–215

Theißen G, Becker A, Di Rosa A, Kanno A, Kim JT, Münster T, Winter K-U, Saedler H (2000) A short history of MADS-box genes in plants. Plant Mol Biol 42:115–149

Yan L, Loukoianov A, Blechl A, Tranquilli G, Ramakrishna W, SanMiguel P, Bennetzen JL, Echenique V, Dubcovsky J (2004) The wheat *VRN2* gene is a flowering repressor downregulated by vernalization. Science 303:1640–1644

Yan L, Loukoianov A, Tranquilli G, Helguera M, Fahima T, Dubcovsky J (2003) Positional cloning of the wheat vernalization gene *VRN1*. Proc Natl Acad Sci USA 100:6263–6268

Yu J, Hu S, Wang J et al. (2002) A draft sequence of the rice genome (*Oryza sativa* L. ssp. *indica*). Science 296:79–91

Zhang H, van Nocker S (2002) The *VERNALIZATION INDEPENDENCE 4* gene encodes a novel regulator of *FLOWERING LOCUS C*. Plant J 31:663–673

Zhang H, Ransom C, Ludwig P, van Nocker S (2003) Genetic analysis of early flowering mutants in *Arabidopsis* defines a class of pleiotropic developmental regulators required for expression of the flowering-time switch *FLOWERING LOCUS C*. Genetics 164:347–358

Prof. Dr. Günter Theißen
Friedrich-Schiller-Universität Jena
Lehrstuhl für Genetik
Philosophenweg 12
07743 Jena, Germany
Tel.: +49-3641-949550
Fax: +49-3641-949552
e-mail: guenter.theissen@uni-jena.de
URL: http://www2.uni-jena.de/biologie/genetik/index.htm

Biotechnology:
Engineered male sterility in plant hybrid breeding

Kerstin Stockmeyer and Frank Kempken[1]

1 Introduction

The exploitation of hybrid crop varieties in agriculture has enabled enormous increases in food productivity because of increased uniformity and hybrid vigour. Because of the hybrid vigour or heterosis, hybrids are characterized by increased resistance to diseases and enhanced performance in different environments compared with the heterozygous hybrid progeny (called F1 hybrids) over the homozygous parents (Lefort-Buson et al. 1987). Heterotic hybrid varieties in major crops such as wheat, cotton and rice show more than 20% yield advantage over the conventional ones under the same cultivation conditions. The increased vigour, uniformity and yield of F1 hybrids have been exploited in most crops where the pollination system allows for economical and convenient cross hybridization (Basra 2000).

In hybrid seed production, one line is designated as the female parent and the other as the male parent. The production of hybrid seeds requires a pollination control system in order to prevent unwanted self-pollination of the female line. Especially for those crop species with hermaphrodite flowers, this can be a great challenge. Many methods exist to prevent self-pollination of the female line during hybrid seed production: application of male-specific gametocides, such as mitomycin and streptomycin (Jan and Rutger 1988), some inter- and intraspecific crosses (Hanson and Conde 1985), mechanical removal of male flowers or anthers, or use of genetic cytoplasmic (CMS) or nuclear-encoded male sterility. Naturally occurring genetically male sterile plants generally maintain fully normal female functions. The phenotypic characteristics of male sterility are very diverse from the complete absence of male organs, the abortion of pollen at any step of its development, the failure to develop normal sporogenous tissues, the absence of stamens dehiscence or the inability of mature pollen to germinate on compatible stigma.

The generation of, mainly nuclear-encoded, male sterility is the basis of new reliable and cost-effective pollination control systems for genetic engineering that have been developed during the past decade. Thereby the propagation of the male-sterile female parent lines is an important aspect for the successful application of these systems for large-scale hybrid seed production.

2 Natural male-sterility systems in plants

In order to prevent self-pollination of the female line, its pollen fertility must be controlled to permit fertilization only by pollen from the male parent. A simple way to establish a female line for hybrid seed production is to identify or create a line that is unable to produce viable pollen, like some maize (Laughnan and Gabay-Laughnan 1983) or rice (Kadowaki et al. 1988) lines. Therefore, this male-sterile line is unable to self-pollinate and seed formation is dependent upon pollen from the male line.

2.1 Cytoplasmic male sterility (CMS)

The mitochondrion serves essential functions as a centre for energy metabolism in the development of eukaryotic organisms. Pollen development in plants appears to be particularly influenced by mitochondrial function. Rearrangements of the mitochondrial DNA leading to unique chimeric genes sometimes result in the inability of the plant to produce fertile pollen (Fig. 1). This process, known as cytoplasmic male sterility, is particularly useful for the production of hybrid varieties with increased crop productivity and has been extensively reviewed previously (Hanson 1991; Schnable and Wise 1998; Kempken and Pring 1999). The association of CMS with abnormal mitochondrial gene expression has been established in many plant species including maize (Levings 1990), petunia (Bino 1985) and sorghum (Pring et al. 1995; Xu et al. 1995a). It is thought that the disruption in pollen development is a consequence of mitochondrial dysfunction associated with the chimeric genes. Incorporation of the derived proteins into the mitochondrial membrane or into multiprotein enzyme complexes may lead to the impairment of mitochondrial function. A unique feature of CMS is that expression of the trait is influenced by nuclear fertility restorer (RF) genes (Schnable and Wise 1998; Kempken and Pring 1999). Nuclear restorer genes can suppress the effect of the sterile cytoplasm and restore fertility to the next generation.

Fig. 1. Fertile and sterile sorghum pollen. Iodine-potassium stain of sorghum pollen from a fertile and sterile line: (**A**) Dark stained fertile pollen indicating starch production. (**B**) Unstained pollen from the sterile line.

Cytoplasmic male sterility has been utilized in some important crops, such as sunflower, rice and sorghum, to prevent unwanted pollinations, but CMS mutants and restorer systems are not available for all crops used in agriculture. In some cases CMS has been associated with increased disease susceptibility. As an example, the T-cytoplasm of maize and its susceptibility to race T of the southern corn leaf blight (*Bipolaris maydis*) led to an epidemic in the USA in 1970 (Wise et al. 1987). CMS is transmitted only maternally and all progeny will be sterile. These CMS lines must be maintained by repeated crossing to a sister line, the maintainer line, that is genetically identical except that it possesses normal cytoplasm and is male fertile. Fertility restoration is essential in crops such as corn or sunflower where the harvested commodity from the F1 generation is a seed.

2.2 Nuclear male sterility

Anther and pollen development and fertilization processes have been the subjects of much investigation (Goldberg et al. 1993). Many nuclear genes involved in pollen development have been identified as mutants leading to pollen abortion and male sterility. This nuclear (or genic) male sterility is useful for hybrid seed production, but it has limitations due to the need to maintain female parent lines as heterozygotes and the segregation of fertile and sterile plants each generation. Nuclear male sterility in plants includes both spontaneous and engineered sterility. Spontaneous mutations leading to nuclear male sterility commonly occur in plants with a high frequency. Such mutations can easily be induced with chemical mutagents or ionising radiation. Nuclear male sterility is usually controlled by a pair of recessive genes (*msms*). These recessive mutations generally affect a huge number of functions and proteins, which for example are involved in male meiosis (Glover et al. 1998). In many crops nuclear male sterility does not permit effective production of population with 100% male sterile plants. This fact seriously limits its use in hybrid seed production.

3 Methods of producing male-sterile plants

Many different strategies to produce male-sterile plants by interfering with the development and the metabolism of the tapetum (Mariani et al. 1990; van der Meer et al. 1992; Hernould et al. 1998) or pollen (Worrall et al. 1992) in transgenic plants have been reported since the first transgenic male sterility system was described. Male sterility was further induced by using sense or antisense suppression to inhibit essential genes (Xu et al. 1995b; Luo et al. 2000) or by expressing aberrant mitochondrial gene products (Hernould et al. 1993; He et al. 1996; Gomez-Casati 2002). However, any of the available strategies has drawbacks such as interference with metabolism or general development or restriction to specific species. Thus, a universal and dominant male sterility system with efficient effect on pollen growth offering the possibility to efficiently restore fertility would be a great advantage for the production of hybrid seeds.

3.1 Selective destruction of tissues important for the production of functional pollen

In these systems a gene encoding a protein that is able to disrupt cell function, for example a ribonuclease that destroys the RNA of the tapetal cells (Mariani et al. 1990; Burgess et al. 2002) or the diphtheria toxin A-chain (Koltunow et al. 1990), is tissue-specific expressed. The tapetum serves as a good target for those expression strategies, because it plays a critical role in the process of pollen formation by secreting important substances for the pollen. In some of these systems, sterility or fertility can be chemically regulated. For example, inducible sterility can be obtained through the expression of a gene encoding a protein catalysing the conversion of a pro-herbicide into a toxic herbicide only in male reproductive tissues. Kriete et al. (1996) induced male-sterility in transgenic *Nicotiana tabacum* plants by tapetum-specific deacetylation of the externally applied non-toxic compound *N*-acetyl-L-phosphinothricin (*N*-ac-Pt). They constructed transgenic tobacco plants expressing the *argE* gene from *Escherichia coli* under the control of the tapetum-specific tobacco TA29 promoter. The *argE* gene product represents an *N*-acetyl-L-ornithine deacetylase, which removes the acetyl-group from *N*-ac-Pt resulting in the cytotoxic compound L-phosphinothricin (Pt, glufosinate). The application of *N*-ac-Pt led to empty anthers, resulting in male-sterile plants. Another example for tissue-specific cell ablation is the use of a bacterial phosphonate monoester hydrolase as a conditional lethal gene (Dotson et al. 1996). The *pehA* gene from *Burkholderia caryophilli*, a glyphosate metabolizing bacterium, was expressed in *Arabidopsis thaliana* using a tapetum-specific promoter. The treatment of transgenic plants with the protoxin glyceryl phosphate led to male sterility, because of the hydrolysis to glyphosate, which is a potent herbicide inhibiting the biosynthesis of aromatic amino acids. Another example for such a chemical control is the inducible expression of a male-sterility gene by the application of a chemical (Mariani et al. 1990; Goff et al. 1999). In order to induce fertility, the expression of a fertility restorer gene that can complement the male sterility or of a repressor of the male sterility gene can be chemically controlled (Cigan and Albertsen 2000). An alternative method for fertility restoration was suggested by Luo et al. (2000). They used a site-specific recombination system FLP/*FRT* from yeast to restore fertility in *Arabidopsis* plants that were male sterile because of the antisense expression of the pollen- and tapetum-specific gene *bcp1*. Mariani et al. (1992) restored fertility of male-sterile plants, which were generated through the use of the bacterial extracellular ribonuclease barnase (Paddon et al. 1989), by expressing a specific inhibitor of barnase, called barstar.

3.2 Changing of levels of metabolites needed for the production of viable pollen

Another approach to induce male sterility in plants is metabolic engineering of the carbohydrate supply. Carbohydrates are important for anther and pollen development. The extracellular invertase Nin88 mediates phloem unloading of carbohydrates via an apoplastic pathway. Tissue-specific antisense repression of *nin88* in tobacco caused male-sterility, because early stages of pollen development were blocked (Goetz et al. 2001). McConn and Browse (1996) demonstrated that *Arabidopsis* triple mutants, that contained negligible levels of trienoic fatty acids, such as jasmonate, were male-sterile and produced no seed. In that case the fertility could be restored through exogenous application of jasmonate.

4 Strategies for the multiplication of male-sterile lines

Although the described systems have provided important informations about anther and pollen development and ways to interfere with it, their potential use for commercial hybrid seed production is often limited because of the lack of cost-effective and efficient methods to multiply the engineered male-sterile plants (for an overview of multiplication strategies, see Perez-Prat und van Lookeren Campagne 2002).

4.1 Herbicide application for selection of male-sterile plants

A strategy for the propagation of male-sterile plants is to combine the gene conferring dominant male sterility to a herbicide resistance gene (e.g. Denis et al. 1993). After crossing the heterozygous male-sterile plants with a wild-type line in the same genetic background the male-sterile progeny can be selected by herbicide application. It is important to eliminate all the fertile plants in order to prevent any self-pollination, because this would lead to impure hybrid seeds.

4.2 Reversible male sterility

One approach to multiply male-sterile plants is to produce plants that are conditionally fertile. During female parent multiplication, male-sterile plants are treated with a fertility-restoring chemical and can be self-fertilized. For the production of hybrid seeds, chemical application is not required and the plants remain sterile. This system has some advantages over the selection of male-sterile plants by herbicide application, for example that the chemical has to be used during female parent multiplication and not during hybrid seed production and it can be applied to a smaller acreage.

Based on conditional male fertility several pollination control systems have been described. An example for the regulation of male fertility is the manipulation of hormones in male reproductive tissues. Huang et al. (2003) induced male-sterile plants through tissue-specific expression of *CKX1* and *gai* genes that are involved in oxidative cytokinin degradation and gibberellin signal transduction. In this dominant male sterility system, the male-sterile phenotype is achieved in transgenic plants that are homozygous for the transgene and it is reversible by exogenous hormone applications.

Alternatively, fertility can be induced through environmental conditions. In rice TGMS (thermo-sensitive genetic male sterility) and PGMS (photoperiod-sensitive genetic male sterility) mutants male sterility is influenced by temperature and photoperiod length (He et al. 1999; Dong et al. 2000). The temperature occurring just after panicle initiation is the most critical in the expression of fertility and sterility. Most rice TGMS lines are male fertile at temperatures under 25°C and sterile at higher temperatures (Sun et al. 1989). The seeds of TGMS lines are multiplied by selfing when exposed to the right temperature at the critical growth stage. PGMS lines are fertile under natural short day and male sterile under long day conditions. In this system, the male-sterile female line can be propagated by growing it under environmental conditions that restore fertility. This approach requires no restorer lines and no chemical treatment. However, controlled environmental conditions are needed to avoid the plants to be constantly challenged by unfavourable fluctuations in their environment.

Other conditional male fertility systems are based on a repressor of the male sterility gene or on the inducible expression of a fertility restorer gene that complements the defect (Cigan and Albertsen 2000).

4.3 Use of maintainer lines

The propagation of nuclear male-sterile plants can also be achieved through a crossbreed with a maintainer plant, that is male fertile but produces 100% male-sterile progeny when used to pollinate male-sterile plants. Perez-Prat and van Lookeren Campagne (2002) developed pollen lethality and colour maintainer lines that are useful for propagating both dominant and recessive male-sterile lines. The maintainer plants are genetically identical to the nuclear male-sterile plants except for a transgenic maintainer locus that renders it male fertile.

This system does not need chemical application, but otherwise a fertility restorer gene is required and, in the case of colour maintainers, seed sorting might also be needed.

5 Conclusion

The use of hybrid crops is a very important advance in agriculture in recent years, because hybrids have increased yield, wider environmental adaptability and are more insect and disease resistant. One strategy that has been utilized for hybrid crop production is male sterility. Biotechnology has enabled new methods to obtain male-sterile plants and the development of several new pollination control systems that could be useful for hybrid seed production. However, the inability to propagate the male-sterile female parent line in a cost-effective and efficient way limits the potential application for the commercial production of hybrids. Future research should take into account the importance of developing solutions for this propagation, because for many crops it is the limiting factor for the production of hybrids on a large scale.

References

Basra AS (ed) (2000) Hybrid seed production in vegetables: rationale and methods in selected crops. Food Products Press, Binghamton, New York

Bino RJ (1985) Histological aspects of microsporogenesis in fertile, cytoplasmic male-sterile and restored fertile *Petunia hybrida*. Theor Appl Genet 69:423–428

Burgess DG, Ralston EJ, Hanson WG, Heckert M, Ho M, Jenq T, Palys JM, Tang K, Gutterson N (2002) A novel, two-component system for cell lethality and its use in engineering nuclear male-sterility in plants. Plant J 31:113–125

Cigan AM, Albertsen MC (2000) Reversible nuclear genetic system for male sterility in transgenic plants. US patent number 6072102

Denis M, Delourme R, Gourret JP, Mariani C, Renard M (1993) Expression of engineered nuclear male sterility in *Brassica napus*. Plant Physiol 101:1295–1304

Dong NV et al. (2000) Molecular mapping of a rice gene conditioning thermosensitive genic male sterility using AFLP, RFLP and SSR techniques. Theor Appl Genet 100:727–734

Dotson SB, Lanahan MB, Smith AG, Kishore GM (1996) A phosphonate monoester hydrolase from *Burkholderia caryophilli* PG2982 is useful as a conditional lethal gene in plant. Plant J 10:383–392

Glover J, Grelon M, Craig S, Chaudury A, Dennis L (1998) Cloning and characterisation of *MS5* from *Arabidopsis*: a gene critical in male meiosis. Plant J. 15:345–356

Goetz M, Godt DE, Guivarc'h A, Kahmann U, Chriqui D, T. R (2001) Induction of male sterility in plants by metabolic engineering of the carbohydrate supply. Proc Natl Acad Sci USA 98:6522–6527

Goff SA, Crossland LD, Privalle LS (1999) Control of gene expression in plants by receptor mediated transactivation in the presence of a chemical ligand. US patent number 5880333

Goldberg RB, Beals TP, Sanders PM (1993) Anther development: basic principles and practical applications. Plant Cell 5:1217–1229

Gomez-Casati D, Busi MV, Gonzalez-Schain N, Mouras A, Zabaleta EJ, Araya A (2002) A mitochondrial dysfunction induces the expression of nuclear-encoded complex I genes in engineered male sterile *Arabidopsis thaliana*. FEBS Lett 532:70–74

Hanson M (1991) Plant mitochondrial mutations and male sterility. Annu Rev Genet 25:461–486

Hanson MR, Conde MF (1985) Function and variation of cytoplasmic genomes: lessons from cytoplasmic-nuclear interactions affecting male sterility in plants. Int Rev Cytol 94:213–267

He S, Abad AR, Gelvin SB, Mackenzie SA (1996) A cytoplasmic male sterility-associated mitochondrial protein causes pollen disruption in transgenic tobacco. Proc Natl Acad Sci USA 93:11763–11768

He YQ, Yang J, Xu CG, Zhang ZG, Zhang Q (1999) Genetic bases of instability of male sterility and fertility reversibility in photoperiod-sensitive genic male-sterile rice. Theor Appl Genet 99:683–693

Hernould M, Suharsono S, Litvak S, Araya A, Mouras A (1993) Male-sterility induction in transgenic tobacco plants with an unedited atp9 mitochondrial gene from wheat. Proc Natl Acad Sci USA 90:2370–2374

Hernould M, Suharsono, Zabaleta E, Carde JP, Litvak S, Araya A, Mouras A (1998) Impairment of tapetum and mitochondria in engineered male-sterile tobacco plants. Plant Mol Biol 36:499–508

Huang S, Cerny RE, Qi Y, Bhat D, Aydt CM, Hanson DD, Malloy KP, Ness LA (2003) Transgenic studies on the involvement of cytokinin and gibberellin in male development. Plant Physiol 131:1270–1282

Jan CC, Rutger JN (1988) Mitomycin C- and streptomycin-induced male sterility in cultivated sunflower. Crop Science Madison: Crop Science Society of America 28:792–795

Kadowaki K, Osumi T, Nemoto H, Harada K, Shinjyo C (1988) Mitochondrial DNA polymorphism in male-sterile cytoplasm of rice. Theor Appl Genet 75:234–236

Kempken F, Pring DR (1999) Male sterility in higher plants—fundamentals and applications. Prog Bot 60:139–166

Koltunow AM, Truettner J, Cox KH, Wallroth M, Goldberg RB (1990) Different temporal and spatial gene expression patterns occur during anther development. Plant Cell 2:1201–1224

Kriete G, Niehaus K, Perlick AM, Pühler A, Broer I (1996) Male sterility in transgenic tobacco plants induced by tapetum-specific deacetylation of the externally applied non-toxic compound N-acetyl-L-phosphinothricin. Plant J 9:809–818

Laughnan JR, Gabay-Laughnan S (1983) Cytoplasmic male sterility in maize. Annu Rev Genet 17:27–48

Lefort-Buson M, Guillot-Lemoine B, Datté Y (1987) Heterosis and genetic distance in rapeseed (Brassica napus L): crosses between European and Asiatic selfed lines. Genome 29:413–418

Levings CS III (1990) The Texas cytoplasm of maize: cytoplasmic male sterility and disease susceptibility. Science 250:942–947

Luo H, Lyznik LA, Gidoni D, Hodges TK (2000) FLP-mediated recombination for use in hybrid plant production. Plant J 23:423–430

Mariani C, de Beuckeleer M, Truettner J, Leemans J, Goldberg RB (1990) Induction of male sterility in plants by a chimaeric ribonuclease gene. Nature 347:737–741.

Mariani C, Gossele V, De Beuckeleer M, De Block M, Goldberg RB, De Greef W, Leemans J (1992) A chimaeric ribonuclease-inhibitor gene restores fertility to male sterile plants. Nature 357:384–387

McConn M, Browse J (1996) The critical requirement for linoleic acid in pollen development, not photosynthesis, in an Arabidopsis mutant. Plant Cell 8:403–416

Paddon CJ, Vasantha N, Hartley RW (1989) Translation and processing of Bacillus amyloliquefaciens extracellular RNase. J Bacteriol 171:1185–1187

Perez-Prat E, van Lookeren Campagne MM (2002) Hybrid seed production and the challenge of propagating male-sterile plants. Trends Plant Science 7:199–203

Pring DR, Tang HV, Schertz KF (1995) Cytoplasmic male sterility and organelle DNAs of sorghum. In: Levings CS III , Vasil IK (eds) The molecular biology of plant mitochondria. Kluwer Academic Publishers, Dordrecht, pp 461–495

Schnable PS, Wise RP (1998) The molecular basis of cytoplasmic male sterility and fertility restoration. Trends Plant Science 3:175–180

Sun ZX, Min SK, Xiong ZM (1989) A temperature-sensitive male sterile line found in rice. Rice Genet Newslett 6:116–117

van der Meer IM, Stam ME, van Tunen AJ, Mol JNM, Stuitje AR (1992) Antisense inhibition of flavonoid biosynthesis in petunia anthers results in male sterility. Plant Cell 4:253–262

Wise RP, Pring DR, Gengenbach BG (1987) Mutation to male fertility and toxin insensitivity in T-cytoplasm maize is associated with a frameshift in a mitochondrial open reading frame. Proc Natl Acad Sci USA 84:2858–2862

Worrall D, Hird DL, Hodge R, Wyatt P, Draper J, Scott R (1992) Premature dissolution of the microsporocyte callose wall causes male sterility in transgenic tobacco. Plant Cell 4:759–771

Xu GW, Cui YX, Schertz KF, Hart GE (1995a) Isolation of mitochondrial DNA sequences that distinguish male-sterility-inducing cytoplasms in *Sorghum bicolor* (L.) Moench. Theor Appl Genet 90:1180–1187

Xu H, Knox RB, Taylor PE, Singh MB (1995b) *Bcp1*, a gene required for male fertility in Arabidopsis. Proc Natl Acad Sci USA 92:2106–2110

Kerstin Stockmeyer and Frank Kempken[1]

Abteilung für Botanik mit Schwerpunkt Genetik und Molekularbiologie, Botanisches Institut und Botanischer Garten, Christian-Albrechts-Universität zu Kiel, Olshausenstr. 40, D-24098 Kiel, Germany

[1]Correspondence: e-mail: fkempken@bot.uni-kiel.de

Physiology

Membrane turnover in plants

Ulrike Homann

1 Introduction

The plasma membrane of plant cells has long been viewed as a rather static system. However, recent studies have uncovered the importance of exo- and endocytosis and membrane cycling for physiological functioning in a variety of different plant cell types. It is now clear that the plasma membrane is a dynamic system and subject to a constant exchange with intracellular compartments. This review attempts to summarize recent results on membrane turnover in plant cells and to introduce possible regulatory mechanisms of this process.

2 Membrane turnover during polarized cell growth

Cell growth is associated with irreversible increase in cell volume and surface area. It requires secretion of cell wall components and addition of new plasma membrane material. During non-polarized cell growth, this process occurs uniformly over the whole cell whereas in tip growing cells it is restricted to the apex. Polarized cell growth can be extremely fast with extension rates of more than 200 nm/s for pollen tubes. It is obvious that such a rapid growth relies on a highly active secretory system. However, it also requires fast membrane retrieval. Estimations from electron microscopy studies clearly demonstrated that delivery of secretory vesicle membrane to the tip may well exceed the amount of membrane material necessary for cell growth (Picton and Steer 1983). The total vesicle production rate in pollen tubes is maintained under similar conditions even when the growth rate changes (Picton and Steer 1983). In extreme cases, only 10% of the delivered membrane material is used for extension of the plasma membrane. Thus, there is a large demand for endocytosis during tip growth. Similar results were found for growing coleoptile cells. By comparing the area of vesicle membrane delivered to the cell surface with the membrane area required for growth Phillips et al. (1988) found that at least 65% of the delivered membrane must be recycled. Membrane turnover is therefore most likely an integral part of cell growth in general.

2.1 Regulation of exocytosis

Studies on growth of pollen tubes and root hairs have so far mainly focused on the regulation of vesicle delivery. From these investigations, two major regulatory components have emerged: polarized ion gradients and fluxes, in particular of calcium and protons and a dynamic cytoskeleton (for review, see Hepler et al. 2001). Briefly, a tip focused Ca^{2+} gradient, which is established in polarized plant cells, is essential for growth. It presumably participates in exocytosis of secretory vesicles, although the exact mechanism and interacting proteins have not yet been identified. The Ca^{2+} gradient oscillates with the same frequency as the oscillation in growth rate, but phase shifted. In pollen tubes, an internal pH gradient that oscillates in relation to growth has also been observed (Feijó et al. 1999). In addition, GTP-binding proteins and kinases that act along the secretory pathway of tip growing cells have been identified (Moutinho et al. 1998; Molendijk et al., 2001; Cheung et al. 2002; Šamaj et al. 2002; Preuss et al. 2004).

Actin microfilaments are crucial for the delivery of secretory vesicles to the apex of the cell. They are organized in longitudinally oriented bundles up to the so-called "apical-clear zone" where secretory vesicles accumulate prior to fusion with the plasma membrane (Hepler et al. 2001). In addition, actin is supposed to be involved in the establishment of cell polarity and the regulation of tip extensibility (Miller et al. 1999; Vidali and Hepler 2001). In agreement with the central role of actin in the control of tip growth, numerous actin binding proteins have been identified in pollen tubes and root hairs (Chen et al. 2002; Šamaj et al. 2002; Fan et al. 2004; Limmongkon et al. 2004).

2.2 Endocytosis during tip growth

In contrast to the detailed information on regulatory components of the secretory pathway much less is known about endocytosis in tip growing plant cells. Recent investigations have led to the conclusion that clathrin plays an important role in endocytosis, not only in animal, but also in plant cells (for review, see Holstein 2002). During clathrin-dependent endocytosis, a complex structure of clathrin and associated proteins is assembled at specialized sites, called coated pits, in order to concentrate surface proteins and to drive the invagination of the plasma membrane. This finally leads to the formation of clathrin-coated endocytic vesicles. Clathrin-coated pits and clathrin-coated vesicles have been observed in a variety of plants (Robinson 1996), including the subapical region of growing root hairs and pollen tubes (Emons and Trass 1986). In addition, components of the clathrin coat could be immunolocalized to the tip of pollen tubes (Blackbourn and Jackson 1996). These observations suggest that clathrin-dependent endocytosis operates in polarized growth. A number of proteins that may be involved in both cytoskeleton dynamics and endocytosis have recently been localized in tips of root hairs (for review, see Šamaj et al. 2004). However, their exact role in endocytosis remains largely speculative.

New insights into the dynamics of membrane turnover in growing pollen tubes have recently been gained from the application of FM dyes, mainly FM4-64 and FM1-43. These dyes are increasingly used to study endocytosis and vesicle trafficking in living plant cells. They partition into the outer leaflet of the membrane and are believed not to penetrate through the membrane. FM-dyes fluoresce significantly only after incorporation into membranes. This makes the dye an ideal candidate for the investigation of endocytosis in living cells. However, recent studies also demonstrated that precautions have to be taken when the dye is used to probe for vesicle trafficking in plant cells and analysis of FM-stained images is not always straightforward. Depending on the cell type and time of incubation, different organelles along the secretory and endocytic pathway may become stained by the dye (Bolte et al. 2004). In guard cells, FM1-43 and FM2-10 were found to label not only endocytic structures but also mitochondria (Meckel et al. 2004). In addition, long incubation times and/or high concentrations of FM-dyes may result in irreversible damage to cells (Meckel et al. 2004). Nevertheless, FM dyes provide a valuable tool for dissecting vesicle trafficking. When applied to growing pollen tubes, FM4-64 resulted in a distinct staining pattern in the tube apex (Parton et al. 2001). This staining pattern corresponds to the previously identified cone-shaped "apical-clear zone" where secretory vesicles accumulate prior to fusion with the plasma membrane. A similar staining pattern was found in *Arabidopsis* root hairs incubated in FM4-64 (Ketelaar et al. 2003). In pollen tubes and root hairs the cone-shaped distribution of FM4-64 was visible after about 5 and 15 min, respectively. Considering that cytosolic structures stained by FM4-64 are supposed to correspond to endocytic vesicles or endosomes, these results seem at first glance difficult to explain. However, taking into account a fast recycling of secretory vesicles, the result can be interpreted as follows. During tip growth, a large amount of membrane material is retrieved via endocytosis. Most of these FM-stained vesicles are not transported along the endocytic pathway to the vacuole, but are quickly recycled to deliver new cell wall material. This will then lead to the accumulation of FM-stained secretory vesicles found in the tip of pollen tubes and root hairs.

2.3 Coupling between exo- and endocytosis

The delivery of cell wall material and retrieval of excess membrane material during growth requires a coordinated regulation of the rate of exo- and endocytosis. How this coupling is achieved in tip growing cells is still unknown. It is worth noting that a tight coupling between exo- and endocytosis is most likely not only important for tip growing cells, but for cell growth in general and during maturation of cells. After the cell has stopped enlarging, the building of a thick cell wall still requires delivery of large amounts of cell wall material and consequently membrane material. This membrane material must be retrieved via endocytosis.

Possible mechanisms for the coupling between vesicle fusion and recycling can be derived from studies on neurotransmission. During synaptic transmission a large amount of vesicles filled with neurotransmitter fuse with the plasma membrane. In order to sustain neurotransmitter release, these vesicles need to be recycled. There are two routes for cycling of synaptic vesicles, a slow and a fast track (for review, see Galli and Hauker 2004). Recycling via the slow track takes about 40–60 s and involves clathrin-mediated endocytosis. The other fast recycling pathway allows cycling of vesicles within less than a second. This pathway may require similar mechanisms of exo- and endocytosis as the slow track but function at a higher rate. Alternatively, the fast track could involve the so-called "kiss-and-run" mechanism. This mechanism is characterized by successive rounds of transient opening and closing of the fusion pore without complete incorporation of the vesicular membrane into the plasma membrane. Such a mechanism is not only found in neurons, but has been identified in a variety of cell types (Schneider 2001), including plant cells (Weise et al. 2000). Using patch-clamp capacitance measurements, Weise et al. (2000) observed transient and permanent fusion of vesicles in *Zea mays* coleoptile protoplasts. It has been suggested that delivery of cell wall components occurs mainly via transient fusion, while tension-driven exocytosis (see also below) accommodates for surface area increase (Morris and Homann 2001).

The advantage of transient fusion is a rapid cycling between a fusion and a non-fusion state. During transient opening of the fusion pore secretory products can be released, while the vesicular matrix is retained for re-use in subsequent cycles. Slow and fast cycling of vesicles generally exist in one cell. How the cell shifts from the slow to the fast track is not fully understood. In synapses, high cytosolic Ca^{2+} concentration has been shown to alter the mode of exocytosis to the kiss-and-run mechanism (Ales et al. 1999).

Despite the detailed understanding of the mechanisms involved in exo- and endocytosis in neurons the question of how the rate of exocytosis is adjusted to an increase or decrease in endocytosis has not been fully answered. Recent investigations show that synaptotagmin, a Ca^{2+}-binding protein that is required for Ca^{2+}-stimulated exocytosis, is also involved in the control of endocytosis (Nicholson-Tomishima and Ryan 2004). Synaptotagmin interacts with both the SNARE (soluble N-ethylmaleimide-sensitive fusion protein attachment protein receptor) complex, which is essential for membrane fusion, and the clathrin adaptor complex AP-2. This suggests that synaptotagmin plays an important role in coupling the rate of exocytosis to endocytosis. SNARE proteins have also been identified in different plants and are supposed to function at all fusion steps along the secretory and endocytic pathway. In addition, homologs of synaptotagmin can be found in the *Arabidopsis* genome (Craxton 2001), even though their role in exo-and/or endocytosis is not clear. Thus, all components necessary for fusion and fission of vesicles in neurons can be found in plant cells. One may therefore suggest that the cycling of vesicles in tip-growing plant cells may involve similar mechanisms as described for neurons. However, in plant cells the rates of exo- and endocytosis have to be adjusted in relation to cell growth, whereas in neurons endocytosis ensures that the pool of secretory vesicles is not depleted during repetitive excitation and that cells do not grow during neurotransmitter release. In addition, vesicle cycling in neurons and tip-growing plant cells happens on different time scales, i.e. seconds versus minutes. It is therefore most likely that further investigation on the membrane turnover during polarized growth will also reveal completely new regulatory mechanisms. Membrane tension, which has so far only been suggested to be involved in surface area changes of guard cells and osmotically treated protoplasts, may be one of these regulatory components (see below).

3 Guard cell functioning and tension modulated exo- and endocytosis

Guard cells mediate opening and closing of the stomatal pores, which regulate gas exchange in plants. During stomatal movement, guard cells undergo over a period of minutes large osmotically driven changes in cell volume and consequently surface area. These large changes in surface area of up to 40%

(Raschke 1979) cannot result from stretching of the existing membrane, as the maximum possible stretching of membranes is limited to about 2% (Wolfe et al. 1986). In addition, the large turgor pressure of up to 4 MPa (Raschke 1979) prevents plant cell plasma membrane from maintaining infoldings that could provide excess surface area. Alterations in surface area must therefore be accomplished by addition and removal of membrane material to and from the plasma membrane, respectively.

3.1 Exo- and endocytosis during osmotically driven surface area changes

To investigate osmotically driven changes in surface area, patch-clamp capacitance measurements have been applied to guard cell protoplasts. This technique allows the examination of exo- and endocytosis in single living protoplasts in real time. Results from these measurements demonstrated that osmotically induced swelling and shrinking of guard cell protoplasts are associated with incorporation and removal of membrane material into and out of the plasma membrane, respectively (Homann 1998). High resolution capacitance measurements that allow the detection of single exo- and endocytotic events indicated that fusion and fission of vesicles with a diameter of around 300 nm mainly accommodates for changes in surface area (Homann and Thiel 1999). These results were confirmed by fluorescence imaging of guard cell protoplasts with confocal laser scanning microscopy (CLSM). Staining of the plasma membrane with the fluorescent membrane probe FM1-43 revealed fast vesicular retrieval of plasma membrane into the cytoplasm during shrinking of the cells (Kubitscheck et al. 2000).

Adjustment of surface area in response to osmotic imbalance also plays an important role in cold acclimation of plants. During freezing ice formation occurs initially only extracellularly. The chemical potential of ice is lower than that of water, which leads to water efflux and large reduction in cell volume. During thawing, the process is reversed and cells swell. The fast adjustment of surface area in response to osmotic imbalance is a prerequisite for cold acclimation. Wolfe et al. (1985) found that cold-acclimated plants can swell without rupture to about twice the volume of those from non-acclimated plants. Measurements of membrane tension in these cells led to the hypothesis of tension-sensitive surface area regulation (Wolfe et al. 1985). This is in agreement with results derived from membrane capacitance measurements of guard cell protoplasts. An increase in membrane tension as a result of osmotically induced water influx or evoked by application of hydrostatic pressure resulted in an increase in exocytotic activity which was Ca^{2+} independent (Homann 1998; Bick et al. 2001).

3.2 Surface area regulation and membrane tension

Tension-sensitive exo- and endocyotis have been implicated to be an important component of surface area regulation not only in plant but also in animal cells (Morris and Homann 2001). In general, cells seem to detect and respond to derivations around a membrane tension set point. An increase above this set point results in addition of membrane material to the plasma membrane until the membrane tension set point is restored. Upon decrease in membrane tension excess plasma membrane material is retrieved to re-establish the resting tension. More detailed investigations revealed that the effective membrane tension is not identical to the bilayer tension alone but is strongly influenced by the interaction between the cortical cytoskeleton and the plasma membrane (Raucher and Sheetz 1999). Moreover, measurements on guard cell protoplasts implicated that stretching of the plasma membrane leads to the reinforcement of the actin cytoskeleton (Bick et al. 2001). This in turn results in stabilisation of the plasma membrane and desensitisation of pressure-driven vesicle fusion. The mechanisms by which cells sense changes in membrane tension are not yet known. Neither have the signal transduction pathways been identified that link changes in membrane tension to changes in the rate of exo- or endocytosis.

Considering the striking similarities between tension-sensitive exo- and endocytosis in plant and animal cells (Morris and Homann 2001), it seems likely that this mechanism was developed early in evolution. One may speculate that cells were exposed to large changes in environment and therefore had to establish a mechanism that would prevent rupturing of the plasma membrane under hypoosmotic conditions.

An important but yet unresolved question in tension modulated surface area changes is the origin and quality of the membrane material which is added and removed in the course of this process. In guard cells the addition of membrane material could often be detected immediately after application of hydrostatic pressure (see Fig. 1 in Bick et al. 2001). This indicates the existence of an intracellular reservoir of membrane material that is instantly available for incorporation into the plasma membrane. Guard cell protoplasts can undergo several cycles of swelling and shrinking. It is therefore most likely that the membrane material that is retrieved from the plasma membrane during surface area decrease is reused in subsequent cell swelling. To address the question of the quality of the membrane that is added and removed during surface area changes the fate of two types of K^+ channels, the K^+ inward and K^+ outward rectifier, was investigated in guard cell protoplasts (Homann and Thiel 2002; Hurst et al. 2004). The K^+ channels are localized in the plasma membrane of guard cells and play an important role in guard cell functioning. Parallel measurements of membrane capacitance and conductance in guard cell protoplasts implied that the vesicular membrane, which is inserted and retrieved during pressure-driven changes in surface area, carries active K^+ channels (Homann and Thiel 2002; Hurst et al. 2004). From the parallel increase in current and capacitance, the number of K^+ channels added for a given increase in surface area could be estimated (Homann and Thiel 2002). This led to the conclusion that only about one of nine vesicles that fuse with the plasma membrane contains a K^+ channel (Homann and Thiel 2002). Together, the results imply that membrane turnover during guard cell functioning is associated with turnover of plasma membrane material and membrane proteins, in particular K^+ channels.

3.3 Role of tension modulated exo- and endocytosis

Work on maize coleoptile cells provides evidence that tension driven exocytosis and Ca^{2+} stimulated exocytosis can exist as two independent exocytotic pathways in one cell (Thiel et al. 2000). Ca^{2+} recruits membrane from a small pool, whereas the tension driven pathway is Ca^{2+} independent and draws membrane from a much larger reservoir. Ca^{2+}-stimulated exocytosis is proposed to be a key element in hormone stimulated cell growth (Thiel et al. 1994), but the osmotically evoked increase in surface area suggests that tension-driven exocytosis also plays a role in cell elongation (Thiel et al. 2000). Recent studies on surface area changes in turgid guard cells demonstrate that reversible internalization of membrane material is not limited to plant protoplasts. Using the plasma membrane marker FM4-64, Shope et al. (2003) demonstrated that a decrease in surface area under hyperosmotic conditions is correlated with the internalization of the membrane dye. The internalization of plasma membrane material was reversed upon swelling of guard cells in hypo-osmotic conditions. Endocytic vesicles could not be identified, suggesting that these vesicles may be too small to incorporate a sufficient amount of dye for detection with the fluorescent microscope. The regulatory mechanism underlying the reversible internalization of plasma membrane material in turgid guard cells may be similar to that found in swelling and shrinking protoplasts. Even though a large expansion or even rupturing of the plasma membrane is prevented by the cell wall the rather flexible wall of guard cells (Willmer and Fricker 1996) may still allow for stretching of the plasma membrane. During opening of the stomatal pore, osmotically driven water influx and subsequent increase in cell volume may result in sufficient stretching of the plasma membrane to stimulate tension-driven exocytosis (Fig. 1). In the process of stomatal closure, loss of water and decrease in cell volume would be accompanied by a decrease in membrane tension that would stimulate endocytosis (Fig. 1).

Changes in membrane tension may also occur at the very tip of growing pollen tubes or root hairs. The cell wall at the tip of these cells is mainly composed of pectins, and is thus highly plastic (Li et al. 1996). The plasma membrane at the tip may therefore be subject to a much higher tension than in the rest of the cell. This could directly contribute to an increased rate of exocytosis at the tip. It would also explain why endocytosis in pollen tubes is believed to occur mainly in an area just behind the apex and not at the very tip (Parton et al. 2001).

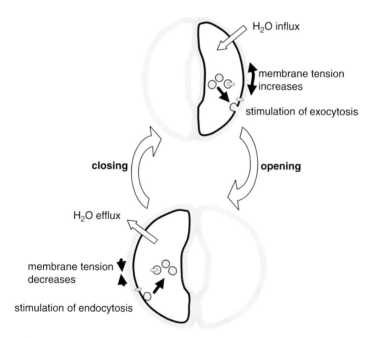

Fig. 1. Model for tension-driven exo- and endocytosis during stomatal movement. Alterations in surface area during stomatal movement must be accomplished by addition and removal of membrane material to and from the plasma membrane. This may be achieved via tension driven exo- and endocytosis: During opening of the stomatal pore, osmotically driven water influx causes stretching of the plasma membrane which stimulates exocytosis. In the process of stomatal closure loss of water leads to a decrease in membrane tension and stimulation of endocytosis

4 Constitutive exo- and endocytosis

4.1 Constitutive turnover of K^+ channels

Recent studies on membrane trafficking in plant cells have demonstrated that the plasma membrane of non-growing and un-stimulated cells is not static but subject to continuous membrane turnover. Investigations of membrane turnover in turgid guard cells revealed a constitutive internalization of small vesicles with a diameter of ≤ 270 nm (Meckel et al. 2004). Endocytosis of these vesicles occurred under constant osmotic pressure within a timescale of minutes. In guard cells transiently expressing the K^+ channel KAT1 a subset of endocytic vesicles carried the ion channel (Meckel et al. 2004). This points to a constitutive turnover of K^+ channels in guard cells. KAT1 was also found in putative endosomal compartments, which might correspond either to prevacuolar compartments or recycling endosomes (Meckel et al. 2004). In contrast to KAT1, the protein TM23, a plasma membrane protein with a 23 amino acid long transmembrane domain of the human lysosomal protein LAMP1, was not constitutively internalized (Meckel et al. 2005). This difference in constitutive endocytosis may be explained by a lack of a motif that is recognized by the endocytic machinery. This would imply the existence of a corresponding motif in the K^+ channel KAT1. However, such an endocytic signal motif has not yet been identified. The physiological function of a constitutive turnover of the K^+ channel is not clear. It may act as a quality control ensuring that only functional channels stay in the plasma membrane. Changes in constitutive turnover could also affect the ion channel concentration and thus have a strong impact on membrane transport properties. In animal cells, variation in the stability of ion channels in the plasma membrane were, for example, found to affect water absorption of epithelia cells (Rotin et al. 2001) and have been implicated in functional plasticity of neurotransmission (Luscher and Keller 2004).

4.2 Cycling of PIN proteins

Recent investigation on the functioning of the putative auxin export carrier PIN1 have highlighted the importance of cycling for physiological functioning of membrane proteins in plants. The plant hormone auxin plays a central role in the regulation of plant growth and development. Its function depends on a polar transport of the phytohormone. This polar transport is mediated by the asymmetric localization of PIN proteins. How the polarity of PIN proteins is established is not yet fully understood. PIN1 has been shown to cycle between the plasma membrane and an internal compartment (Steinmann et al. 1999; Geldner et al. 2001). Cycling of PIN was found to be important for auxin transport, and the polarized localization of PIN1 (Geldner et al. 2001, 2003). It is actin-dependent and can be blocked by inhibitors which were thought to block specifically auxin transport, such as TIBA (Geldner et al. 2001). The auxin transport inhibitors were also found to inhibit trafficking of other membrane proteins. This led to the hypothesis that cycling of PIN1 is essential for effective auxin transport. It is not yet clear how cycling of PIN is connected to its function. One model suggests that auxin is accumulated in PIN1 carrying vesicles. Fusion of these vesicles with the plasma membrane would then lead to release of auxin (Baluska et al. 2003). According to this model, cycling of PIN1 would allow vesicles to be refilled with auxin. Alternatively, cycling may be associated with reversible modifications of components of the auxin efflux machinery that are essential for auxin transport. So far, there is no direct evidence for either of these models. The second big question arising from investigations of PIN localization is how the polar distribution is established. Targeted trafficking to subdomains of the plasma membrane may occur via sorting of proteins at the Golgi and direct transport to the specific domains. Alternatively, establishment of polarity may involve recycling of proteins from the endosome. The latter model is supported by studies that demonstrated that cycling of PIN1 is not only important for maintaining auxin efflux but also for the polar distribution of the protein. Inhibitors such as brefeldin A (BFA) or mutation of GNOM, an exchange factor for ARF GTPases (ARF-GEFs) that regulate vesicle trafficking in a variety of organisms, resulted in the loss of polar PIN localization (Geldner et al. 2001, 2003). BFA is known to inhibit function of ARF-GEFs, which then results in a block of recruitment of vesicle coat components and thus inhibition of vesicle budding from intracellular compartments (Robineau et al. 2000). The *Arabidopsis* protein GNOM is a BFA sensitive ARF-GEF and is localized to endosomes (Geldner et al. 2003). Mutations in GNOM were found to result in developmental defects that resembled those of BFA treated plants. Together with investigations of BFA-resistant variants of GNOM, this led to the conclusion that GNOM is required for PIN1 recycling between the plasma membrane and endosomal compartments (Geldner et al. 2003).

Recently, another component that acts in the polar sorting of PIN proteins has been identified. The serine–threonine kinase PINOID (PID) was found to be essential for the polarized targeting of PIN proteins (Friml et al. 2004). The results suggest that PID acts as a switch, leading to apical or subapical distribution of PIN depending on its presence above or below a certain threshold level. So far, it is unknown which proteins are targeted by the PID kinase.

Judging from investigations on animal cells, it seems likely that a large number of different factors which act on various points along the secretory and endocytic pathway contribute to the distribution of proteins in the plasma membrane. In animal cells the trafficking of the cystic fibrosis transmembrane conductance regulator (CFTR), which localized to the apical membrane of epithelial cells and vesicular targeting of the glutamate receptor GLUT 4, has been extensively studied (for review, see Kleizen et al. 2000; Bryant et al. 2002; Bertrand and Frizzell 2003). A number of regulatory steps along the trafficking pathway of CFTR have been found to contribute to normal epithelia functioning and control of polar CFTR channel density. These regulatory mechanisms include control of ER and Golgi export as well as exo- and endocytosis and cycling between the plasma membrane and the recycling endosome (Betrand and Frizzell 2003). Sequence homologies and analogous function between components involved in GLUT 4 targeting and those implicated in auxin transport regulation have led to the hypothesis that similar mechanisms function in cycling of transporters in animal and plant systems (Muday et al. 2003). Future investigation will reveal which of the proposed mechanisms is indeed related to cycling of proteins in plant cells and will certainly also uncover plant specific control factors.

Acknowledgement. I thank Professor G. Thiel for helpful comments on the manuscript.

References

Ales E, Tabares L, Poyato JM, Valero V, Lindau M, Alvarez de Toledo G (1999) High calcium concentrations shift the mode of exocytosis to the kiss-and-run mechanism. Nat Cell Biol 1:40–44

Baluška F, Šamaj J, Menzel D (2003) Polar transport of auxin: carrier-mediated flux across the plasma membrane or neurotransmitter-like secretion? Trends Cell Biol 13:282–285

Bertrand CA, Frizzell RA (2003) The role of regulated CFTR trafficking in epithelial secretion. Am J Physiol Cell Physiol 285:C1–18

Bick I, Thiel G, Homann U (2001) Cytochalasin D attenuates the desensitisation of pressure-stimulated vesicle fusion in guard cell protoplasts. Eur J Cell Biol 80:521–526

Blackbourn HD, Jackson AP (1996) Plant clathrin heavy chain: Sequence analysis and restricted localisation in growing pollen tubes. J Cell Sci 109:777–787

Bolte S, Talbot C, Boutte Y, Catrice O, Read ND, Satiat-Jeunemaitre B (2004) FM-dyes as experimental probes for dissecting vesicle trafficking in living plant cells. J Microsc 214:159–173

Bryant NJ, Govers R, James DE (2002) Regulated transport of the glucose transporter GLUT4. Nat Rev Mol Cell Biol 3:267–277

Chen CY, Wong EI, Vidali L, Estavillo A, Hepler PK, Wu HM, Cheung AY (2002) The regulation of actin organization by actin-depolymerizing factor in elongating pollen tubes. Plant Cell 14:2175–2190

Cheung AY, Chen CY, Glaven RH, de Graaf BH, Vidali L, Hepler PK, Wu HM (2002) Rab2 GTPase regulates vesicle trafficking between the endoplasmic reticulum and the Golgi bodies and is important to pollen tube growth. Plant Cell 14:945–962

Craxton M (2001) Genomic analysis of synaptotagmin genes. Genomics 77:43–49

Emons AMC, Traas JA (1986) Coated pits and coated vesicles on the plasma membrane of plant cells. Eur J Cell Biol 41:57–64

Fan X, Hou J, Chen X, Chaudhry F, Staiger CJ, Ren H (2004) Identification and characterization of a Ca^{2+}-dependent actin filament-severing protein from lily pollen. Plant Physiol 136:3979–3989

Feijó JA, Sainhas J, Hackett GR, Kunkel JG, Hepler PK (1999) Growing pollen tubes possess a constitutive alkaline band in the clear zone and a growth-dependent acidic tip. J Cell Biol 144:483–496

Friml J, Yang X, Michniewicz M, Weijers D, Quint A, Tietz O, Benjamins R, Ouwerkerk PB, Ljung K, Sandberg G, Hooykaas PJ, Palme K, Offringa R (2004) A PINOID-dependent binary switch in apical-basal PIN polar targeting directs auxin efflux. Science 306:862–865

Galli T, Hauker V (2004) Cycling of synaptic vesicles: How far? How fast! Sci. STKE re 19

Geldner N, Friml J, Stierhof YD, Jurgens G, Palme K (2001) Auxin transport inhibitors block PIN1 cycling and vesicle trafficking. Nature 413:425–428

Geldner N, Anders N, Wolters H, Keicher J, Kornberger W, Muller P, Delbarre A, Ueda T, Nakano A, Jurgens G (2003) The *Arabidopsis* GNOM ARF-GEF mediates endosomal recycling, auxin transport, and auxin-dependent plant growth. Cell 112:219–230

Hepler PK, Vidali L, Cheung AY (2001) Polarized cell growth in higher plants. Annu Rev Cell Dev Biol 17:159–187

Holstein SEH (2002) Clathrin and plant endocytosis. Traffic 3:614–620

Homann U (1998) Fusion and fission of plasma-membrane material accommodates for osmotically induced changes in the surface area of guard-cell protoplasts. Planta 206:329–333

Homann U, Thiel G (1999) Unitary exocytotic and endocytotic events in guard-cell protoplasts during osmotic-driven volume changes. FEBS Lett 460:495–499

Hurst AC, Meckel T, Tayefeh S, Thiel G, Homann U (2004) Trafficking of the plant potassium inward rectifier KAT1 in guard cell protoplasts of *Vicia faba*. Plant J 37:391–397

Ketelaar T, de Ruijter NC, Emons AM (2003) Unstable F-actin specifies the area and microtubule direction of cell expansion in *Arabidopsis* root hairs. Plant Cell 15:285–292

Kleizen B, Braakman I, de Jonge HR (2000) Regulated trafficking of the CFTR chloride channel. Eur J Cell Biol 79:544–556

Kubitscheck U, Homann U, Thiel G (2000) Osmotic evoked shrinking of guard cell protoplasts causes retrieval of plasma membrane into the cytoplasm. Planta 210:423–431

Li YQ, Zhang HQ, Pierson ES, Huang FY, Linskens HF, Hepler PK, Cresti M (1996) Enforced growth-rate fluctuation causes pectin ring formation in the cell wall of *Lilium longiflorum* pollen tubes. Planta 200:41–49

Luscher B, Keller CA (2004) Regulation of GABAA receptor trafficking, channel activity, and functional plasticity of inhibitory synapses. Pharmacol Ther 102:195–221

Limmongkon A, Giuliani C, Valenta R, Mittermann I, Heberle-Bors E, Wilson C (2004) MAP kinase phosphorylation of plant profilin. Biochem Biophys Res Commun 324:382–386

Meckel T, Hurst AC, Thiel G, Homann U (2004) Endocytosis against high turgor: intact guard cells of Vicia faba constitutively endocytose fluorescently labelled plasma membrane and GFP-tagged K^+-channel KAT1. Plant J. 39:182–193

Meckel T, Hurst AC, Thiel G, Homann U (2005) Guard cells undergo constitutive and pressure driven membrane turnover. Protoplasma (in press)

Miller DD, de Ruijter NCA, Bisseling T, Emons AMC (1999) The role of actin in root hair morphogenesis: studies with lipochito-oligosaccharide as a growth stimulator and cytochalasin as an actin perturbing drug. Plant J 17:141–154

Molendijk AJ, Bischoff F, Rajendrakumar CS, Friml J, Braun M, Gilroy S, Palme K (2001) *Arabidopsis thaliana* Rop GTPases are localized to tips of root hairs and control polar growth. EMBO J 20:2779–2788

Morris CE, Homann U (2001) Cell surface area regulation and membrane tension. J Membr Biol 179:79–102

Moutinho A, Trewavas AJ, Malho R (1998) Relocation of a Ca²⁺-dependent protein kinase activity during pollen tube reorientation. Plant Cell 10:1499–1510

Muday GK, Peer WA, Murphy AS (2003) Vesicular cycling mechanisms that control auxin transport polarity. Trends Plant Sci 8:301–304

Nicholson-Tomishima K, Ryan TA (2004) Kinetic efficiency of endocytosis at mammalian CNS synapses requires synaptotagmin I. Proc Natl Acad Sci USA 101:16648–16652

Parton RM, Fischer-Parton S, Watahiki MK, Trewavas AJ (2001) Dynamics of the apical vesicle accumulation and the rate of growth are related in individual pollen tubes. J Cell Sci 114:2685–2695

Picton JM, Steer MW (1983) Membrane recycling and control of secretory activity in pollen tubes. J Cell Sci 63:303–310

Phillips GD, Preshaw C, Steer MW (1988) Dictyosome vesicle production and plasma membrane turnover in auxin-stimulated outer epidermal cells of coleoptile segments from *Avena sativa* (L.). Protoplasma 145:59–65

Preuss ML, Kovar DR, Lee YR, Staiger CJ, Delmer DP, Liu B (2004) A plant-specific kinesin binds to actin microfilaments and interacts with cortical microtubules in cotton fibers. Plant Physiol 136:3945–3955

Raucher D, Sheetz MP (199) Characteristics of a membrane reservoir buffering membrane tension. Biophys J 77:1992–2002

Raschke K (1979) Movements of stomata. In: Haupt W, Feinleb ME (eds) Encyclopedia of plant physiology, vol 7, physiology of movements. Springer, Berlin, pp 382–441

Robineau S, Chabre M, Antonny B (2000) Binding site of brefeldin A at the interface between the small G protein ADP-ribosylation factor 1 (ARF1) and the nucleotide-exchange factor Sec7 domain. Proc Natl Acad Sci USA 97:9913–9918

Robinson DG (1996) Clathrin-mediated trafficking. Trends Plant Sci 1:349–355

Rotin D, Kanelis V, Schild L (2001) Trafficking and cell surface stability of ENaC. Am J Physiol Renal Physiol 281:F391–399

Šamaj J, Ovecka M, Hlavacka A, Lecourieux F, Meskiene I, Lichtscheidl I, Lenart P, Salaj J, Volkmann D, Bogre L, Baluska F, Hirt H (2002) Involvement of the mitogen-activated protein kinase SIMK in regulation of root hair tip-growth. EMBO J 21:3296–3306

Šamaj J, Baluska F, Voigt B, Schlicht M, Volkmann D, Menzel D. (2004) Endocytosis, actin cytoskeleton, and signaling. Plant Physiol 135:1150–1161

Schneider SW (2001) Kiss and run mechanism in exocytosis. J Membr Biol 181:67–76

Shope JC, DeWald DB, Mott KA (2003) Changes in surface area of intact guard cells are correlated with membrane internalization. Plant Physiol 133:1314–1321

Steinmann T, Geldner N, Grebe M, Mangold S, Jackson CL, Paris S, Galweiler L, Palme K, Jurgens G (1999) Coordinated polar localization of auxin efflux carrier PIN1 by GNOM ARF GEF. Science 286:316–318

Thiel G, Rupnik M, Zorec R (1994) Raising the cytosolic Ca²⁺ concentration increases the membrane capacitance of maize coleoptile protoplasts: evidence for Ca²⁺-stimulated exocytosis. Planta 195:305–308

Thiel G, Sutter J-U, Homann U (2000) Ca²⁺-sensitive and Ca²⁺-insensitive exocytosis in maize coleoptile protoplasts. Pflügers Arch – Eur J Physiol 439 [Suppl.]:R152–R153.

Vidali L, Hepler PK (2001) Actin and pollen tube growth. Protoplasma 215:64–76

Weise R, Kreft M, Zorec R, Homann U, Thiel G (2000) Transient and permanent fusion of vesicles in Zea mays coleoptile protoplasts measured in the cell-attached configuration. J Membr Biol 174:15–20

Willmer CM, Fricker M (1996) Stomata, 2nd edn. Chapmann and Hall, London

Wolfe J, Dowgert MF, Steponkus PL (1985) Dynamics of membrane exchange of the plasma membrane and the lysis of isolated protoplasts during rapid expansion in area. J Membr Biol 86:127–138

Wolfe J, Dowgert MF, Steponkus PL (1986) Mechanical study of the deformation of the plasma membranes of protoplasts during osmotic expansions. J Membr Biol 93:63–74

Ulrike Homann
Institute of Botany,
Darmstadt University of Technology,
64287 Darmstadt,
Germany.
e-mail: homann-u@bio.tu-darrmstadt.de

Besides water:
Functions of plant membrane intrinsic proteins and aquaporins

Ralf Kaldenhoff

1 Aquaporins in plants

Our view on the mechanism of membrane water transport has been renewed by the molecular and functional characterization of aquaporins. Further analysis of additional conductivities of the so-called MIPs (membrane intrinsic proteins), in which aquaporins form a subclass, could easily improve many concepts of plant physiology.

Functional assays of MIPs revealed conductivity to and specificity for water and/or small solutes. The general biochemical mechanism of transport and selectivity was understood after pore structure analysis of aquaporin proteincrystals (Borgnia et al. 1999a,b). Aquaporins exhibit a characteristic conserved arrangement with six transmembrane helices linked by three extra- and two intracellular loops, N- and C-terminal domains protruding into the cytoplasm and a highly conserved amino acid motif, asparagine-proline-alanine (NPA), occurring twice in the pore region. Due to a supposedly greater necessity for fine tuned water control, plant aquaporins are particularly abundant with a greater diversity than the paralogs in metazoans (Johanson et al. 2001). In *Arabidopsis*, for example, 35 MIP like isoforms were predicted from genome analysis. Some of them are, however, assumed to be pseudogenes (Quigley et al. 2001). The plant aquaporins are classified into four major subfamilies: plasma membrane intrinsic proteins (PIPs), tonoplast intrinsic proteins (TIPs), Nodulin-26-like intrinsic proteins (NLMs, NIPs), and small basic intrinsic proteins (SIPs) (Johanson and Gustavsson 2002). The NLMs were the first aquaporins identified in plants, and are located in the peribacteroid membrane of symbiotic root nodules. Here they are believed to control transport of metabolites between the host cytosol and bacteria (Fortin et al. 1987). However, these proteins are also found in non-legume plants, and exhibit glycerol transport activity when expressed heterologously in *Xenopus* oocytes (Dean et al. 1999; Weig and

Jakob 2000; Ciavatta et al. 2001). The PIP subfamily can be further subdivided into PIP1, PIP2 and PIP3. The proteins differ in the lengths of the N- and C-termini, the former being longer in the PIP1s. The TIP subfamily can be split into five subgroups (TIP1 to TIP5), in several higher plant species (Johanson et al. 2001). With the exception of TIPs, analysis of *Physcomitrella patens* ESTs showed a similar diversification into the PIP, TIP, NIP and SIP as aquaporins of higher plants, indicating that gene duplication events leading to the subfamilies must have occurred early during land plant evolution (Borstlap 2002).

2 Plant aquaporins and water transport

2.1 Characterization of aquaporin function in *Xenopus* oocytes

For an initial characterization, aquaporin function is usually assessed in single cell expression systems like *Xenopus* oocytes (Preston et al. 1992). These cells possess relatively low intrinsic water permeability, are large in size, manipulation by injection with capillary needles or isolation of membranes is feasible, and oocytes are available in amounts of several hundred per surgery. For an oocyte assay, a cDNA of a putative aquaporin is cloned into an expression vector with a start site for an RNA polymerase close to *Xenopus* translation consensus sites. In-vitro transcription of a cloned cDNA generates a cRNA coding for the target protein. The cRNA is injected into the oocyte and after an appropriate incubation time, the putative aquaporin could be inserted into the oocyte plasma membrane, probably without an aquaporin specific integration mechanism and simply because of the hydrophobic nature of the proteins. Even aquaporins targeted to other membranes, such as the plant tonoplast, were functionally integrated into the oocyte plasma membrane (Maurel et al. 1995). An oocyte expressing the putative aquaporin can now be subjected to hypo-osmotic conditions in order to determine the membrane water permeability by following the cell volume increase and determination of the swelling rate. In comparison to controls that do not express further aquaporins, additional to the intrinsic frog aquaporins, the contribution of the extra protein can be determined. Regarding the initial swelling rate and the osmotic gradient, a membrane water permeability coefficient can be calculated (Zhang and Verkman 1991). It is given as P_{os} (osmotic permeability) or converted into P_f (diffusive permeability). If the calculated permeability value is increased in comparison to the values obtained for water-injected oocytes, the protein under investigation facilitates membrane water transport and fulfils one criterion of an aquaporin. A representative permeability coefficient for a control oocyte is 1×10^{-3} cm/s and 2×10^{-3} cm/s, e.g. for PIP1, up to 50×10^{-3} cm/s, e.g. for a PIP2 from *Samanea saman* expressing in oocytes (Moshelion et al. 2002). Major characteristics for an aquaporin mediated membrane water transport are a low activation energy, a linear kinetic of water transport and a water flux towards the lower water potential, which is independent of the chemical nature of the osmotic gradients.

Since all aquaporins are structurally related and share highly similar consensus regions, particularly in the pore area, a similar function could be postulated. However, the osmotic water permeability (P_f) as well as the selectivity of aquaporins varies considerably. A combination of water permeability measurements in frog oocytes with quantitative immunoprecipitation provided data on the single channel water permeability (P_f) for human aquaporins, e.g. 0.25 $cm^3/s\times10^{-14}$ for AQP0 and 24 $cm^3/s\times10^{-14}$ for AQP4 (Yang and Verkman 1997). Plant plasma membrane aquaporins also displayed different aquaporin activity in oocytes (Chaumont et al. 2000; Bots et al. 2005). Coexpression of ZmPIP1;2 and ZmPIP2;1 isoforms induced a P_f increase above that obtained for expression of a single aquaporin-species. A function of the C-terminal part of loop E in PIP interaction was indicated by results of mutational analysis (Fetter et al. 2004). Also in the plant, interaction or heteromer formation could be important, as was concluded after analysis of PIP1 and PIP2 double antisense *Arabidopsis* plants (Martre et al. 2002).

If the cells were kept in isoosmotic conditions and instead a gradient for other substances such as urea, gases or glycerol was applied, facilitated transport for the specific compound can be studied as long as it could be directly determined or a correlated indirect effect on the molecules' membrane transport could be measured.

2.2 Other single cell systems

In a similar experimental set-up as for the *Xenopus* oocytes, other different single cell systems were used to identify MIP function. For example, some yeast laboratory strains do not express functional aquaporins and can be transformed with a cDNA sequence coding for an aquaporin in a commercially available expression vector. After digestion of the cell wall, yeast spheroblasts were used for a swelling respective shrinking assay according to the principle of the oocyte system or yeast plasma membrane vesicles were analyzed in a stopped flow device (Laize et al. 1997; Lagree et al. 1998; Suga and Maeshima 2004). In addition, *E. coli* has been adopted for aquaporin functional analysis

(Delamarche et al. 1999). Bacteria cells expressing the protein of interest were subjected to a hyperosmotic shock and the change in cell shape was recorded by electron microscopy. If an aquaporin was functionally expressed, the cytoplasm retracted due to rapid water loss and plasmolytic spaces were obtained.

2.3 Plant protoplasts for functional analysis of aquaporins

Employing plant protoplasts for the functional analysis has the advantage that plant derived aquaporins can be studied in a homologous or at least in a more similar system than oocytes, yeasts or *E. coli*. Various technical setups were developed for determination of plant protoplast swelling kinetics. From the technical point of view the most challenging task is the fixation of the fragile protoplasts during the transfer to hypo-osmotic conditions. This was achieved either by attachment to a pipette, gluing to glass slides or transfer with the help of an oil coverage (Kaldenhoff et al. 1995; Ramahaleo et al. 1999; Moshelion et al. 2002; Suga et al. 2002). Protoplast analysis was used to determine the water permeability of specific plant tissues (Suga et al. 2002; Siefritz et al. 2004) or to characterize the effects of the inhibition of aquaporin expression by mutation, respectively, after transformation. In this regard the pressure probe is also a tool for analysis of hydraulic conductivity of cells (Tournaire-Roux et al. 2003).

However, all plant techniques were restricted by the laborious techniques to introduce a permanent transformation into plants, or screening for a mutant. The fact that no aquaporin null plant system is at hand, such as, e.g. those in yeast or *E. coli*, is also impeding a simple technical set-up for water permeability analysis of specific aquaporins.

Inhibitor studies, e.g. with mercury, were used as a loophole for this dilemma. However, interpretation of the obtained results remain difficult, no matter which of the above mentioned techniques is applied to measure water permeability. This is due to non-specific side effects of heavy metal ions on metabolism, protein synthesis, or ion channels (Zhang and Tyerman 1999). In addition, plant aquaporins differ in sensitivity to mercury (Biela et al. 1999) and heavy metal ions penetrate certain plant tissues and cells imperfectly (Hill et al. 2004).

3 Effects of aquaporin water conductivity in plants

Despite the difficulties given by a comparable high diversity and large copy number of aquaporin homologues in plants, the function of some plant aquaporins appears to be well studied with regard to water conductivity and its relevance for plant physiology. It is extensively reviewed elsewhere (Maurel 1997; Kaldenhoff and Eckert 1999; Tyerman et al. 1999; Luu and Maurel 2005) and just adumbrated here.

Clear indications for a physiological relevance were obtained in transgenic plants with a modified expression of certain aquaporins or by analysis of aquaporin mutations. First evidence for a function in cellular and whole plant water transport came from plasma membrane aquaporin antisense *Arabidopsis* plants. These developed a larger root system than comparable controls (Kaldenhoff et al. 1998; Grote et al. 1999). Findings from experiments with double antisense constructs targeted to a PIP1 and a PIP2 aquaporin indicated that these aquaporins play an important role in the recovery from a water-deficient condition (Martre et al. 2002). Using antisense and overexpression in tomato, a fruit ripening associated aquaporin was shown to be important for organic acids and sugar composition of tomatoes (Chen et al. 2001). The tobacco plasma membrane aquaporin NtAQP1 was found to be important for root hydraulic conductivity and water stress resistance (Siefritz et al. 2002). Results that imply a role of plasma membrane aquaporins in plant water management were also obtained from monocotyledon species like barley (Katsuhara et al. 2003) or rice (Lian et al. 2004). Studies with *Arabidopsis* PIP2 mutants demonstrated that besides the role for water uptake (Javot et al. 2003), the regulation of aquaporins by pH is important for a developmental adaptation to flooding (Tournaire-Roux et al. 2003). Some reports specify a role of aquaporins during leaf movement, a process with high rates of cellular water transport (Otto and Kaldenhoff 2000; Moshelion et al. 2002; Siefritz et al. 2004). Constitutive overexpression of a plasma membrane aquaporin from *Arabidopsis* in tobacco resulted in increased growth rates under optimal irrigation (Aharon et al. 2003), which was interpreted as the sum of effects on water uptake and photosynthesis due to increased cellular water permeability.

4 Permeability to small non-ionic molecules

4.1 Glycerol

The aquaporin selectivity filter is not always producing exclusive strict water conductivity. Some aquaporins exhibit an additional conductivity for glycerol. A mechanism of water and glycerol transport by the *E. coli* GlpF glycerol facilitator was obtained by cryo-electron microscopy and X-ray studies (Fujiyoshi et al. 2002; de Groot et al. 2003). The data were used for molecular dynamics simulation similar to that performed for Aqp1 (de Groot and Grubmuller 2001). GlpF also contains the conserved NPA motifs at comparable positions to the water selective aquaporins. However, the preference for glycerol is achieved by aromatic amino acids at the periplasmatic side. Tryptophan at position 48, phenylalanine at position 200 and arginine at position 206 form a constriction, and the arginine residue builds hydrogen bonds with two hydroxyl groups of the glycerol molecule. In this way, the glycerol carbon backbone faces into the cavity assembled by the aromatic amino acids mentioned above. Glycerol is separated from other linear polyols and passes the pore in a single file. The GlpF pore is completely amphipathic with polar residues opposed in a hydrophobic wall.

Plant NIPs exhibit similar selectivity to the GlpF. However, the NIP selectivity filter differs (Biswas 2004). Computational analyses of this filter region, the aromatic/Arg [ar/R] filter, lead to a classification of plant MIPs according to pore size and amphiphilicity into eight subfamilies: one PIP, three TIP, two NIP and two SIP. It was speculated that the specific features of the subfamilies result in divergent transport selectivity (Wallace and Roberts 2004).

The first plant aquaporins with glycerol permeability were found in soybean root nodules (Rivers et al. 1997; Dean et al. 1999). The soybean nodulin 26 and *Lotus japonicus* LMIP2, showed a low intrinsic osmotic water permeability and conductivity to uncharged polyols such as glycerol (Wallace et al. 2002). In addition, the PsNIP-1 from *Pisum* was shown to be an aquaglyceroporin (Schuurmans et al. 2003). NIPs expressed in nodules were thought to be part of a metabolite transfer system between the plant cytoplasm and the symbiotic bacterioids (Rivers et al. 1997). The function of other NIPs, which are expressed in non-symbiotic tissue or plants that generally do not interact with N_2-fixing bacteria remains rather speculative (Ciavatta et al. 2002).

Heterologous expression of two NIPs from *Arabidopsis thaliana* (AtNLM1 and AtNLM2) in baker's yeast demonstrated a glycerol permease activity. The transport was non-saturable up to 100 mM extracellular glycerol concentration (Weig and Jakob 2000). The urea-transporter mutant Dur3p was used for complementation by CpNIP1, a Nod26-like protein from zucchini (*Cucurbita pepo* L.), and by *Arabidopsis thaliana* delta-TIP or gamma-TIP (Klebl et al. 2003).

4.2 CO_2

It is generally accepted that gases, such as CO_2 or NH_3, easily cross cell membranes by dissolving in the membrane lipid. However, it has been suggested that their transport is facilitated by aquaporins because the gas permeability of some cell membranes seems quite high, of others relatively low (Prasad et al. 1998; Terashima and Ono 2002) and can be altered by inhibitors of aquaporins (Cooper and Boron 1998; Niemietz and Tyerman 2000; Terashima and Ono, 2002). Theoretically, these gas molecules are small enough to traverse the central pore. Due to very high calculated CO_2 permeability values for biological membranes (48 cm/s) and supposedly existing unstirred layer effects, some authors think a gas transport facilitated by aquaporins is unlikely (Hill et al. 2004). Experimental data, e.g. of plant plasma membranes (Gimmler et al. 1990) or a cholesterol:lecithin bilayer, were by a factor 10–100 lower, but still relatively high in comparison to water for instance (Gutknecht et al. 1977). The opinion that gas transport by proteins is rather improbable was supported by experimental data showing no differences, in animal lungs for example, if AQP1, AQP5 or both in combination were expressed or not expressed (Fang et al. 2002). In contrast, there are experimental evidences from studies with *Xenopus* oocytes indicating that the human AQP1 or the tobacco NtAQP1 increases CO_2 permeability (Nakhoul et al. 1998a; Uehlein et al. 2003). This facilitated CO_2 transport could be attributed to the gas channel function of aquaporins, effects on the membrane lipid composition, or expression pattern of oocyte intrinsic genes that could modify oocyte CO_2 permeability. Due to results of inhibitor studies and analysis of a mercury insensitive aquaporin mutant, these factors could be excluded and a CO_2 conductivity for hAQP1 was suggested (Cooper and Boron 1998). Physiological consequences of AQP1 facilitated CO_2 transport are still a matter of debate (Cooper et al. 2002), because animals with an aquaporin gene knock-out did not show differences in CO_2 exchange rates (Sun et al. 2001; Fang et al. 2002). On the other hand, results obtained with human erythrocytes at low chemical gradients for CO_2, demonstrated that nearly the entire CO_2 transport across the membrane was mediated by AQP1 and the HCO_3^-–Cl^- transporter (Blank and Ehmke 2003). It was concluded that these proteins might function as high affinity sites for CO_2 transport across the erythrocyte membrane. Taken together, the situation in animals appears quite confusing. Diverse tissues, cells and membranes with differing physiological functions were analysed and diverse experimental set-ups were applied. As a consequence, the scientific discussion with this regard is rather controversy.

In contrast to the situation in animals, data from physiological studies for plant aquaporins indicate that a P_{CO2} is of physiological relevance. When *Vicia faba* or *Phaseolous vulgaris* leaf discs were treated with sub-mM concentrations of $HgCl_2$, the hydraulic permeability of the plasma membrane was decreased by 70–80% as well as photosynthetic CO_2 fixation and the conductance for CO_2 from the intercellular spaces to the chloroplast stroma (40% and 30% of controls). Although the $HgCl_2$ treatment should be considered with the same carefulness as in experiments investigating water conductivity, it was assumed that the photosynthetic CO_2 uptake across the plasma membrane of the mesophyll cells was facilitated by mercury-sensitive aquaporins (Terashima and Ono 2002).

Plants with aquaporin knock-out or aquaporin overexpression show several differences not only in water transport (Siefritz et al. 2002), but also in CO_2 limited processes like photosynthesis. When grown under favourable growth conditions, transgenic tobacco overexpressing an *Arabidopsis* PIP1b showed significantly higher transpiration and photosynthetic rates than non-transformed control plants (Aharon et al. 2003). However, these authors did not relate the effects to an increase in CO_2 transport rates, but to a facilitated water transport. Another study, also using an aquaporin from a different species, was conducted with rice in order to confirm the hypothesis that a PIP2 contributes to facilitated CO_2 transport (Hanba et al. 2004). The transgenics overexpressed the barley aquaporin HvPIP2;1. Mesophyll conductance (g_i) was determined for intact leaves by concurrent measurements of gas exchange and carbon isotope ratio. The level of HvPIP2;1 was found to be strongly related to g_i and the results were interpreted to suggest that HvPIP2;1 has a role in CO_2 diffusion in rice leaves. A molecular characterization of HvPIP2;1 was, however, not provided in this study. It remained to be determined whether the correlation between aquaporin expression and CO_2 permeability increase was just a side effect or causal to HvPIP2;1 CO_2 conductance.

A molecular characterization of CO_2 conductance for a plant aquaporin was performed in *Xenopus* oocytes with the tobacco NtAQP1 (Uehlein et al. 2003). Oocytes were injected with a NtAQP1 cRNA and a solution of carbonic anhydrase, an enzyme that accelerates the conversion of CO_2 to HCO_3^-. In this experimental setup, CO_2 membrane transport rather than the conversion reaction to HCO_3^- is rate limiting for HCO_3^- accumulation in the oocyte (Gutknecht et al. 1977; Nakhoul et al. 1998b). Consequently, CO_2 transport into the cells generated a decrease in intracellular pH. It was found that CO_2 uptake rates in oocytes expressing NtAQP1 were 45% higher and comparable with those of the human AQP1. It was concluded that NtAQP1 has a function as a CO_2 membrane transport facilitator in the oocyte expression system. The role for CO_2 transport in plants was studied in tobacco expressing an NtAQP1 antisense construct or in plants with an NtAQP1 coding region under the control of a tetracycline-inducible promoter (Gutknecht et al. 1977; Gatz 1995) (TET-NtAQP1). Photosynthetic ^{14}C incorporation from $^{14}CO_2$ was found to be dependent on NtAQP1 expression: a low NtAQP1 level resulted in lower incorporation rates and an increased NtAQP1 level resulted in higher incorporation rates compared to controls. Other processes that are known to be regulated by CO_2 steady state concentrations, such as stomatal movement and net photosynthesis in intact and detached leaves, were found to be dependent on the level of NtAQP1 expression in these plants. It was concluded that if a relatively low CO_2 concentration gradient exists, as is the case between a photosynthesising plant cell and the atmosphere, CO_2 permeability of NtAQP1 is of substantial physiological significance. Additional yet unpublished studies (J. Flexas, M. Ribas-Carbó, J. Bota, J. Cifre, H. Medrano, Palma d. Mallorca, Spain; B. Otto, R. Kaldenhoff Darmstadt, Germany) on the NtAQP1 antisense and NtAQP1 overexpressing plants showed that other intrinsic factors that might change CO_2 transport rates, like rubisco or carboanhydrase content respectively, were not changed. However, leaf mesophyll conductance varied by about 30%, depending on the level of NtAQP1 expression.

4.3 NH_3

Like the debate about CO_2 conductivity of aquaporins and its implementation on physiology, there is a great deal of lively discussion on NH_3 permeation. Reports taking sides on aquaporin mediated NH_3 permeability (Nakhoul et al. 2001; Cooper et al. 2002) or providing facts against it were published (Zeidel et al. 1994).

NH_4 uptake at low extracellular concentration in plants is catalyzed by members of the ammonium transporter/methylammonium permease (AMT/Mep) family (Ninnemann et al. 1994). Yet, no specific NH_4/NH_3 transporter, operating at elevated concentrations, has been isolated in any organism. There is, however, evidence from inhibitor studies in plants in favour of NH_3 permeability by aquaporins (Niemietz and Tyerman 2000). Using functional complementation of a yeast ammonium transport mutant (Dmep1–3), three wheat (*Triticum aestivum*) TIP2 aquaporins were characterized, which complement the effect of the deletion mutation on reduced ammonium supply (2 mM). When expressed in oocytes, an additional conductivity for the NH_4^+ analogues methylammonium and formamide was registered. Homology modelling of the TIP2 combined with data from site directed mutagenesis and electrically measurements suggested that NH_3 enters the pore, is protonated and released as NH_4^+ (Jahn et al. 2004). The specific TIP2 seems to fulfil the requirements for the predicted low affinity NH_4^+ transporter. A physiological function has so far not been circumstantiated.

4.4 Boron

Boron is important as a micronutrient. The boron permeability of purified plasma membrane vesicles obtained from squash (*Cucurbita pepo*) roots was found to be 6 times higher than the permeability of microsomal vesicles and boron permeation was partially inhibited by mercuric chloride or phloretin. Expression of a PIP1 in oocytes increased the boron permeability by about 30% (Dordas et al. 2000).

References

Aharon R, Shahak Y, Wininger S, Bendov R, Kapulnik Y, Galili G (2003) Overexpression of a plasma membrane aquaporin in transgenic tobacco improves plant vigor under favorable growth conditions but not under drought or salt stress. Plant Cell 15:439–447

Biela A, Grote K, Otto B, Hoth S, Hedrich R, Kaldenhoff R (1999) The Nicotiana tabacum plasma membrane aquaporin NtAQP1 is mercury-insensitive and permeable for glycerol. Plant J 18:565–570

Biswas S (2004) Functional properties of soybean nodulin 26 from a comparative three-dimensional model. FEBS Lett 558:39–44

Blank ME, Ehmke H (2003) Aquaporin-1 and HCO_3^-–Cl^- transporter-mediated transport of CO_2 across the human erythrocyte membrane. J Physiol 550:419–429

Borgnia M, Nielsen S, Engel A, Agre P (1999a) Cellular and molecular biology of the aquaporin water channels. Annu Rev Biochem 68:425–458

Borgnia MJ, Kozono D, Calamita G, Maloney PC, Agre P (1999b) Functional reconstitution and characterization of AqpZ, the *E.-coli* water channel protein. J Mol Biol 291:1169–1179

Borstlap AC (2002) Early diversification of plant aquaporins. Trends Plant Sci 7:529–530

Bots M, Feron R, Uehlein N, Weterings K, Kaldenhoff R, Mariani T (2005) PIP1 and PIP2 aquaporins are differentially expressed during tobacco anther and stigma development. J Exp Bot 56:113–121

Chaumont F, Barrieu F, Jung R, Chrispeels MJ (2000) Plasma membrane intrinsic proteins from maize cluster in two sequence subgroups with differential aquaporin activity. Plant Physiol 122:1025–1034

Chen GP, Wilson ID, Kim SH, Grierson D (2001) Inhibiting expression of a tomato ripening-associated membrane protein increases organic acids and reduces sugar levels of fruit. Planta 212:799–807

Ciavatta VT, Morillon R, Pullman GS, Chrispeels MJ, Cairney J (2001) An aquaglyceroporin is abundantly expressed early in the development of the suspensor and the embryo proper of loblolly pine. Plant Physiol 127:1556–1567

Ciavatta VT, Egertsdotter U, Clapham D, von Arnold S and Cairney J (2002) A promoter from the loblolly pine PtNIP1;1 gene directs expression in an early-embryogenesis and suspensor-specific fashion. Planta 215: 694–698.

Cooper GJ, Boron WF (1998) Effect of PCMBS on CO_2 permeability of Xenopus oocytes expressing aquaporin 1 or its C189S mutant. Am J Physiol Cell Physiol 44:C1481–C1486

Cooper GJ, Zhou Y, Bouyer P, Grichtchenko II, Boron WF (2002) Transport of volatile solutes through AQP1. J Physiol 542:17–29

de Groot BL, Grubmuller H (2001) Water permeation across biological membranes: Mechanism and dynamics of aquaporin-1 and GlpF. Science 294:2353–2357

de Groot BL, Engel A, Grubmuller H (2003) The structure of the aquaporin-1 water channel: a comparison between cryo-electron microscopy and X-ray crystallography. J Mol Biol 325:485–493

Dean RM, Rivers RL, Zeidel ML, Roberts DM (1999) Purification and functional reconstitution of soybean nodulin 26. An aquaporin with water and glycerol transport properties.
Biochemistry 38:347–353

Delamarche C, Thomas D, Rolland JP, Froger A, Gouranton J, Svelto M, Agre P, Calamita G
(1999) Visualization of AqpZ-mediated water permeability in Escherichia coli by cryo-
electron microscopy. J Bacteriol 181:4193–4197

Dordas C, Chrispeels MJ, Brown PH (2000) Permeability and channel-mediated transport of
boric acid across membrane vesicles isolated from squash roots. Plant Physiol
124:1349–1362

Fang XH, Yang BX, Matthay MA, Verkman AS (2002) Evidence against aquaporin-1-depenent
CO_2 permeability in lung and kidney. J Physiol Lond 542:63–69

Fetter K, Van Wilder V, Moshelion M, Chaumont F (2004) Interactions between plasma
membrane aquaporins modulate their water channel activity. Plant Cell 16:215–228

Fortin MG, Morrison NA, Verma DP (1987) Nodulin-26, a peribacteroid membrane nodulin
is expressed independently of the development of the peribacteroid compartment. Nucleic
Acids Res 15:813–824

Fujiyoshi Y, Mitsuoka K, de Groot BL, Philippsen A, Grubmuller H, Agre P, Engel A (2002)
Structure and function of water channels. Curr Opin Struct Biol 12:509–515

Gatz C (1995) Novel inducible/repressible gene expression systems. Meth Cell Biol
50:411–424

Gimmler H, Weiss C, Baier M, Hartung W (1990) The Conductance of the Plasmalemma for
CO_2. J Exp Bot 41:785–795

Grote K, Gimmler H, Kaldenhoff R (1999) Coulter counter cell size determination of protoplasts from *Arabidopsis thaliana* PIP1b aquaporin antisense lines under iso- and hypo-
osmotic conditions. Protoplasma 210:31–35

Gutknecht J, Bisson MA, Tosteson FC (1977) Diffusion of carbon-dioxide through lipid
bilayer membranes—effects of carbonic-anhydrase, bicarbonate, and unstirred layers.
J Gen Physiol 69:779–794

Hanba YT, Shibasaka M, Hayashi Y, Hayakawa T, Kasamo K, Terashima I, Katsuhara M
(2004) Overexpression of the barley aquaporin HvPIP2;1 increases internal CO_2 conductance and CO_2 assimilation in the leaves of transgenic rice plants. Plant Cell Physiol
45:521–529

Hill AE, Shachar-Hill B, Shachar-Hill Y (2004) What are aquaporins for? J Memb Biol 197:1–32

Jahn TP, Moller ALB, Zeuthen T, Holm LM, Klaerke DA, Mohsin B, Kuhlbrandt W
Schjoerring JK (2004) Aquaporin homologues in plants and mammals transport ammonia. FEBS Lett 574:31–36

Javot H, Lauvergeat V, Santoni V, Martin-Laurent F, Guclu J, Vinh J, Heyes J, Franck KI,
Schaffner AR, Bouchez D, Maurel C (2003) Role of a single aquaporin isoform in root
water uptake. Plant Cell 15:509

Johanson U, Karlsson M, Johansson I, Gustavsson S, Sjovall S, Fraysse L, Weig AR, Kjellbom P
(2001) The complete set of genes encoding major intrinsic proteins in *Arabidopsis* provides a framework for a new nomenclature for major intrinsic proteins in plants. Plant
Physiol 126:1358–1369

Johanson U, Gustavsson S (2002) A new subfamily of major intrinsic proteins in plants. Mol
Biol Evol 19:456–461

Kaldenhoff R, Kolling A, Meyers J, Karmann U, Ruppel G, Richter G (1995) The blue light-
responsive AthH2 gene of *Arabidopsis thaliana* is primarily expressed in expanding as well
as in differentiating cells and encodes a putative channel protein of the plasmalemma.
Plant J 7:87–95

Kaldenhoff R, Grote K, Zhu JJ, Zimmermann U (1998) Significance of plasmalemma aquaporins for water-transport in *Arabidopsis thaliana*. Plant J 14:121–128

Kaldenhoff R, Eckert M (1999) Features and function of plant aquaporins. J Photochem
Photobiol B-Biol 52:1–6

Katsuhara M, Koshio K, Shibasaka M, Hayashi Y, Hayakawa T, Kasamo K (2003) Over-expresion of a barley aquaporin increased the shoot/root ratio and raised salt sensitivity in transgenic rice plants. Plant Cell Physiol 44:1378–1383

Klebl F, Wolf M and Sauer N (2003) A defect in the yeast plasma membrane urea transporter Dur3p is complemented by CpNIP1, a Nod26-like protein from zucchini (*Cucurbita pepo* L.), and by *Arabidopsis thaliana* delta-TIP or gamma-TIP. FEBS Lett 547:69–74

Lagree V, Pellerin I, Hubert JF, Tacnet F, LeCaherec F, Roudier N, Thomas D, Gouranton J Deschamps S (1998) A yeast recombinant aquaporin mutant that is not expressed or mistargeted in *Xenopus* oocyte can be functionally analyzed in reconstituted proteoliposomes. J Biol Chem 273:12422–12426

Laize V, Ripoche P, Tacnet F (1997) Purification and functional reconstitution of the human CHIP28 water channel expressed in *Saccharomyces cerevisiae*. Protein Expression Purification 11:284–288

Lian HL, Yu X, Ye Q, Ding XS, Kitagawa Y, Kwak SS, Su WA, Tang ZC (2004) The Role of aquaporin RWC3 in drought avoidance in rice. Plant Cell Physiol 45:481–489

Luu DT, Maurel C (2005) Aquaporins in a challenging environment: molecular gears for adjusting plant water status. Plant Cell Environ 28:85–96

Martre P, Morillon R, Barrieu F, North GB, Nobel PS, Chrispeels MJ (2002) Plasma membrane aquaporins play a significant role during recovery from water deficit. Plant Physiol 130:2101–2110

Maurel C (1997) Aquaporins and water permeability of plant membranes. Ann Rev Plant Physiol Plant Mol Biol 48:399–429

Maurel C, Kado RT, Guern J, Chrispeels MJ (1995) Phosphorylation regulates the water channel activity of the seed-specific aquaporin alpha-TIP. EMBO J 14:3028–3035

Moshelion M, Becker D, Biela A, Uehlein N, Hedrich R, Otto B, Levi H, Moran N, Kaldenhoff R (2002) Plasma membrane aquaporins in the motor cells of *Samanea saman*: diurnal and circadian regulation. Plant Cell 14:727–739

Nakhoul NL, Davis BA, Romero MF, Boron WF (1998a) Effect of expressing the water channel aquaporin-1 on the CO_2 permeability of *Xenopus* oocytes. Am J Physiol Cell Physiol 43:C543–C548

Nakhoul NL, Davis BA, Romero MF, Boron WF (1998b) Effect of expressing the water channel aquaporin-1 on the CO_2 permeability of *Xenopus* oocytes. Am J Physiol-Cell Physiol 43:C543–C548

Nakhoul NL, Hering-Smith KS, Abdulnour-Nakhoul SM, Hamm LL (2001) Transport of NH3/NH4+ in oocytes expressing aquaporin-1. Am J Physiol-Renal Physiol 281:F255–F263

Niemietz CM, Tyerman SD (2000) Channel-mediated permeation of ammonia gas through the peribacteroid membrane of soybean nodules. FEBS Lett 465:110–114

Ninnemann O, Jauniaux JC, Frommer WB (1994) Identification of a high-affinity Nh4+ transporter from plants. Embo J 13:3464–3471

Otto B, Kaldenhoff R (2000) Cell-specific expression of the mercury-insensitive plasma-membrane aquaporin NtAQP1 from *Nicotiana tabacum*. Planta 211:167–172

Prasad GVR, Coury LA, Finn F, Zeidel ML (1998) Reconstituted aquaporin 1 water channels transport CO_2 across membranes. J Biol Chem 273:33123–33126

Preston GM, Carroll TP, Guggino WB, Agre P (1992) Appearance of water channels in *Xenopus* oocytes expressing red cell CHIP28 protein. Science 256:385–387

Quigley F, Rosenberg J, Shachar-Hill Y, Bohnert H (2001) From genome to function: the *Arabidopsis* aquaporins. Gen Biol 3:research0001

Ramahaleo T, Morillon R, Alexandre J, Lassalles JP (1999) Osmotic water permeability of isolated protoplasts. Modifications during development. Plant Physiol 119:885–896

Rivers RL, Dean RM, Chandy G, Hall JE, Roberts DM, Zeidel ML (1997) Functional analysis of nodulin 26, an aquaporin in soybean root nodule symbiosomes. J Biol Chem 272:16256–16261

Schuurmans JA, van Dongen JT, Rutjens BP, Boonman A, Pieterse CM, Borstlap AC (2003) Members of the aquaporin family in the developing pea seed coat include representatives of the PIP, TIP, and NIP subfamilies. Plant Mol Biol 53:633–645

Siefritz F, Tyree MT, Lovisolo C, Schubert A, Kaldenhoff R (2002) PIP1 plasma membrane aquaporins in tobacco: from cellular effects to function in plants. Plant Cell 14:869–876

Siefritz F, Otto B, Bienert GP, van der Krol A, Kaldenhoff R (2004) The plasma membrane aquaporin NtAQP1 is a key component of the leaf unfolding mechanism in tobacco. Plant J 37:147–155

Suga S, Maeshima M (2004) Water channel activity of radish plasma membrane aquaporins heterologously expressed in yeast and their modification by site-directed mutagenesis. Plant Cell Physiol 45:823–830

Suga S, Komatsu S, Maeshima M (2002) Aquaporin isoforms responsive to salt and water stresses and phytohormones in radish seedlings. Plant Cell Physiol 43(10):1229–1237

Sun XC, Allen KT, Xie Q, Stamer WD, Bonanno JA (2001) Effect of AQP1 expression level on CO_2 permeability in bovine corneal endothelium. Invest Ophthalmol Vis Sci 42:417–423

Terashima I, Ono K (2002) Effects of $HgCl_2$ on CO_2 dependence of leaf photosynthesis: evidence indicating involvement of aquaporins in CO_2 diffusion across the plasma membrane. Plant Cell Physiol 43:70–78

Tournaire-Roux C, Sutka M, Javot H, Gout E, Gerbeau P, Luu DT, Bligny R, Maurel C (2003) Cytosolic pH regulates root water transport during anoxic stress through gating of aquaporins. Nature 425:393–397

Tyerman SD, Bohnert HJ, Maurel C, Steudle E, Smith JAC (1999) Plant aquaporins: their molecular biology, biophysics and significance for plant water relations. J Exp Bot 50 Spec Iss. SI:1055–1071

Uehlein N, Lovisolo C, Siefritz F, Kaldenhoff R (2003) The tobacco aquaporin NtAQP1 is a membrane CO_2 pore with physiological functions. Nature 425:734–737

Wallace IS, Roberts DM (2004) Homology modeling of representative subfamilies of *Arabidopsis* major intrinsic proteins. Classification based on the aromatic/arginine selectivity filter. Plant Physiol 135:1059–1068

Wallace IS, Wills DM, Guenther JF, Roberts DM (2002) Functional selectivity for glycerol of the nodulin 26 subfamily of plant membrane intrinsic proteins. FEBS Lett 523:109–112

Weig AR, Jakob C (2000) Functional identification of the glycerol permease activity of *Arabidopsis thaliana* NLM1 and NLM2 proteins by heterologous expression in *Saccharomyces cerevisiae*. FEBS Lett 481:293–298

Yang B, Verkman AS (1997) Water and glycerol permeabilities of aquaporins 1-5 and mip determined quantitatively by expression of epitope-tagged constructs in *Xenopus* pocytes. J Biol Chem 16140–16146

Zeidel ML, Nielsen S, Smith BL, Ambudkar SV, Maunsbach AB, Agre P (1994) Ultrastructure, pharmacological inhibition, and transport selectivity of aquaporin channel-forming integral protein in proteoliposomes. Biochemistry 33:1606–1615

Zhang R, Verkman AS (1991) Water and urea permeability properties of *Xenopus* oocytes: expression of mRNA from toad urinary bladder. Am J Physiol 260:C26–C34

Zhang WH, Tyerman SD (1999) Inhibition of water channels by $HgCl_2$ in intact wheat root cells. Plant Physiol 120:849–857

Prof. Dr. R. Kaldenhoff
Applied Plant Sciences
Institute of Botany
Darmstadt University of Technology
Schnittspantr. 10
D-64287 Darmstadt
Germany
e-mail: kaldenhoff@bio.tu-darmstadt.de

New insight into auxin perception, signal transduction and transport

May Christian, Daniel Schenck, Michael Böttger, Bianka Steffens, Hartwig Lüthen

1 Growth stimulation: the classical effect of auxin

The auxin problem has been a classical question of plant physiology, ever since the hormone was first identified (Went 1928) and isolated in the 1930s (Kögl et al. 1934). Generations of researchers have trained their skills on the mechanism of auxin action. However, the classical auxin effect is still not completely understood. Obviously, micromolar concentrations trigger a cellular programme that induces an enormous rise in rates of elongation growth in shoots and coleoptiles. This effect is rapid, occurring after lag phases of only 10–20 min (Dela Fuente and Leopold 1970). It is generally accepted that the growth response is caused by a loosening of the cell wall (e.g. Heyn 1931; Cleland 1967; for review, see Cosgrove 1999). But how does the plant cell perceive the auxin signal? What is the nature of the signalling chain? What kind of mechanism is responsible for gene expression and cell wall loosening?

In recent years, there has been significant progress in our understanding of auxin-induced gene expression. Having now a patchy idea of how auxin turns on genes, the question is which of these genes are relevant for growth control. Another field of rapid progress is the investigation of auxin transport, which is a crucial prerequisite for the control of tropisms. This review will focus on the new advances and will, in the end, try to define some open problems.

2 Auxin receptors

One paradigm of auxin research is the concept of an auxin receptor detecting the auxin signal. Although auxin binding protein 1 (ABP1) has been identified as one relevant auxin sensor, there is still space for other binding proteins involved in auxin perception.

Progress in Botany, Vol. 67
© Springer-Verlag Berlin Heidelberg 2006

2.1 Auxin binding protein 1 (ABP1)

Specific auxin binding to subcellular fractions of plant membranes was first
shown in the early 1970s in the pioneering work of Hertel (Hertel et al. 1972;
Dohrmann et al. 1978). The activity was localized in the endoplasmic reticulum and was first purified to homogeneity by Löbler and Klämbt (1985).
Auxin binding protein 1 (ABP1) was sequenced and cloned (Hesse et al.
1989). Meanwhile, sequence data from a variety of species are available (for
review, see Napier et al. 2002).

Figure 1 shows the structure of ABP1 and a number of synthetic oligopep-
tides and antibodies that were used in electrophysiological studies. It was
shown that anti-ABP1-antibodies inhibited auxin-induced membrane hyperpolarization in tobacco protoplasts (Barbier-Brygoo et al. 1991) and also an
ATP-driven transmembrane current in maize protoplasts (Rück et al. 1993).
It was speculated from these data that extracellular ABP1 was an auxin receptor for these electrophysiological effects. There were several problems with
this theory, however. Firstly, the amino acid sequence of ABP1 has a KDEL
motif at the C-terminus. This signal sequence marks the protein to be
retained in the lumen of the endoplasmic reticulum. How can a protein with
this cellular "return to sender" ZIP-code be functional at the outer surface of
the PM? Secondly, no transmembrane domains are predicted from the amino

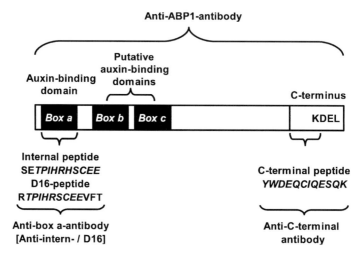

Fig. 1. Molecular structure of ABP1 and source of peptides and antibodies used in studies of
auxin signal transduction. The boxes *a*, *b* and *c* are conserved domains involved in auxin
binding, whereas the C-terminus transmits the auxin signal to a yet unidentified trans plasma
membrane protein. Anti-ABP1-antibodies inhibit auxin action in protoplast systems. The
anti-box a-antibodies have auxin agonist activities, as have synthetical C-terminal peptides

acid sequence of ABP1. ABP1 does not feel, look or smell like a typical hormone receptor.

In order to overcome these conceptual problems the so-called docking protein hypothesis was proposed (Fig. 2). Although most of the ABP1 is trapped inside the cell, a small amount escaping the KDEL recycling may be excreted to the apoplastic side of the plasma membrane. Upon binding extracellular auxin, ABP1 is thought to undergo a conformational change, facilitating binding to a hypothetical trans-membrane docking protein, from where the signalling chain is started. This general concept was first outlined by Klämbt as early as 1990, but the docking protein is still not identified. A detailed description of Klämbt's original model and several variations is included in our earlier review (Lüthen et al. 1999). In this review, we will focus on some more recent findings.

2.1.1 3-D structure of ABP1

ABP1 has been crystallized and a 1.9 Å resolution structure has recently been published (Woo et al. 2002) (Fig. 3). As was suggested earlier, ABP1 was found to be a dimer. N- and C-terminus are linked by disulfide bridges

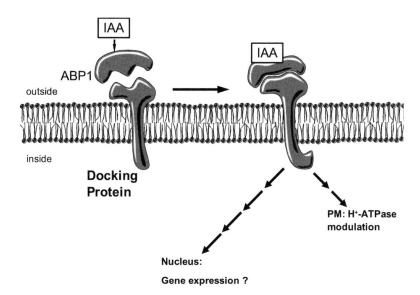

Fig. 2. Docking protein hypothesis. Upon binding auxin, the ABP1-auxin-complex attaches to a not yet identified transmembrane docking protein, transmitting the auxin signal into the cell

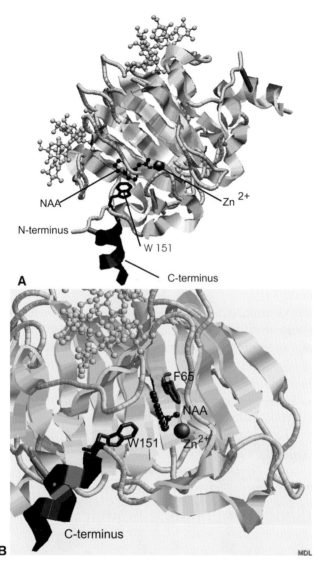

Fig. 3. Three-dimensional structure of ABP1, binding the synthetic auxin NAA, as derived from X-ray diffraction data (Wu et al. 2002; PDB ID code: 1LRH). The figure was created using Protein Explorer in conjunction with Chime 2.6 (Elsevier MDL) and Netscape as a browser. **A** ABP1 dimer binding NAA, one monomer is hidden in this view. Note the auxin binding site consisting of a Zn^{2+} ion and the tryptophan residue 151 (W151). The C-terminus sticks out of the structure and is coupled to the N-terminus via a disulfide bridge. **B** Detailed view of the auxin binding domain of ABP1. The aromatic ring of auxin is in close proximity of both W151 and the phenylalanine residue F65

between cysteines (Fig. 3A). A Zn^{2+} ion is coordinatively bound to several histidine residues, and also accounts for the binding to the carboxylic group of the auxin molecule (Fig. 3B). The short helical structure at the C-terminus, thought to link ABP1 to the docking protein, sticks out of the protein complex.

Astonishingly, structures for the protein in complex and in the absence of auxin do not differ very much. This may indicate that auxin binding causes no dramatic conformational change in the molecule, probably because the intermolecular movements are hampered by the disulfide bridges. It has been suggested that crystallization prefers the bound state of ABP1, and that in the native protein the C-terminal alpha-helix acts as a rigid rod conveying movement to the protein's surface. Auxin binding may pull the C-terminal helix into the core of the protein, thus transferring the signal to the docking protein (Fig. 3B). A role of a tryptophan residue close to the C-terminus (W151) in this process has been proposed. The phenylalanine residue F65 might also be a part of the aromatic binding site. A possibility may also be that auxin inserts between F65 and W151, thereby pushing the C-terminus outward, facilitating its interaction with the docking protein. The KDEL sequence alone is not essential for auxin binding or PM interaction (David et al. 2001).

2.1.2 ABP1 mutants

An analysis of ABP1 knockout plants or overexpressors could decide the much-debated question if ABP1 is the auxin receptor relevant for growth control by simply putting them into an auxanometer. Aside from the fact that building auxanometers for the tiny *Arabidopsis* hypocotyls is not an easy task (Christian and Lüthen 2000), this approach has not yet given a conclusive answer.

A number of ABP1 mutants and transgenic ABP1 plants have been described (Chen et al. 2001, 2003). In *Arabidopsis*, a knockout mutant was created by T-DNA transformation (Chen et al. 2001). It was found that homozygous plants were embryolethal. Although this demonstrates that ABP1 is an important protein, it makes a direct investigation of auxin-induced growth in a classical growth test impossible. Chen et al. therefore investigated the development of wild type and *abp1* embryos and found that *abp1* develops normally until the early globular stages. During the mid-globular stages, several tier cells start to elongate only in the wild type. After that, *abp1* embryo tier cells form misoriented cross walls and are arrested in the globular stage. The fact that critical cell elongation steps do not occur in *abp1* embryos is a telltale sign that ABP1 has to be somehow relevant for cell expansion, but rapid growth responses to auxin have not yet been tested in these mutants.

In order to overcome the problem of embryolethality, Jones and coworkers used DEX-inducible antisense plants (Chen et al. 2001). Treatment with dexamethasone (DEX) drastically reduced the levels of *ABP1* expression in BY-2 cells. Wild type cells develop an elongated shape, whereas DEX-induced transformed cells do not elongate. DEX-induced cell cultures also show a suppressed increase in fresh weight. These results also suggest a role of ABP1 in growth control.

DEX-inducible *Arabidopsis* ABP1 antisense plants have also been created. Using a CCD-auxanometer, our group investigated their auxin-induced growth response, but found no difference to the wild type. However, it was shown by the Jones group (Chen et al. 2001, 2003) that the ABP1 protein level was near normal, although the ABP1 mRNA was dramatically reduced in DEX-treated plants.

2.1.3 Indirect evidence for ABP1 as a growth relevant receptor at the single cell level

The most convincing results for the receptor role of ABP1 came from studies of electrophysiological auxin responses of protoplasts (see above). The "protoplast membrane hyperpolarization assay" (Barbier-Brygoo et al. 1991) responds to auxin, but is technically demanding and the physical nature of the measured parameter is not clear. The situation is much better for the whole cell patch clamp studies by Rück et al. (1993), which basically confirmed the activity of anti-ABP1 antibodies on an auxin-induced transmembrane current. Both responses occur very early, long before the end of the lag phase of the rapid growth response. Thus, it was unclear whether the parameters measured in the electrophysiological systems were really linked to the classical growth responses.

We investigated another single cell system, this time based on protoplast swelling. Protoplasts were long known to swell upon auxin treatment (Keller and Van Volkenburgh 1996). The time scale of this response is very similar to that of the growth effect on organ level (Steffens and Lüthen 2000). It has to be stressed that the physical nature of protoplast swelling, which is due to subtle changes in osmoregulation, is probably very different to turgor driven growth.

We could, however, show that protoplast swelling was generally inhibited by pretreatment with anti-ABP1 antibodies. Antibodies directed against box a, the putative auxin-binding pocket of ABP1 (Fig. 1), had auxin agonist activity. Peptides with the C-terminal sequence of ABP1, which are supposed to bind to the docking protein, induced an auxin-like effect (Steffens et al.

2001). Taken together with earlier similar results from the electrophysiological responses, these data clearly demonstrate that ABP1 is a receptor relevant for auxin-induced protoplast swelling responses. But does that mean that ABP1 has a role in growth control?

To address this question, our group used an auxin-insensitive signal transduction mutant in tomato, *diageotropica* (Christian et al. 2003). Hypocotyls of these plants do not respond to exogenously applied auxin, and gravitropism is severely impaired (Muday et al. 1995). Protoplasts prepared from *dgt* hypocotyls did not respond with a swelling to auxin, agonistic anti-ABP1-antibodies (Christian et al. 2003) or the C-terminal peptide (Schenck and Lüthen, unpublished data). This shows that the mutation in DGT interrupts a signalling chain that uses ABP1 as a receptor, as it impairs the growth response. Although other interpretations are possible, one can hardly escape the impression that ABP1 is somehow involved in the machinery controlling growth, if it is not the only relevant auxin receptor.

On the other hand, there are studies of auxin-induced growth in the influence of inhibitors of the auxin efflux carrier that indicate that growth can occur in the absence of extracellular auxin (Davies et al. 1974; Vesper and Kuss 1990; Claussen et al. 1996). A similar line of evidence is based on an analysis of mutants of putative auxin influx carriers (*aux1, lax*). They show reduced sensitivity for those auxins that are taken up by carriers rather than by diffusion (Bennett et al. 1996, 1998; Marchant et al. 2002). Therefore, all available evidence supports models proposing a concerted action of extracellular ABP1 and some kind of intracellular receptor. The decision in the life of a cell to undergo an enormous irreversible increase in volume is a very serious one; it may well be that it is secured by more than one lock. Recent data on protoplast swelling in pea protoplasts indicate that there are two distinct pathways regulating this process, one of them being ABP1-dependent, the other not (Yamagami et al. 2004).

The research in the last decade made clear that ABP1 is more than just a red herring (Hertel 1995; Venis 1995), and made it very probable that it is involved in growth control, perhaps without being the only player. An open question is still how the signal is transmitted into the cell; the mysterious docking protein is still not identified.

2.2 Other receptor candidates

2.2.1 ABP57

A Korean group recently isolated a soluble auxin binding protein from rice (Kim et al. 2001) directly interacting with the plasma membrane ATPase.

In assays of ATPase activity of isolated plasma membrane vesicles, this protein was able to stimulate ATPase activity in an auxin-dependent manner. The very rapid direct stimulation of proton pumping observed in protoplasts may be in part mediated by this protein. On the other hand, ABP57 appears to be related to other proteins binding IAA like bovine serum albumin (in fact it was purified on anti-BSA affinity columns). BSA is known to bind various aromatic substances like IAA and tryptophan. In addition, ABP57 does not stimulate ATPase activity when the preparation is subject to active synthetic auxins such as 2,4-D and NAA, which makes the relevance of this candidate as a primary auxin receptor quite doubtful. It may however have a function in modulating auxin action.

2.2.2 Receptor-like kinases (RLKs) – novel players in auxin perception?

In animal signalling, receptor tyrosine kinases (RTKs) play a pivotal role (see Fantl et al. 1993 for review). Growth factors and hormones such as PDGF, EGF and insulin are perceived by this class of receptors. In plant genome databases, similar genes have been identified and were termed "receptor-like kinases". A gene family of such RLKs identified in *Arabidopsis* are the *TMK* genes (Chang et al. 1992; Shiu and Bleecker 2001). Plant RLKs surprisingly do not possess tyrosine kinase activity, but are serine–threonine kinases (Zhang 1998). An unpublished observation from the Bleecker group links TMKs to auxin action. They created *TMK*-knockout mutants in *Arabidopsis* using T-DNA-insertions. *Tmk1*, *tmk3*, *tmk4* triple mutants showed abnormal hypocotyl and root growth. Root growth in these plants was completely insensitive to auxin at concentrations of up to 30 μmol/l IAA (Dai et al. 2004). Auxin binding data of TMKs have yet not been reported, and their role in auxin signalling is not yet clearly defined. The available evidence, however, makes them promising candidates for important players in auxin action.

3 Auxin-induced gene expression

While there are at least some ideas how auxin is perceived by the cell, the nature of the subsequent signalling steps is not known. Recent years have brought, however, considerable progress in the understanding of the final step of auxin signalling, the regulation of gene expression.

As in many other cases, mutant analysis has been used as a powerful tool to identify key players in auxin signalling (Hobbie et al. 1994; Leyser 1997). Several loci were found conferring auxin resistance and defining genes essential for a proper response. Many are supposed to be involved in, or were transcriptionally dependent on, ubiquitin-mediated repressor degradation (Gray et al. 2001; Ward and Estelle 2001; Zenser et al. 2001). In awareness of this fact, the hunt for transcriptional regulators and mechanisms of auxin-triggered proteolysis began to become a central point of auxin research (Gray and Estelle 2000; Eckardt 2001; for review, see Kepinski and Leyser 2002; Leyser 2002).

3.1 Transcriptional regulators

Auxin mediates its several effects by controlling transcription of auxin-induced genes. The helpers to keep expression under control are repressor proteins (Kepinski and Leyser 2002; Leyser 2002) identified as members of the Aux/IAA protein family (Abel et al. 1994; Ulmasov et al. 1997; Tiwari et al. 2001). The products of the *Aux/IAA* genes are metabolically unstable nuclear proteins (Abel and Theologis 1996).

Most of the Aux/IAA proteins contain four highly conserved domains (Guilfoyle et al. 1998) (Fig. 4). Domain I is a potent transcriptional repression

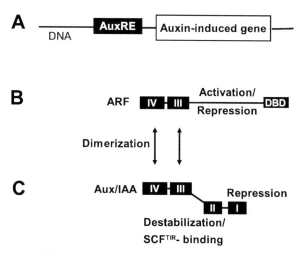

Fig. 4. Aux/IAAs and ARFs are auxin-dependent transcription factors. **A** Auxin-induced gene with an upstream auxin response element (AuxRE) in the promoter. **B** The ARFs (auxin response factors) are activators or repressors of transcription, depending on composition of their middle region. ARFs bind to AuxREs through a DNA-binding domain (DBD). **C** Aux/IAAs are dominant repressors of ARF-enhanced transcription. Heterodimerization between Aux/IAAs and ARFs occurs through domain III and IV. Domain II of Aux/IAAs contains a degron sequence that is responsible for instability and high turnover rate of the protein. Additionally, this region facilitates SCF[TIR1]-binding, which is a prerequisite for protein degradation

domain (Tiwari et al. 2004), domain II confers instability to the protein through a special degron sequence (Ramos et al. 2001). Gain-of-function mutations caused an amino acid substitution within domain II, stabilizing the respective protein by interrupting the turnover. This in turn leads to an auxin-insensitive phenotype (Worley et al. 2000; Gray et al. 2001; Ouellet et al. 2001; Ramos et al. 2001; Tiwari et al. 2001; Tian et al. 2002; Zenser et al. 2003). Domain II mutations in several *Aux/IAA* genes stabilized the corresponding proteins (for review, see Reed 2001). Thus, the instability of Aux/IAA proteins is essential for their function.

Domains III and IV are involved in homo- and heterodimerization with other Aux/IAAs or auxin response factors (ARFs) and are required for auxin responsiveness (Kim et al. 1997; Ouellet et al. 2001; Tiwari et al. 2003).

ARFs are transcriptional regulators that act in concert with Aux/IAAs. They bind directly to DNA in the promoter region of auxin-induced genes through a DNA-binding domain (DBD) and either activate or repress transcription, depending on the central region of the ARF (Liscum and Reed 2002). A model of auxin transcription factor functionality has been developed (Fig. 5): ARF activators bind to a TGTCTC-sequence in the promoter region upstream of auxin-induced genes, the so-called auxin response elements (AuxREs) (Ulmasov et al. 1999; Hagen and Guilfoyle 2002). If auxin concentrations are low, Aux/IAA-repressors heterodimerize (Ulmasov et al. 1997; Tiwari et al. 2001) with ARFs and prevent transcription of relevant genes. Elevated auxin levels cause instability of Aux/IAAs via an unknown mechanism. They dissociate from DNA-bound ARFs, interact with SCF[TIR1] and are degraded by the 26S proteasome (Hellmann and Estelle 2002). ARF activators then homodimerize with free ARFs, enhancing transcription of auxin-induced genes and transcription factors themselves, forming a feed-back loop (Leyser 2002).

This hypothetical mechanism would allow rapid fine tuning of auxin responses. Additionally, the endless possible combinations of Aux/IAA-repressors and ARF-activators as well as ARF-repressors allow a highly accurate control of transcription at any time during the development and in different tissues (Kim et al. 1997). This model also provides an acceptable explanation for the fact that auxin has opposite effects in the aerial parts of plants and roots. Beyond this, it implicates that auxin acts as a morphogen (Liscum and Reed 2002; Bhalerao and Bennett 2003) and that auxin concentration is the crucial signal, starting signal transduction. Evidence came from auxin transport mutant studies. It turned out that both functional auxin influx and efflux carriers, and therefore the direction of auxin flux, are essential for a proper auxin response (Bennett et al. 1998; Marchant et al. 1999, 2002).

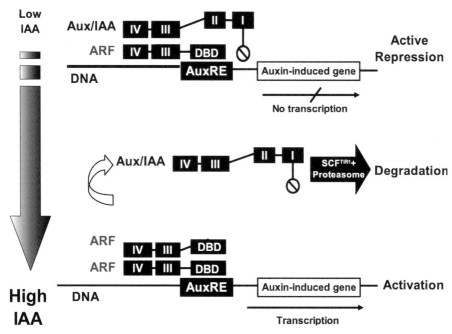

Fig. 5. Model for auxin-induced gene expression. Auxin response factors (ARFs) bind to auxin response elements (AuxREs) in the promoter region of auxin-induced genes. At low auxin concentrations, Aux/IAA proteins bind to the ARFs, blocking the expression of auxin-induced genes. Increasing auxin concentration induces degradation of the Aux/IAA repressors, thus triggering auxin-induced gene expression by removal of the blockade. Formation of ARF-ARF homodimers might further increase the activation

Twenty-nine different Aux/IAAs and 23 ARFs have been found in *Arabidopsis* ("*Arabidopsis* Genome Initiative 2000"; Hagen and Guilfoyle 2002). It remains unclear which of these proteins facilitate regulation of cell elongation. The careful characterization of the rapidly increasing number of relevant mutants or new biochemical approaches (Hayashi et al. 2003; Armstrong et al. 2004) will help to deepen our understanding.

Loss-of-function mutants in the *Aux/IAA* gene family members showed no visible phenotype, whereas most mutations in the *ARF* gene family members result in visible phenotypes. It is suggested that the Aux/IAA proteins have broader overlapping functions than the ARFs.

3.2 Protein degradation—an essential step in auxin signalling

Recent findings indicate that members of the Aux/IAA protein family interact directly with SCF^{TIR1}, which is a part of the protein degradation machinery of

the plant (Gray et al. 2001; Kepinski and Leyser 2004). Auxin appears to promote this interaction. Indeed, two members of the Aux/IAA family,
AXR3/IAA17 and AXR2/IAA7, are stabilized by mutations in parts of SCF^{TIR1}
or related proteins (Gray et al. 2001).

Ubiquitination labels a target protein for subsequent degradation by the
26S proteasome. In general, ubiquitin is initially activated by enzyme E1
(Fig. 6, step A). In the following, it is transferred to the ubiquitin conjugat-
ing enzyme E2 (Fig. 6, step B) which acts in concert with the SCF-type ubiquitin ligase complex E3 SCF^{TIR1} (Zheng et al. 2002) to covalently bind
ubiquitin to the target protein (Fig. 6, step C). Gray et al. (1999) showed that
SCF^{TIR1} is essential for auxin response in *Arabidopsis*.

In *Arabidopsis* SCF^{TIR1} contains an F-box receptor subunit (Fig. 6), TIR1,
which recognizes and interacts with substrate proteins (del Pozo and Estelle

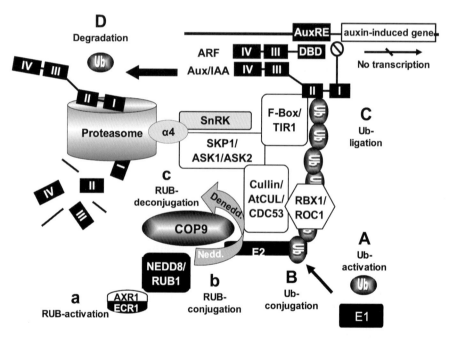

Fig. 6. Ubiquitin-mediated degradation of an Aux/IAA repressor through the SCF^{TIR1} complex and the proteasome. The substrate (Aux/IAA) protein has to be labelled with a multiu-
biquitin-chain for recognition by the proteasome. *A* Activation of ubiquitin (Ub). *B*
Conjugation of Ub to the ubiquitin-chain. *C* Ligation of Ub to the substrate. *D* Passing on of
labelled repressor to the proteasome and following degradation. To fascilitate efficient activity of the SCF^{TIR1} complex, it has to be modified at the cullin subunit (neddylation). *a* The
modifier (Nedd8), a RUB-like protein, is first activated by the heterodimer AXR1/ECR1. *b*
The activity of SCF^{TIR1} is regulated by neddylation and *c* deneddylation. For explanations for
SCF^{TIR1} subunits, see text

2000). TIR1 was identified to be important for auxin signalling by mutant analysis. The auxin resistant phenotype of *tir1* and mutants of other components of the SCF^{TIR1} linked ubiquitination to auxin signalling. Overexpression studies of *TIR1* resulted in enhancement of auxin-induced gene expression (Ruegger et al. 1998; Gray et al. 1999; del Pozo et al. 2002).

In addition, a cullin subunit (AtCUL) of SCF^{TIR1} and the RING-finger protein RBX1 form a dimer, catalyzing ubiquitin chain formation (Seol et al. 1999). The Skp1-like proteins ASK1 or ASK2, together with the cullin AtCUL and RBX1, build the core of the SCF complex. The core components associate with TIR1 through ASK1 or ASK2, bringing together the F-box-bound substrate and the ubiquitin chain building/conjugating cullin/RBX unit (Kishi and Yamao 1998; Wang et al. 2002). ASK1 together with SnRK and α_4 is involved in proteasomal binding of SCF^{TIR1} (Farrás et al. 2001).

The cullin is involved in another step of auxin-mediated proteolysis, the conjugation of SCF^{TIR1} to the ubiquitin-related protein NEDD8/RUB1. This so-called neddylation (Osaka et al. 1998; del Pozo and Estelle 1999; Deshaies 1999; Gray et al. 2000) and subsequent deneddylation (Fig. 6, steps b and c) are required for regulation of SCF^{TIR1} activity (Schwechheimer et al. 2001; del Pozo et al. 2002; Eckardt 2003; Serino and Deng 2003; Wolf et al. 2003), perhaps by modulating the binding and positioning of E2 or E2-ubiquitin conjugate (Kawakami et al. 2001; Wu et al. 2002) or connection of SCF^{TIR1} to the proteasome (von Arnim 2001). The COP9 signalosome (CNS), a highly conserved complex cleaves the cullin-NEDD8/RUB conjugate (Lyapina et al. 2001).

Like ubiquitin, NEDD8/RUB1 first has to be activated by a special enzyme (Fig. 6, step a) (del Pozo et al. 2002). The RUB-activating enzyme is a heterodimer with homology to E1. The amino-terminal half is called AXR1 (**AUXIN RESISTANT** 1), because loss of function confers auxin resistance to the plants (Leyser et al. 1993; Timpte et al. 1995). ECR1 was identified to be the missing C-terminal part of this enzyme (del Pozo et al. 1998). Mutants of *AXR1* show slightly more severe auxin-related defects than *tir1* mutants. The role of AXR1 in auxin-mediated growth promotion is supported by the fact that AXR1 is accumulated in dividing and elongating cells, but not present in mature non-growing cells (del Pozo et al. 1998, 2002). These findings brought the AXR1-part upstream of ubiquitination into play (del Pozo et al. 2002; Schwechheimer et al. 2002). In addition to the morphological phenotype of *axr1*, members of the *Aux/IAA* family of auxin-regulated genes are not expressed normally in the mutant (Abel et al. 1994; Timpte et al. 1995). It was obvious to assume that ubiquitination is involved in auxin-induced gene expression.

3.3 How does auxin regulate gene expression?

It is still unclear which of these steps of the complex degradation is controlled by auxin. The repressor could be phosphorylated by an auxin-
dependent protein kinase promoting association with SCF[TIR1] (Deshaies
et al. 1999; Christensen et al. 2000; Colón-Carmona 2000; Harari-
Steinberg and Chamowitz 2004), or auxin-regulated neddylation could be
responsible for auxin-triggered proteolysis. A hypothesis bringing this
together was suggested by Harari-Steinberg and Chamovitz (2004). They
see the COP9 signalosome as a master docking station. The COP9 signalosome would organize a kinase, its substrate and a specific SCF complex
that would ubiquitinate the phosphorylated substrate, and pass it on for
degradation by the proteasome. However, recent findings by Kepinski and
Leyser (2004) and Dharmasiri et al. (2003) question the role of phosphorylation in auxin-induced repressor degradation. Alternatively, the
involvement of a peptidyl-prolyl *cis/trans* isomerase (PPIase) of the parvu-
line type, isomerizing proline residues within domain II, was proposed
(Dharmasiri et al. 2003). At this point, attention has to be directed towards
the tomato mutant *diageotropica* (*dgt*). *Dgt* is insensitive to exogenously
applied auxin (Kelly and Bradford 1986). The *DGT* gene encodes a
cyclophilin (LeCYP1) (Oh et al. 2002, 2003; Ivanchenko and Lomax 2004)
that has intrinsic peptidyl-prolyl *cis/trans* isomerase activity. Interestingly,
the *dgt* mutation affects the expression of a subset of auxin-regulated genes
(Nebenführ et al. 2000). In contrast, Kepinski and Leyser (2004) found no
clues for an involvement of PPIases. They propose that auxin acts through
modification of TIR1 or a tightly bound protein rather than Aux/IAA
modification. In any case repressor degradation is followed by ARF-
enhanced gene expression.

The characterization of mutants using combined genetic, physiological
and genomic approaches is a promising way to unravel the clue of interacting signalling pathways. One good example is again the *diageotropica*
mutant. Nemhauser et al. (2004) examined the interdependency of brassinosteroid (BR) and auxin signalling. They discovered that the ARF-binding
TGTCTC sequence, previously identified as an auxin-responsive promoter
element, is also enriched in some genes expressed following BR-treatment
(Goda et al. 2004). Interestingly, the auxin-insensitivity of *dgt* is recovered by
combined treatment with auxin and BR (Park 1998). Looking at reduced
gene expression in *dgt*, the attention falls on *IAA5*. We found *IAA5* on the list
of genes induced by both auxin and BR (Nakamura et al. 2003). Thus, *IAA5*
must be an intersection between brassinosteroid and auxin signalling,
involved in growth control.

4 Auxin-upregulated genes and their functions

4.1 Plasma membrane ATPase

According to the acid growth theory of auxin action, auxin causes a stimulation of the plasma membrane proton pump, leading to cell wall acidification, cell wall loosening, and turgor-driven growth (Ruge 1937; Hager et al. 1971). The debate on this classical theory of plant physiology has been discussed in our earlier review and elsewhere. At least a significant fraction of auxin-induced growth is, without any doubt, triggered by acid-induced wall loosening (Lüthen et al. 1990; Cleland et al. 1991; Rayle and Cleland 1992). Expansins, the biochemical basis of cell wall loosening, have been discovered (Cosgrove 1998). In any case, auxin-induced proton secretion has been detected in many systems. To increase proton pumping, ATPase activity can be stimulated, or more ATPase protein can be incorporated into the plasma membrane. Both mechanisms have been experimentally demonstrated. Rück et al. (1993) could detect in a whole cell patch clamp configuration an ATP-dependent current reflecting proton pump activity. They could stimulate this current with IAA treatment or by application of anti-ABP1-antibodies. The fact that this current occurred without any delay indicates that gene expression was not involved. Hence, stimulation of the ATPase by auxin via a very short signalling chain must have taken place.

On the other hand, Hager and co-workers (Hager et al. 1991) could demonstrate immunologically a rapid *de novo* synthesis of plasma membrane ATPase protein in response to auxin. Apparently the cell increases the density of proton pump molecules at the membrane. However, there are isoforms of ATPase that are not induced by auxin (Jahn et al. 1995). Frias et al. (1996) could clone the ATPase isoform MHA2 and demonstrate auxin induction both on the mRNA and the protein level. At least some promoters of PM-ATPase genes carry AuxREs (Kirschke et al. 2000).

4.2 K⁺-inward channels

Compared with H^+-ATPase, the auxin induction of K^+-channel genes is much better characterized. The discovery of our group that auxin-induced growth of maize coleoptiles was K^+-dependent (Claussen et al. 1997) and was reversibly inhibited by K^+-channel blockers triggered the molecular investigation of K^+-channel gene expression by the Hedrich group (Philippar et al. 1999). They could clone from maize coleoptiles a potassium inward channel ZMK1. It could be shown that auxin treatment induced a marked increase of *ZMK1* mRNA on time scales consistent with the growth response (Philippar et al. 1999), and an increase in channel density was monitored by means of electrophysiology (Thiel and Weise 1999). The potential promoter region of the *ZMK1* gene contains an AuxRE.

ZMK1 expression appears to have a function in the control of gravi- and phototropism. Phototropic bending was not only accompanied by a differential redistribution of auxin across the coleoptile, but also by a differential expression of *ZMK1* in the shaded flank of the organ (Philippar et al. 1999; Fuchs et al. 2003). This is very much in line with the classical Cholodny–Went theory of phototropism (Cholodny 1924). Similar results were also obtained after gravistimulation (Philippar et al. 1999; Fuchs, personal communication).

In *Arabidopsis* the K^+-inward channel gene *KAT1* was shown to be induced by auxin in a similar manner as *ZMK1* in maize (Philippar et al. 2004). In *KAT1*-knockout mutants, K^+-currents after auxin stimulation were characterized by reduced amplitudes. Thus, this change in the electrical properties of the K^+-uptake channel in hypocotyl protoplasts resulted from an auxin-induced increase of active KAT1 proteins. However, the growth responses, measured by a CCD-auxanometer at high temporal resolution, were not changed from the wild type to the mutant. Thus it remains unclear whether KAT1 is involved in growth control. It may well be that potassium sensitivity of growth differs from monocots to dicots. This can, however be only investigated using ZMK1 overproducers or knockouts in a growth test.

4.3 Others

There are a number of families of classical auxin-induced genes. In most cases, their function is not yet clarified. Since auxin action is a rapid process, the "early" auxin response genes in particular are relevant. They fall into three major classes: *Aux/IAAs*, *SAURs* and *GH3* genes. A recent review by Hagen and Guilfoyle (2002) covers them in detail, so that only a brief discussion will be sufficient here.

Aux/IAAs are key players in controlling auxin-induced gene expression, and, in a kind of feedback regulation, are auxin-induced genes by themselves. They have already been discussed in detail in section 3.3, "How does auxin regulate gene expression?", in this article. Small auxin-upregulated genes (*SAURs*) code for 20–35kDa proteins and are expressed rapidly after auxin treatment, mRNAs show up within 2–5 min. Their function is still unknown, but may be related in some way to calcium signalling, as their amino terminal has been reported to bind calmodulin (Yang and Poovaiah 2000). On the other hand, auxin-induced growth in wheat coleoptiles is not accompanied by transients of cytosolic calcium, as recently demonstrated by Nagel-Volkmann (unpublished data).

5 Polar auxin transport

The ability of cells to respond to auxin is not the only prerequisite of auxin action. For auxin to be able to control crucial steps of plant development, concentration gradients across tissues have to be established. In 1924 the Cholodny–Went hypothesis proposed a differential distribution of auxin in lateral direction after light or gravity stimulus of shoots and coleoptiles (see Went 1974 for a personal retrospect on these pioneering days of auxin physiology). This is one of the classical theories of plant physiology and, although having been challenged in all the years, recent molecular evidence appears to support it quite convincingly.

For plant growth and other developmental processes, the directed transport from the site of biosynthesis, the growing shoot tip, to the site of auxin action, the elongation zone is important. Auxin moves between cells in a polar fashion. Old studies showed that auxin movement in shoots was normally strictly basipetal (reviewed in Goldsmith 1977). Auxin also moves in an acropetal direction, as in new leaf primordia (Reinhardt et al. 2003). In roots, the Cholodny–Went theory had serious problems, since it was known that the gravity stimulus was perceived in the root cap, and that a root cap inhibitor had to be postulated to explain gravitropism. Although auxin is known to inhibit root growth, how could auxin fit to this role, being predominantly transported in the opposite direction? To find a way out of this problem, Evans, Hasenstein and others suggested that auxin is transported acropetally through the parenchyma of the stele, is gravitropically redistributed in the root cap and transported basispetally in the cortex through the elongation zones, where it mediates root differential growth in root gravitropism (for review, see Moore and Evans 1986; Evans 1991). Among the workers in the field, this version of the Cholodny–Went theory was termed the "orange juice cooler model" of auxin transport (Hasenstein, personal communication).

But how does auxin transport work and which transport mechanisms exist? A first breakthrough was the isolation of probable elements of the auxin efflux carrier complex from *Arabidopsis thaliana*, the PIN proteins. Other proteins believed to be involved in auxin efflux belong to the class of the ABC transporters, whereas AUX/LAX proteins are candidates for an auxin influx carrier. Taken together, these proteins constitute the polar auxin transport machinery.

5.1 How does auxin efflux work?

5.1.1 The PINs

In the 1970s and 1980s, it was proposed that auxin was taken up by diffusion, but excreted by an auxin efflux carrier (Rubery and Sheldrake 1974). In order to explain the polar direction of auxin efflux, it was suggested that these carriers are asymmetrically distributed across the cell. Auxin would be taken up at the upper side of a cell and excreted at the lower. Thus, it would move in a polar fashion slowly from cell to cell, precisely what has been shown experimentally. The effect of phytotropins and other inhibitors of polar auxin transport and the fact that they bind to plasma membrane preparations strongly supported this electrochemical theory of auxin transport.

In the 1980s and 1990s, the auxin efflux carrier resisted all attempts of isolation by methods of protein biochemistry; it appeared that it was a multi-protein complex that falls into pieces upon solubilization. Molecular biology brought new ways to analyse auxin efflux. There are some mutants that show defects in auxin transport and are therefore excellent tools for identifying auxin carrier candidates. One mutant *pinformed 1* (*pin1*) has a "pin-like" inflorescence. The phenotype resembles the effect of phytotropins on *Arabidopsis* development. Therefore, PIN proteins were soon postulated to play an important role in auxin transport. The most popular idea was that PIN1 and other members of the PIN protein family were identical to the auxin efflux carrier or (more probably) a component of it (Friml and Palme 2002). This concept was underscored dramatically by the localization of the PIN1 gene product, which was found at the lower cell surface of shoot cells, and at the lower side of parenchyma cells in the root central cylinder (Gälweiler et al. 1998) (Fig. 7). Auxin transport inhibitors block rapid actin-dependent cycling of PIN1 between the plasma membrane and endosomal compartments (Geldner et al. 2001) and inhibit trafficking of membrane proteins that are unrelated to auxin transport. PIN1 cycling is of central importance for directed auxin transport, and auxin transport inhibitors affect efflux by generally interfering with membrane-trafficking processes.

Another PIN family member, PIN4, is localized in developing and mature root meristems. *Pin4* mutants are defective in establishing and maintaining endogenous auxin gradients. They fail to canalize externally applied auxin and display various patterning defects in both embryonic and seedling roots. PIN4 plays a role in generating a sink for auxin below the quiescent centre of the root meristem that is essential for auxin distribution and patterning (Friml et al. 2002b).

PIN3 is expressed in gravity-sensing tissues of the root tip. Most PIN3 protein is accumulated in the lateral cell surface. In the root columella, PIN3 is positioned symmetrically at the plasma membrane. When cells are gravis-timulated, it is rapidly relocalized laterally (Friml et al. 2002a). PIN3 is a regulator of auxin efflux and especially for lateral auxin transport which is important for tropic growth.

The role of PIN2 is to regulate basipetal auxin transport and gravitropism of roots (Rashotte et al. 2000). In terms of the "orange juice cooler model", it can be speculated that PIN1 is important for transporting the auxin down the shoot and acropetally to the root tip, PIN4 for focussing it to the gravitationally sensitive cells, PIN3 for redistributing it, and PIN2 for transporting it basipetally through the root cortex to the elongtion zone. In this model auxin acts as the root cap inhibitor (Fig. 7). The resulting distribution of

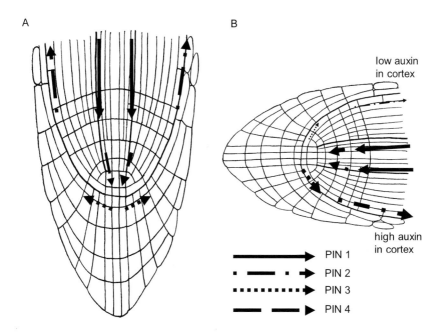

Fig. 7. Model of auxin transport in a root before and after gravistimulation. **A** Auxin flow in a vertical root according to the "orange juice cooler model" of auxin transport. Acropetal auxin transport occurs in the stele and is mediated by the PIN1 carrier. PIN4 focuses the auxin stream to the region around the quiescent center. In the gravistimulated cell, auxin is distributed laterally by PIN3, and then transported basipetally through the cortex cells, passing the elongation zone of the root. **B** Upon gravistimulation auxin is redistributed by PIN3, and more auxin is directed to the lower half of the root than to the upper part. Cell elongation is inhibited in the lower half of the root, resulting in a positive gravitropic curvature

auxin across the root may not only affect and mediate gravitropism, but, given the strong effect of IAA on the proton pump, also differentially control ion uptake in various root tissues. More than 40 years ago, Lüttge and Weigl (1962) showed that distribution of radioactive sulfate across root tips displays a very distinct pattern: sulfate can penetrate the root only up to the plerome-peribleme border, i.e. to the central cylinder, although in this zone of the root there are no anatomical limitations (e.g. Casparian bands) for apoplastic entry into the inner root tissues. It would be interesting to know if such putative ionic intake patterns are changed by auxin treatment or by gravistimulation of the roots.

5.1.2 ABC transporters as efflux carrier candidates

Besides (or in conjunction with) the PINs, members of another family of putative efflux carriers are ATP-binding cassette (ABC) transporters. They were shown to be responsible for the control of hypocotyl cell elongation under certain light conditions (Sidler et al. 1998). It has been speculated that they are the real efflux carriers and that the PIN protein, being easily redistributed and repositioned, regulate their activity. Analysis of mutants defective in *Arabidopsis* ABC proteins indicates that primary active transport might participate in the control of auxin homeostasis (Luschnig 2002).

It is clear that the direction of auxin transport in stems results from basal PIN1 localization within cells (Noh et al. 2003). Mutations in two genes homologous to those encoding ABC transporters were recently shown to block polar auxin transport in the hypocotyls of *Arabidopsis* seedlings. Noh et al. (2004) showed that *mdr* (*multi-drug resistance*) mutants display enhanced gravitropism and phototropism. These phenotypes resulted from a disruption of the normal accumulation of PIN1 protein along the basal end of hypocotyl cells. Lateral auxin conductance is increased as a result, enhancing growth differentials responsible for the two tropic responses.

5.2 How does auxin influx work?

The influx of auxin is the other part of the auxin transport machinery. Originally, it has been supposed that auxin is exclusively taken up passively by diffusion. It was suggested that only the free acid of IAA can penetrate the plasma membrane. In the cytoplasm (pH=7, pK_a auxin=4.8) most of the auxin occurs as the IAA$^-$ anion, which can leave the cell only via the efflux carrier. Auxin is thus accumulated by the so-called ion trap mechanism.

Some early electrophysiological data (Rubery and Sheldrake 1974) suggested the involvement of carriers in auxin uptake operating in concert with the ion trap mechanism. As in the case of the efflux carriers, mutant studies paved the way to the identification of promising candidates for such transporters. One of them, AUX1 (Bennett et al. 1996), is located at the plasma membrane (Swarup et al. 2004). It belongs to a family of amino acid/auxin permeases. AUX1 represents a polytopic membrane protein composed of 11 transmembrane spanning domains. In addition, a large *aux1* allelic series containing null, partial-loss-of-function, and conditional mutations was characterized to identify the functionally important domains and amino acid residues within the polypeptide. Almost all partial-loss-of-function and null alleles cluster in the core permease region, with one notable exception, *aux1-7*, which modifies the function of the external C-terminal domain (Swarup et al. 2004). AUX1 activity appears to be required for polar and phloem-based auxin transport in root and shoot tissues. For example, basipetal transport via lateral root cap and epidermal tissues requires AUX1 for auxin influx, and PIN2 for auxin efflux, as discussed above (Rashotte et al. 2000; Swarup et al. 2001). Also, processes other than gravitropism have been linked to the interactions of auxin transport proteins. Acropetal auxin transport, which is dependent on AUX1 and PIN1, leads to the positioning of newly formed leaf primordia (Reinhardt et al. 2003).

As there are several PIN genes for controlling auxin efflux, there are several auxin influx carrier genes: AUX1 is only one member of a recently discovered familiy of related genes termed *LAX* (*Like Aux1*), coding for a number of additional auxin influx carrier candidate proteins (Parry et al. 2001). This diversity of genes controlling auxin transport gives the organism the necessary degrees of freedom to fine-tune the delicate auxin distribution across the plant.

AUX and LAX homologues have also been found in other plants. Recently, Schnabel and Frugoli (2004) found ten PIN homologs in the legume *Medicago truncatula* (*MtPINs*) and five LAX homologs (*MtLAXs*), which appear to have a functional role in nodule formation.

5.3 Auxin transport depends on K⁺

Auxin-regulated processes are often dependent on the presence of cations. In roots of *Arabidopsis* auxin transport requires external potassium (Vicente-Agullo et al. 2004). This finding was observed by disrupting a potassium transporter (TRH1). Disruption of TRH1 affects root hair development and the gravitropic root response. Rescue of the observed morphological defects by exogenous auxin indicates a link between TRH1 activity and auxin transport. This hypothesis is supported by the findings that the rate of auxin translocation from shoots to roots and efflux of labelled IAA in isolated root segments were reduced in the *trh1* mutant. The TRH1 carrier seems to be an important part of auxin transport system in roots of *Arabidopsis* (Vicente-Agullo et al. 2004).

6 Prospects

Despite the enormous recent progress, our understanding of auxin action and transport remains patchy at best. Targets of future research could be the signalling chain between perception and gene expression. Physiological data also suggest that pinpointing the precise role of auxin in gravitropism, especially its interaction with ethylene and cytokinins (Aloni et al. 2004) may be a promising field of research.

Note added in Proof

After Submission of this review two groups (Kepinski and Leyer 2005, Dharmasiri et al. 2005) independently demonstrated that the F-Box protein TIR1, a key regulator of auxin induced gene expression, binds auxin and apparently acts as nuclear auxin receptor. This major discovery suggests that auxin can directly control gene expression by binding to TIR1. It will be a fascinating perspective for auxin research to explore possible crosstalk between ABP1 and TIR1 signalling in growth control

References

Abel S, Oeller PW, Theologis A (1994) Early auxin-induced genes encode short-lived nuclear proteins. Proc Natl Acad Sci USA 91:326–330

Abel S, Theologis A (1996) Early genes and auxin action. Plant Physiol 111:9–17

Armstrong JI, Yuan S, Dale JM, Tanner VN, Theologis A (2004) Identification of inhibitors of auxin transcriptional activation by means of chemical genetics in *Arabidopsis*. Proc Natl Acad Sci USA 101:14978–14983

Aloni R, Langhans M, Aloni E, Ullrich CE (2004) Role of cytokine in the regulation of root gravitropism. Planta 220:177–182

Barbier-Brygoo H, Ephritikhine G, Klämbt D, Maurel C, Palme K, Schell J, Guern J (1991) Perception of the auxin signal at the plasma membrane of tobacco mesophyll protoplasts. Plant J 1:83–93

Bennett MJ, Marchant A, Green HG, May ST, Ward SP, Millner PA, Walker AR, Schulz B, Feldmann KA (1996) *Arabidopsis AUX1* gene: a permease-like regulator of root gravitropism. Science 273:948–950

Bennett MJ, Marchant A, May ST, Swarup R (1998) Going the distance with auxin: unravelling the molecular basis of auxin transport. Philos Trans R Soc London B 353:1511–1515

Bhalerao RP, Bennett MJ (2003) The case of morphogens in plants. Nature Cell Biol 5:939–943

Chang C, Schaller GE, Patterson SE, Kwok SF, Meyerowitz EM, Bleecker AB (1992) The *TMK1* gene from *Arabidopsis* codes for a protein with structural and biochemical characteristics of a receptor protein kinase. Cell 4:1263–1271

Chen JG, Ullah H, Young JC, Sussman MR, Jones AM (2001) ABP1 is required for organized cell elongation and division in *Arabidopsis* embryogenesis. Genes Dev 15:902–911

Chen JG, Shimomura S, Sitbon F, Sandberg G, Jones AM (2003) The role of auxin-binding protein 1 in the expansion of tobacco leave cells. Plant J 28:607–617

Cholodny N (1924) Über die hormonale Wirkung der Organspitze bei der geotropischen Krümmung. Ber dt Bot Ges 42:356–360

Christian M, Lüthen H (2000) New methods to analyse auxin-induced growth I: Classical auxinology goes *Arabidopsis*. Plant Growth Regulation 32:107–114

Christian M, Steffens B, Schenck D, Lüthen H (2003) The *diageotropica* mutation of tomato disrupts a signalling chain using extracellular auxin binding protein 1 as a receptor. Planta 218:309–314

Christensen SK, Dagenais N, Chory J, Weigel D (2000) Regulation of auxin response by the protein kinase PINOID. Cell 100(4):469–478

Claussen M, Lüthen H, Böttger M (1996) Inside or outside? Localization of the auxin receptor relevant to auxin-induced growth. Physiol Plant 98:861–867

Claussen M, Lüthen H, Blatt M, Böttger M (1997) Auxin-induced growth and its linkage to potassium channels. Planta 201:227–234

Cleland RE (1967) Extensibility of isolated cell walls: measurement and changes during cell elongation. Planta 74:197–209

Cleland RE, Buckley G, Nowbar S, Lew NM, Stinemetz C, Evans ML, Rayle DL (1991) The pH profile for acid induced growth of coleoptile and epicotyl sections is consistent with the acid-growth-theory of auxin action. Planta 186:70–74

Colón-Carmona A, Chen DL, Yeh KC, Abel S (2000) Aux/IAA proteins are phosphorylated by phytochrome in vitro. Plant Physiol 124:1728–1738

Cosgrove D (1998) Cell wall loosening by expansins. Plant Physiol 118:333–339

Cosgrove D (1999) Enzymes and other agents that enhance cell wall extensibility. Annu Rev Plant Physiol Plant Mol Biol 50:391–417

Dai N, Wang W, Shui S, Bleecker AB (2004) Auxin associated growth is mediated by the TMK subfamily of receptor-like kinases from *Arabidopsis*. In: Auxin 2004, May 22–27, 2004, Kolympari, Crete, Greece, Abstract No. 15. http://pgec-genome.ars.usda.gov/Auxin2004/

David K, Carnero-Diaz E, Leblanc N, Monestiez M, Grosclaude J, Perrot-Rechenmann C (2001) Conformational dynamics underlie the activity of the auxin binding protein, Ntabp1. J Biol Chem 276:34517–34523

Davies PJ (1974) The uptake and elution of indoleacetic acid by pea stem section in relation to auxin induced growth. In: Plant Growth Substances 1973, Tokyo (Hirokawa Publishing Company, Inc.), pp 767–778.

del Pozo JC, Estelle M (1999) Function of the ubiquitin-proteasome pathway in auxin response. Trends in Plant Science 4:107–112

del Pozo JC, Estelle M (2000) F-box proteins and protein degradation: an emerging theme in cellular regulation. Plant Mol Biol 44:123–128

del Pozo JC, Timpte C, Tan S, Callis J, Estelle M (1998) The ubiquitin-related protein RUB1 and auxin response in *Arabidopsis*. Science 280:1760–1763

del Pozo JC, Dharmasiri S, Hellmann H, Walker L, Gray WM, Estelle M (2002) AXR1-ECR1-dependent conjugation of RUB1 to the *Arabidopsis* cullin AtCul1 is required for auxin response. Plant Cell 14:421–433

Dela Fuente RK, Leopold AC (1970) Time course of auxin stimulation of growth. Plant Physiol 46: 186–189

Deshaies RJ (1999) SCF and Cullin/RING H2-based ubiquitin ligases. Annu Rev Cell Dev Biol 15:435–467

Dharmasiri N, Dharmasiri S, Jones AM, Estelle M (2003) Auxin action in a cell free system. Current Biology 13:1418–1422

Dharmasiri N, Dharmasiri S, Estelle M (2005) The F-Box protein TIR1 is an auxin receptor. Nature 435:441–445

Dohrmann U, Hertel R, Kowalik H (1978) Properties of auxin binding sites in different sub-cellular fractions from maize coleoptiles. Planta 140:97–106

Eckardt NA (2001) Auxin and the power of the proteasome in plants. Plant Cell 13:2161–2163

Eckardt NA (2003) Characterization of the last subunit of the *Arabidopsis* COP9 signalosome. The Plant Cell 15:580–581

Evans ML (1991) Gravitropism: Interaction of sensitivity modulation and effector redistribution. Plant Physiol 95:1–5

Fantl WJ, Johnson DE, Williams LT (1993) Signalling by receptors of tyrosine kinase. Annu Rev Biochem 62:453–481

Farrás R, Ferrando A, Jásik J, Kleinow T, Ökrész L, Tiburcio A, Salchert K, del Pozo C, Schell J, Koncz C (2001) SKP1-SnRK protein kinase interactions mediate proteasomal binding of a plant SCF ubiquitin ligase. EMBO J 20:2742–2756

Frias I, Caldeira MT, Perez-Castineira JR, Navarro-Avino JP, Culianez-Macia FA, Kuppinger O, Stransky H, Pages M, Hager A, Serrano R (1996) A major isoform of the maize plasma membrane H+-ATPase. Characterization and induction by auxin in coleoptiles. Plant Cell 8:1533–1544

Friml J, Palme K (2002) Polar auxin transport-old questions and new concepts? Plant Mol Biol 49:273–284

Friml J, Wisniewska J, Benkova E, Mendgen K, Palme K (2002a) Lateral relocation of auxin efflux regulator PIN3 mediates tropism in *Arabidopsis*. Nature 415:806–809

Friml J, Benkova E, Blilou I, Wisniewska J, Hamann T, Ljung K, Woody S, Sandberg G, Scheres B, Jurgens G, Palme K (2002b) AtPIN4 mediates sink-driven auxin gradients and root patterning in *Arabidopsis*. Cell 8:108:661–673

Fuchs I, Philippar K, Ljung K, Sandberg G, Hedrich R (2003) Blue light regulates an auxin induced K+-channel gene in the maize coleoptile. Proc Natl Acad Sci USA 100(20):11795–11800

Gälweiler L, Guan C, Müller A, Wisman E, Mendgen K, Yephremov A, Palme K (1998) Regulation of polar auxin transport by AtPIN1 in *Arabidopsis* vascular tissue. Science 282:2226–2230

Geldner N, Friml J, Stierhof YD, Jurgens G, Palme K (2001) Auxin transport inhibitors block PIN1 cycling and vesicle trafficking. Nature 413:425–428

Goda H, Sawa S, Asami T, Fujioka S, Shimada Y, Yoshida S (2004) Comprehensive comparison of auxin-regulated and brassinosteroid-regulated genes in *Arabidopsis*. Plant Physiology 134:1555–1573

Goldsmith MHM (1977) The polar transport of auxin. Annu Rev Plant Physiol 28:439–478

Gray W, del Pozo JC, Walker L, Hobbie L, Risseeuw E, Banks T, Crosby WL, Yang M, Ma H, Estelle M (1999) Identification of an SCF ubiquitin-ligase complex required for Auxin response in *Arabidopsis thaliana*. Genes Dev 13:1678–1691

Gray W, Kepinski S, Rouse D, Leyser O, Estelle M (2001) Auxin regulates SCF[TIR1]-dependent degradation of AUX/IAA proteins. Nature 414:271–276

Gray WM, Estelle M (2000) Function of the ubiquitin-proteasome pathway in auxin response. Trends Biochem Sci 25:133–138

Guilfoyle T, Hagen G, Ulmasov T, Murfett J (1998) How does auxin turn on genes? Plant Physiol 118:341–347

Hagen G, Guilfoyle T (2002) Auxin-responsive gene expression: genes, promoters, and regulatory factors. Plant Mol Biol 49:373–385

Hager A, Menzel H, Krauss A (1971) Versuche und Hypothese zur Primärwirkung des Auxins beim Streckungswachstum. Planta 100:47–75

Hager A, Debus G, Edel HG, Stransky H, Serrano R (1991) Auxin induces exocytosis and a rapid synthesis of a high-turnover pool of plasma membrane H^+-ATPase. Planta 185:527–537

Harari-Steinberg O, Chamowitz DA (2004) The COP9 signalosome: mediating between kinase signalling and protein degradation. Current Protein and Peptide Science 5:185–189

Hayashi K, Jones AM, Ogino K, Yamazoe A, Oono Y, Inoguchi M, Kondo H, Nozaki H (2003) Yokonolide B, a novel inhibitor of auxin action, blocks degradation of AUX/IAA factors. J Biol Chem 278:23797–23806

Hellmann H, Estelle M (2002) Plant Development: regulation by protein degradation. Science 297:793–797

Hertel R (1995) Auxin binding protein 1 is a red herring. J Exp Bot 46:461–462

Hertel R, Thompson KS, Russo VEA (1972) In vitro auxin binding to particulate fractions from corn coleoptiles. Planta 107:325–340

Hesse T, Feldwisch J, Balshusemann D, Bauw G, Puype M, Vanderkerckhove J, Löbler M, Klämbt D, Schell J, Palme K (1989) Molecular cloning and structural analysis of a gene from Zea mays (L.) coding for a putative receptor for the plant hormone auxin. EMBO J 8:2453–2461

Heyn AJN (1931) Der Mechanismus der Zellstreckung. Rec Trav Bot Neerl 28:113–114

Hobbie L, Estelle M (1994) Genetic approaches to auxin action. Plant Cell Environ 17:525–540

Ivanchenko M, Lomax T (2004) Roles of a cyclophilin, LeCYP1/DGT, in auxin-mediated tomato development. In: Auxin 2004, May 22–27, 2004, Kolympari, Crete, Greece, Abstract No. 20. http://pgec-genome.ars.usda.gov/Auxin2004/

Jahn T, Johannson S, Lüthen H, Volkmann D, Larsson C (1996) Reinvestigation of the auxin and fusicoccin stimulation of the plasma membrane ATPase. Planta 199:359–365

Kawakami T, Chiba T, Suzuki T, Iwai K, Yamanaka K, Minato N, Suzuki H, Shimbara N, Hidaka Y, Osaka F, Omata M, Tanaka K (2001) NEDD8 recruits E2-ubiquitin to SCF E3 ligase. EMBO J 20:4003–4012

Keller CP, Van Volkenburgh E (1996) Osmoregulation by oat coleoptile protoplasts. Plant Physiol 110:1007–1016

Kelly MO, Bradford KJ (1986) Insensitivity of the diageotropica tomato mutant to auxin. Plant Physiol 82:713–717

Kepinski S, Leyser O (2002). Ubiquitination and auxin signaling. A degrading story. Plant Cell 14:81–95

Kepinski S, Leyser O (2004) Auxin-induced SCF^{TIR1}-Aux/IAA interaction involves stable modification of the SCF^{TIR1} complex. Proc Natl Acad Sci USA 101:12381–12386

Kepinski S, Leyser O (2005) The Arabidopsis F-Box protein TIR1 is an auxin receptor. Nature 435:446–451

Kim J, Harter K, Theologis A (1997) Protein-protein interactions among the Aux/IAA proteins. Proc Natl Acad Sci USA 94:11786–11791

Kim YS, Min JK, Kim D, Min JK, Kim D, Jung J (2001) A soluble auxin-binding protein, ABP 57. J Biol Chem 276:10730–10736

Kirschke CP, Seungil R, Bradshaw KG, Sayna A, Dubrovsky J, Ewing NN (2000) The promoter of tomato plasma membrane H^+-ATPase LHA2 is auxin responsive in roots and shoots and drives expression during lateral root initiation and in abscission zones. Annual Meeting of the American Society of Plant Physiologists, Abstract 723. http://216.133.76.127/pb2000/public/P40/0885.html

Kishi T, Yamao F (1998) An essential function of Grr1 for the degradation of Cln2 is to act as binding core that links Cln2 to Skp1. J Cell Sci 111:3655–3661

Klämbt D (1990) A view about the function of the auxin-binding proteins at the plasma membrane. Plant Mol Biol 14:1045–1050

Kögl F, Hagen-Smit J, Erxleben H (1934) Über ein neues Auxin (Heteroauxin) aus Harn. Physiol Chem 228:104–112

Leyser O (1997) Lessons from a mutant weed. Physiologia Plantarum 100:407–414

Leyser O (2002) Molecular genetics of auxin signalling. Annu Rev Plant Biol 53:377–398

Leyser OHM, Lincoln C, Timpte C, Turner JC, Lammer D, Estelle MA (1993) The hormone-resistance gene *AXR1* of *Arabidopsis* is related to ubiquitin-activating enzyme E1. Nature 364:161–164

Liscum E, Reed JW (2002) Genetics of Aux/IAA and ARF action in plant growth and development. Plant Mol Biol 49:387–400

Löbler M, Klämbt D (1985) Auxin-binding protein from coleoptile membranes from corn (*Zea mays* L.). I. Purification by immunological methods and characterization. J Biol Chem 260:9848–9853

Luschnig C (2002) Auxin transport: ABC proteins join the club. Trends Plant Sci 7:329–332

Lüthen H, Bigdon M, Böttger M (1990) Reexamination of the acid growth theory of auxin action. Plant Physiol 93:931–939

Lüthen H, Claussen M, Böttger M (1999) Growth: Progress in auxin research. Prog Bot 60:315–340

Lüttge U, Weigl J (1962) Mikroautoradiographische Untersuchungen der Aufnahme und des Transports von$^{35}SO_4{}^-$ und$^{45}Ca^{++}$ in Keimwurzeln von *Zea mays* L. und *Pisum sativum* L. Planta 58:113-126

Lyapina S, Cope G, Shevchenko A, Serino G, Tsuge T, Zhou C, Wolf DA, Wei N, Shevchenko A, Deshaies RJ (2001) Promotion of NEDD8-CUL1 conjugate cleavage by COP9 signalosome. Science 292:1382–1385

Marchant A, Kargul K, May ST, Muller P, Delbarre A, Perrot-Rechenmann C, Bennett MJ (1999) AUX1 regulates root gravitropism in *Arabidopsis* by facilitating auxin uptake within root apical tissues. EMBO J 18:2066–2073

Marchant A, Bhalerao R, Casimiro I, Eklof J, Casero PJ, Bennett M, Sandberg G (2002) AUX1 promotes lateral root formation by facilitating indole-3-acetic acid distribution between sink and source tissues in the *Arabidopsis* seedling. Plant Cell 14:589–597

Moore R, Evans ML (1986) How roots perceive and respond to gravity. Am J Bot 73:574–587

Muday KG, Lomax TL, Rayle DL (1995) Characterization of the growth and auxin physiology of the tomato mutant *diageotropica*. Planta 195:548–553

Nakamura A, Higuchi K, Goda H, Fujiwara MT, Sawa S, Koshiba T, Shimada Y, Yoshida S (2003) Brassinolide induces *IAA5*, *IAA19* and DR5, a synthetic auxin response element in *Arabidopsis*, implying a cross talk point of brassinosteroid and auxin signalling. Plant Physiol 133:1–11

Napier RM, David KM, Perrot-Rechenmann C (2002) A short history of auxin-binding proteins. Plant Mol Biol 49:339–348

Nebenführ A, White TJ, Lomax TL (2000) The *diageotropica* mutation alters auxin induction of a subset of the *Aux/IAA* gene family in tomato. Plant Mol Biol 44:73–84

Nemhauser JL, Mockler TC and Chory J (2004) Interdependency of brassinosteroid and auxin signalling in *Arabidopsis*. PLOS Biology 2: e258. http://biology.plosjournals.org/plosonline/

Noh B, Bandyopadhyay A, Peer WA, Spalding EP, Murphy AS (2003) Enhanced gravi- and phototropism in plant *mdr* mutants mislocalizing the auxin efflux protein PIN1. Nature 423:999–1002

Oh K, Hardeman K, Ivanchenko MG, Ellard-Ivey M, Nebenführ A, White TJ., Lomax Tl (2002) Fine mapping in tomato using microsynteny with the *Arabidopsis* genome: the *Diageotropica* (*Dgt*) locus. Genome Biology 3(9). http://genomebiology.com/2002/3/9/research/0049

Oh KC, Ivanchenko MG, White TJ, Lomax T (2003) The *Diageotropica* gene of tomato encodes a novel player, a cyclophilin, in auxin signaling. http://abstracts.aspb.org/pb2003/public/P46/0680.html

Osaka F, Kawasaki H, Aida N, Saeki M, Chiba T, Kawashima S, Tanaka K, Kato S (1998) A new NEDD8-ligating system for cullin-4A. Genes Dev 12:2263–2268

Ouellet F, Overvoorde PJ, Theologis A (2001) IAA17/AXR3: Biochemical insight into an auxin mutant phenotype. Plant Cell 13:829–841

Park WJ (1998) Effect of epibrassinolide on hypocotyl growth of the tomato mutant *diageotropica*. Planta 207:120–124

Parry G, Marchant A, May S, Swarup R, Swarup K, James N, Graham N, Allen T, Martucci R, Yemm A, Napier R, Manning K, King G, Bennett M (2001) Quick on uptake: characterization of a family of plant auxin influx carriers. J Plant Growth Reg 20:217–225

Philippar K, Fuchs I, Lüthen H, Hoth S, Bauer CS, Haga K, Thiel G, Edwards K, Ljung K, Sandberg G, Böttger M, Becker D, Hedrich R (1999) Auxin-induced K$^+$-channel expression represents an essential step in coleoptile growth and gravitropism. Proc Natl Acad Sci USA 96:2186–12191

Philippar K, Ivashikina N, Ache P, Christian M, Lüthen H, Palme K, Hedrich R (2004) Auxin activates *KAT1* and *KAT2*, two K$^+$-channel genes expressed in seedlings of *Arabidopsis thaliana*. Plant J 37: 815–827

Ramos J, Zenser N, Leyser O, Callis J (2001) Rapid degradation of auxin/indoleacetic acid proteins requires conserved amino acids of domain II and is proteasome dependent. Plant Cell 13:2349–2360

Rashotte AM, Brady SR, Reed RC, Ante SJ, Muday GK (2000) Basipetal auxin transport is required for gravitropism in roots of *Arabidopsis*. Plant Physiol 122(2):481–490

Rayle DL, Cleland RE (1992) The acid growth theory of auxin-induced cell elongation is alive and well. Plant Physiol 99:1271–1274

Reed JW (2001) Roles and activities of Aux/IAA proteins in *Arabidopsis*. Trends Plant Sci 6:420–425

Reinhardt D, Pesce ER, Stieger P, Mandel T, Baltensperger K, Bennett M, Traas J, Friml J, Kuhlemeier C (2003) Regulation of phyllotaxis by polar auxin transport. Nature 426:255–260

Rubery PH, Sheldrake AR (1974) Carrier-mediated auxin transport. Planta 118:101–121

Rück A, Palme K, Venis MA, Napier RM, Felle HH (1993) Patch clamp analysis establishes a role for an auxin binding protein in the auxin stimulation of plasma membrane current in *Zea mays* protoplasts. Plant J 4:41–46

Ruegger M, Dewey E, Gray WM, Hobbie L, Turner J, Estelle M (1998) The TIR1 protein of *Arabidopsis* functions in auxin response and is related to human SKP2 and yeast Grrp1. Genes Dev 12:198–207

Ruge U (1937) Untersuchungen über den Einfluss des Heteroauxins auf das Streckungswachstum von *Helianthus annuus*. Z f Botanik 31:1–56

Schnabel EL, Frugoli J (2004) The *PIN* and *LAX* families of auxin transport genes in *Medicago truncatula*. Mol Genet Genom 272:420–432

Schwechheimer C, Serino G, Callis J, Crosby WL, Lyapina S, Deshaies RJ, Gray WM, Estelle M, Deng XW (2001) Interactions of the COP9 signalosome with the E3 ubiquitin ligase SCF^{TIR1} in mediating auxin response. Science 292:1379–1382

Schwechheimer C, Serino G, Deng X-W (2002) Multiple ubiquitin ligase-mediated processes require COP9 signalosome and AXR1 function. Plant Cell 14:2553–2563

Seol JH, Feldman RMR, Zachariae W, Shevchenko A, Correll CC, Lyapina S, Chi Y, Galova M, Claypool J, Sandmeyer S, Nasmyth K, Shevchenko A, Deshaies RJ (1999) Cdc53/cullin and the essential Hrt1 RING-H2 subunit of SCF define a ubiquitin ligase module that activates the E2 enzyme Cdc34. Genes Dev 13:1614–1626

Serino G, Deng XW (2003) The COP9 signalosome: regulating plant development through the control of proteolysis. Annu Rev Plant Biol 54:165–182

Shiu SH, Bleecker AB (2001) Receptor-like kinases from *Arabidopsis* form a monophyletic gene family related to animal receptor kinases. Proc Natl Acad Sci USA 98:10763–10768

Sidler M, Hassa P, Hasan S, Ringli C, Dudler R (1998) Involvement of an ABC transporter in a developmental pathway regulating hypocotyl cell elongation in the light. Plant Cell10:1623–1636

Steffens B, Lüthen H (2000) New methods to analyse auxin-induced growth II: The swelling reaction of protoplasts—a model system for the analysis of auxin signal transduction? Plant Growth Regul 32:115–122

Steffens B, Feckler C, Palme K, Christian M, Böttger M, Lüthen H (2001) The auxin signal for protoplast swelling is perceived by extracellular ABP1. Plant J 27:591–599

Swarup R, Friml J, Marchant A, Ljung K, Sandberg G, Palme K, Bennett M (2001) Localization of the auxin permease AUX1 suggests two functionally distinct hormone transport pathways operate in the *Arabidopsis* root apex. Genes Dev 15:2648–2653

Swarup R, Kargul J, Marchant A, Zadik D, Rahman A, Mills R, Yemm A, May S, Williams L, Millner P, Tsurumi S, Moore I, Napier R, Kerr ID, Bennett MJ (2004) Structure-function analysis of the presumptive *Arabidopsis* auxin permease AUX1. Plant Cell 16:3069–3083

Thiel G, Weise R (1999) Auxin augments K+ inward rectifier in coleoptiles. Planta 208:38–45

Tian Q, Uhlir NJ, Reed JW (2002) *Arabidopsis* SHY2/IAA3 inhibits auxin-regulated gene expression. Plant Cell 14: 301–319

Timpte C, Lincoln C, Pickett FB, Turner J, Estelle M (1995) The *AXR1* and *AUX1* genes of *Arabidopsis* function in separate auxin-response pathways. Plant J 8:561–569

Tiwari SB, Wang XJ, Hagen G, Guilfoyle TJ (2001) AUX/IAA proteins are active repressors, and their stability and activity are modulated by auxin. Plant Cell 13:2809–2822

Tiwari SB, Hagen G, Guilfoyle TJ (2003) The roles of auxin response factor domains in auxin-responsive transcription. Plant Cell 15:533–543

Tiwari SB, Hagen G, Guilfoyle TJ (2004) Aux/IAA proteins contain a potent transcriptional repression domain. Plant Cell 16:533–543

Ulmasov T, Murfett J, Hagen G, Guilfoyle TJ (1997) Aux/IAA proteins repress expression of reporter genes containing natural and highly active synthetic auxin response elements. Plant Cell 9:1963–1971

Ulmasov T, Hagen G, Guilfoyle TJ (1999) Dimerization and DNA binding of auxin response factors. Plant J 19:309–319

Venis M (1995) Auxin binding protein 1 is a red herring? Oh no it isn't! J Exp Bot 46:463–465

Vesper MJ, Kuss CL (1990) Physiological evidence that the primary site of auxin action is an intracellular site. Planta 182:486–491

Vicente-Agullo F, Rigas S, Desbrosses G, Dolan L, Hatzopoulos P, Grabov A (2004) Potassium carrier TRH1 is required for auxin transport in *Arabidopsis* roots. Plant J 40:523–535

von Arnim AG (2001) A hitchhiker's guide to the proteasome. Science's stake. http://fp.bio.utk .edu/vonarnim/BOT404/STKE-von%20Arnim.pdf

Wang H, Huang J, Lai Z, Xue Y (2002) F-box proteins in flowering plants. Chinese Science Bulletin 47:1497–1501

Ward S, Estelle M (2001) Auxin signalling involves regulated protein degradation by the ubiquitin-proteasome pathway. J. Plant Growth Regul 20:265–273

Went FW (1928) Wuchsstoff und Wachstum. Rec Trav Bot Neerl 25:1–116

Went FW (1974) Reflections and speculations. Annu Rev Plant Physiol 25:1–26

Wolf DA, Zhou C, Wee S (2003) The COP9 signalosome: an assembly and maintenance platform for cullin ubiquitin ligases? Nature Cell Biol 5:1029–1033

Woo EJ, Marshall J, Bauly J, Chen JG, Venis M, Napier RM, Pickersgill RW (2002) Crystal structure of auxin binding protein 1 in complex with auxin. EMBO J 21:2877–2885

Worley CK, Zenser N, Ramos J, Rouse D, Leyser O, Theologis A, Callis J (2000) Degradation of Aux/IAA proteins is essential for normal auxin signaling. Plant J 21:553–562

Wu K, Chen A, Tan P, Pan ZQ (2002) The Nedd8-conjugated ROC1-CUL1 core ubiquitin ligase utilizes Nedd8 charged surface residues for efficient polyubiquitin chain assembly catalyzed by Cdc34. J Biol Chem 277:516–527

Yamagami M, Haga K, Napier RM, Iino M (2004) Two distinct signaling pathways participate in auxin-induced swelling of pea epidermal protoplasts. Plant Physiol 134:735–747

Yang T, Poovaiah BW (2000) Molecular, biochemical evidence for the involvement of calcium/calmodulin in auxin action. J Biol Chem 275:3137–3143

Zenser N, Ellsmore A, Leasure C, Callis, J. (2001) Auxin modulates the degradation rate of Aux/IAA proteins. Proc Natl Acad Sci USA 98:11795–11800

Zenser N, Dreher KA, Edwards SR, Callis J (2003) Acceleration of Aux/IAA proteolysis is specific for auxin and independent of AXR1. Plant J 35:285–294

Zhang X (1998) Leucine-rich repeat receptor-like kinases in plants. Plant Mol Biol Reporter 16:301–311

Zheng N, Schulman BA, Song L, Miller JJ, Jeffrey PD, Wang P, Chu C, Koepp DM, Elledge SJ, Pagano M, Conaway JW, Harper JW, Pavletich NP (2002) Structure of the Cul1-Rbx1-Skp1-F-box[Skp2] SCF ubiquitin ligase complex. Nature 416:703–709

May Christian
Daniel Schenck
Michael Böttger
Hartwig Lüthen
Biozentrum Klein Flottbek und Botanischer Garten
Universität Hamburg
Ohnhorststr. 18
22609 Hamburg
Fax: +49-40-42816254
e-mail: h.luthen@botanik.uni-hamburg.de

Bianka Steffens
Botanisches Institut und Botanischer Garten
Christian-Albrechts-Universität zu Kiel
Olshausenstr. 40
24098 Kiel

New insights into abiotic stress signalling in plants

Margarete Baier, Andrea Kandlbinder, Karl-Josef Dietz, Dortje Golldack

Plants respond and adapt to variable environmental conditions with a wide range of cellular and metabolic changes that are triggered by signalling and regulatory pathways. Recently, new insights into signalling networks involved in abiotic stress adaptation have been gained by transcriptome analyses that suggest the existence of both specific signalling and of cross-talk between signal transduction pathways in response to environmental changes. In the following, selected studies on cellular signalling induced by abiotic stresses as high light intensities and elevated temperature, UV-B and ozone, low temperature, salinity, and heavy metals focussing on transcriptome studies will be reviewed.

1 Light and elevated temperature

In their natural habitat, plants experience changes in light intensity and temperature that range from a limitation to an excess status. Long-term physiological studies in wheat and radish (Zavorueva and Ushakova 2004) demonstrated that the adaptability of plants to elevated temperature is best in high light and worst in low light. Plants have evolved to the combination of the two stresses by responding synergistically to heat and high light. This hypothesis is supported by transcriptome analysis comparing *Arabidopsis thaliana* mRNA patterns in high light and heat-filtered high light (Rossel et al. 2002). Most of the 66 co-regulated genes, including AP×2 (encoding a cytosolic ascorbate peroxidase), GP×6 (encoding a glutathione peroxidase), an early light inducible protein (ELIP) and several heat shock proteins, showed higher transcript level variation with unfiltered light.

The focus of this chapter is specifically on plant responses to moderate and excess high light. For light quality sensing we would like to refer to excellent reviews, e.g. by Schepens et al. (2004), Lin (2002) and Chamovitz and Deng (1996). Light sensors such as cryptochromes and phytochromes are important in quantitative light sensing in low light environments (Devlin et al. 2003). However, in high light, their responsiveness is overwhelmed (Bailey et al. 2001). A more sensitive perception system for light intensity is

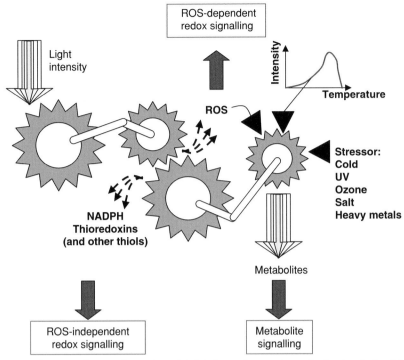

Fig. 1. Light controls ROS-dependent, ROS-independent and metabolic signals by driving photosynthetic electron transport. The generation of photosynthates depends mainly on the flux of photosynthetic electrons into reductive metabolism, while initiation of putative redox signals like ROS-signals, thiol signals and signals linked to the redox state of $NADP^+/NADPH$ is controlled by the interface of the photosynthetic light and dark reaction. Abiotic stressors, e.g. high and low temperature, salt and heavy metals, affect signal initiation by imbalancing the system.

photosynthesis. Depending on the absolute photosynthetic electron transport rate as well as on photosynthetic electron transport relative to downstream reductive metabolism various ROS-dependent and -independent redox signals and metabolite signals are induced to control nuclear gene expression and transcript stability (Fig. 1).

1.1 Saturating light intensities and moderate excess light

At saturating light intensities, photosynthesis leads to accumulation of photosynthates, which are sensed and initiate metabolite-specific signalling cascades (Coruzzi and Zhou 2001; Rook and Bevan 2003). Supported by intersystem photosynthetic redox signals, like the redox state of the plastoquinone pool (Escoubas et al. 1995; Oswald et al. 2001), they control expression of genes involved in photosynthetic electron transport (e.g. PetE; Oswald et al. 2001), light harvesting (e.g. cab) and CO_2-fixation (rbcS) (Krapp et al. 1993; Escoubas et al. 1995). In carbohydrate sensing, hexokinase is involved (Moore et al. 2003), which triggers SNF1-like kinases similar to the carbohydrate signalling pathway in yeast (Halford et al. 2003). However, in plants, the overlap between carbon and light regulated genes is high indicating photosynthesis-dependent signal initiation. The extent of co-regulation relative to specific regulation became evident in the genome wide cDNA array study performed by Thum et al. (2004) with *Arabidopsis thaliana*. Compared to 1247 transcripts co-regulated by light and carbon, only 201 transcripts were selectively regulated by carbon and 77 by light. Promoter profiling led to identification of 16 different *light-and-carbon-responsive* cis-elements (LCR), responding to independent or linked signal transduction pathways (Thum et al. 2004).

Besides photosynthate production, photosynthetic electron transport regulates the redox poise of redox carrier proteins like thioredoxin (Fig. 1). It could trigger thiol-based signalling cascades involved in light-dependent mRNA stabilization in polyribosomes as observed, e.g. for ferredoxin-1 mRNA (Tang et al. 2003). A third signalling pathway was shown recently by analysis of 2-Cys peroxiredoxin-A promoter regulation (Baier et al. 2004). It is driven by the acceptor availability at photosystem II. Like reduction of thioredoxin, it is regulated by photosynthetic electron transport relative to reductive metabolism, pointing out an important signalling function of the interface between photosynthetic electron transport and downstream reductive metabolism.

The dovetailing of light and dark reaction makes photosynthetically controlled signal initiation dependent on various abiotic stressors (Fig. 1). For example, by speeding up or slowing down reductive metabolism, temperature affects the regeneration of photosynthetic electron acceptors. In elevated temperature besides inactivation of enzymes, photoinhibition takes place, which reduces the threshold for the light intensity in photooxidative ROS-formation (Lavorel 1975) and might be a reason for the synergism of high light and temperature effects observed by cDNA array analysis (Rossel et al. 2002).

1.2 Excess high light

In excess high light, the regulatory capacities of photosynthetic electron transport and chloroplast biochemistry are overtaxed (Niyogi 2000). In response, protection mechanisms are activated. First of all, the energy uptake capacity is decreased by down-regulation of the transcript levels of genes encoding light harvesting antenna (Kimura et al. 2003). Further suppression of photosynthetic electron transport results from decreasing the transcript levels for the two photosystems and the cytochrome b_6f complex (Kimura et al. 2003). Under the same conditions, the transcript levels for ferredoxin and ferredoxin-NADP$^+$ reductase, whose corresponding proteins control the interface between photosynthetic electron transport and the reductive chloroplast biochemistry, only respond slightly (Kimura et al. 2003), indicating that their regulation is not primarily responsive to high light.

In all experiments using light intensities above 800 μmol quanta m^{-2} s^{-1} (Rossel et al. 2002; Kimura et al. 2003; Vandenabeele et al. 2004), consistently the transcripts encoding antioxidant enzymes, PR-proteins and enzymes of the phenylpropanoid metabolism accumulated. The individual results however strongly vary. Kimura et al. (2003) reported that the transcript levels for chloroplast antioxidant enzymes hardly respond to high light, in contrast to those for extra-plastidic antioxidant enzymes. The observation indicates that the gene expression response of nuclear encoded chloroplast antioxidant enzymes follows independent regulatory forces and is not sensitive to high light. However, in its generalized form this conclusion contradicts previous observations, e.g. the increase of the transcript amounts for chloroplast superoxide dismutases (Kliebenstein et al. 1998) and all four chloroplast peroxiredoxins (Horling et al. 2003).

Most transcriptome analysis performed so far showed that the overlap of high light responses with regulation by other stresses is high. For example, RD29A, ERD27, ERD10, KIN1, COR15a and Lea14 are up-regulated by any of the high light, drought, cold and high-salinity stresses (Kimura et al. 2003). They are like the heat-shock protein HSP70 biomarkers for environmental stress (Cho et al. 2004; Ireland et al. 2004), which can provide systemic resistance against various physically different stressors. Putative transcription factors involved are heat-shock factors (HSFs), which were originally described in respect of their importance in heat responses. In plants, HSFs stimulate, e.g. expression of genes for antioxidant enzymes, like APx1 and APx2 (Panchuk et al. 2002). The various HSFs can either interact synergistically or interfere with activators (Bharti et al. 2004). In contrast to the general stress markers, e.g. four of the *light-stress regulated*-genes described by Dunaeva and Adamska (2001) and the high-light induced genes HSC70-G7, APx1, ELIP, GST1, CHS and HY5 identified by Rossel et al. (2002) differentially respond to various stressors, demonstrating that multiple high light signal transduction cascades exist.

Many high light induced genes, e.g. the genes for antioxidant enzymes and PR-proteins, are redox-regulated, suggesting that formation of reactive oxygen species (ROS) are involved in signal initiation (summarized in Foyer et al. 1997a; Baier and Dietz 1998; Rodermel 2001). To analyse the importance of photorespiratory H_2O_2 in high-light-regulated gene expression, cDNA-array hybridizations were performed in wild-type and catalase-deficient A. thaliana in high light (Vandenabeele et al. 2004). Several transcripts, e.g. encoding heat shock proteins and PR-proteins such as dihydroflavonol-4-reductase or proteolytic enzymes, were more induced in the catalase-deficient plants than in wild-type. The difference in the mRNA pattern confirmed that H_2O_2 is involved in triggering the high light-response. A function is especially indicated for down-regulation of the transcript levels for genes involved in the oxidative burst, including AtrbohC (Torres et al. 2002) and a Ca^{2+}-ATPase (Hrabak et al. 1996), and for the induction of the translation and proteolytic machinery by F-box and WD40 proteins (Vandenabeele et al. 2004). Because high-light driven expression of antioxidant enzymes, hypersensitive response related and anthocyanin and phenylpropanoid biosynthetic genes (PAL, CHS and dihydroflavonol-4-reductase) were reversed or at least mitigated, Vandenabeele and coworkers (2004) proposed that elevated peroxisomal H_2O_2 levels (as in catalase-deficient plants) reverse the masterswitch for a subset of genes. An alternative explanation for this response, however, could be hardening of catalase-deficient plants. In the antisense lines, the low catalase activity could have resulted in induction of catalase-independent defences, as observed for the induction of peroxiredoxin-independent antioxidant protection in 2-Cys peroxiredoxin antisense lines (Baier et al. 2000). On the other hand, the existence of H_2O_2-repressed besides H_2O_2-induced high-light signalling pathways may reflect alternative, antagonistically triggered signalling cascade (with presumably different thresholds). The regulatory mechanisms could be similar to the mechanisms leading either to induction of antioxidant defences or of oxidative bursts following pathogen attack. In this context, it is interesting to note, that in the high light response the transcript levels of many transcription factors known to be involved in pathogen responses, such as WRKY, Myb, AP-2 type transcription factors and NAC, are induced (Vandenabeele et al. 2004).

Although in the response to high light, the precise signals, the signal transduction pathways and targets are still not understood in detail, the transcriptome analysis performed in recent years supports the hypothesis that photooxidative ROS formation is involved in signal initiation. Due to the reactivity of most ROS in combination with the redox buffering capacity of chloroplasts and the cytosol and the sink strength of mesophyll peroxisomes, however, long distance signal transduction by chloroplast ROS/H$_2$O$_2$ is unlikely. Putative second messengers are, e.g. mobile homo- and heterodithiols (Foyer et al. 1997b; Baier and Dietz 1998) and oxolipids which derive, e.g. from oxidation of membrane lipids (Montillet et al. 2004). Alternatively, protochlorophyllids have been discussed as signalling molecules in regulation of cab-gene expression (Strand et al. 2003; Rodermel 2001). The idea is based on work on the gun5 mutant, which is defective in the H-subunit of Mg^{2+}-chelatase and shows de-repressed expression of cab genes (Mochizuki et al. 2001; Strand et al. 2003). However, in high light, negative regulation, including excess light dependent expression of a helix-loop-helix transcription factor *Phytochrome-Interacting Factor 1* (PIF1), avoids accumulation of protochlorophyllides (Huq et al. 2004). Recently, special attention has been paid to the products of the 13-LOX-pathway. 13-LOX products accumulate early in response to many stressors, while the 9-LOX pathway is more specific as, e.g. induced by cadmium stress (Montillet et al. 2004). Consistent with the hypothesis on a signalling function of 13-LOX products, transcriptome analysis with plant material treated with the oxolipid jasmonic acid showed accumulation of transcripts for PR proteins and antioxidative enzymes (Schenk et al. 2000), similar to high light treated plants. However, AP×2, which is strongly responsive to excess light (Fryer et al. 2003), cannot be induced by jasmonic acid (Chang et al. 2004). Its constitutive induction in a low glutathione background (Ball et al. 2004) hints at sensing of other signals related to the cellular redox homeostasis, presumably to the thiol redox poise. Identification of the precise signals triggering nuclear gene expression in response to high light, and especially identification of the mobile signals passing the chloroplast envelope and transmitting photosynthetic signals, will be a great challenge for the next years.

2 Perception, signalling and transcriptional regulation in response to UV-B and ozone

Effects of UV-B radiation and ozone on plants range from changes in growth and development to more specific effects on primary metabolic functions, such as decrease in photosynthetic activity, changes in pigment composition and enzyme activities. These responses depend on the perception of the specific environmental condition, signal transduction mechanisms, and modification of gene expression.

In general, before environmental factors trigger a cellular response, they have to be perceived by receptor(s) and the information has to be transduced via a signalling pathway. So far, no specific photoreceptor molecule has been identified that perceives the UV-B signal. This is made more complex as UV-B is strongly absorbed by a wide range of biologically active molecules such as nucleic acids, aromatic amino acids, lipids and phenolic compounds (Jordan 2002). Numerous authors have suggested the existence of a specific UV-B photoreceptor, although the exact nature of such a receptor is still unknown (Nagy and Schäfer 2000; Wade et al. 2001; Brosché and Strid 2003; Gyula et al. 2003). Recently, Stratmann (2003) proposed that UV-B co-opts other stress signalling pathways by activating membrane-bound receptors like the wound signalling receptor SR160 in plants. As ozone degradates rapidly in the apoplast to form various reactive oxygen species (ROS), e.g. H_2O_2, $O_2^{.-}$, and $OH^.$. (Langebartels et al. 2002), only its primary reaction products might be sensed by still unknown plasma membrane bound receptors (Baier et al. 2005).

It has been reported that exposure to ozone and UV-B induces changes in gene expression, and recent data suggest that these stresses share many common features via the generation of ROS (Surplus et al. 1998; A-H-Mackerness et al. 1999). For instance, levels of ROS increase in response to UV-B pulses (Allan and Fluhr 1997), as well as ozone, degrades rapidly in the apoplast to form various ROS species (Kangasjärvi et al. 1994). Therefore, ROS might act as a signal inducer for the expression of certain UV-B responsive genes (A-H-Mackerness et al. 1999; 2001; Green and Fluhr 1995).

In addition to ROS, salicylic acid, jasmonic acid and ethylene are involved as signal transduction intermediates in *A. thaliana* (A-H-Mackerness et al. 1999). The involvement of jasmonic acid in UV-B and ozone responses was shown using the *A. thaliana* mutant jar1, which is insensitive to jasmonic acid. In this mutant, less PDF1.2 mRNA was accumulated in response to UV-B irradiation. Furthermore, UV-B treatment increased the level of jasmonic acid in *A. thaliana*, and the jar1 mutant was more sensitive to UV-B exposure than wild-type plants (A-H-Mackerness et al. 1999). Based on analysis of jar1 and the ozone-tolerant ethylene-insensitive *A. thaliana* mutant ein2 after ozone-treatment, Tuominen et al. (2004) postulated that early accumulation of ethylene stimulates spreading of cell death and suppresses protection by jasmonic acid. Late accumulation of jasmonic acid, however, inhibits the ethylene pathway and the propagation of cell death (Tuominen et al. 2004). In NahG plants, which are salicylate hydroxylase over-expressing *A. thaliana* transformants that do not accumulate salicylic acid, UV-B induced increases in PR-1, whereas PR-2 and PR-5 mRNA accumulation were blocked, indicating that salicylic acid is required for this response (Surplus et al. 1998). Further analysis of this mutant revealed that ozone-induced ethylene production depends on salicylic acid (Rao et al. 2002). Ethylene has also been shown to be involved in UV-B responses by using the ethylene-insensitive *A. thaliana* mutant etr1. This mutant failed to up-regulate both PR-1 and PDF1.2 genes in response to UV-B (A-H-Mackerness et al. 1999). The sensitivity to UV-B shown in the jar1 and etr1 mutants was only partly mimicked in ozone-fumigated plants: jar1 was more sensitive than wild type, whereas etr1 plants were as tolerant as the wild type (Overmyer et al. 2000).

In the transmission of secondary signals such as ROS, jasmonic acid, salicylic acid and ethylene mitogen activated protein kinases (MAPK) are involved. *A. thaliana* encodes 10 MAPKKKK, 80 MAPKKK, 10 MAPKK and 23 MAPK (Jonak et al. 2002; Yu et al. 2004), which form complex signalling networks characterized by convergence and divergence at the level of the MAPKs and their upstream activating kinases. Cardinale et al. (2002) suggested that one stress might activate several MAPKs, and a particular MAPK might be activated by multiple stress signals. In *A. thaliana*, the MAP kinases MPK3, MPK4 and MPK6 are activated by various abiotic stresses (Ichimura et al. 2000; Kovtun et al. 2000). In a recent study, Moon et al. (2003) showed that a nucleotide diphosphate kinase (AtNDPK2) is involved in ROS signalling and specifically interacts with the MPK3 and MPK6. MPK6 and MPK3 are activated by the MAP kinase kinases MKK4 and MKK5 (Asai et al. 2002) and the MAPKKK ANP1 (Kovtun et al. 2000). In *Lycopersicon peruvianum* suspension-cultured cells, specific MAPKs, LeMPK1, LeMPK2 and LeMPK3 were transcriptionally activated after UV-B treatment as well as after wounding, but the activation kinetics of LeMPK1, LeMPK2 and LeMPK3 were different in response to wounding and UV-B (Holley et al. 2003). Thus, differences in MAPK-mediated responses might be determined by a combined effect of different active MAPKs, different activation kinetics and additional MAPK-independent signalling components.

2.1 Ozone and UV-B induced gene expression

New insights into the responses to UV-B and ozone signalling and the specificity of secondary messengers have been provided recently by transcriptome analysis investigating responses to oxidative stress in wild type plants of *A. thaliana*, tobacco and maize (Desikan et al. 2001; Brosché et al. 2002; Izaguirre et al. 2003; Tamaoki et al. 2003a; Casati and Walbot 2004; Ulm et al. 2004), the analysis of transgenic plants and mutants with various genetic backgrounds (Vranova et al. 2002; Casati and Walbot 2003; Tamaoki et al.

2003b) and PCR-based suppression subtractive hybridization (Mahalingham et al. 2003). Ulm et al. (2004) monitored the gene expression profile of UV-B irradiated *A. thaliana* seedlings by using microarrays comprising almost the full *A. thaliana* genome (>24,000 genes on the array). A set of early low-level UV-B responsive genes, 100 activated and 7 repressed, was identified. 64% of these genes are currently annotated as encoding proteins of known or putative functions. More than 30% of these UV-B-responsive genes encode transcription regulators, including genes encoding transcription factors implicated in response to abiotic stress (DREB2A, ABF3, ZAT10, ZAT12), during development (CIA2, COL1, MYB13), in light responses (HY5, HYH) and unknown functions (MYB44, MYB111, bHLH34, bHLH149, bHLH150 and two NAM-related proteins). Brosché et al. (2002) performed a microarray experiment with 5000 ESTs from *A. thaliana* and identified 70 UV-B-responsive genes. These encode photosynthesis-related proteins, pathogenesis-related proteins (e.g. PR-1), antioxidant enzymes, enzymes involved in flavonoid and lignin biosynthesis and signal transducers. Izaguirre et al. (2003) analysed a set of approximately 250 wound- and UV-responsive genes in field grown *Nicotiana longiflora*. The array hybridization experiment revealed that 20% of insect herbivory-responsive genes such as photosynthesis-related genes and a WRKY transcription factor were also regulated by UV-B. Recently, Casati and Walbot (2003) examined changes in transcript abundance for approximately 2500 maize ESTs after UV irridation treatments in leaves of four maize genotypes that differ in flavonoid and anthocyan content. They identified 304 genes that were responsive to UV-B radiation, 268 gene transcripts being upregulated. These genes encode, e.g. proteins involved in protein synthesis like cytoplasmic ribosomal proteins, initiation and elongation factors. In contrast, only 36 genes were downregulated after the treatment like transcripts encoding proteins related to photosynthesis and CO_2 fixation, such as Rubisco and proteins of both photosystems I and II. Over the last decade, UV-related promoter elements and candidate transcription factors have also been identified like ACE (ACGT-containing elements that recognize common plant regulatory factors), MRE (Myb-recognition elements) (Jordan 2002) and a 11 bp GC-rich promoter element found in SAD genes that are upregulated rapidly by UV-B (Gittins et al. 2002).

By hybridization of *A. thaliana* cDNA-macroarrays, Tamaoki et al. (2003a) identified 205 ozone-responsive transcripts after 12 h exposure to O_3 and comprehensively compared the involvement of ethylene, jasmonic and salicylic acid signalling pathways on ozone-responsive gene expression. Approximately 75% of the ozone-responsive transcripts were induced and 48 of 205 genes were suppressed by O_3. Among the 109 transcripts with known functions, 11 are involved in signal transduction, for example, calmodulin-3, calmodulin-like protein, cyclophilin ROC7, GTP-binding protein GB3, putative serine threonine kinase and a putative MAP3K epsilon protein kinase, 33 in metabolism like monodehydroascorbate reductase, glutaredoxin and pyruvate kinase, 24 in cellular organization and biogenesis, 25 in cell rescue/defence (e.g. glutathione S-transferase, PR4), six in energy, five in protein synthesis and degradation, three in transcription and two in transport.

Utilizing a DNA array with 5000 ESTs and cDNAs from *A. thaliana*, Brosché et al. (2002) found six genes (MEB5.2, PyroA, Ubq3, Lhcb6, F5D21.10 and the gene for a RNA polymerase II subunit) that were regulated under UV-B and ozone: PyroA, Ubq3 and RNA polymerase II subunit were specifically increased by UV-B, MEB5.2 increased and Lhcb6, F5D21.10 decreased under both treatments. The PyroA is putatively required for resistance towards harmful singlet oxygen radicals (Osmani et al. 1999), but the functions of MEB5.2 and F5D21.10 have not yet been identified.

Via the generation of reactive oxygen species ozone and UV-B share common features (A-H-Mackerness et al. 1999). For instance, several authors reported that the pattern of gene expression for UV-B-induced stress is similar to ozone (Matsuyama et al. 2002; Sävenstrand et al. 2002), but different to drought or wounding. In addition, by the comparison of 205 ozone-responsive transcripts in response to drought, salinity, UV-B, low temperature, high temperature and acid rain, Tamaoki et al. (2004) confirmed these findings.

Detailed functional characterization of transcripts and proteins identified in transcriptomic and proteomic studies will provide further information on physiological responses to ozone and UV-B in plants elucidating the steps of signal perception and transduction.

3 Signalling and transcriptional regulation in response to drought, cold, and salt stress

Drought, low temperature, and salinity are abiotic environmental factors that greatly influence plant growth and development, and complex adaptational responses are induced by these stresses on the physiological, biochemical, and molecular level in plants. Freezing temperatures can cause membrane damage, dehydration of cells that is associated with osmotic stress and that may lead to generation of active oxygen species (Thomashow 2001; Xiong et al. 2002). Mechanisms of plant cold acclimation include, e.g. alterations in lipid composition, accumulation of osmoprotective compounds as proline and sucrose, as well as synthesis of hydrophilic and cryoprotective, respectively, polypeptides as COR15 that decreases lamellar- to hexagonal II phase transitions (e.g. Steponkus et al. 1998; Gilmour et al. 2000; Cook et al. 2004). Salinity affects intracellular ion homeostasis and water balance of plants, and may also induce oxidative damage as a secondary stress effect (Golldack 2004). Strategies of salt adaptation include, e.g. regulation of Na^+-influx by transcriptional control of HKT1-type Na^+-transporters and AKT1-type K^+-channels as well as vacuolar sequestration of Na^+ that is regulated by the vacuolar H^+-ATPase and the tonoplast NHX1-type Na^+/H^+-antiporter (Apse et al. 1999; Golldack and Dietz 2001; Rus et al. 2001; Golldack et al. 2002, 2003). Salt-induced synthesis and cytoplasmic accumulation of compatible solutes as polyols and sugars, proline, and quaternary ammonium compounds, respectively, have a function in preventing cellular dehydration and in structural stabilization of proteins and membranes (Popova et al. 2002, 2003).

Several shared transcriptional changes but also stress-specific responses to drought, low temperature, and hyperosmotic treatment have been reported indicating both cross-talks as well as specificity of stress sensing, signal transduction, and regulation of *cis*- and *trans*-acting factors (Shinozaki et al. 2003; Chinnusamy et al. 2004). *Cis*-acting elements that have been found in promoters of drought-, cold-, and salt-responsive genes are the ABA-responsive element (ABRE) and the dehydration-responsive element (DRE; C-repeat element (CRT); Yamaguchi-Shinozaki and Shinozaki 1994; Thomashow 1999; Shinozaki et al. 2003). DRE-binding proteins that belong to the ERF/AP2-type transcription factors have been identified with CBF1 (DREB1B), CBF2 (DREB1C), and CBF3 (DREB1A) that are responsive to cold stress, and DREB2 that is induced by drought and salt stress (Shinozaki et al. 2003). Over-expression of the cold-inducible rice homologue OsDREB1A in *A. thaliana* induced expression of target genes of *A. thaliana* DREB1A and increased tolerance to cold, drought, and salt stress in the transgenic plants indicating the existence of conserved DREB1A-homologous pathways in distant plant species (Dubouzet et al. 2003). In *A. thaliana*, regulation of intracellular Na^+-homeostasis under salt stress includes the SOS-pathway with the Ca^{2+}-sensor SOS3 and the serine/threonine protein kinase SOS2 that activate the plasma membrane Na^+/H^+ antiporter SOS1 (Chinnusamy et al. 2004).

Novel elements involved in signalling and transcriptional activation that are induced by drought, cold, and salinity have been identified recently by large-scale transcriptome analyses. Microarray-based studies led to identification of transcription factors and signalling elements that had not been related to these environmental stresses before, and enabled comparative analyses of the signalling networks regulating transcriptional responses to these factors. In addition, by monitoring transcript profiles of transgenic plants over-expressing stress-responsive signal transduction elements and transcription factors, respectively, down-stream target genes of these adaptive regulating pathways could be identified, accordingly, new insights into regulation of abiotic stress responses emerge.

3.1 bZIP transcription factors responsive to abiotic stress treatment

A. thaliana ABF/AREB proteins that are members of group A bZIP transcription factors bind to the ABRE element and have been isolated by yeast-one hybrid screening (Choi et al. 2000; Uno et al. 2000; Jakoby et al. 2002). Transcript levels of these ABRE binding factors increased by ABA treatment, drought, and NaCl, and AREB1 and AREB2 activated ABRE-dependent reporter gene activity in a transient assay in *A. thaliana* protoplasts (Choi et al. 2000; Uno et al. 2000). Over-expression of ABF3 and ABF4 (AREB2) caused enhanced drought tolerance and ABA hypersensitivity, and altered transcript levels of ABA-responsive genes, e.g as rd29B and rab18 in *A. thaliana* (Kang et al. 2002). Using *A. thaliana* full-length microarrays with 7000 cDNAs obtained from unstressed and stressed plants, Seki et al. (2002) compared transcript profiles in response to drought, cold, and salinity. ABF3 that was included in the microarray showed increased transcript levels in response to drought, NaCl and cold treatment, thus confirming and extending previous knowledge on this stress response factor (Seki et al. 2002). Kreps et al. (2002) used a GeneChip microarray with approximately 8000 probes for monitoring abiotic stress responses and reported increased transcript levels of ABF3 at cold treatment, hyperosmotic mannitol stress and salt in *A. thaliana* roots. Moreover, by use of microarray hybridizations stress-induced transcription of other members of the family of bZIP transcription factors could be detected that had not been reported before. Thus, expression of the group S bZIP transcription factor AtbZIP60 increased by cold treatment, drought and salinity (Table 1). In contrast, group S ATbZIP1 was specifically up-regulated by drought and NaCl but was not regulated by low temperature (Seki et al. 2002). Using a rice microarray including 1700 cDNAs, Rabbani et al. (2003) compared transcript profiles under ABA-treatment, salt, drought, and low temperature. The rice bZIP transcription factor OSE2, which is homologous to the *A. thaliana* group A factor AtbZIP13, was induced by drought stress and ABA, whereas a homologue to *A. thaliana* group S AtbZIP53 was specifically induced by cold stress (Rabbani et al. 2003). In addition, function of AtbZIP53 as a transcriptional activator of

Table 1. Transcription factors with stress-induced regulation of transcript levels. *C* cold treatment, *D* drought, *S* salt stress, *M* hyperosmotic mannitol treatment

Gene name	Species	AGI and *Arabidopsis* orthologue	Transcrip-tional regulation	Transgenic traits	Reference
OsNAC6	Rice	At1g01720	C, D, S, ABA		Rabbani et al. (2003)
ANAC002	*Arabidopsis*	At1g01720	D, S		Seki et al. (2002)
STZ	*Arabidopsis*	At1g27730	C, D, S	Drought tolerance	Seki et al. (2002), Fowler and Thomashow (2002), Sakamoto et al. (2004)
ANAC013	*Arabidopsis*	At1g32870	D, S		Seki et al. (2002)
AtbZIP60	*Arabidopsis*	At1g42990	C, D, S		Seki et al. (2002)
ANAC019	*Arabidopsis*	At1g52890	D, S, ABA	Drought tolerance	Tran et al. (2004)
ANAC029	*Arabidopsis*	At1g69490	C		Fowler and Thomashow (2002)
ANAC041	*Arabidopsis*	At2g33480	C		Fowler and Thomashow (2002)
ANAC055	*Arabidopsis*	At3g15500	D, S, ABA	Drought tolerance	Tran et al. (2004)
lip19	Rice	At3g62420	C		Rabbani et al. (2003)
ANAC072 (RD26)	*Arabidopsis*	At4g27410	C, D, S, ABA	Hypersen-sitivity to ABA, drought tolerance	Seki et al. (2002), Fujita et al. (2004), Tran et al. (2004)
ABF3 (AtbZIP37)	*Arabidopsis*	At4g34000	C, D, S, M	Hypersen-sitivity to ABA	Seki et al. (2002), Kreps et al. (2002), Kang et al. (2002)
ANAC092	*Arabidopsis*	At5g39610	D, S		Seki et al. (2002)
OSE2	Rice	At5g44080	D, ABA		Rabbani et al. (2003)
ATbZIP1	*Arabidopsis*	At5g49450	D, S		Seki et al. (2002)
ANAC102	*Arabidopsis*	At5g63790	C, D, S		Seki et al. (2002)
SCOF-1	Soybean		C, ABA	Cold tolerance	Kim et al. (2001)
ZPT2-3	Petunia		C, D	Drought tolerance	Sugano et al. (2003)

hypoosmolarity-induced reporter gene expression in *A. thaliana* has been shown indicating a regulatory role of members of bZIP transcription factors in adaptation to hypo-osmotic conditions as well (Satoh et al. 2004).

3.2 Stress-induced regulation of NAC transcription factors and zinc finger proteins

Transcription factors of the NAC-type family are well known for involvement, e.g. in plant development, auxin signalling, and responses to biotic stresses whereas less evidence had been presented for regulation by abiotic stresses (Riechmann et al. 2000; Xie et al. 2000; Collinge and Boller 2001; Ooka et al. 2003). Using GeneChip arrays representing approximately 8000 probes, Fowler and Thomashow (2002) reported transient decrease of transcript levels for two NAC-type proteins in cold-treated *A. thaliana*. In rice, OsNAC6 was induced by ABA, cold, drought, and salt stress, whereas the *A. thaliana* homologue ANAC002 was up-regulated under drought and salinity (Rabbani et al. 2003; Seki et al. 2002). In *A. thaliana*, the NAC-protein RD26 was induced by ABA, drought, and salt (Fujita et al. 2004). Over-expression of RD26 in *A. thaliana* resulted in hypersensitivity to ABA and stimulation of ABA-inducible transcripts suggesting a role of RD26 as a transcriptional activator in ABA-dependent gene expression (Fujita et al. 2004). Expression, e.g. of the *A. thaliana* NAC transcription factors ANAC019 and ANAC055 was induced by ABA, drought, and salt stress, and conferred increased drought tolerance to over-expressing plants by up-regulation of stress inducible genes (Tran et al. 2004).

Another group of transcription factors that have been shown recently to control abiotic stress-induced regulatory pathways are Cys2/His2 zinc finger proteins of the TFIIIA-type. The soybean C2H2-type zinc finger protein SCOF-1 is specifically regulated by cold and ABA but not by drought and salt stress (Kim et al. 2001). SCOF-1 induced ABRE-dependent gene expression by enhancing the DNA binding activity of the bZIP transcription factor SGBF-1 to ABRE (Kim et al. 2001). Expression of the petunia C2H2-type zinc finger protein ZPT2-3 was induced by wounding, jasmonic acid, cold, and drought, and constitutive over-expression of the gene increased drought tolerance in the transgenic plants (Sugano et al. 2003). Using microarray hybridizations, Fowler and Thomashow (2002) detected transient up-regulation of the C2H2-type zinc finger protein STZ by cold, and Seki et al. (2002) also reported enhanced expression in response to salt stress and drought. With a microarray-based approach, Maruyama et al. (2004) identified up-regulation of STZ as a downstream gene of DREB1A/CBF3 in

DREB1A/CBF3 over-expressing plants that were characterized by improved tolerance to drought, salt, and freezing. Over-expression of STZ in *A. thaliana* caused tolerance to drought stress and function of STZ and also of the related C2H2-type zinc finger proteins AZF1, AZF2, and AZF3 as transcriptional repressors under drought, cold, and salt stress has been suggested (Sakamoto et al. 2004).

4 Heavy metal toxicity and tolerance

Excess uptake into the symplast of (heavy) metal ions is deleterious to most plant species (Clemens 2001; Hall 2002). Metal elements either play an essential role as micronutrients, for example Cu, or lack an essential function, for example Cd. Originally the group of heavy metals was defined on the basis of specific weight of the metal that should be above 5 g/cm^3. However, other metals with lower specific weight such as aluminium also are toxic to plants, and the term heavy metal toxicity is occasionally used with rather diffuse meaning. Al^{3+}, Al^{2+}, AsO_2^-, AsO_4^{3-}, Au^+, Cd^{2+}, Cu^+, Cu^{2+}, Co^{2+}, Cr^{3+}, CrO_4^{2-}, Fe^{2+}, Fe^{3+} Hg^{2+}, Mn^{2+}, Ni^{2+}, Pb^{2+}, SeO_3^{2-}, SeO_4^{2-}, Sn^{2+}, W^{6+} and Zn^{2+} represent a selection of toxic metal ions with diverse chemical properties. It should be noted that some of them exist in various states of oxidation, as cations and/or as anions and may pose serious threat to man's health if excessively taken up with the diet. In addition, particularly Cu^+ and Fe^{2+} are strong oxidants and are involved in Fenton-type reactions where the extraordinarily reactive hydroxyl radical is produced. The length of the above incomplete list of metal ions also implicates that a comprehensive treatise of metal induced signalling, modification of metabolism and gene expression, damage development and adaptation is impossible here, and in fact the biochemical analysis of toxicity mechanisms is just beginning for many of these metal ions. Furthermore, only very few metals have been analysed in respect to gene expression in any depth, for example Cd^{2+}, Al^{3+}, Cu^{2+} and Pb^{2+}. Toxic metal ion concentrations for sensitive species begin in the below or low micromolar range, for example in case of Cd, Cu and Ni, as measured by root growth inhibition. Many metals are strong ligands to functional groups of biomolecules. For example, Cd and Cu have a high affinity to thiol groups, while Ni coordinates with carboxylates and histidine residues (Hall 2002). Metal binding alters the structure of target molecules and the binding frequently modifies and inhibits, respectively, their function in cell metabolism. As a consequence, disregulation of metabolism including liberation of ROS occurs in response to metal exposure. Signalling in context of metal ions can either be initiated from specific sensors or from imbalances in metabolism. Although metals interact with defined chemical groups of proteins such as thiol, carboxyl, amino and imido groups, binding to specific targets as primary event in signalling in plants has not been described so far.

In a converse manner, metabolic disorder following metal application has been studied in detail, often with focus on oxidative stress. Metal-dependent depletion of the glutathione pool and generation of reactive oxygen species are considered to induce damage to cells and to trigger adaptive responses not only in case of redox active transition metals but as a rather general mechanism (Dietz et al. 1999; Schützendübel et al. 2001).

In contrast to the highly toxic metals, other metal ions such as MoO_4^{2-} may be administered at close to millimolar concentrations without many adverse effects on plant biomass production (Brune and Dietz 1995). Tolerant ecotypes that repeatedly evolved locally under selective pressure of contaminated sites have adapted to otherwise toxic concentrations and tolerate two to four orders of magnitude higher concentrations of the metals. The best studied example on the level of gene expression is *Arabidopsis halleri* (Becher et al. 2004; Weber et al. 2004). Interestingly, tolerant ecotypes often exhibit a hyperaccumulation phenotype and can accumulate metals to levels above 1% of dry mass.

Although not belonging to the group of heavy metals in *sensu strictu*, Al and arsenate should also be mentioned in this context. Particularly in acid soils, high levels of free Al^{3+}-ions interfere with plant root metabolism and are among the most important abiotic stressors limiting crop productivity on a global scale (Kochian 1995). In case of Al, various hypotheses have been proposed to explain rapid root growth inhibition in response to low Al doses; (i) inhibition of cell proliferation by Al-binding to DNA (Matsumoto et al. 1976), (ii) inhibition of root cell elongation through apoplastic Al (Horst 1995) and (iii) Al-induced inhibition of auxin transport (Kollmeier et al. 2000). Each of these processes is likely to contribute to the overall Al syndrome. In a recent study, Doncheva et al. (2005) observed a change in cell division pattern within the maize root tissue 5 min after addition of 17 μM Al to the rooting medium indicating rapid initiation of a signalling chain. The specific change of pattern with strong inhibition of cell division in the central part of the root tip but not in the periphery cannot be explained through a direct contact of the cells to Al. Such a type of direct inhibition should be most pronounced in the periphery. The data are more in line with Al-dependent activation of a developmental programme involving sensing and signalling processes. Al-stress alters the gene expression of plant tissues. From 50 cDNAs identified in Al-stressed sugarcane using suppression subtractive hybridisation, 14 were related to signalling (Watt 2003), among them a putative histone deacetylase, a Ser/Thr kinase and a RNA binding protein.

The advantage of the approach was the specific addressing of rare transcripts. In particular from the group of rare transcripts, new insight into stress-dependent signalling and regulation can be expected. However, but not unexpectedly, more than half of the identified transcripts coded for proteins with unknown function. Recently, using a cDNA-amplified fragment length polymorphism (cDNA-AFLP) approach to identify Al-regulated gene expression in a tolerant and sensitive rice line, Mao et al. (2004) identified a set of up-regulated genes involved in lignin and cell wall synthesis. Two genes mapped on the common QTL (quantitative trait locus) interval for Al-tolerance in rice, one encodes an unknown protein, the other a small ubiquitin-like modifier 1 (SUMO 1) that is involved in a pathway of protein modification also often implicated in stress adaptation (Kurepa et al. 2003; Mao et al. 2004).

Metals other than Al also affect long distance signalling: Cadmium, but neither Zn nor Al, inhibited systemic spreading of posttranscriptional gene silencing (PTGS) from old to young leaves in tobacco (Ueki and Citovsky 2001). The inhibition was only seen in experiments where low non-toxic Cd concentration had been administered. Other metals than Al also affect hormone homeostasis of plants, for example up-regulation of lipid transfer protein expression in Cd-stressed barley correlated with increasing ABA concentrations in the shoot (Hollenbach et al. 1997).

AsO_4^{3-} acts as a phosphate analogue and binds to protein domains with four oxygen coordination sites often containing arginine. Depending on speciation, toxicity starts at soil levels of >1 mg/l (Carbonell-Barrachina et al. 1999). Tolerance is achieved by reduction of AsO_4^{3-} to the more toxic AsO_2^{2-} and subsequent efficient binding to thiols such as glutathione and phytochelatins. The stability of the glutathione–arsenite complex is high between pH 1.5 and 7.5. For example, the fern *Pteris vittata* rapidly imports and hyperaccumulates arsenic. Prior or following translocation to the shoot, arsenate is reduced to arsenite that is complexed by thiol compounds for detoxification and deposited mainly in epidermal cells (Zhao et al. 2003; Gumaelius et al. 2004). In a biotechnological project, arsenate reductase was introduced into *A. thaliana*. As expected, the transgenic plants were more sensitive to arsenate. However, following simultaneous transformation with γ-glutamylcysteine synthetase (γ-ECS) under control of a strong promoter, the plants accumulated more arsenic and were significantly more tolerant to arsenate (Dhankher et al. 2002). γ-ECS is the key enzyme in synthesis of glutathione needed for arsenite complexation and detoxification. The authors suggested using this strategy to create As-accumulating plants for phytoremediation of As-polluted soils.

Recently, mitogen activated protein kinase (MAPK) cascades have moved into the focus of metal toxicity research. Following elicitation of *Medicago sativa* with Cu and Cd, respectively, specific MAPK isoforms were immuno-precipitated and measured for activity by an in-gel method with γ-[^{32}P]ATP (Jonak et al. 2004). Activity states of specific MAPK as well as upstream MAPKK differed between both metals in dependence of metal concentration and exposure time. From the results, it was concluded that Cu mostly acts via reactive oxygen generation that activates MAPK-pathways to trigger genetic responses, whereas Cd ions induce different cellular signalling pathways that at least partly integrate into the MAPK networks (Jonak et al. 2004). In a parallel study and by using similar methods, Yeh et al. (2004) showed the involvement of a MAPK in Cd response in rice. Pretreatment with glutathione strongly suppressed Cu^{2+}-induced activation of MAPK in rice, and to a much lesser extent Cd^{2+}-induced activation. From that, the authors concluded that oxidative stress and ROS are involved in Cd-stress signalling (Yeh et al. 2004). Despite differences in interpretation, both studies have proven that MAP kinases participate in cell signalling in response to metal stress and that different metals trigger distinct signalling pathways.

Hyperaccumulation of metal ions is linked to a common set of biochemical events. Following (accelerated) uptake into root cells, the metal ions are efficiently loaded into the xylem, translocated to the shoots and deposited in safe stores such as epidermal cells, leaf hairs or mesophyll vacuoles. Transport and sequestration of metal ions apparently are key elements of the hyperaccumulation phenotype. Transcripts encoding zinc-regulated transporter, P-type ATPase, cation diffusion facilitator and nicotianamine synthase were upregulated in response to Zn stress in two studies where the Zn responses of Zn-sensitive *A. thaliana* and Zn-tolerant and hyperaccumulating *A. halleri* were compared on a transcriptome-wide scale (Becher et al. 2004; Weber et al. 2004). Transcripts of MYB and bHLH transcription factors and also for other signalling elements were more induced in *A. halleri* than in *A. thaliana* (Becher et al. 2004). The functional significance of these findings on signal transducers and regulators still needs to be assessed.

Synthesis of phytochelatins from glutathione precursors is induced upon administration of metal ions such as Cu^{2+}, Cd^{2+} and Zn^{2+}. Until recently, the regulation was believed to depend on posttranslational activation of the phytochelatin synthase (Grill et al. 1989). Late studies have shown that depending on isoform, developmental stage and growth condition, transcriptional activation can occur (Lee and Korban 2002; Finkemeier et al. 2003). This arbitrarily chosen example shows that sensing and signalling pathways involved in metal stress and adaptation are far from being understood, particularly in the light of the diversity of metal stressors (see above) and the complex dependency of the stress response on endogenous and exogenous parameters.

5 Perspectives

Novel signalling and regulatory elements involved in adaptation to abiotic stresses as high light intensities, ozone, cold, salinity, and heavy metals have been identified using the technology of microarray hybridization indicating that stress responses in plants appear to be less diverse than the stresses and the stress receptors. Overlaps among regulatory pathways extends from the transcriptional level to intracellular signalling that regulates gene expression. Stress responses frequently involve ROS and phytohormones such as salicylic acid, ethylene and abscisic acid among other signalling components. In the transmission of these signals MAPK kinase cascades often play a central role as they form complex signalling networks characterized by convergence and divergence at the level of the MAPKs and their upstream activating kinases. In addition to the important findings gained from studies in the plant models *A. thaliana* and rice, comparative transcriptomics analyses of species with natural tolerance to abiotic stresses will be a powerful tool to further dissect regulatory networks of stress adaptation. These investigations combined with forward and reverse genetics approaches will further elucidate our knowledge on regulatory networks that trigger environmental adaptation, and will help to identify key regulators that may be suitable to confer improved abiotic stress tolerance to sensitive species as crops.

References

A-H-Mackerness S, John CF, Jordan BR, Thomas B (2001) Early signalling components in ultraviolet-B responses: distinct roles for different reactive oxygen species and nitric oxide. FEBS Lett 489:237–242

A-H-Mackerness S, Surplus SL, Blake P, John CF, Buchanan-Wollaston V, Jordan BR, Thomas B (1999) Ultraviolet-B-induced stress and changes in gene expression in *Arabidopsis thaliana*: role of signalling pathways controlled by jasmonic acid, ethylene and reactive oxygen species. Plant Cell Environ 22:1413–1423

Allan A, Fluhr R (1997) Two distinct sources of elicited reactive oxygen species in tobacco epidermal cells. Plant Cell 9:1559–1572

Apse MP, Aharon GS, Snedden WA, Blumwald E (1999) Salt tolerance conferred by overexpression of a vacuolar Na^+/H^+ antiport in *Arabidopsis*. Science 285:1256–1258

Asai T, Tena G, Plotnikova J, Willmann MR, Chiu W-L, Gomez-Gomez L, Boller T, Ausubel FM, Sheen J (2002) MAP kinase signalling cascade in *Arabidopsis* innate immunity. Nature 415:977–983

Baier M, Dietz K-J (1998) The costs and benefits of oxygen for photosynthesizing cells. Progress Botany 60:282–314

Baier M, Noctor G, Foyer CH, Dietz KJ (2000) Antisense suppression of 2-Cys peroxiredoxin in *Arabidopsis* specifically enhanced activities and expression of enzymes associated with ascorbate metabolism but not glutathione metabolism. Plant Physiol 124:823–832

Baier M, Ströher E, Dietz KJ (2004) The acceptor availability at photosystem I and ABA control nuclear expression of 2-cyst peroxiredoxin-A in *Arabidopsis thaliana*. Plant Cell Physiol 45:997–1006

Baier M, Kandlbinder A, Golldack D, Dietz KJ (2005) Oxidative stress and ozone: perception, signalling and response. Plant Cell Environ 28:1012–1020

Bailey S, Walters RG, Jansson S, Horton P (2001) Acclimation of *Arabidopsis thaliana* to the light environment: the existence of separate low light and high light responses. Planta 213:794–801

Ball L, Accotto GP, Bechtold U, Creissen G, Funk D, Jimenez A, Kular B, Leyland N, Mejia-Carranza J, Reynolds H, Karpinski S, Mullineaux PM (2004) Evidence for a direct link between glutathione biosynthesis and stress defense gene expression in *Arabidopsis*. Plant Cell 16:2448–2462

Becher M, Talke IN, Krall L, Krämer U (2004) Cross-species microarray transcript profiling reveals high constitutive expression of metal homeostasis genes in shoots of the zinc hyperaccumulator *Arabidopsis halleri*. Plant J 37:251–268

Bharti K, von Koskull-Döring P, Bharti S, Kumar P, Tintschl-Körbitzer A, Treuter E, Nover L (2004) Tomato heat stress transcription factor HsfB1 represents a novel type of general transcription coactivator with a histone-like motif interacting with the plant CREB binding protein ortholog HAC1. Plant Cell 16:1521–1535

Brosché M, Strid A (2003) Molecular events following perception of ultraviolet-B radiation by plants. Physiol Plant 117:1–10

Brosché M, Schuler MA, Kalbina I, Connor L, Strid A (2002) Gene regulation by low level UV-B radiation: Identification by DNA array analysis. Photochem Photobiol Sci 1:656–664

Brune A, Dietz KJ (1995) A comparative analysis of element composition of barley roots and leaves under cadmium-, molybdenum-, nickel- and zinc-stress. J Plant Nutr 18:853–868

Carbonell-Barrachina AA, Burlo F, Valero D, Lopez E, Martinez-Romero D, Martinez-Sanchez F (1999) Arsenic toxicity and accumulation in turnip as affected by arsenic chemical speciation. J Agric Food Chem 47:2288–2294

Cardinale F, Meskiene I, Ouaked F, Hirt H (2002) Convergence and divergence of stress induced mitogen-activated protein kinasee signaling pathways at the level of two distinct mitogen-activated protein kinase kinases. Plant Cell 14:703–711

Casati P, Walbot V (2003) Gene expression profiling in response to ultraviolet radiation in maize genotypes with yarying flavonoid content. Plant Physiol 132:1739–1754

Casati P, Walbot V (2004) Rapid transcriptome responses of maize (*Zea mays*) to UV-B in irridiated and shielded tissues. Genome Biol 5:R16

Chamovitz DA, Deng XW (1996) Light signaling in plants. Crit Rev Plant Sci 15:455–478

Chang CC, Ball L, Fryer MJ, Baker NR, Karpinski S, Mullineaux PM (2004) Induction of ascorbate peroxidase 2 expression in wounded *Arabidopsis* leaves does not involve known wound-signalling pathways but is associated with changes in photosynthesis. Plant J 38:499–511

Chinnusamy V, Schumaker K, Zhu JK (2004) Molecular genetic perspectives on cross-talk and specificity in abiotic stress signalling in plants. J Exp Bot 55:225–236

Cho EK, Hong CB (2004) Molecular cloning and expression pattern analyses of heat shock protein 79 genes from *Nicotiana tabacum*. J Plant Biol 47:149–159

Choi H, Hong J, Ha J, Kang J, Kim SY (2000) ABFs, a family of ABA-responsive element binding factors. J Biol Chem 275:1723–1730

Clemens S (2001) Molecular mechanisms of plant metal tolerance and homeostasis. Planta 212:475–486

Collinge M, Boller T (2001) Differential induction of two potato genes, Stprx2 and StNAC, in response to infection by *Phytophthora infestans* and to wounding. Plant Mol Biol 46:521–529

Cook D, Fowler S, Fiehn O, Thomashow MF (2004) A prominent role for the CBF cold response pathway in configuring the low-temperature metabolome of *Arabidopsis*. Proc Natl Acad Sci USA 101:15243–15248

Coruzzi GM, Zhou L (2001) Carbon and nitrogen sensing and signaling in plants:emerging "matrix effects". Curr Opin Plant Biol 4:247–253

Desikan R, A-H-Mackerness A, Hancock JT, Neill SJ (2001) Regulation of the *Arabidopsis* transcriptome by oxidative stress. Plant Physiol 127:159–172

Devlin PF, Yanovsky MJ, Kay SA (2003) A genomic analysis of the shade avoidance response in *Arabidopsis*. Plant Physiol 133:1617–1629

Dhankher OP, Li YJ, Rosen BP, Shi J, Salt D, Senecoff JF, Sashti NA, Meagher RB (2002) Engineering tolerance and hyperaccumulation of arsenic in plants by combining arsenate reductase and gamma-glutamylcysteine synthetase expression. Nature Biotechnol 20:1140–1145

Dietz KJ, Krämer U, Baier M (1999) Free radicals and reactive oxygen species as mediators of heavy metal toxicity. In: Prasad J, Hagemeyer MNV (eds) Heavy metal stress in plants: from molecules to ecosystems. Springer Verlag, Berlin Heidelberg New York, pp 73–97

Doncheva S, Amenos M, Poschenrieder C, Barcelo J (2005) Root pattern—a primary target for aluminium toxicity in maize. J Exp Bot (in press)

Dubouzet JG, Sakuma Y, Ito Y, Kasuga M, Dubouzet EG, Miura S, Seki M, Shinozaki K, Yamaguchi-Shinozaki K (2003) OsDREB genes in rice, *Oryza sativa* L., encode transcription activators that function in drought-, high-salt- and cold-responsive gene expression. Plant J 33:751–763

Dunaeva M, Adamska I (2001) Identification of genes expressed in response to light stress in leaves of *Arabidopsis thaliana* using RNA differential display. Eur J Biochem 268:5521–5529

Escoubas J-M, Lomas M, LaRoche J, Falkowski PG (1995) Light intensity regulation of cab gene transcription is signalled by the redox state of the plastoquinone pool. Proc Natl Acad Sci USA 92:10237–10241

Finkemeier I, Kluge C, Metwally A, Georgi M, Grotjohann N, Dietz KJ (2003) Alterations in Cd-induced gene expression under nitrogen deficiency in *Hordeum vulgare*. Plant Cell Environ 26:821–833

Fowler S, Thomashow MF (2002) *Arabidopsis* transcriptome profiling indicates that multiple regulatory pathways are activated during cold acclimation in addition to the CBF cold response pathway. Plant Cell 14:1675–1690

Foyer CH, Lelandais M, Kundert K-J (1997a) Photooxidative stress in plants. Physiol Plant 92:696–717

Foyer CH, Lopez-Delgando H, Dat JF, Scott IM (1997b) Hydrogen peroxide- and glutathione-associated mechanisms of acclamatory stress tolerance and signalling. Physiol Plant 100:241–254

Fryer MJ, Ball L, Oxborough K, Karpinski S, Mullineaux PM, Baker NR (2003) Control of *Ascorbate Peroxidase 2* expression by hydrogen peroxide and leaf water status during excess light stress reveals a functional organisation of *Arabidopsis* leaves. Plant J 33:691–705

Fujita M, Fujita Y, Maruyama K, Seki M, Hiratsu K, Ohme-Takagi M, Tran LS, Yamaguchi-Shinozaki K, Shinozaki K (2004) A dehydration-induced NAC protein, RD26, is involved in a novel ABA-dependent stress-signaling pathway. Plant J 39:863–876

Gilmour SJ, Sebolt AM, Salazar MP, Everard JD, Thomashow MF (2000) Overexpression of the *Arabidopsis* CBF3 transcriptional activator mimics multiple biochemical changes associated with cold acclimation. Plant Physiol 124:1854–1865

Gittins JR, Schuler MA, Strid A (2002) Identification of a novel nuclear factor-binding site in the *Pisum sativum sad* gene promoters. Biochim Biophys Acta 1574:231–244

Golldack D (2004) Molecular responses of halophytes to high salinity. Prog Bot 65:219–234

Golldack D, Dietz KJ (2001) Salt-induced expression of the vacuolar H^+-ATPase in the common ice plant is developmentally controlled and tissue specific. Plant Physiol 125:1643–1654

Golldack D, Su H, Quigley F, Kamasani UR, Munoz-Garay C, Balderas E, Popova OV, Bennett J, Bohnert HJ, Pantoja O (2002) Characterization of a HKT-type transporter in rice as a general alkali cation transporter. Plant J 31:529–542

Golldack D, Quigley F, Michalowski CB, Kamasani UR, Bohnert HJ (2003) Salinity stress-tolerant and -sensitive rice (*Oryza sativa* L.) regulate AKT1-type potassium channel transcripts differently. Plant Mol Biol 51:71–81

Green R, Fluhr R (1995) UV-B-induced PR-1 accumulation is mediated by active oxygen species. Plant Cell 7:203–212

Grill E, Löffler S, Winnacker EL, Zenk MH (1989) Phytochelatins, the heavy metal binding peptides of plants, are synthesized from glutathione by a specific gamma-glutamylcysteine dipeptidyl transpeptidase. Proc Natl Acad Sci USA 86:6838–6842

Gumaelius L, Lahner B, Salt DE, Banks JA (2004) Arsenic hyperaccumulation in gametophytes of *Pteris vittata*. A new model system for analysis of arsenic hyperaccumulation. Plant Physiol 136:3198–3208

Gyula P, Schäfer E, Nagy F (2003) Light perception and signalling in higher plants. Curr Opin Plant Biol 6:446–452

Halford NG, Hey S, Jhurreea D, Laurie S, McKibbin RS, Paul M, Zhang Y (2003) Metabolic signalling and carbon partitioning: role of Snf1-related (SnRK1) protein kinase. J Exp Bot 54:467–475

Hall JL (2002) Cellular mechanisms for heavy metal detoxification and tolerance. J Exp Bot 53:1–11

Hollenbach B, Schreiber L, Hartung W, Dietz KJ (1997) Cadmium leads to stimulated expression of a lipid transfer protein (*ltp*) in barley: Implications for the involvement of LTP in wax assembly. Planta 203:9–19

Holley SR, Yalamanchili RD, Moura DS, Ryan CA, Stratmann JW (2003) Convergence of signaling pathways induced by systemin, oligosaccharide elicitors, and ultraviolet-B radiation at the level of mitogen-activated protein kinases in *Lycopersicon peruvianum* suspension-cultured cells. Plant Physiol 132:1728–1738

Horling F, Lamkemeyer P, König J, Finkemeier I, Kandlbinder A, Baier M, Dietz KJ (2003) Divergent light-, ascorbate-, and oxidative stress-dependent regulation of expression of peroxiredoxin gene family in *Arabidopsis*. Plant Physiol 131:317–325

Horst WJ (1995) The role of the apoplast in aluminium toxicity and resistance of higher plants. Zeitschrift für Pflanzenernährung und Bodenkunde 158:419–428

Hrabak EM, Dickmann LJ, Satterlee JS Sussman MR (1996) Characterization of eight members of the calmodulin-like domain protein kinase gene family from *Arabidopsis thaliana*. Plant Mol Biol 31:405–412

Huq E, Al-Sady B, Hudson M, Kim CH, Apel M, Quail PH (2004) Phytochrome-Interacting-Factor 1 is a critical regulator of chlorophyll biosynthesis. Science 305:1937–1941

Ichimura K, Mizoguchi T, Yoshida R, Shinozaki K (2000) Various abiotic stresses rapidly activate *Arabidopsis* MAP kinases ATMPK4 and ATMPK6. Plant J 24:655–665

Ireland HE, Harding SJ, Bonwick GA, Jones M, Smith CJ, Williams JHH (2004) Evaluation of heat shock protein 70 as a biomarker of environmental stress in *Fucus serratus* and *Lemna minor*. Biomarkers 9:139–155

Izaguirre MM, Scopel AL, Baldwin IT, Ballaré CL (2003) Convergent responses to stress. Solar UV-B radiation and *Manduca sexta* herbivory elicit overlapping transcriptional responses in field-grown plants of *Nicotiana longiflora*. Plant Physiol 132:1755–1767

Jakoby M, Weisshaar B, Droge-Laser W, Vicente-Carbajosa J, Tiedemann J, Kroj T, Parcy F (2002) bZIP transcription factors in *Arabidopsis*. Trends Plant Sci 7:106–111

Jonak C, Nakagami H, Hirt H (2004) Heavy metal stress. Activation of distinct mitogen-activated protein kinase pathways by copper and cadmium. Plant Physiol 136:3276–3283

Jonak C, Ökresz L, Bögre L, Hirt H (2002) Complexity, cross talk and integration of plant MAP kinase signalling. Curr Opin Plant Biol 5:415–424

Jordan BR (2002) Molecular response of plant cells to UV-B stress. Funct Plant Biol 29:909–916

Kang JY, Choi HI, Im MY, Kim SY (2002) *Arabidopsis* basic leucine zipper proteins that mediate stress-responsive abscisic acid signaling. Plant Cell 14:343–357

Kangasjärvi J, Talvinen J, Utriainen M, Karjalainen R (1994) Plant defense systems induced by ozone. Plant Cell Environ 17:783–794

Kim JC, Lee SH, Cheong YH, Yoo CM, Lee SI, Chun HJ, Yun DJ, Hong JC, Lee SY, Lim CO, Cho MJ (2001) A novel cold-inducible zinc finger protein from soybean, SCOF-1, enhances cold tolerance in transgenic plants. Plant J 25:247–259

Kimura M, Yamamoto YY, Seki M, Sakurai T, Sato M, Abe T, Yoshida S, Manabe K, Shinozaki K, Matsui M (2003) Identification of *Arabidopsis* genes regulated by high light-stress using cDNA microarrays. Photochem Photobiol 77:226–233

Kliebenstein DJ, Monde R-A, Last RL (1998) Superoxide dismutase in *Arabidopsis*: an eclectic enzyme family with disparate regulation and protein localization. Plant Physiol 118:637–650

Kochian LV (1995) Cellular mechanisms of aluminium toxicity and resistance of higher plants. Ann Rev Plant Physiol Plant Mol Biol 46:237–260

Kollmeier M, Felle HH, Horst WJ (2000) Genotypical differences in aluminium resistance of maize are expressed in the distal part of the transition zone. Is reduced basipetal auxin flow involved in inhibition of root elongation by aluminium? Plant Physiol 122:945–956

Kovtun Y, Chiu WL, Tena G, Sheen J (2000) Functional analysis of oxidative stress-activated mitogen-activated protein kinase cascades in plants. Proc Natl Acad Sci USA 97:2940–2945

Krapp A, Hofmann B, Schäfer C, Stitt M (1993) Regulation of the expression of rbcS and other photogenic genes by carbohydrates: a mechanism for the "sink regulation" of photosynthesis. Plant J 3:817–828

Kreps JA, Wu Y, Chang HS, Zhu T, Wang X, Harper JF (2002) Transcriptome changes for *Arabidopsis* in response to salt, osmotic, and cold stress. Plant Physiol 130:2129–2141

Kurepa J, Walker JM, Smalle J, Gosink MM, Davis SJ, Durham TL, Sung DY, Vierstra RD (2003) The small ubiquitin-like modifier (SUMO) protein modification system in *Arabidopsis*–accumulation of SUMO1 and -2 conjugates is increased by stress. J Biol Chem 278 (9):6862–6872

Langebartels C, Wohlgemuth H, Kschieschan S, Grun S, Sandermann H (2002) Oxidative burst and cell death in ozone-exposed plants. Plant Physiol Biochem 40:567–575

Lavorel J (1975) Luminescence. In: Govindjee (ed) Bioenergetics of photosynthesis. Academic Press, New York, pp 223–317

Lee S, Korban SS (2002) Transcriptional regulation of *Arabidopsis thaliana* phytochelatin synthase (AtPCS1) by cadmium during early stages of plant development. Planta 215:689–693

Lin C (2002) Blue light receptors and signal transduction. Plant Cell 14 (Suppl):207–225

Mahalingham R, Gomez-Buitrago A, Eckardt N, Shah N, Guevara-Garcia A, Day P, Raina R, Fedoroff NV (2003) Characterizing the stress/defense transcriptome of *Arabidopsis*. Genome Biol 4:R20

Mao CZ, Yi K, Yang L, Zheng BS, Wu YR, Liu FY, Wu P (2004) Identification of aluminium-regulated genes by cDNA-AFLP in rice (*Oryza sativa* L.): aluminium-regulated genes for the metabolism of cell wall components. J Exp Bot 55:137–143

Maruyama K, Sakuma Y, Kasuga M, Ito Y, Seki M, Goda H, Shimada Y, Yoshida S, Shinozaki K, Yamaguchi-Shinozaki K (2004) Identification of cold-inducible downstream genes of the *Arabidopsis* DREB1A/CBF3 transcriptional factor using two microarray systems. Plant J 38:982–993

Matsumoto H, Hirasawa E, Morimura S, Takahashi E (1976) Localization of absorbed aluminium in pea root and its binding to nucleic acid. Plant Cell Physiol 17:627–631

Matsuyama T, Tamaoki M, Nakajima N, Aono M, Kubo A, Moriya S, Ichihara T, Suzuki O, Saji H (2002) cDNA microarray assessment for ozone-stressed *Arabidopsis thaliana*. Environ Poll 117:191–194

Mochizuki N, Brusslan JA, Larkin R, Nagatan A, Chory J (2001) *Arabidopsis* genomes uncoupled 5 (GUN5) mutant reveals the involvement of Mg-chelatase H subunit in plastid-to-nucleus signal transduction. Proc Natl Acad Sci USA 98:2053–2058

Montillet JL, Cacas JL, Garnier L, Montane MH, Douki T, Bessoule JJ, Polkowska-Kowalczyk L, Maciejewska U, Agnel JP, Vial A, Triantaphyllides C (2004) The upstream oxolipid profile of *Arabidopsis thaliana*: a tool to scan for oxidative stresses. Plant J 40:439–451

Moon H, Lee B, Choi G, Shin D, Prasad DT, Lee O, Kwak SS, Kim DH, Nam J, Bahk J, Hong JC, Lee SY, Cho MJ, Lim CO, Yun DJ (2003) NDP kinase 2 interacts with two oxidative stress-activated MAPKs to regulate cellular redox state and enhances multiple stress tolerance in transgenic plants. Proc Natl Acad Sci USA 100:358–363

Moore B, Zhou L, Rolland F, Hall Q, Cheng WH, Liu YX, Hwang I, Jones T, Sheen J (2003) Role of the *Arabidopsis* glucose sensor HXK1 in nutrient, light, and hormonal signalling. Science 300:332–336

Nagy F, Schäfer E (2000) Nuclear and cytosolic events of light-induced, phytochrome-regulated signaling in higher plants. EMBO J 19:157–163

Niyogi KK (2000) Safety valves for photosynthesis. Curr Opin Plant Biol 3:455–460

Ooka H, Satoh K, Doi K, Nagata T, Otomo Y, Murakami K, Matsubara K, Osato N, Kawai J, Carninci P, Hayashizaki Y, Suzuki K, Kojima K, Takahara Y, Yamamoto K, Kikuchi S (2003) Comprehensive analysis of NAC family genes in *Oryza sativa* and *Arabidopsis thaliana*. DNA Res 31:239–247

Osmani H, May GS, Osmani SA (1999) The extremely conserved pyroA gene of *Aspergillus nidulans* is required for pyridoxine synthesis and is required indirectly for resistance to photosensitizers. J Biol Chem 274:23565–23569

Oswald O, Martin T, Dominy PJ, Graham IA (2001) Plastid redox state and sugars. Interactive regulators of nuclear-encoded photosynthetic gene expression. Proc Natl Acad Sci USA 98:2047–2052

Overmyer K, Tuominen H, Kettunen R, Betz C, Langebartels C, Sandermann H Jr, Kangasjarvi J (2000) Ozone-sensitive *Arabidopsis rcd1* mutant reveals opposite roles for ethylene and jasmonate signaling pathways in regulating superoxide-dependent cell death. Plant Cell 12:1849–1862

Panchuk II, Volkov RA, Schöffl F (2002) Heat stress- and heat shock transcription factor-dependent expression and activity of ascorbate peroxidase in *Arabidopsis*. Plant Physiol 129:838–853

Popova OV, Ismailov SF, Popova TN, Dietz KJ, Golldack D (2002) Salt-induced expression of NADP-dependent isocitrate dehydrogenase and ferredoxin-dependent glutamate synthase in *Mesembryanthemum crystallinum*. Planta 215:906–913

Popova OV, Dietz KJ, Golldack D (2003) Salt-dependent expression of a nitrate transporter and two amino acid transporter genes in *Mesembryanthemum crystallinum*. Plant Mol Biol 52:569–578

Rabbani MA, Maruyama K, Abe H, Khan MA, Katsura K, Ito Y, Yoshiwara K, Seki M, Shinozaki K, Yamaguchi-Shinozaki K (2003) Monitoring expression profiles of rice genes under cold, drought, and high-salinity stresses and abscisic acid application using cDNA microarray and RNA gel-blot analyses. Plant Physiol 133:1755–1767

Rao MV, Lee H, Davis KR (2002) Ozone-induced ethylene production is dependent on salicylic acid, and both salicylic acid and ethylene act in concert to regulate ozone-induced cell death. Plant J 32:447–456

Riechmann JL, Heard J, Martin G, Reuber L, Jiang C, Keddie J, Adam L, Pineda O, Ratcliffe OJ, Samaha RR, Creelman R, Pilgrim M, Broun P, Zhang JZ, Ghandehari D, Sherman BK,

Yu G (2000) *Arabidopsis* transcription factors: genome-wide comparative analysis among eukaryotes. Science 290:2105–2110

Rodermel S (2001) Pathways of plastid-to-nucleus signalling. Trends Plant Sci 6:471–478

Rook F, Bevan M (2003) Genetic approaches to understanding sugar response pathways. J Exp Bot 54:1–7

Rossel JB, Wilson IW, Pogson BJ (2002) Global changes in gene expression in response to high light in *Arabidopsis*. Plant Physiol 130:1109–1120

Rus A, Yokoi S, Sharkhuu A, Reddy M, Lee BH, Matsumoto TK, Koiwa H, Zhu JK, Bressan RA, Hasegawa PM (2001) AtHKT1 is a salt tolerance determinant that controls Na^+ entry into plant roots. Proc Natl Acad Sci USA 98:14150–14155

Sakamoto H, Maruyama K, Sakuma Y, Meshi T, Iwabuchi M, Shinozaki K, Yamaguchi-Shinozaki K (2004) *Arabidopsis* Cys2/His2-type zinc-finger proteins function as transcription repressors under drought, cold, and high-salinity stress conditions. Plant Physiol 136:2734–2746

Satoh R, Fujita Y, Nakashima K, Shinozaki K, Yamaguchi-Shinozaki K (2004) A novel subgroup of bZIP proteins functions as transcriptional activators in hypoosmolarity-responsive expression of the ProDH gene in *Arabidopsis*. Plant Cell Physiol 45:309–317

Sävenstrand H, Brosché M, Strid A (2002) Regulation of gene expression by low levels of ultraviolet-B radiation in *Pisum sativum*: isolation of novel genes by suppression subtractive hybridisation. Plant Cell Physiol 43:402–410

Schenk PM, Kazan K, Wilson I, Anderson JP, Richmond T, Somerville SC, Manners JM (2000) Coordinated plant defense responses in *Arabidopsis* revealed by microarray analysis. Proc Natl Acad Sci USA 97:11655–11660

Schepens I, Duek P, Fankhauser C (2004) Phytochrome-mediated light signalling in *Arabidopsis*. Curr Opin Plant Biol 7:564–569

Schützendübel A, Schwanz P, Teichmann T, Gross K, Langenfeld-Heyser R, Godbold DL, Polle A (2001) Cadmium-induced changes in antioxidative systems, hydrogen peroxide content, and differentiation in Scots pine roots. Plant Physiol 127:887–898

Seki M, Narusaka M, Ishida J, Nanjo T, Fujita M, Oono Y, Kamiya A, Nakajima M, Enju A, Sakurai T, Satou M, Akiyama K, Taji T, Yamaguchi-Shinozaki K, Carninci P, Kawai J, Hayashizaki Y, Shinozaki K (2002) Monitoring the expression profiles of 7000 *Arabidopsis* genes under drought, cold and high-salinity stresses using a full-length cDNA microarray. Plant J 31:279–292

Shinozaki K, Yamaguchi-Shinozaki K, Seki M (2003) Regulatory network of gene expression in the drought and cold stress responses. Curr Opin Plant Biol 6:410–417

Steponkus PL, Uemura M, Joseph RA, Gilmour SJ, Thomashow MF (1998) Mode of action of the COR15a gene on the freezing tolerance of *Arabidopsis thaliana*. Proc Natl Acad Sci USA 95:14570–14575

Strand A, Asami T, Alonso J, Ecker JR, Chory J (2003) Chloroplast-to-nucleus communication triggered by accumulation of Mg-protoporphyrin IX. Nature 421:79–83

Stratmann J (2003) Ultraviolet-B radiation co-opts defense signaling pathways. Trends Plant Sci 8:526–533

Sugano S, Kaminaka H, Rybka Z, Catala R, Salinas J, Matsui K, Ohme-Takagi M, Takatsuji H (2003) Stress-responsive zinc finger gene ZPT2-3 plays a role in drought tolerance in petunia. Plant J 36:830–841

Surplus SL, Jordan BR, Murphy AM, Carr JP, Thomas B, A-H-Mackerness S (1998) Ultraviolet-B-induced responses in *Arabidopsis thaliana*: role of salicylic acid and reactive oxygen species in the regulation of transcripts encoding photosynthetic and acidic pathogenesis-related proteins. Plant Cell Environ 21:685–694

Tamaoki M, Nakajima N, Kubo A, Aono M, Matsuyama T, Saji H (2003a) Transcriptome analysis of O_3-exposed *Arabidopsis* reveals that multiple signal pathways act mutually antagonistically to induce gene expression. Plant Mol Biol 53:443–456

Tamaoki M, Matsuyama T, Kanna M, Nakajima N, Kubo A, Aono M, Saji H (2003b) Differential O_3-sensitivity among *Arabidopsis* accessions and its relevance to ethylene synthesis. Planta 216:552–560

Tamaoki M, Matsuyama T, Nakajima N, Aono M, Kubo A, Saji H (2004) A method for diagnosis of plant environmental stresses by gene expression profiling using a cDNA macroarray. Environ Poll 131:137–145

Tang L, Bhat S, Petracek ME (2003) Light control of nuclear gene mRNA abundance and translation in tobacco. Plant Physiol 133:1979–1990

Thomashow MF (1999) Plant cold acclimation: Freezing tolerance genes and regulatory mechanisms. Annu Rev Plant Physiol Plant Mol Biol 50:571–599

Thomashow MF (2001) So what's new in the field of plant cold acclimation? Lots! Plant Physiol 125:89–93

Thum KE, Shin MJ, Palenchar PM, Kouranov A, Coruzzi GM (2004) Genome-wide investigation of light and carbon signalling interactions in *Arabidopsis*. Genome Biol 5:R10

Torres MA, Dangl JL, Jones JDG (2002) *Arabidopsis* gp91phox homologues *AtrbohD* and *AtrbohF* are required for accumulation of reactive oxygen intermediates in the plant defense responses. Proc Natl Acad Sci USA 99:517–522

Tran LS, Nakashima K, Sakuma Y, Simpson SD, Fujita Y, Maruyama K, Fujita M, Seki M, Shinozaki K, Yamaguchi-Shinozaki K (2004) Isolation and functional analysis of *Arabidopsis* stress-inducible NAC transcription factors that bind to a drought-responsive cis-element in the early responsive to dehydration stress 1 promoter. Plant Cell 16:2481–2498

Tuominen H, Overmyer K, Keinanen M, Kollist H, Kangasjarvi J (2004) Mutual antagonism of ethylene and jasmonic acid regulates ozone-induced spreading cell death in *Arabidopsis*. Plant J 39:59–69

Ueki S, Citovsky V (2001) Inhibition of systemic onset of post-transcriptional gene silencing by non-toxic concentrations of cadmium. Plant J 28:283–291

Ulm R, Baumann A, Oravecz A, Máté Z, Ádám E, Oakeley EJ, Schäfer E, Nagy F (2004) Genome-wide analysis of gene expression reveals the function of the bZIP transcription factor HY5 in the UV-B response of *Arabidopsis*. Proc Natl Acad Sci USA 101:1397–1402

Uno Y, Furihata T, Abe H, Yoshida R, Shinozaki K, Yamaguchi-Shinozaki K (2000) *Arabidopsis* basic leucine zipper transcription factors involved in an abscisic acid-dependent signal transduction pathway under drought and high-salinity conditions. Proc Natl Acad Sci USA 97:11632–11637

Vandenabeele S, Vanderauwera S, Vuylsteke M, Rombauts S, Langebartels C, Seidlitz HK, Zabeau M, Van Montagu M, Inzé D, Van Breusegem F (2004) Catalase deficiency drastically affects gene expression induced by high light in *Arabidopsis thaliana*. Plant J 39:45–58

Vranova E, Atichartpongkul S, Villarroel R, Van Montagu M, Inze D, Van Camp W (2002) Comprehensive analysis of gene expression in *Nicotiana tabacum* leaves acclimated to oxidative stress. Proc Natl Acad Sci USA 99:10870–18075

Wade HK, Bibikova TN, Valentine WJ, Jenkins GI (2001) Interactions within a network of phytochrome, cryptochrome and UV-B phototransduction pathways regulate chalcone synthase gene expression in *Arabidopsis* leaf tissue. Plant J 25:675–685

Watt DA (2003) Aluminium-responsive genes in sugarcane: identification and analysis of expression under oxidative stress. J Exp Bot 54:1163–1174

Weber M, Harada E, Vess C, von Roepenack-Lahaye E, Clemens S (2004) Comparative microarray analysis of *Arabidopsis thaliana* and *Arabidopsis halleri* roots identifies nicotianamine synthase, a ZIP transporter and other genes as potential metal hyperaccumulation factors. Plant J 37:269–281

Xie Q, Frugis G, Colgan D, Chua NH (2000) *Arabidopsis* NAC1 transduces auxin signal downstream of TIR1 to promote lateral root development. Genes Dev 14:3024–3036

Xiong L, Schumaker KS, Zhu JK (2002) Cell signaling during cold, drought, and salt stress. Plant Cell 14:S165–183

Yamaguchi-Shinozaki K, Shinozaki K (1994) A novel cis-acting element in an *Arabidopsis* gene is involved in responsiveness to drought, low-temperature, or high-salt stress. Plant Cell 6:251–264

Yeh CM, Hsiao LJ, Huang HJ (2004) Cadmium activates a mitogen-activated protein kinase gene and MBP kinases in rice. Plant Cell Physiol 45:1306–1312

Yu SW, Tang KX (2004) MAP kinase cascades responding to environmental stress in plants. Acta Bot Sinica 46:127–136

Zavorueva EN, Ushakova SA (2004) Characteristics of slow induction of chlorophyll fluorescence and CO_2-exchange for the assessment of plant heat tolerance at various levels of light intensity. Russian J Plant Physiol 51:194–301

Zhao FJ, Wang JR, Barker JHA, Schat H, Bleeker PM, McGrath SP (2003) The role of phytochelatins in arsenic tolerance in the hyperaccumulator *Pteris vittata*. New Phytologist 159:403–410

Margarete Baier
Andrea Kandlbinder
Karl-Josef Dietz
Dortje Golldack
Department of Physiology and Biochemistry of Plants,
Faculty of Biology,
University of Bielefeld,
33615 Bielefeld,
Germany
Fax: +49-521-106-6039

Genetically transformed root cultures – generation, properties and application in plant sciences

Dr. Inna N. Kuzovkina and Dr. Bernd Schneider

1 Introduction

A unique characteristic of the Gram-negative soil bacterium *Agrobacterium rhizogenes*—the ability to insert T-DNA into the genome of cells of wounded plant tissue—contains a fundamental process that has been utilized to develop a valuable technique for genetically transforming plants. Neoplastic root tissue growing at the infected plant site is the morphological result of inserting T-DNA of the Ri (root-inducing) plasmid of *A. rhizogenes* into the plant genome. Like the formation of "crown galls" after the introduction of T-DNA from Ti (tumor-inducing) plasmid of *Agrobacterium tumefaciens*, Ri transformation has been considered a plant disease (Riker et al. 1930; Hildebrand 1934). However, in contrast to the tumor-inducing T-DNA of the Ti-plasmid (pTi), which is an abnormal process in plant development, the active formation of a more branched root system induced in planta by pRi T-DNA often benefits the plant, since it contributes to improved water and nutrient supply. Plants regenerated from pRi-transformed roots often show a characteristically altered phenotype ("T phenotype") (Tepfer 1984), which is not necessarily a plant disease syndrome. Thus, pRi T-DNA has been assumed to possess an adaptive function in nature (Tepfer 1983). Recently *A. rhizogenes* was found to be involved in transformation during the early evolution of *Nicotiana* plants (Suzuki et al. 2002).

Root tips dissected from pRi-transformed plants are capable of growing *in vitro* under sterile conditions. These roots have been referred to as "hairy root cultures" or "transformed root cultures" and used in plant science for more than 2 decades. During that period, hairy roots have become an essential tool in plant physiology, biochemistry, and molecular biology. In this review, we intend to provide a general overview on this field of research rather than focusing on particular themes that are the subject of more detailed reviews, e.g. the contributions to the book edited by Doran (1997).

2 Some genetic and historical remarks

The history of early discoveries of *Agrobacterium*-plant interactions was retrospectively reviewed by Chilton (2001), and a recent overview (Zupan et al. 2000) has updated our knowledge of the genetics of *Agrobacterium*-mediated gene transfer. Although they focus in part to *Agrobacterium tumefaciens*, these and other review articles (Rhodes et al. 1990; Weising and Kahl 1996) provide an excellent survey. Therefore, these aspects will be discussed here only briefly.

In the first step of contact between the plant and *Agrobacterium rhizogenes*, the virulence (*vir*) gene of the Ri-plasmid is induced by specific phenolic metabolites, e.g. acetosyringone (Stachel et al. 1985), which are released by the injured plant cell. During transformation, one or two T-DNA sections (T_L and T_R) (Jouanin 1984) are transferred to the plant genome but not the *vir*-gene. Introducing the T_R gene section, which shows some similarity with T-DNA of the Ti plasmid, into the plant genome, includes transfer of loci involved in auxin biosynthesis. Thus, the auxin level of these transformants is immediately enhanced, and simultaneously their requirement for exogenous auxin diminishes. Roots, which have been transformed only with T_R-DNA, retain apical dominance and morphological traits of the normal plant root.

In addition to auxin biosynthetic genes, genetic information transferred from the *Agrobacterium* Ri (or Ti) plasmid also includes genes in the T_R section that encode for biosynthesis of specific opines (Rhodes et al. 1990). These *N*-(carboxyalkyl)-amino acids (Thompson and Donkersloot 1992), agropine, mannopine, and cucumopine, are utilized as a carbon and nitrogen source by free-living bacteria of *A. rhizogenes*. Wild strains of *A. rhizogenes* have been classified by their opine type: agropine strains (e.g. A4, 15834, and 1855) produce agropine, mannopine and agropinic acid; mannopine strains (e.g. 8196) contain only mannopine, and cucumopine strains contain only cucumopin (Petit et al. 1983). Ri plasmids of mannopine strains transfer only the T_L-DNA section containing the *rol* genes, which cause of the typical hairy root phenotype (Petersen et al. 1989; Dessaux et al. 1993). Thus, there is a specific relationship between the bacterium strain and the structure of the characteristic opine produced by the infected tissue. Specific degradation pathways indicate that the purpose of opine production is to redirect plant metabolism for bacterial nutrition.

T_L-DNA of the Ri plasmid does not show homology with T-DNA of the Ti plasmid. Four genes, *rol* A, B, C and D, whose major functions in root development have been worked out by White et al. (1985), Schmülling et al. (1988), and Petersen et al. (1989), are included in T_L. The *rol* genes are not involved in auxin biosynthesis, but alter perception of auxin receptor proteins and, in so doing, enhance the hormone susceptibility of plant cells. Thus, they are responsible for the characteristic phenotypic properties of hairy roots such as rapid plagiotropic growth rate and extensive root branching. Recent studies regarding the function of *rol* genes have been reviewed in detail by Christey (2001).

Typically, pRi T_R-DNA and T_L-DNA are only integrated into the genome of dicotyledons. Monocots have been demonstrated to be highly resistant to *A. rhizogenes*, and special techniques are required to transform them genetically with pRi T-DNA. Moreover, even among dicots, there is some specificity of various *Agrobacterium* strains and the host range is frequently limited. After the plant is infected with pRi, each transformed cell results in an individual clone, which keeps the diploid character of the parent cell (Aird et al. 1988).

Hairy roots were first obtained from *Agrobacterium*-infected carrot tissue (Chilton et al. 1982) and initial axenic hairy root cultures were obtained from carrot and potato disks (Willmitzer et al. 1982) and *Hyoscyamus* (Flores and Filner 1985). The latter also provided evidence that transformed root cultures not only show stable growth properties but also keep the capacity of the parent tissue to synthesize root-specific alkaloids.

After the foundations in *Agrobacterium*-based transformation techniques were prepared, a considerable number of transformed root cultures were established and used in various research directions. By 1987, hairy roots had been obtained from 29 plant species and this number quickly rose in subsequent years. By the year 1988, for example, 49 species from 16 families had been established from only one group (Mugnier 1988), and in 1990 a compilation of hairy root cultures showed 116 species (Tepfer 1990). By the middle of 2004, we counted genetically transformed root cultures of 185 plants belonging to 41 families. Solanaceae have been transformed most successfully (at least 40 species), followed by Leguminosae (26) and Compositae (24). Although these numbers mainly reflect the scientific interest in the plants being transformed, they also depend on the virulence of the *Agrobacterium* strain and the resistance of the particular plant species to infection.

3 Transformation process and cultivation conditions

Sterile seedlings are the preferred objects for genetic transformation with *Agrobacterium rhizogenes*, but other plant parts such as surface-sterilized hypocotyl segments, cotyledons, stems, leaf pieces, or leaf stems can also be used for inoculation. Protocols that have been used successfully in our laboratory include slightly wounding the hypocotyl of intact seedlings, inoculating the wounded site with diluted *Agrobacterium* suspension (10^9–10^{11} cells ml^{-1} in MS or B5 medium) for 24-48 h, flushing the plant with sterile medium, and incubation on an agar dish containing an antibiotic (e.g. carbenicillin or carboforan). After an incubation of approximately 2–4 weeks, roots start to emerge on the infected tissue (Fig. 1). Freshly grown root tips are excised and transferred to liquid hormone-free medium containing an antibiotic in order to eliminate residual bacteria. The antibiotic frequently inhibits the growth rate of the roots to some extent. After completely removing free-living *Agrobacteria*, the hairy roots are cultivated in the usual manner (Fig. 1). Similar procedures have been described in more detail, for example, by Hamill et al. (1987a) and Hamill and Lidget (1997).

Some plants are highly resistant to infection by *Agrobacterium rhizogenes*, and, vice versa, various bacterial strains exhibit different levels of virulence to particular plant species. Agropine strain A4, for example, is highly virulent to

a   b

Fig. 1. Adventitious roots appearing after infection of **a** carrot tissue and **b** the stem of a *Peganum harmala* plant with *Agrobacterium rhizogenes*. Mature transformed root cultures of **c** carrot and **d** *P. harmala*

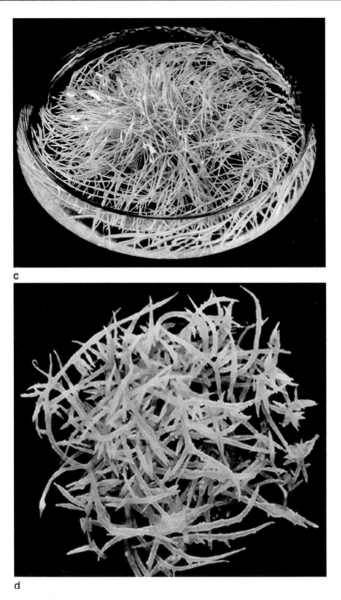

c

d

Fig. 1. cont'd

most dicotyledonous plants but virulence of mannopine strain 8196 is limited. In problematic cases, isolated protoplasts have been incubated with *A. rhizogenes*, or pRi T-DNA was transferred by microinjection or electroporation (Grisebach 1983; Fromm et al. 1985). Electroporation was also used to transform the first monocotyledonous species (Matsuki et al. 1989; Dommisse et al. 1990). Sonication for 60–100 s of the explants after inoculation followed by co-cultivation for another 1–2 days improved penetration of agrobacteria into the plant tissue (Trick and Finer 1997). A "difficult-to-transform" plant, *Papaver somniferum*, was transformed successfully by employing this sonication-assisted method (Le Flem-Bonhomme et al. 2004). Successful transformation with pRi is usually obvious from the accelerated plagiotropic and hormone-autotrophic growth of the hairy roots. Plants regenerated from transformed roots sometimes lose their apical dominance and form an enhanced number of adventitious roots. In addition to morphological characteristics, transformation with pRi can be determined by biochemical or molecular methods. The determination of opines was frequently used for this purpose (Petit et al. 1983), until it was found that opine synthesis is not always stable (Kamada et al. 1986). A more reliable method is the determination of T-DNA by southern blot analysis (Payne et al. 1987) or PCR (Edwards et al. 1991). Modified Ri plasmids containing selection marker genes in their T-DNA make it possible to discriminate transformed roots from untransformed explants. Examples include the antibiotic resistance marker genes such as the catalase (CAT) gene, coding for chloramphenicol resistance, and the neomycin phosphotransferase II (NPT II) gene from a bacterial transposable element for kanamycin resistance (Hamill et al. 1987b). The β-glucuronidase (GUS) gene has also been used as a selection marker (Jefferson 1989; Puddephat et al. 2001). A new marker for establishing Leguminosae hairy roots has been developed recently by Widholm's group (Cho et al. 2004). Feedback-insensitive anthranilate synthase (ASA2) cDNA was isolated from a 5-methyltryptophan-resistant tobacco cell line and introduced into *Astragalus sinicus* and *Glycine max* using *Agrobacterium rhizogenes* strains DC-AR2 or K599, respectively. The results demonstrated that the tobacco ASA2 gene is useful as a selectable marker for transforming legumes.

The effect of growth conditions, medium composition, and exogenous hormones has been thoroughly investigated and reviewed (Toivonen et al. 1991; Rhodes et al. 1994; Nussbaumer et al. 1998; Morgan et al. 2000; Sevón and Oksman-Caldentey 2002) and will not be considered in detail here. Although hairy roots in general are hormone-autotrophic, supplementary auxin (IAA, NAA or IBA) has proved beneficial in some cases for root growth and production of secondary metabolites, e.g. for *Linum flavum* (Lin et al. 2003), *Papaver somniferum* (Park and Facchini 2000), *Catalpa ovata* (Wysokinska et al. 2001), and *Panax ginseng* × *P. quinquifolium* (Washida et al. 2004).

4 Morphological and physiological aspects of hairy root cultures

Transformed root cultures possess some properties that make them very attractive in comparison with untransformed root cultures and undifferentiated cell suspension and tissue cultures. They show constant and rapid growth characteristics in hormone-free media of relatively simple composition. Doubling time after inoculation of various hairy root cultures was reported to be 24–90 h (Payne et al. 1991). Inoculates do not require conditioning of the medium, making the amount of the inoculum unimportant (Rhodes et al. 1990; Toivonen 1993).

The growth rate of transformed root cultures mainly depends on two physiological parameters: (1) the linear extension of the root tip, and (2) the exponential formation of lateral roots. Thus, the growth curve of hairy root cultures is characterized by a short lag-phase followed by exponential increase of biomass. The euploid chromosome number represents an important basis for the high genetic and growth stability of hairy roots compared to undifferentiated cell and tissue cultures with their typical somatic variability (Aird et al. 1988).

Transformed root cultures do not show apical dominance but rapidly form adventitious and acropetally growing roots. The epidermis of most transformed root cultures is covered by a large number of root hairs. In some cases, e.g. *Lithospermum erythrorhizon*, *Lupinus hartwegii* and *Ruta graveolens*, the root tips of transformed root cultures, when grown in liquid medium, retain the calyptra, which plays an important role in the formation of border cells and secretion of secondary compounds into the medium (Hawes et al. 2000, 2003). Hairy root physiology is further characterized by limited secondary growth and the occurrence of simple idioblasts, which are not significantly different from parenchyma cells. They do not possess differentiated depots such as oil bodies or lactifers but often store secondary metabolites in lipophilic vesicles or in the vacuoles. The increase in diameter of the basal root segments is due to cell extension and thickness of the primary root cortex rather than cell division. The intercellular space in these root parts can be filled with exudates of the cortex cells (Payne et al. 1991). From a physiological point of view, hairy roots are comparable in many respects to germinating roots. Another morphological peculiarity of hairy root cultures is their plagiotropic growth, which is connected to the loss of gravitational sensitivity of the root tips and due to the reduced amount of amyloplasts in the root tips (Kim and Soh 1996).

Lanoue et al. (2004) recently grew *Datura innoxia* hairy roots in a bioreactor equipped with on-line data analyses in the dark at a constant temperature. Their observations revealed regular oscillations of the CO_2 evolution rate, O_2 uptake rate, pH, and conductivity. The imposition of an external day/night cycle did not change this oscillation, indicating a circadian rhythm in hairy roots of this species.

Root hairiness is the most conspicuous property of transformed "hairy" root cultures. However, especially in large-scale bioreactor cultivation, root hairs have a detrimental effect on the performance of root cultures, specifically, transport processes within the reactor and resistance of oxygen transfer to the roots (Bordonaro and Curtis 2000; Shiao and Doran 2000). Thus, controlling root hair formation genetically or by growth conditions is desirable for biotechnological applications and has been achieved in transformed root cultures of *Arabidopsis thaliana* (Shiao and Doran 2000) and *Hyoscyamus muticus* (Bordonaro and Curtis 2000).

5 Secondary metabolites from hairy root cultures

5.1 Secondary products extracted from hairy root tissue

The occurrence of secondary metabolites in various types of plant *in vitro* cultures has been reviewed several times (Ellis 1988; Stöckigt et al. 1995; Gräther and Schneider 2001). The genetic stability of transformed root cultures is a major advantage for the production of root-specific secondary compounds. More than 500 hairy root lines of *Datura stramonium* were checked for long-term genetic stability (Maldonado-Mendoza et al. 1993). Stable alkaloid production was demonstrated over a period of 5 years.

A number of reviews (Rhodes et al. 1990; Tepfer 1990; Toivonen 1993; Wysokinska and Chmiel 1997; Bajaj and Ishimaru 1999; Shanks and Morgan 1999; Giri and Narasu 2000; Christey 2001; Facchini 2001; Sevón and Oksman-Caldentey 2002) have focused on the potential of hairy roots to accumulate a large variety of natural products. The compounds obtained from hairy roots include, for example, acridones (Kuzovkina et al. 2004), anthraquinones (Mantrova et al. 1999), coumarins (Bais et al. 2000), diterpenes (Hu and Alfermann 1993), flavonoids (Li et al. 2002a) including flavones (Kuzovkina et al. 2005), glucosinolates (Wielanek and Urbanek 1999), indole alkaloids (Falkenhagen et al. 1993; Sheludko et al. 2002a,b), lignans (Lin et al. 2003), naphthoquinones (Brigham et al. 1999), nicotine alkaloids (Rhodes et al. 1986), polyamines (Biondi et al. 2000), sesquiterpenes (Komaraiah et al. 2003), steroids (Fujimoto et al. 2000), and tropane alkaloids (Mano et al. 1989). Valuable plant natural products produced by hairy roots are frequently subject to biotechnological production studies (Kim et al. 2002; Wysokinska and Chmiel 1997).

Many of these compounds are biologically active and have, for example, antifungal, antibacterial, insecticidal, nematicidal and phytotoxic properties. Many alkaloids, especially tropanes from various species of the Solanaceae family (Rhodes et al. 1990; Canto-Canché and Loyola-Vargas 1999), morphinanes and benzophenanthridines from *Papaver* species (Le Flem-Bonhomme et al. 2004), terpenoid indole alkaloids from Apocynaceae (Toivonen et al. 1991; Bhadra and Shanks 1997), and monoterpene indole-derived quinoline alkaloids such as campthothecin and related alkaloids (Lorence and Nessler 2004) from Rubiaceae (*Ophiorrhiza pumila*) (Saito et al. 2001; Kitajima et al. 2002; Yamazaki et al. 2003a) and Nyssaceae (*Camptotheca acuminata*) (Lorence et al. 2004), which are highly active as medicinal drugs or useful as model compounds for drug development, have been produced at reasonable levels in hairy roots. Several new natural products were detected first in hairy root cultures (Gräther and Schneider 2001). The first naturally occurring alkaloid of the raumacline group, 10-hydroxy-$N(\alpha)$-demethyl-19,20-dehydroraumacline, which was isolated as a mixture of *E*- and *Z*-isomers from *Rauvolfia serpentina* (Sheludko et al. 2002a) and other monoterpenoid indole alkaloids (Sheludko et al. 2002b) represent striking examples. In several cases, transformed root cultures of endemic or endangered plants attracted attention for isolating pharmaceutically or otherwise commercially interesting natural products because of the limited availability of the whole plants. For example, ajmaline and ajmalicine were produced by hairy root cultures of *Rauvolfia micrantha*, a rare species endemic to the Western Ghats of India (Sudha et al. 2003).

The occurrence of N-(carboxyalkyl)-amino acids (opines) is a characteristic feature of *Agrobacterium*-infected plant tissue, which is why these microbial metabolites have been used as indicators of plant-microbe interaction (Dessaux et al. 1993). Nevertheless, opines may be considered secondary products with potential biological activity (Thompson and Donkersloot 1992).

In addition to alkaloid-producing species, plants containing root-specific nitrogen-free compounds were also used to establish transformed root cultures. pRi T-DNA transformed root cultures of *Lithospermum erythrorhizon* synthesize the root-specific naphthoquinone shikonin when grown in the dark on ammonium-free medium (Yazaki et al. 1998, 2001). Hairy roots of the endangered plant *Valeriana wallichii* were shown to produce enhanced levels of valepotriates (Banerjee et al. 1998). Transformed root cultures of *Scutellaria baicalensis* were initiated by several groups (Zhou et al. 1997; Nishikawa et al. 1999; Stojakowska and Malarz 2000; Kuzovkina et al. 2005) and proved to be efficient producers of flavones (e.g. baicalein, wogonin and respective glucuronides). These natural products showed pronounced antioxidative, antimicrobial, and cytostatic effects (Ciesielska et al. 2002; Zhang et al. 2003), and activity on vascular diseases (Schramm and German 1998).

Inducing plant secondary metabolism by elicitors, e.g. fungal cell wall components, inorganic ions, or intermediates of signaling pathways such as jasmonates often enhanced levels of biologically active natural products (phytoalexins) (Benhamou 1996; Nimchuk et al. 2003). Since the fundamental studies of Zenk (Zenk 1988, 1991; Gundlach et al. 1992) on jasmonic acid as a signal transducer of secondary metabolism in plant cell cultures, many applications of elicitors to hairy root cultures have been published (Sevón and Oksman-Caldentey 2002). The response of alkaloid levels to elicitation with pectinase and jasmonic was reported for hairy root cultures of *Catharanthus roseus* (Rijwani and Shanks 1998). Enhanced flux to several branches of the indole alkaloid pathway was demonstrated after treatment with jasmonic acid. Transiently increasing levels were found for lochnericine and tabersonine, and specific yields of a number of compounds reached several time the amount of the control, e.g. 500% for horhammericine. However, results regarding the elicitation of indole alkaloids are contradictory. In some studies using hairy root cultures of *Catharanthus roseus*, jasmonate did not elicit the formation of indole alkaloids (Morgan and Shanks 2000). Methyl jasmonate and salicylic acid, respectively, did not enhance camptothecin level of an *Ophiorrhiza pumila* hairy root culture, but led to secretion to the medium (Saito et al. 2001).

In contrast, elicitation of genetically transformed madder (*Rubia tinctorum*) roots with methyl jasmonate resulted in a 5- to 8-fold increase in the anthraquinone content without altering the pigment composition (Mantrova et al. 1999). Methyl jasmonate has been suggested to induce the biosynthesis of one of the key enzymes responsible for the first stages of anthraquinone biosynthesis in madder roots.

A specific 70 kD polypeptide accumulates in *R. tinctorium* hairy roots, which confirms this suggestion. Elicitation by methyl jasmonate of *Scutellaria baicalensis* hairy roots, which produce antimicrobial flavones, resulted in a 2-fold increase in flavone levels (Kuzovkina et al. 2005). Hairy root cultures of *Ocimum basilicum* transformed with *Agrobacterium* strain ATCC-15834 showed 3-fold increases in the growth and production of rosmarinic acid compared to untransformed roots (Bais et al. 2002b). Elicitation of hairy root cultures with fungal cell wall elicitors from *Phytophthora cinnamoni* enhanced the production of rosmarinic acid significantly compared to the untreated roots. The use of salicylic acid, yeast extract, and different abiotic elicitors led to enhanced concentrations of tropane alkaloids (hyoscyamine, scopolamine) in hairy root cultures and exudates of *Brugmansia candida* (Pitta-Alvarez et al. 2000). In part, preference for scopolamine synthesis has been observed in these cultures. Elicitor treatments of transformed root cultured of *Datura stramonium* in the order MeJa>fungal elicitor>oligogalacturonide increased alkaloid accumulation (Zabetakis et al. 1999). An additional effect of elicitor treatment, permeabilization of cell walls, has been observed in several cases (Sevón et al. 1992), as a result of which metabolites were secreted into the growth medium (Pitta-Alvarez et al. 2000; Bais et al. 2002a). The content of shikonin (Shimomura et al. 1991) and camptothecin (Saito et al. 2001) in the medium was increased by the presence of a polystyrene resin, which absorbed the secreted metabolites. An equilibrium has been posited to exist with respect to the metabolite levels between the inside and the outside of cells. A reduced level outside the cells as a result of absorption by the polystyrene resin might enhance secretion (Saito et al. 2001).

Mycelial and medium filtrates of *Pythium aphanidermatum* and *Phytopthora parasitica* var. nicotiana were used to elicit coumarin production in a *Cichorium intybus* hairy root culture (Bais et al. 2000). After treatment with *P. parasitica*, esculin and esculetin yields were approximately four-fold higher than in the untreated control. In addition, the growth rate of hairy roots of *C. intybus* was significantly enhanced after treatment with these fungal elicitors. Short-term treatment (24–48 h) with fungal cell wall extracts during the logarhythmic growth phase significantly enhanced levels of thiophene and thiarubine A in hairy roots of *Tagetes patula* (Buitelaar et al. 1993) and *Ambrosia artemisiifolia* (Bhagwath and Hjotso 2000).

Transformed roots of *Lithospermum erythrorhizon* are highly responsive to microbial treatment. A shikonin concentration up to 30 times that of the untreated control was observed in these cultures (Brigham et al. 1999). In addition to the normal storage of shikonin in root hairs, active formation was induced by microbial treatment in root tips and epidermal and border cells of *L. erythrorhizon*. Despite many successful applications, elicitation by jasmonates and other signaling compounds cannot be generalized. It has been speculated that only formation of typical phytoalexins and phytoanticipins, which are involved in response to wounding, herbivore attack, and pathogen infection, is affected by elicitors (Verpoorte 2000). Elicitation of hairy roots of *Datura stramonium* with heavy metals led to enhanced levels of sesquiterpenoids (lubimin and hydroxylubimin) but did not stimulate tropane alkaloid biosynthesis (Furze et al. 1991). Biondi et al. (2000) reported a dramatic increase of free and conjugated amines (putrescine, spermine, spermidine) after treatment of genetically transformed root cultures of *Hyoscyamus muticus* with jasmonates. However, the levels of tropane alkaloids were enhanced only slightly in these experiments. Formation of calystegines in transformed root cultures of *Atropa belladonna* was stimulated by enhanced sucrose content of the medium but not by jasmonic acid or chitosan (Rothe et al. 2001).

Hairy root cultures of two genotypes of *Glycine max* showing different susceptibility to the "sudden death syndrome" disease, which is caused by the fungus *Fusarium solani* f. sp. glycines, have been used to study the role of isoflavonoids and their conjugates in plant defense (Lozovaya et al. 2004). Rapid formation of enhanced levels of glyceollin in response to infection by *F. solani* seems to be important in providing partial resistance to this fungus.

Treatment of hairy root cultures of *Salvia miltiorrhiza* with an abiotic elicitor, Ag^+ ($Ag_2S_2O_3$, 15–40 µM), between days 12 and 22 of the culture period resulted in a dose-dependent increase in the levels of tanshinone diterpenoids (Zhang C et al. 2004). Ag^+ also inhibited root growth significantly, which effect was neutralized by supplementing the medium with sucrose or renewing the medium before elicitation. Combined elicitation and medium renewal enhanced tanshinone yield more than 6-fold.

5.2 Secondary products in hairy root exudates

Secondary products of the above-ground plant parts and root/rhizomes are not necessarily identical and do not even always belong to the same structural type. Moreover, secreted compounds do not always reflect the extractable metabolic pattern of the roots. Most of the compounds representing up to 10% of photosynthetically fixed carbon secreted from the roots (Johansson 1992; Shepherd and Davies 1993) remain to be assigned to particular structures. The biological functions of compounds secreted from the roots of intact plants are thought to be connected to defense against soil organisms, chemical competition with other plants (allelopathy), enhancing phytoavailability of nutrients and trace element, or preparing the environment for root extension. Hairy roots, like untransformed root cultures, in general possess a fully functional system of vascular bundles. However, due to the missing transpiration stream of above-ground plant parts, the metabolite flow is not comparable with that of intact plants.

Hairy roots are frequently capable of secreting metabolites into the nutrient medium. Although rhizosecretion of genetically transformed roots has not been studied systematically, it seems to depend mainly on the special cultivation conditions. This makes them interesting for both the production of valuable compounds and the study of root biological processes, as shown, for example, for *Hyoscyamus muticus* hairy roots, which secrete hyoscyamine and scopolamine (Sevón et al. 1992). The productivity of tropane alkaloid-forming hairy roots in some cases was shown to be higher compared to whole plants and up to 20% of the total amount of produced alkaloids was secreted into the medium, perhaps because of altered or missing storage and transport mechanisms in the cultured roots (Rhodes et al. 1990).

Fluorescent β-carbolines were recently identified as the source of a phenomenon called "root fluorescence" which has been observed in roots of *Oxalis tuberosa* (oca) (Bais et al. 2002a). Exudation of harmine and harmaline was demonstrated due to their fluorescent properties. The biological function of root fluorescence, which so far has remained speculative, is of considerable interest because it seems not to be restricted to *Oxalis* but has been observed in other plant species as well, including *Arabidopsis thaliana*. Recently, hairy roots of the oca plant have been established which, upon treatment with fungal cell wall elicitors, secrete harmine and harmaline in even higher doses than do intact plants (Bais et al. 2003). The transformed root cultures of *Oxalis tuberosa* are considered a suitable model system for studying the mechanism of secretion and evaluating the biological significance of the exudation of fluorescent metabolites.

Plants of the *Tropeolaceae* contain benzylisothiocyanates, which are glucosinolate degradation products formed by the action of myrosinase. Hairy roots of *Tropaeolum majus* and *Tropaeolum tuberosum* produced glucosinolates and/or their degradation products both in roots and exudates (Flores et al. 2003). In hairy roots of *T. majus* the content of the benzylisothiocyanate glucotropaeolin was higher in comparison with callus, cell suspension, and leaves of intact plants (Wielanek and Urbanek 1999). Myrosinase activity was found to be enhanced accordingly. Glucotropaeolin was secreted into the medium, which makes these hairy roots interesting for biotechnological application.

Moreno-Valenzuela et al. (2003) studied the involvement of calcium in the accumulation of indole alkaloids in hairy roots of *Catharanthus roseus* and release of alkaloids into the medium. The use of different inhibitors of internal and external Ca^{2+} flux for this purpose stimulated alkaloid production and secretion into the medium, indicating that the secretion of secondary compounds from hairy roots depends strongly on external conditions.

An interesting example of how elicitation leads to secretion of coumarine-type phytoalexins and an altered glucoside-to-aglykone ratio was found in a *Pharbitis nil* hairy root culture. Umbelliferone, presumably formed from its glucoside skimmin upon elicitation with $CuSO_4$ and MeJA, and scopoletin were secreted into the medium. This typical phytoalexin formation was accompanied by root-growth-inhibiting activity of increasing levels of umbelliferone. Thus, detoxification of umbelliferone results in re-glucosylation and accumulation of plant-growth-non-inhibiting skimmin in the root tissue (Yaoya et al. 2004).

6 Use of hairy root cultures in biosynthetic and metabolic studies

6.1 Biosynthesis

The pioneering work of Zenk (1988) demonstrated the biosynthetic potential of plant cell cultures. Using plant *in vitro* systems for labeling studies, the low incorporation rates obtained by feeding experiments with whole plants in the 1950s and 1960s have almost been overcome. The biosynthetic pathways of many secondary metabolites have been elucidated in detail in cell cultures and other *in vitro* systems (Bourgaud et al. 2001) and the enzymes characterized (Zenk 1991). Hairy roots, as a less dedifferentiated plant *in vitro* system, are suitable for labeling studies for the same reasons that cell and organ suspensions are: They are independent from seasonal and other periodic changes of plant growth and development; highly productive clones are selectable; and precursors can be easily absorbed from the liquid medium. In addition, hairy roots seem to be more genetically stable than undifferentiated cell suspensions and growth is possible for almost unlimited time periods (Maldonado-Mendoza et al. 1993).

Many aspects of plant secondary metabolite biosynthesis have been studied using transformed root cultures (Robins 1998). Various species of the Solanaceae and Convolvulaceae have been used to investigate tropane alkaloid biosynthesis. The conversion of tropinone to tropine and pseudotropine was observed directly in living root cultures of *Datura stramonium* by ^{15}N NMR spectroscopy (Ford et al. 1994). Further *in vivo* NMR studies using the same hairy root culture described the transformation of tropine into littorine and hyoscyamine (Ford et al. 1996) and detailed physiological aspects of polyamine and tropane alkaloid metabolism (Ford et al. 1998). Efficient incorporation of labeled littorine into hyoscyamine was demonstrated in transformed root cultures of *Datura stramonium* (Robins et al. 1994). Evidence for intramolecular rearrangement of littorine into hyoscyamine and the mechanism of this rearrangement have mainly been obtained from *D. stramonium* root cultures. These results have been reviewed by Humphrey and O'Hagan (2001). The kinetics of the isomerization of littorine into hyoscyamine were investigated by ^{13}C NMR techniques, again using a transformed root culture of the genus *Datura* fed with labeled tropoyl moiety precursors (Lanoue et al. 2002). Using feeding experiments with ^{15}N precursors and analysis by GC-MS and NMR, Scholl et al. (2001, 2003) established the pathway from tropinone via pseudotropine to calystegines in transformed root cultures of *Calystegia sepium*. These results showed that tropinone is a common precursor of tropane and nortropane alkaloid biosynthesis.

Böhm and Mäck (2004) studied betaxanthin formation, which is assumed to be a spontaneous non-enzymatic condensation between betalamic acid and amino acids, in hairy roots of *Beta vulgaris* upon feeding glutamate or glutamine. The results obtained in this study indicated that the chemical character of the amino acids seems to be less important than the intracellular situation. This might also explain the species-specific patterns of betaxanthins.

In a series of papers, Fujimoto and collaborators reported on the biosynthesis of phytoecdysones (Fujimoto et al. 2000; Hyodo and Fujimoto 2000; Okuzumi et al. 2003) in *Ajuga reptans* var. atropurpurea hairy roots (see also review by Dinan 2001). Feeding ^{13}C- and ^{2}H-labeled precursors to these ecdysone-producing hairy root cultures and determination of labeling positions using NMR spectroscopy led to the elucidation of a variety of biosynthetic steps. Thus, the pathway from cholesterol to 7-dehydrocholesterol to 7-dehydrocholesterol-5α,6α-epoxide to 3β-hydroxy-5β-cholestan-6-one to 2β,3β-dihydroxy-5β-cholestan-6-one was proposed for the early stages of 20-hydroxyecdysone biosynthesis. On the other hand, *Ajuga hairy* root cultures converted labeled 3β-hydroxy-5β-cholestan-6-one and 2β, 3β-dihydroxy-5β-cholestan-6-one to 20-hydroxyecdysone, demonstrating that the Δ^{7}-double bond can be introduced at a later stage in the biosynthesis and that 7-dehydrocholesterol is not an obligatory intermediate.

The discovery of a pathway via 1-deoxy-D-xylulose-5-phosphate (DXP) but independent from the mevalonate pathway and leading to isopentenyl diphosphate (IPP) and finally to isoprenoids raised the question of whether or not this route also operates in hairy roots and, even more interesting, if crosstalk between both pathway exists. Treatment of *Catharanthus roseus* hairy roots with the antibiotic fosmidomycin, a specific inhibitor of DXP reductoisomerase, and feeding experiments with 1-deoxy-D-xylulose, 10-hydroxygeraniol, or loganin suggested that the DXP pathway mainly provides the carbon skeleton of monoterpenoids in *C. roseus* hairy roots (Hong et al. 2003).

The biosynthesis of camptothecin, a strong inhibitor of topoisomerase I, has also been studied in hairy roots. Due to strong lipophilic and cytotoxic properties, camptothecin itself is less useful in clinical applications. However, because its water-soluble semisynthetic derivatives such as irinotecan and topotecan are of special interest in oncological practice, their market potential in 2003 was about $1 billion US (Lorence and Nessler 2004). The limited availability of camptothecin from its major natural sources, the Chinese tree *Camptotheca acuminata* (Nyssaceae) and *Nothapodytes foetida* (Icacinaceae), underlines the need for biotechnological production. Whereas cell cultures of *C. acuminata* and *N. foetida* produced only 0.00017% on a dry-weight basis (Fulzele et al. 2002), up to 0.10% camptothecin and 0.15% 10-hydroxycamptothecin were obtained from hairy roots of *Ophiorrhiza pumila* (Rubiaceae)

(Saito et al. 2001) and *C. acuminata* (Lorence et al. 2004). This yield is comparable with that from roots of intact plants. The faster growth rate of *Ophiorrhiza pumila* hairy roots compared to *C. acuminata* and enhanced camptothecin secretion into the medium upon absorption by polystyrene resin results in a higher total productivity of *O. pumila*. Due to the biotechnological interest in *O. pumila*, this culture has been investigated in more detail with respect to phytochemistry and camptothecin biosynthesis. The formation of camptothecin from the terpenoid pathway via secologanin raised the question of whether the biosynthesis proceeds through the classical mevalonate route or the DXP pathway. Specific inhibitors of both pathways, lovastatin and fosmidomycin, were used in combination with [1-^{13}C]glucose feeding, and computer-based data analysis was performed to reconstruct the biosynthetic pathway (Yamazaki et al. 2004). The results clearly showed that, as in *C. roseus*, secologanin is synthesized via the DXP pathway. However, the regulation mechanism of monoterpenoid biosynthesis seems to be different in *C. roseus* and *O. pumila*. Strictosidine synthase (STR) and tryptophan decarboxylase (TDC), two key enzymes of camptothecin biosynthesis, seem to be coordinately regulated in *O. pumila*. This was concluded on the basis of expression analyses. No expression of STR and TDC was observed by northern blot analysis of elicited hairy roots, and neither enhanced STR and TDC activities nor higher levels of camptothecin were observed upon elicitation (Yamazaki et al. 2003b).

The biosynthetic origin of prenyl side chains of plant metabolites has been studied by Asada et al. 2000. Incorporation of [1-^{13}C]glucose indicates that the biosynthesis of the hemiterpene moiety of glabrol, the main prenylated flavanone in the hairy root cultures of *Glycyrrhiza glabra*, also proceeds via the non-mevalonate pathway.

In addition to elucidating precursor-product relationships by labeling experiments, hairy roots are useful for characterizing biosynthetic enzymes. For example, epoxidation of tabersonine to lochnericine was studied in *Catharanthus roseus* hairy root cultures. This important step in the biosynthesis of horhammericine and other bisindole alkaloids in *C. roseus* was promoted by methyl jasmonate. The enzyme activity catalyzing this epoxidation was located in the microsomal fractions and was inhibited by carbon monoxide and other P450 inhibitors, suggesting a cytochrome P450-dependent monooxygenase (Rodriguez et al. 2003).

6.2 Biotransformation

The potential of plant cells to perform particular biochemical reactions more efficiently than synthetic chemistry is able to has been frequently utilized to produce specific metabolites. Biotransformations, including examples for enhancing the production of pharmacologically interesting compounds from hairy root cultures, have been reviewed by Giri et al. (2001) and Ramachandra and Ravishankar (2002).

Transformed root cultures of *Lobelia sessilifolia* were shown to glucosylate phenolics such as epicatechin and protocatechuic acid (Yamanaka et al. 1995). Although not occurring in *Panax ginseng* hairy root cultures, umbelliferone was transformed to a variety of glycosides, demonstrating the potential of this culture to glycosylate exogenous compounds. The glycosylation was catalyzed by glycosyltransferase rather than by glycosidase, which was demonstrated by the administration of inhibitors (Li et al. 2002b). Glycosidation of low-molecular-weight alcohols was found in *Coleus forskohlii* hairy root cultures (Li et al. 2003), which produce forskolin.

7 Hairy roots for biotechnological production of secondary metabolites

7.1 "Wild-type" hairy roots

The qualitative composition of secondary metabolite patterns of genetically transformed root cultures almost is identical to those of roots from intact plants. Only those secondary products that are formed in photosynthetic tissue and after translocation stored in the roots are not usually accessible by hairy roots. Remarkably, clones obtained from different root tips produced different amounts of secondary products (Jankovic et al. 2002). This variability among transformed root cultures demonstrates the opportunity for enhancing the biosynthetic rate, for example, by selecting highly producing plants, screening clones, and optimizing medium and other growth conditions. However, despite numerous reports of laboratory-scale systems for biotechnological production of secondary metabolites by *in vitro* cultures such as dedifferentiated cell suspensions and organ cultures (Banthorpe 1994; Yeoman and Yeoman 1996; Berlin 1997), only few have been successfully scaled up to commercial levels. Shikonin from cell cultures of *Lithospermum erythrorhizon* was the first example (Tabata and Fujita, 1985) and paclitaxel (Taxol®) currently is the most prominent one. Relatively few research efforts have been translated into commercial application, partly due to costly cultivation equipment and media, slow growth rates, the occasional relative deficiency of secondary products, genetic instability and latent endophytic bacteria (Leifert and Cassells 2001). Some of these drawbacks can be overcome by hairy roots. They are able to grow on hormone-free medium, making cultivation less expensive; their growth is generally faster in comparison to cell suspensions and non-transformed organ cultures; and they exhibit higher genetic stability.

The vinca alkaloids vinblastin and vincristin were the subject of considerable biotechnological efforts due to their pronounced cytostatic activity. However, heterotrophic transformed root cultures of *Catharanthus roseus* produced only small amounts (0.05 mg g^{-1} dry weight) of these alkaloids (Parr et al. 1988) because in whole plants the biosynthesis takes place in the chloroplasts of above-ground green tissue, but not in the roots (De Carolis and De Luca 1993). This typical example indicated that hairy roots are mainly able to produce almost only root-specific secondary metabolites. Exceptions are, for example, illuminated root cultures, which in some cases produced metabolites characteristic of photosynthetic tissue. For example, green cultures were able to produce higher levels of scopolamine than were the dark-grown cultures (Flores et al. 1993) in the case of *Acmella oppositifolia* and *Datura innoxia*.

Camptothecin has already been discussed (section 6.1) with respect to its biosynthesis in transformed root cultures of *Ophiorrhiza pumila*. Sudo et al. (2002) cultivated hairy roots of this plant in a 3-l bioreactor and obtained 22 mg camptothecin within a cultivation period of 8 weeks. A considerable portion (17%) of the total camptothecin was secreted into the medium.

In recent years, the early studies of Yoshikawa and Furuya (1987) on *Panax ginseng* hairy roots have continued to focus research in this field (Mallol et al. 2001), mainly because of the growing worldwide market for preventive and therapeutic herbal drugs in which ginseng is of particular interest. Some ginseng hairy root cultures grow well in fermentors (Jeong et al. 2002; Palazon et al. 2003) and may hold commercial promise.

7.2 Conventional strategies to affect production of secondary metabolites

Some feasible strategies for enhancing the production of secondary products in hairy roots have been already discussed in the context of fundamental research subjects, e.g. elicition (section 5.1). Elicitors can enhance the production of valuable secondary metabolites by transformed roots and other *in vitro* cultures. More conventional methodologies include precursor feeding, strategies for stimulating secretion into the medium, optimization of growth conditions, and selection of high-productive cell lines.

Nicotine production and diffusive release from the intracellular environment into the medium is based on an equilibrium partitioning process (Larsen et al. 1993). Removing nicotine from the culture medium of tobacco cell cultures by applying the surfactant Triton X-100 enhanced both production and release rates. Moreno-Valenzuela et al. (1999), who studied the effect of macerozyme on the accumulation of coumarin and alkaloid production in hairy root cultures of *Catharanthus roseus*, demonstrated that macerozyme treatment affected the activities of biosynthesis-related enzymes such as tryptophan decarboxylase, phenylalanine ammonium lyase, and phosphatidylinositol 4,5-bisphosphate phospholipase. The production of secondary metabolites and phospholipase C activity in *C. roseus* hairy roots was stimulated as well. Further experiments investigated the involvement of calcium in the accumulation of indole alkaloids in hairy roots of the same species and their release into the medium. The use of different inhibitors of internal and external Ca^{2+} flux for this purpose stimulated alkaloid production and secretion into the medium considerably (Moreno-Valenzuela et al. 2003).

Precursor feeding may be used to overcome rate-limiting biosynthetic steps and thus to enhance the level of desired products in hairy root cultures. In *Datura innoxia* transformed root cultures, a combination of precursor feeding (e.g. L-phenylalanine or DL-β-phenyllactic acid) and treatment with Tween 20 amplified the total hyoscyamine content by 40% and 60%, respectively, compared to the control. In contrast, the addition of precursors without Tween 20 was ineffective at stimulating hyoscyamine production (Boitel-Conti et al. 2000). In order to enhance productivity of terpenoid indole alkaloids, Morgan and Shanks (2000) fed various terpenoid and shikimate-derived precursors to *Catharanthus roseus* hairy roots. During the early stationary growth phase, application of terpenoid intermediates resulted in tabersonine accumulation, indicating that the terpenoid pathway is rate limiting and tryptophan pathway may not be. In contrast, neither tryptophan nor terpenoid pathways seem to be limited during the late growth phase.

Precursor feeding did not always enhance production of secondary metabolites in hairy root cultures. Application of phenylalanine to hairy root cultures of *Scutellaria baicalensis*, for example, failed to enhance production of flavones (Kuzovkina et al. 2005). Presumably, amino acids are not the limiting factor in the biosynthesis of flavones in this culture. Exogenous cholesterol even reduced biosynthesis and growth rates of transformed root cultures of *Trigonella foenum-graecum* (Christen 2002).

Supplementing the medium of hairy root cultures of *Cichorium intybus* with DMSO accelerated root growth and enhanced coumarine production. These results were improved substantially by combining the application of DMSO and a filtrate of *Phytophtora parasitica* (Bais et al. 2001). This example demonstrates that permeabilization together with biotic elicititation may boost the production of secondary metabolites in hairy roots.

The effect of carbohydrate composition of the medium on not only root growth but also secondary metabolite production has recently attracted attention. For example, both optimal growth rate and maximal production of aromatic carboxylic acids in hairy roots of *Isatis indigotica* were achieved when maltose was used as a carbon source (Xu et al. 2004). Sucrose concentration of the medium also affected the secondary metabolite production of hairy roots. The flavonoid formation of transformed root cultures of *Scutellaria baicalensis* was highest in B5 medium containing 7% sucrose (Stojakowska and Malarz 2000). In these hairy root cultures, increasing sucrose concentration promoted glucuronide formation, whereas aglykone levels remained almost constant (Kóvacs et al. 2004). The possibility that various carbohydrates play a role as signalling compounds in terpenoid biosynthesis in hairy roots of *Artemisia annua* has been discussed by Weathers et al. (2004).

The effect of light on the formation of root-specific secondary compounds seems due to the specific regulation of metabolism in tissue derived from underground plant parts. In *Lithospermum erythrorhizon*, dark-inducible genes (LeDIs) were identified, which are involved in the downregulation of shikonin biosynthesis (Yazaki et al. 2001). Remarkably, the activities of other shikonin biosynthetic enzymes, such as phenylalanine ammonia lyase and *p*-hydroxybenzoic acid geranyltransferase, were almost unchanged. Although the exact function of LeDIs in *L. erythrorhizon* has not yet been defined, they seem to play an indirect role in biosynthesis, perhaps in regulating transport mechanisms.

The invasion of *Solidago altissima* (Asteraceae), originally a native of North America, of Japan is thought to be due to its content of *cis*-dehydromatricaria ester, an allelopathic polyacetylene. Hairy roots established from this plant grew better under light conditions than in the dark but produced considerably lower levels of the polyacetylene (Inoguchi et al. 2003). Thus, an inhibiting effect of light on the biosynthesis of polyacetylenes was thought to be responsible for this difference.

Another negative effect on the production of secondary metabolites is connected to the undifferentiated growth of root cultures, which was observed after application of exogenous auxins (2,4-D, IAA, NAA) (Rhodes et al. 1994). However, this behavior is reversible and, in the undiffentiated growth state, can be used to select productive culture lines (Payne et al. 1991).

Various growth conditions—carbohydrate composition of the medium and treatment with elicitors—have been monitored for their role in enhancing the production of amarogentin in hairy roots cultures of *Swertia chirata* growing in a bioreactor. A 15-fold enhancement of amarogentin content in the medium was achieved by permeabilizing roots using Tween 20 (Keil et al. 2000).

7.3 Genetic engineering of secondary metabolite biosynthesis

Transgenic plants obtained by *Agrobacterium rhizogenes*-mediated transformation have been compiled by Giri and Narasu (2000). In addition to the *Agrobacterium*-derived Ri T-DNA, they are equipped with foreign genes that have been introduced, for example, by binary vector systems. Some of these hairy roots contain additional biosynthetic genes aimed to enhance secondary metabolite production (Saito et al. 1992; Verpoorte and Memelink 2002). For example, transformed root cultures of *Rubia peregrina* containing a bacterial isochorismate synthase showed enhanced biosynthetic activity and accumulated 20% higher total anthraquinone levels compared to the control cultures (Lodhi et al. 1996). The pharmaceutically important tropane alkaloids are one of the classes of secondary products most frequently explored by hairy root culture techniques, including genetic engineering. *Agrobacterium*-mediated introduction of the tobacco putrescine SAM *N*-methyltransferase gene (*pmt*) into the genomes of *Datura metel* and *Hyoscyamus muticus* resulted in amounts of tropane alkaloids up to 5 times those in control hairy roots. Both hyoscyamine and scopolamine levels were enhanced in *D. metel* hairy roots, whereas in *H. muticus* only hyoscyamine content was increased by *pmt* gene overexpression (Moyano et al. 2003). Thus in different solanaceous plants, the enzyme catalyzing conversion of hyoscyamine to scopolamine, hyoscyamine-6β-hydroxylase (H6H), is differently regulated. Overexpression of the *h6h* gene in hairy roots of *Hyoscyamus muticus* resulted in a 100-fold increase in scopolamine levels compared to the control culture line (Jouhikainen et al. 1999). An excellent example of genetically engineering a plant secondary metabolite pathway recently has been published by Zhang L et al. (2004). A binary expression vector was used to simultaneously introduce genes encoding two enzymes of scopolamine biosynthesis in transgenic *Hyoscyamus niger* hairy root cultures. The upstream putrescine *N*-methyltransferase (PMT), which catalyzes a rate-limiting step, and the downstream enzyme H6H were overexpressed. The effect on scopolamine production was compared with that of wild type and transgenic *H. niger* hairy root lines harboring PMT of H6H as single genes. The best line produced 411 mg l^{-1} scopolamine, which provided the basis for the large-scale commercial production of scopolamine in hairy root culture.

Sense and antisense constructs of genes encoding the berberine bridge enzyme (BBE), a key enzyme in the biosynthesis of benzophenanthridine alkaloids such as sanguinarine, chelirubine, and macarpine, were introduced into *Eschscholzia californica* root cultures (Park et al. 2003). Transgenic roots expressing *BBE* from *Papaver somniferum* displayed higher levels of BBE mRNA, protein and enzyme activity, and increased accumulation of benzophenanthridine alkaloids compared to control roots, but reduced cellular pools of several amino acids. In contrast, roots transformed with an antisense-*BBE* construct from *E. californica* had lower levels of BBE mRNA and enzyme activity, reduced benzophenanthridine alkaloid accumulation and larger cellular pools of certain amino acids compared with controls. These results demonstrate that altering metabolic flux through benzophenanthridine alkaloid pathways can affect the cellular pools of specific amino acids.

The formation of terpenoid indole alkaloids has been discussed several times in this review. Precursor feeding experiments (Morgan and Shanks 2000) were mentioned, for example, in the preceding section 7.2. Terpenoid indole alkaloids represent compounds that are formed from precursors coming from two different biosynthetic pathways. In such cases, biogenetic engineering is still a challenge. Terpenoid indole alkaloids are formed by coupling secologanin from the terpenoid pathway to tryptamine from the indole pathway. An indole pathway in *Catharanthus roseus* hairy roots using an inducible promoter has been reported (Hughes et al. 2004a), and experiments have been carried out in order to study the regulation of processes providing precursors, tryptophan and tryptamine, to the terpenoid indole alkaloid biosynthetic pathway (Hughes et al. 2004b). Hairy root cultures of *C. roseus* were established with a glucocorticoid-inducible promoter controlling the expression of an *Arabidopsis* feedback-resistant anthranilate synthase alpha subunit. Although this culture showed large increases in tryptophan (300-fold) and tryptamine concentration, the levels of most terpenoid indole alkaloids, except lochnerine, remained almost unchanged (Hughes et al. 2004b). Regulation of the biosynthesis of secondary metabolites, especially when precursors from different pathways are involved, is clearly complex.

Engineering metabolic pathways by *Agrobacterium*-mediated gene transfer does not always lead to the desired result. Introducing the microbial *ubiC* gene to encode chorismate pyruvate lyase into hairy roots of *Lithospermum erythrorhizon* did not result in overproduction of the desired product, shikonin, but instead the level of the nitrile glucoside menisdaurin unexpectedly increased. The enhanced formation of menisdaurin demonstrates the complexity of biosynthetic processes and their regulation (Sommer et al. 1999). In addition to the expression of the bacterial *ubiC* gene, 3-hydroxy-3-methylglutaryl-CoA (HMG-CoA) reductase, which is involved in the formation of geranyl diphosphate, has been expressed in *L. erythrorhizon* but so far without positively affecting shikonin production (Köhle et al. 2002).

An altered metabolic flux owing to expression of foreign genes can lead to premature senescence and even cell death (Kholodenko et al. 1998). In conclusion, the genetic transfer of genes encoding the biosynthesis of single enzymes is not always sufficient and control of multiple steps and regulatory processes may be required to enhance production of secondary metabolites without negatively affecting the entire metabolism.

8 Hairy roots for genetic improvement and clonal propagation

8.1 Experimental systems

Agrobacterium-mediated transformation (Horsch et al. 1985; Koncz and Schell, 1986) has become a standard technique for introducing foreign genes into a given plant genome not only in order to enhance secondary metabolite production in hairy root cultures but also to regenerate plants with improved physiological properties and agriculturally or horticulturally useful traits. The transfer and stable integration of a DNA sequence (T-DNA) from the *Agrobacterium rhizogenes* root-inducing (Ri) plasmids into the plant genome has often led to genetically improved root cultures and regenerated plants that carry useful external genetic information. The objective of genetic engineering is to establish hairy root cultures for biotechnological and environmental applications and useful recombinant proteins (Grusak 2002). Moreover, regenerated plants for improved crop yield and functional food production, resistance to total herbicide treatment and pathogens, and clonal propagation of crop and ornamental plants are of particular interest (Christey 2001). Genetically engineered root cultures have often been used as research tools to study the effects of transformation on physiological and morphological parameters. Many studies have explored the relevance of *rol* genes in the development of transformed plants and their use as an instrument to specifically alter plant properties.

Agriculturally important traits have been investigated using hairy root cultures. An *A. rhizogenes*-mediated transformation was applied to identify the plant nematode resistance gene Hs1^{pro-1} from sugar beets (Cai et al. 1997). Using an improved transformation protocol based on co-inoculation with *A. tumefaciens* LBA4404 and *A. rhizogenes* 15834, this nematode resistance gene was introduced into *Beta vulgaris* (Kifle et al. 1999). Stable expression and mitotic stability of the foreign genes were demonstrated by southern blot hybridization and an *in vitro* nematode resistance test. The beet cyst nematode *Heterodera schachtii* was not able to complete its life cycle on the resistant hairy root line. Further investigations were performed on the expression pattern of Hs1^{pro-1} under biotic and abiotic stresses, as well as after nematode infection; Hs1^{pro-1} was shown to be upregulated transcriptionally in resistant *Beta vulgaris* roots after nematode infection (Thurau et al. 2003). Promoter analysis with the GUS reporter gene demonstrated that the Hs1[pro-1] promoter activates expression of a feeding site-specific gene in both *Beta vulgaris* and *Arabidopsis*. This example shows that pathogen responsiveness of a disease resistance gene promoter can be studied at the cellular level using hairy roots.

The expression of other genes potentially involved in the resistance to the beet cyst nematode *Heterodera schachtii* was investigated using sugar beet hairy root cultures. Among a large number of transcripts from nematode-infected resistant and susceptible hairy roots, only a small number of resistance-specific fragments were obtained (Samuelian et al. 2004). Only a single transcript-derived fragment clearly resulted from nematode infection in resistant roots. Sequence analysis of full-length cDNA suggests that the gene encodes a polypeptide of unknown function that shows no homology to any sequence present in the public databases. To elucidate its specific function, this cDNA was transferred into the hairy roots of susceptible sugar beet and clones were inoculated with nematodes. In most clones, the number of developing female nematodes was significantly reduced, suggesting that the gene may be useful for inducing resistance to cyst nematodes in plants.

8.2 Plant regeneration and clonal propagation

Spontaneous shoot regeneration of Ri-transformed tissue proceeds directly from hairy roots without the need for an intermediate callus stage, thus avoiding somaclonal variation. Spontaneous shoot formation was observed, for example, in transformed root cultures of *Nicotiana hesperis* (Hamill and Rhodes 1988), or after illumination of *Brassica napus* (Damgaard and Rasmussen 1991) and *Ruta graveolens* (Kuzovkina, unpublished). The phenomenon of spontaneous regeneration can be observed mainly in newly established hairy root cultures. In other cases, e.g. with hairy root cultures of *Armoracia lapathifolia*, the ability to perform organogenesis was sustained over a period of 10 years (Kuzovkina, unpublished data). The "old" cultures required media supplemented with kinetin, BAP, and auxines to induce shoot formation. Optimal hormone concentrations were determined for each of the cultures individually (Stiller et al. 1997; Wang et al. 2001; Hyeon-Je and Widholm 2002).

Hairy roots have been used to study physiological response to herbicides. Phosphinothricin-induced effects, such as lower glutamine synthase activity, enhanced ammonium accumulation, and slower growth, were overcome by engineering the cytosolic glutamine synthetase gene from soybean into *Brassica napus* (rape). This effect was observed both in hairy root cultures and regenerated shoots (Downs et al. 1994). Subtoxic concentrations of bialaphos, a phosphinothricin-derived herbicide, were shown to stimulate shoot regeneration of *Agrobacterium*-transformed snapdragon from hairy roots (Hoshino and Mii 1998).

Citrus species are important fruit crops but conventional breeding strategies have occasionally proven difficult. Regeneration of *Citrus aurantifolia* (Mexican lime) from *A. rhizogenes*-transformed explants was achieved both with and without intermediary hairy root phase (Pérez-Mophe-Balch and Ochoa-Alejo 1998). Some transgenic plants were successfully adapted to being grown in soil.

Regeneration of plants from hairy roots sometimes leads to characteristic phenotypic alterations, such as reduced apical growth, wrinkled leaves, and diminished germination capacity (Tepfer 1983; Giri and Narasu 2000). This phenomenon is due to the hairy root syndrome, i.e. each of the *rol* oncogenes located in Ri plasmids, which are expressed in the regenerated plants, causes typical phenotypic changes. In some cases, dwarfed or otherwise altered phenotype characteristics are desired in clonal propagation of some ornamental plants. Several examples have been reviewed by Christey (2001).

Angelonia salicariifolia (Scrophulariaceae), a perennial plant species native to South America, is of interest as an ornamental plant because of its beautiful blue to white flowers. Therefore, phenotypic alterations, especially dwarfness, are desired in genetic improvement of angelonia. *A. rhizogenes*-mediated transformation using a mikomipine-containing wild-type strain was used to introduce a dwarf trait into *A. salicariifolia* (Koike et al. 2003). Plants from two transformed lines exhibiting phenotypic alterations such as dwarfness and smaller leaves flowered 4–6 months after transplantation. No apparent alterations were observed in the number, shape, or size of the flowers.

A study of *Agrobacterium rhizogenes*-mediated DNA transfer to *Aesculus hippocastanum* (horse chestnut) investigated whether desirable morphological modifications of transgenic plants could be obtained in this way (Zdravković-Korać et al. 2004). Stable incorporation of TL-DNA into the *A. hippocastanum* genome and subsequent regeneration of transformed plants was achieved. At least one line was useful in horticulture, namely that for bonsai production.

Cultured hairy roots of *Catharanthus roseus* can produce adventitious shoots that subsequently develop in mature plants (Choi et al. 2004). Plants derived from different hairy root lines differed in their morphological properties or petal colors. The genetic transformation system described in this study seems to be useful for metabolically engineering *C. roseus* in order to enhance the productivity of valuable terpenoid indole alkaloids at the whole-plant level.

Aiming to altered physiological properties, especially xylem vessel structure, of *Ulmus procera* (English ulm), which provide potentially enhanced resistance to the Dutch elm disease fungus, *Ophiostoma novo-ulmi*, Gartland et al. (2001) investigated the effects of pRi A4b plasmid-mediated transformation on regeneration and development. Following an extended culture period, regenerants from the hairy roots of *U. procera* exhibited typical characteristic morphological modifications such as altered leaf shape and growth habit.

The regenerative ability of hairy roots also has been applied to the production and utilization of "artificial seeds" for clonal propagation (Uozumi and Kobayashi 1995). Root fragments of 1–10 mm containing meristematic tissue, encapsulated in alginate and coated with a membrane, were able to regenerate whole plants. These systems are useful especially for micropropagating transgenic plants.

9 Proteins from hairy root cultures

In addition to low-molecular compounds, hairy roots offer a useful system for producing proteins. In this field, plants compete with microbial and animal systems but provide additional several advantages, some of which have been recently reviewed (Gleba et al. 1999; Fischer and Emans 2000; Stockmeyer and Kempken 2004). Many post-translational modifications of proteins, which are required for full expression of biological activity, are so far possible only in eukaryotic systems. Thus, the ability of plant roots to secrete not only small molecules but also functional proteins into the environmental medium increasingly is recognized as a means of producing recombinant proteins for pharmaceutical use as therapeutic or diagnostic agents. Proteins of plant origin are devoid of mammalian-pathogenic viruses, giving them a considerable advantage compared with proteins from animal sources. Hairy roots genetically transformed by transfer of custom-tailored Ri plasmid are becoming accessible in order to produce desired recombinant proteins. Hairy roots established from transgenic tobacco plants were reported to produce IgG monoclonal antibodies (Wongsamuth and Doran 1997). Properties such as accumulation, secretion, and long-term stability of antibody expression have already been investigated. Recently rhizosecretion of a recombinant protein, a human secreted alkaline phosphatase, by plant hairy roots of transgenic tobacco has been reported (Gaume et al. 2003).

Production and rhizosecretion by continuously growing hairy roots under aseptic conditions will undoubtedly develop into a sub-field within biotechnology. First, however, existing problems, such as bioreactor design for cultivation of hairy roots, have to be solved. Nevertheless, these issues, at least in Europe, stand a chance of being solved more easily than those subject to public acceptance and legal restrictions on transgenic plants used in field cultivation or molecular farming.

The secretion of defense proteins plays an important role in below-ground ecological interaction and plant communication. Because of experimental difficulties in handling plant and soil systems, hairy roots are interesting as models in exudation studies. As part of their root exudates, hairy roots of *Phytolacca americana* are able to secrete a ribosome-inactivating protein (RIP) (Park et al. 2002). The RIP is located in the cell walls and root border cells of the hairy roots and secreted constitutively, enhanced by a mechanism mediated by ethylene.

Phytase (myo-inositol hexakisphosphate phosphohydrolase) catalyzes the sequential dephosphorylation of myo-inositol hexakisphosphate (phytate) to inositol and orthophosphate. This enzyme has been used as a feed supplement for monogastric animals. The functional active recombinant fungal phytase was expressed in sesame hairy roots and secreted into the liquid culture medium (Jin et al 2004). The properties of the purified phytase protein were almost similar to those of native fungal phytase. Thus, the sesame hairy root system seems to be useful not only as a model system to study the recombinant phytase but also for biotechnological production.

10 Phytoremediation and environmental detoxification

Agrobacterium-transformed plant roots are considered promising systems for remediation of soil, groundwater, and biowastes (Pletsch et al. 1999). Research is focusing on the extraction of both inorganic and organic pollutants from environmental samples and further metabolic detoxification by transformed root tissue. The highly branched nature and large surface of hairy roots seems to make them suitable for filtering contaminants from soil and water samples. The removal of polychlorinated biphenyls from the culture medium by *Solanum nigrum* hairy roots has been reported (Kas et al. 1997), and its potential demonstrated. Hairy root cultures of *Catharanthus roseus* were shown to be able to take up and transform 2,4,6-trinitrotoluene (TNT) to its dinitroamino derivatives, 4-amino-2,6-dinitrotoluene and 2-amino-4,6-dinitrotoluene, and finally to unidentified metabolites, which were associated with the plant tissue (Hughes et al. 1997). Heavy-metal-hyper-accumulating plant species are of special interest for biosorption studies. Nedelkoska and Doran (2000) reported biosorption properties and the effects of heavy metal concentrations on the growth of hairy roots of various Ni, Cd, Cu, and Zn hyper-accumulating plants. The rapid accumulation of Cd and Cu observed in short-term experiments among all the plant species tested, irrespective of their hyper-accumulator status, led to the conclusion that cell surface mechanisms play a dominant role in the uptake of these metals by hairy roots. A maximum of 10.6 mg g^{-1} dry weight Cd was accumulated when *Thlaspi caerulescens* hairy roots were grown in liquid phase containing an external Cd concentration of 100 ppm. Further studies using hairy root cultures of *T. caerulescens* indicated strong catalase induction in response to Cd exposure, suggesting that antioxidative defenses may play a role in this species' tolerance for heavy metals (Boominathan and Doran 2003). The literature on Cd accumulation by hairy root cultures has been reviewed by Macek et al. 1997.

The absorption rates of Ni from the medium by hairy roots of Ni-resistant and hyper-accumulating plants (*Alyssum bertolonii, A. tenium, A. troodii*) were significantly higher than those of Ni-susceptible plants such as *Nicotiana tabacum* (Nedelkoska and Doran 2001). The absorption by hairy roots and roots of intact plants also differed. Regenerated plants of *A. tenium* were much more tolerant of Ni and capable of accumulating higher Ni concentrations than were hairy roots of this species. Following this demonstration that hairy roots of *Alyssum bertolonii* can hyper-accumulate Ni in the absence of shoots or leaves (Nedelkoska and Doran 2001), hairy roots of this species were used as a model system for generating a metal-enriched product from the harvested plant biomass. After exposure to Ni in liquid medium to give biomass concentrations of 1.9–7.7% Ni based on dry weight, incineration of hairy root tissues in a laboratory-scale horizontal tube furnace yielded a "bio-ore" with crystalline morphology containing up to 82% of the metal (Boominathan et al. 2004). Provided that quality limiting of Ca can be minimized, this procedure might be useful for processing metal-enriched plant material harvested from phytomining operations.

Transformed root cultures of *Brassica juncea* and *Chenopodium amaranticolor* were established and used for removing uranium from solutions containing up to 5 mM uranyl nitrate. Both hairy root cultures were able to absorb considerable amounts of uranium from the solution (Eapen et al. 2003). In addition to potential practical applications, hyper-accumulating hairy roots are useful in physiological studies for understanding the role of roots in heavy metal and radionuclide uptake and accumulation by plants.

11 Conclusions

During the last 20 years, *Agrobacterium rhizogenes*-transformed root cultures have evolved to very promising systems in many areas of plant science.

Certainly, hairy roots are one of the most universally useful plant *in-vitro* systems currently available. In recent years reports on establishing and optimizing transformed root cultures and using them as tools in various disciplines have resulted in rapid progress in the field. Hairy roots traditionally have been employed in plant physiology, natural product chemistry, biotechnology, and plant propagation. While commercial utilization in classical biotechnological production of low-molecular pharmaceuticals and chemicals has so far been of rather limited success, new applications are emerging from the possibilities of genetic engineering and combinatorial biosynthesis. In addition, hairy roots seem to be becoming useful in the production of proteins, in environmental biotechnology for phytoremediation of pollutants from waste water, and for the regeneration of genetically altered plants. The number of applications is anticipated to increase in future. In summary, the history of hairy roots has demonstrated the beneficial utilization of natural gene transfer for both fundamental research and practical applications.

Acknowledgements. Emily Wheeler, Jena, is gratefully acknowledged for linguistic support in the preparation of the manuscript.

References

Aird EH, Hamill JD, Rhodes MJC (1988) Cytogenetic analysis of hairy root cultures from a number of plant species transformed with *Agrobacterium rhizogenes*. Plant Cell Tiss Org 15:47–57

Asada Y, Li W, Yoshikawa T (2000) Biosynthesis of the dimethylallyl moiety of glabrol in *Glycyrrhiza glabra* hairy root cultures via a non-mevalonate pathway. Phytochemistry 55:323–326

Bais HP, Govindaswamy S, Ravishankar GA (2000) Enhancement of growth and coumarin production in hairy root cultures of witloof chicory (*Cichorium intybus* L. cv. Lucknow local) under the influence of fungal elicitors. J Biosci Bioeng 90:648–653

Bais HP, Sudha G, Suresh B, Ravishankar GA (2001) Permeabilization and in situ adsorption studies during growth and coumarin production in hairy root cultures of *Cichorium intybus*. Ind J Exp Biol 39:564–571

Bais HP, Park S-W, Stermitz FR, Halligan KM, Vivanco JM (2002a) Exudation of fluorescent β-carbolines from *Oxalis tuberosa* L. roots. Phytochemistry 61:539–543

Bais HP, Walker TS, Schweizer HP, Vivanco JM (2002b) Root specific elicitation and antimicrobial activity of rosmarinic acid in hairy root cultures of *Ocimum basilicum*. Plant Physiol Biochem 40:983–995

Bais HP, Vepachedu R, Vivanco JM (2003) Root specific elicitation and exudation of fluorescent β-carbolines in transformed root cultures of *Oxalis tuberosa*. Plant Physiol Biochem 41:345–353

Bajaj YPS, Ishimaru K (1999) Genetic transformation of medicinal plants. In: Bajaj YPS (ed) Biotechnology and forestry. transgenic medicinal plants. Springer, Berlin, pp 1–29

Banerjee S, Rahman L, Uniyal GC, Ahuja PS (1998) Enhanced production of valepotriates by *Agrobacterium rhizogenes* induced hairy root cultures of *Valeriana wallichii* DC. Plant Sci 131:203–208

Banthorpe DV (1994) Secondary metabolism in plant tissue culture: scope and limitations. Nat Prod Rep 11:303–328

Benhamou N (1996) Elicitor-induced plant defence pathways. Trends Plant Sci 1:233–240

Berlin J (1997) Secondary products from plant cell cultures. In: Kleinkauf H, von Döhren H (eds) Products of secondary metabolism. VCH, Weinheim, pp 593–640

Bhadra R, Shanks JV (1997) Transient studies of nutrient uptake, growth, and indole alkaloid accumulation in heterotrophic cultures of hairy roots of Catharanthus roseus. Biotechnol Bioeng 55:527–534

Bhagwath SG, Hjortso MA (2000) Statistical analysis of elicitation strategies for thiarubrine A production in hairy root cultures of Ambrosia artemisiifolia. J Biotechnol 80:159–167

Biondi S, Fornalé S, Oksman-Caldentey KM, Eeva M, Agostani S, Bagni N (2000) Jasmonates induce over-accumulation of methylputrescine and conjugated polyamines in Hyoscyamus muticus L. root cultures. Plant Cell Rep 19:691–697

Böhm H, Mäck G (2004) Betaxanthin formation and free amino acids in hairy roots of Beta vulgaris var. lutea depending on nutrient medium and glutamate or glutamine feeding. Phytochemistry 65:1361–1368

Boitel-Conti M, Laberche JC, Lanoue A, Ducrocq C, Sangwan-Norreel BS (2000) Influence of feeding precursors on tropane alkaloid production during an abiotic stress in Datura innoxia transformed roots. Plant Cell Tissue Org 60:131–137

Boominathan R, Doran PM (2003) Cadmium tolerance and antioxidative defenses in hairy roots of the cadmium hyperaccumulator, Thlaspi caerulescens. Biotechnol Bioeng 83:158–167

Boominathan R, Saha-Chaudhury NM, Sahajwalla V, Doran PM (2004) Production of nickel bio-ore from hyperaccumulator plant biomass: applications in phytomining. Biotechnol Bioeng 86:243–250

Bordonaro JL, Curtis WR (2000) Inhibitory role of root hairs on transport within root culture bioreactors. Biotechnol Bioeng 70:176–186

Bourgaud F, Gravot A, Milesi S, Gontier E (2001) Production of plant secondary metabolites: a historical perspective. Plant Sci 161:839–851

Brigham LA, Michaels PJ, Flores HE (1999) Cell-specific production and antimicrobial activity of naphthoquinones in roots of Lithospermum erythrorhizon. Plant Physiol 119:417–428

Buitelaar RM, Leenen EJTM, Geurtsen G, de Groot AE, Tramper J (1993) Effects of the addition of XAD-7 and of elicitor treatment on growth, thiophene production, and excretion by hairy roots of Tagetes patula. Enzyme Microb Tech 15:670–676

Cai DG, Kleine M, Kifle S, Harloff HJ, Sandal NN, Marcker KA, KleinLankhorst RM, Salentijn EMJ, Lange W, Stiekema WJ, Wyss U, Grundler FMW, Jung C (1997) Positional cloning of a gene for nematode resistance in sugar beet. Science 275:832–834

Canto-Canché B, Loyola-Vargas VM (1999) Chemicals from roots, hairy roots, and their application. Adv Exp Med Biol 464:235–275

Chilton MD (2001) Agrobacterium. A memoir. Plant Physiol 125:9–14

Chilton MD, Tepfer DA, Petit A, Casse-Delbart F, Tempé J (1982) Agrobacterium rhizogenes inserts T-DNA into the genomes of the host plant root cells. Nature 295:432–432

Cho HJ, Brotherton JE, Widholm JM (2004) Use of the tobacco feedback-insensitive anthranilate synthase gene (ASA2) as a selectable marker for legume hairy root transformation. Plant Cell Rep 23:104–113

Choi PS, Kim YD, Choi KM, Chung HJ, Choi DW, Liu JR (2004) Plant regeneration from hairy-root cultures transformed by infection with Agrobacterium rhizogenes in Catharanthus roseus. Plant Cell Rep 22:828–831

Christen P (2002) Trigonella species: In vitro culture and production of secondary metabolites. In: Nagata T, Ebizuka Y (eds) Biotechnology in agriculture and forestry; medicinal and aromatic plants XII, vol 51. Springer, Berlin, pp 306–327

Christey MC (2001) Use of Ri-mediated transformation for production of transgenic plants. In Vitro Cell Dev—Pl 37:687–700

Ciesielska E, Gwardys A, Metodiewa D (2002) Anticancer, antiradical and antioxidative actions of novel Antoksyd S and its major components, baicalin and baicalein. Anticancer Res 22:2885–2891

Damgaard O, Rasmussen O (1991) Direct regeneration of transformed shoots in *Brassica napus* from hypocotyl infection with *Agrobacterium rhizogenes*. Plant Mol Biol 17:1–8

De Carolis E, De Luca V (1993) Purification, characterization, and kinetic analysis of a 2-oxoglutarate-dependent dioxygenase involved in vindoline biosynthesis from *Catharanthus roseus*. J Biol Chem 268:5504–5511

Dessaux Y, Petit A, Tempé J (1993) Chemistry and biochemistry of opines, chemical mediators of parasitism. Phytochemistry 34:31–38

Dinan L (2001) Phytoecdysteroids: biological aspects. Phytochemistry 57:325–339

Dommisse EM, Leung DWM, Shaw ML, Conner AJ (1990) Onion is a monocotyledonous host for *Agrobacterium*. Plant Sci 69:249–257

Doran PM (ed) (1997) Hairy roots: culture and applications. Harwood, Amsterdam

Downs CG, Christey MC, Davies KM, King GA, Seelye JF, Sinclair BK, Stevenson DG (1994) Hairy roots of *Brassica napus*. 2. Glutamine-synthetase overexpression alters ammonia assimilation and the response to phosphinothricin. Plant Cell Rep 14:41–46

Eapen S, Suseelan KN, Tivarekar S, Kotwal SA, Mitra R (2003) Potential for rhizofiltration of uranium using hairy root cultures of *Brassica juncea* and *Chenopodium amaranticolor*. Environ Res 91:127–133

Edwards K, Johnstone C, Thompson C (1991) A simple and rapid method for preparation of plant genomic DNA for PCR analysis. Nucleic Acids Res 19:1349–1349

Ellis BE (1988) Natural products from plant tissue culture. Nat Prod Rep 1988:581–612

Facchini PJ (2001) Alkaloid biosynthesis in plants: biochemistry, cell biology, molecular regulation, and metabolic engineering applications. Annu Rev Plant Phys 52:29–66

Falkenhagen H, Stöckigt J, Kuzovkina IN, Alterman IE, Kolshorn H (1993) Indole alkaloids from hairy roots of *Rauwolfia serpentina*. Can J Chem 71:2201–2203

Fischer R, Emans N (2000) Molecular farming of pharmaceutical proteins. Transgenic Res 9:279–299

Flores HE, Filner P (1985) Metabolic relationships of putrescine, GABA and alkaloids in cell and root cultures of *Solanaceae*. In: Neumann KH, Barz W, Reinhard E (eds) Primary and secondary metabolism of plant cell cultures. Springer, Berlin, pp 174–185

Flores HE, Dai Y-R, Cuello JL, Maldonado-Mendoza IE, Loyola-Vargas VM (1993) Green roots: Photosynthesis and photoautotrophy in an underground plant organ. Plant Physiol 101:363–371

Flores HE, Walker TS, Guimaraes RL, Bais HP, Vivanco JM (2003) Andean root and tuber crops: underground rainbows. HortScience 38:161–167

Ford YY, Fox GG, Ratcliffe RG, Robins RJ (1994) *In-vivo* ^{15}N NMR studies of secondary metabolism in transformed root cultures of *Datura stramonium* and *Nicotiana tabacum*. Phytochemistry 36:333–339

Ford YY, Ratcliffe RG, Robins RJ (1996) *In-vivo* NMR analysis of tropane alkaloid metabolism in transformed root and dedifferenciated cultures of *Datura stramonium*. Phytochemistry 43:115–120

Ford YY, Ratcliffe RG, Robins RJ (1998) *In vivo* nuclear-magnetic-resonance analysis of polyamine and alkaloid metabolism in transformed root cultures of *Datura stramnium* L.: evidence for the involvement of putrescine in phytohormone-induced de-differentiation. Planta 205:205–213

Fromm ME, Taylor LP, Walbot V (1985) Expression of genes transferred into monocot and dicot plant cells by electroporation. Proc Natl Acad Sci USA 82:5824–5828

Fujimoto Y, Ohyama K, Nomura K, Hyodo R, Takahashi K, Yamada J, Morisaki M (2000) Biosynthesis of sterols and ecdysteroids in *Ajuga* hairy roots. Lipids 35:279–288

Fulzele DP, Satdive RK, Pol BB (2002) Untransformed root cultures of *Nothapodytes foetida* und production of camptothecin. Plant Cell Tissue Org 69:285–288

Furze JM, Rhodes MJC, Parr AJ, Robins RJ, Whitehead IM (1991) Abiotic factors elicit sesquiterpenoid phytoalexin production but not alkaloid production in transformed root cultures of *Datura stramonium*. Plant Cell Rep 10:111–114

Gartland JS, Brasier CS, Fenning TM, Birch R, Gartland KMA (2001) Ri-plasmid mediated transformation and regeneration of *Ulmus procera* (English Elm). Plant Growth Regul 33:123–129

Gaume A, Komarnytsky S, Borisjuk N, Raskin I (2003) Rhizosecretion of recombinant proteins from plant hairy roots. Plant Cell Rep 21:1188–1193

Giri A, Narasu ML (2000) Transgenic hairy roots: recent trends and applications. Biotechnol Adv 18:1–22

Giri A, Dhingra V, Giri CC, Singh A, Ward OP, Narasu ML (2001) Biotransformations using plant cells, organ cultures and enzyme systems: current trends and future prospects. Biotechnol Adv 19:175–199

Gleba D, Borisjuk NV, Borisjuk LG, Kneer R, Poulev A, Skarzhinskaya M, Dushenko S, Logendra S, Gleba YY, Raskin I (1999) Use of plant roots for phytoremediation and molecular farming. Proc Natl Acad Sci USA 96:5973–5977

Gräther O, Schneider B (2001) The metabolic diversity of plant cell and tissue cultures. In: Esser K, Lüttge U, Kadereit JW, Beyschlag W (eds) Progress in botany, vol 62. Springer, Berlin, pp 266–304

Grisebach RJ (1983) Protoplast microinjection. Plant Mol Biol Rep 1:32–37

Grusak MA (2002) Phytochemicals in plants: genomics-assisted plant improvement for nutritional and health benefits. Curr Opin Biotech 13:508–511

Gundlach H, Müller MJ, Kutchan TM, Zenk MH (1992) Jasmonic acid is a signal transducer in elicitor-induced plant cell cultures. Proc Natl Acad Sci USA 89:2389–2393

Hamill JD, Lidgett AJ (1997) Opportunities and key protocols for studies in metabolic engineering. In: Doran PM (ed) Hairy roots: culture and applications. Harwood, Amsterdam, pp 1–29

Hamill JD, Rhodes MJC (1988) A spontaneous, light independent and prolific plant regeneration response from hairy roots of *Nicotiana hesperis* transformed by *Agrobacterium rhizogenes*. J Plant Physiol 133:506–509

Hamill JD, Parr AJ, Rhodes MJC, Robins RJ, Walton NJ (1987a) New routes to plant secondary products. Bio/Technology 5:800–804

Hamill JD, Prescott A, Martin C (1987b) Assessment of the efficiency of cotransformation of the T-DNA of disarmed binary vectors derived from *Agrobacterium tumefaciens* and the T-DNA of *Agrobacterium rhizogenes*. Plant Mol Biol 9:573–584

Hawes MC, Gunawardena U, Mizasaka S, Zhao X (2000) The role of root border cells in plant defense. Trends Plant Sci 5:128–133

Hawes MC, Bengough G, Cassab G, Ponce G (2003) Root caps and rhizosphere. J Plant Growth Regul 21:352–367

Hildebrand E (1934) Life history of the hairy-root organism in relation to its pathogenesis on nursery apple trees. J Agric Res 48:857–885

Hong SB, Hughes EH, Shanks JV, San KY, Gibson SI (2003) Role of the non-mevalonate pathway in indole alkaloid production by *Catharanthus roseus* hairy roots. Biotechnol Prog 19:1105–1108

Horsch R, Fry J, Hoffmann N, Eichholtz D, Rogers S, Fraley R (1985) A simple and general method for transferring genes into plants. Science 227:1229–1231

Hoshino Y, Mii M (1998) Bialaphos stimulates shoot regeneration from hairy roots of snapdragon (*Antirrhinum majus* L.) transformed by *Agrobacterium rhizogenes*. Plant Cell Rep 17:256–261

Hu ZB, Alfermann AW (1993) Diterpenoid production in hairy root cultures of *Salvia miltiorrhiza*. Phytochemistry 32:699–703

Hughes JB, Shanks JV, Vanderford M, Lauritzen J, Bhadra R (1997) Transformation of TNT by aquatic plants and plant tissue culture. Environ Sci Technol 31:266–271

Hughes EH, Hong SB, Gibson SI, Shanks JV, San KY (2004a) Metabolic engineering of the indole pathway in *Catharanthus roseus* hairy roots and increased accumulation of tryptamine and serpentine. Metab Eng 6:268–276

Hughes EH, Hong SB, Gibson SI, Shanks JV, San KY (2004b) Expression of feedback-resistant anthranilate synthase in *Catharanthus roseus* hairy roots provides evidence for tight regulation of terpenoid indole alkaloid levels. Biotechnol Bioeng 20:718–727

Humphrey AJ, O'Hagan D (2001) Tropane alkaloid biosynthesis. A century old problem unresolved. Nat Prod Rep 18:494–502

Hyeon-Je C, Widholm JM (2002) Improved shoot regeneration protocol for hairy roots of the legume *Astragalus sinicus*. Plant Cell Tissue Org 69:259–269

Hyodo R, Fujimoto Y (2000) Biosynthesis of 20-hydroxyecdysone in *Ajuga* hairy roots: the possibility of 7-ene introduction at a late stage. Phytochemistry 53:733–737

Inoguchi M, Ogama S, Furukawa S, Kondo H (2003) Production of an allelopathic polyacetylene in hairy root cultures of goldenrod (*Solidago altissima* L.). Biosci Biotech Biochem 67:863–868

Jankovic T, Kristic D, Savikin-Fodulovic K, Menkov N, Grubisic D (2002) Xanthones and secoiridoids from hairy root cultures of *Centaurium erythraea* and *C. pulchellum*. Planta Med 68:944–946

Jefferson RA (1989) The GUS reporter gene system. Nature 342:837–838

Jeong GT, Park DH, Hwang B, Park K, Kim SW, Woo J (2002) Studies on mass production of transformed *Panax ginseng* hairy roots in bioreactor. Appl Biochem Biotech 98:1115–1127

Jin UH, Chun JA, Lee JW, Young-Byung-Yi, Lee SW, Chung CH (2004) Expression and characterization of extracellular fungal phytase in transformed sesame hairy root cultures. Protein Expres Purif 37:486–492

Johansson G (1992) Release of organic-C from growing roots of meadow fescue (*Festuca pratensis* L.). Soil Biol Biochem 24:427–433

Jouanin L (1984) Restriction map of an agropine-type Ri plasmid and its homologies with Ti plasmids. Plasmid 12:91–102

Jouhikainen LL, Jokelainen T, Hiltunen R, Teeri TH, Oksman-Caldentey KM (1999) Enhancement of scopolamine production in *Hyoscyamus muticus* L. hairy root cultures by genetic engineering. Planta 208:545–551

Kamada H, Okamura N, Satake M, Harada H, Shimomura K (1986) Alkaloid production by hairy root cultures in *Atropa belladonna*. Plant Cell Rep 5:239–242

Kas J, Burkhard J, Demnerova K, Kostal J, Macek T, Mackova M, Pazlarova J (1997) Perspectives in biodegradation of alkanes and PCBs. Pure Appl Chem 69:2357–2369

Keil M, Härtle B, Guillaume A, Psiorz M (2000) Production of amarogentin in root cultures of *Swertia chirata*. Planta Med 66:452–457

Kholodenko BN, Cascante M, Hoek JB, Westerhoff HV, Schwaber J (1998) Desired metabolite concentrations and fluxes. Biotechnol Bioeng 59:239–247

Kifle S, Shao, M, Jung C, Cai D (1999) An improved transformation protocol for studying gene expression in hairy roots of sugar beet (*Beta vulgaris* L.). Plant Cell Rep 18:514–519

Kim YS, Soh WY (1996) Amyloplast distribution in hairy roots induced by infection with *Agrobacterium rhizogenes*. Biol Sci Space 10:102–104

Kim Y, Wyslouzil BE, Weathers PJ (2002) Secondary metabolism of hairy root cultures in bioreactors. *In Vitro* Cell Dev—Pl 38:1–10

Kitajima M, Yoshida S, Yamagata K, Nakamura M, Takayama H, Saito K, Seki H, Aimi N (2002) Camptothecin-related alkaloids from hairy roots of *Ophiorrhiza pumila*. Tetrahedron 58:9169–9178

Köhle A, Sommer S, Yazaki K, Ferrer A, Boronat A, Li SM, Heide L (2002) High level expression of chorismate pyruvate-lyase (*UbiC*) and *HMG*-CoA reductase in hairy root cultures of *Lythospermum erythrorhizon*. Plant Cell Physiol 43:894–902

Koike Y, Hoshino Y, Mii M, Nakano M (2003) Horticultural characterization of *Angelonia salicariifolia* plants transformed with wild-type strains of *Agrobacterium rhizogenes*. Plant Cell Rep 21:981–987

Komaraiah P, Reddy GV, Reddy PS, Raghavendra1 AS, Ramakrishna SV, Reddanna P (2003) Enhanced production of antimicrobial sesquiterpenes and lipoxygenase metabolites in elicitor-treated hairy root cultures of *Solanum tuberosum*. Biotechnol Lett 25:593–597

Koncz C, Schell J (1986) The promotor of T_L-DNA gene 5 controls the tissue-specific expression of chimeric gene carried by a novel type of *Agrobacterium* binary vector. Mol Gen Genet 204:383–396

Kóvacs G, Kuzovkina I, Szöke E, Kursinszki L (2004) HPLC Determination of flavonoids in hairy root cultures of *Scutellaria baicalensis* Georgi. Chromatographia 60:S81–S85

Kuzovkina IN, Alterman I, Schneider B (2004) Specific accumulation and revised structures of acridone alkaloid glucosides in the tips of transformed roots of *Ruta graveolens*. Phytochemistry 65:1095–1100

Kuzovkina IN, Guseva AV, Kovacs G, Szöke E, Yu M, Vdovitchenko (2005) Flavones in genetically transformed *Scutellaria baicalensis* roots and induction of their synthesis by elicitation with methyl jasmonate. Russ J Plant Physiol 52:77–82

Lanoue A, Boitel-Conti M, Portais JC, Laberche JC, Barbotin JN, Christen P, Sangwan-Norreel B (2002) Kinetic study of littorine rearrangement in *Datura innoxia* hairy roots by C-13 NMR spectroscopy. J Nat Prod 65:1131–1135

Lanoue A, Shakourzadeh K, Marison I, Laberche JC, Christen P, Sangwan-Norreel B, Boitel-Conti M (2004) Occurrence of circadian rhythms in hairy root cultures grown under controlled conditions. Biotechnol Bioeng 88:722–729

Larsen WA, Hsu JT, Flores HE, Humphrey AE (1993) A study of nicotine release from tobacco hairy roots by transient technique. Biotechnol Tech 7:557–562

Le Flem-Bonhomme V, Laurain-Mattar D, Fliniaux MA (2004) Hairy root induction of *Papaver somniferum* var. album, a difficult-to-transform plant, by *A. rhizogenes* LBA 9402. Planta 218:890–893

Leifert C, Cassells AC (2001) Microbial hazards in plant tissue and cell cultures. *In Vitro* Cell Dev–Pl 37:133–138

Li W, Koike K, Asada Y, Hirotani M, Rui H, Yoshikawa T, Nikaido T (2002a) Flavonoids from *Glycyrrhiza pallidiflora* hairy root cultures. Phytochemistry 60:351–355

Li W, Koike K, Asada Y, Yoshikawa T, Nikaido T (2002b) Biotransformation of umbelliferone by *Panax ginseng* root cultures. Tetrahedron Lett 43:5633–5635

Li W, Koike K, Asada Y, Yoshikawa T, Nikaido T (2003) Biotransformation of low-molecular-weight alcohols by *Coleus forskohlii* hairy root cultures. Carbohydrate Res 338:729–731

Lin HW, Kwok KH, Doran PM (2003) Development of *Linum flavum* hairy root cultures for production of coniferin. Biotechnol Lett 25:521–525

Lodhi AH, Bongaerts RJM, Verpoorte R, Coomber SA, Charlwood BV (1996) Expression of bacterial isochorismate sythase (EC 5.4.99.6) in transgenic root cultures of *Rubia peregrina*. Plant Cell Rep 16:54–57

Lorence A, Nessler CL (2004) Camptothecin, over four decades of surprising findings. Phytochemistry 65:2735–2749

Lorence A, Medina-Bolivar F, Nessler CL (2004) Camptothecin and 10-hydroxycamptothecin from *Camptotheca acuminata* hairy roots. Plant Cell Rep 22:437–441

Lozovaya VV, Lygin AV, Zernova OV, Li S, Hartman GL, Widholm JM (2004) Isoflavonoid accumulation in soybean hairy roots upon treatment with *Fusarium solani*. Plant Physiol Biochem 42:671–679

Macek T, Kotrba P, Ruml T, Skacel F, Mackova M (1997) Accumulation of cadmium ions by hairy root cultures. In: Doran PM (ed) Hairy roots: culture and applications. Harwood, Amsterdam, pp 133–138

Maldonado-Mendoza IE, Ayora-Talavera T, Loyola-Vargas, VM (1993) Establishment of root cultures of *Datura stramonium*. Plant Cell Tissue Org 33:321–329

Mallol A, Cusido RM, Palazon J, Bonfill M, Morales C, Pinol MT (2001) Ginsenoside production in different phenotypes of *Panax ginseng* transformed roots. Phytochemistry 57:365–371

Mano Y, Ohkawa H, Yamada Y (1989) Production of tropane alkaloids by hairy root cultures of *Duboisia leichhardtii* transformed by *Agrobacterium rhizogenes*. Plant Sci 59:191–201

Mantrova OV, Dunaeva MV, Kuzovkina IN, Schneider B, Müller-Uri F (1999) Effect of methyl jasmonate on anthraquinone biosynthesis in transformed madder roots. Russ J Plant Physiol 46:248–251

Matsuki R, Onodera H, Yamauchi T, Uchimiya H (1989) Tissue-specific expression of the *rol*C promotor of the Ri plasmid in transgenic rice plants. Mol Gen Genet 22:12–16

Moreno-Valenzuela OA, Monforte-Gonzalez M, Munoz-Sanchez JA, Mendez-Zeel M, Loyola-Vargas VM, Hernandez-Sotomayor SMT (1999) Effect of macerozyme on secondary metabolism plant product production and phospholipase C activity in *Catharanthus roseus* hairy roots. J Plant Physiol 155:447–452

Moreno-Valenzuela OA, Minero-Garcia Y, Chan W, Mayer-Geraldo E, Carbajal E, Loyola-Vargas VM (2003) Increase in the indole alkaloid production and its excretion into the culture medium by calcium antagonists in *Catharanthus roseus* hairy roots. Biotechnol Lett 25:1345–1349

Morgan JA, Shanks JV (2000) Determination of metabolic rate-limitations by precursor feeding in *Catharanthus roseus* hairy root cultures. J Biotechnol 78:137–145

Morgan JA, Barney CS, Penn AH, Shanks JV (2000) Effects of buffered media upon growth and alkaloid production of *Catharanthus roseus* hairy roots. Appl Microbiol Biot 53:262–265

Moyano E, Jouhikainen, K, Tammela P, Palazon J, Cusido RM, Piñol MT, Teeri TH, Oksman-Caldentey KM (2003) Effect of *pmt* gene overexpression on tropane alkaloid production in transformed root cultures of *Datura metel* and *Hyoscyamus muticus*. J Exp Bot 54:203–211

Mugnier J (1988) Establishment of new axenic hairy root lines by inoculation with *Agrobacterium rhizogenes*. Plant Cell Rep 7:9–12

Nedelkoska TV, Doran PM (2000) Characteristics of heavy metal uptake by plant species with potential for phytoremediation and phytomining. Miner Eng 13:549–561

Nedelkoska TV, Doran PM (2001) Hyperaccumulation of nickel by hairy roots of *Alyssum* species: comparison with whole regenerated plants. Biotechnol Prog 17:752–759

Nimchuk Z, Eulgem T, Holt BE, Dangl JL (2003) Recognition and response in the plant immune system. Annu Rev Genet 37:579–609

Nishikawa K, Furukawa H, Fujioka T, Fujii H, Mihashi K, Shimomura K, Ishimaru K (1999) Flavone production in transformed root cultures of *Scutellaria baicalensis* Georgi. Phytochemistry 52:885–890

Nussbaumer P, Kapetanidis I, Christen P (1998) Hairy roots of *Datura candida* x *D. aurea*: effect of culture medium and composition on growth and alkaloid biosynthesis. Plant Cell Rep 17:405–409

Okuzumi K, Hara N, Fujimoto Y, Yamada J, Nakamura A, Takahashi K, Morisaki M (2003) Biosynthesis of phytoecdysteroids in *Ajuga* hairy roots: clerosterol as a precursor of cyasterone, isocyasterone and 29-norcyasterone. Tetrahedron Lett 44:323–326

Palazon J, Mallol A, Eibl R, Lettenbauer C, Cusid RM, Pinol MT (2003) Growth and ginsenoside production in hairy root cultures of *Panax ginseng* using a novel bioreactor. Planta Med 69:344–349

Park SU, Facchini PJ (2000) *Agrobacterium rhizogenes*-mediated transformation of opium poppy, *Papaver somniferum* L., and California poppy; *Eschscholzia californica* Cham., root cultures. J Exp Bot 51:1005–1016

Park SU, Yu M, Facchini PJ (2003) Modulation of berberine bridge enzyme levels in transgenic cultures of California poppy alters the accumulation of benzophenanthridine alkaloids. Plant Mol Biol 51:153–164

Park SW, Lawrence CB, Linden JC, Vivanco JM (2002) Isolation and characterization of a novel ribosome-inactivating protein from root cultures of pokeweed and its mechanism of secretion from roots. Plant Physiol 130:164–178

Parr AJ, Peerless ACJ, Hamill JD, Walton NJ, Robins RJ, Rhodes MJC (1998) Alkaloid production by transformed root cultures of *Catharanthus roseus*. Plant Cell Rep 7:309–312

Payne GF, Bringi V, Prince C, Shuler ML (eds) (1991) Plant cell and tissue culture in liquid systems. Hanser, Munich

Payne J, Hamill JD, Robins RJ, Rhodes MJC (1987) Production of hyoscyamine by hairy root cultures of *Datura stramonium*. Planta Med 53:474–478

Pérez-Mophe-Balch E, Ochoa-Alejo N (1998) Regeneration of transgenic plants of Mexican lime from *Agrobacterium rhizogenes*-transformed tissues. Plant Cell Rep 17:591–596

Petersen SG, Stummann BM, Olesen P, Henningsen KW (1989) Structure and function of root-inducing (Ri) plasmids and their relation to tumor-inducing (Ti) plasmids. Physiol Plant 77:427–35

Petit A, David C, Dahl DA, Ellis JG, Guyon P, Casse-Delbart F, Tempé J (1983) Further extension of the opine concept: plasmids in *Agrobacterium rhizogenes* cooperate for opine degradation. Mol Gen Genet 190:204–214

Pitta-Alvarez SI, Spollanky TC, Giulietti AM (2000) The influence of different biotic and abiotic elicitors on the production and profile of tropane alkaloids in hairy root cultures of *Brugmansia candida*. Enzyme Microb Tech 26:252–258

Pletsch M, Araujo BS, Charlwood BV (1999) Novel biotechnological approaches in environmental remediation research. Biotechnol Adv 17:679–687

Puddephat IJ, Robinson HT, Fenning TM, Barbara DJ, Morton A, Pink DAC (2001) Recovery of phenotypically normal transgenic plants of *Brassica oleracea* upon *Agrobacterium rhizogenes*-mediated co-transformation and selection of transformed hairy roots by GUS assay. Mol Breeding 7:229–242

Ramachandra RS, Ravishankar GA (2002) Plant cell cultures: chemical factories of secondary metabolites. Biotechnol Adv 20:101–153

Rhodes MJC, Hilton MG, Parr AJ, Hamill JD, Robins RJ (1986) Nicotine production by hairy root cultures of *Nicotiana rustica*—fermentation and product recovery. Biotechnol Lett 8:415–420

Rhodes MJC, Robins RJ, Hamill JD, Parr AJ, Hilton MG, Walton NJ (1990) Properties of transformed root cultures. In: Charlwood BV, Rhodes MJC (eds) Secondary products from plant tissue culture. Clarendon Press, Oxford, pp 201–225

Rhodes MJC, Parr AJ, Giulietti A, Aird ELH (1994) Influence of exogenous hormones on the growth and secondary metabolite formation in transformed root cultures. Plant Cell Tissue Org 38:143–151

Rijwani S, Shanks JV (1998) Effect of elicitor dosage and exposure time on biosynthesis of indole alkaloids by *Catharanthus roseus* hairy root cultures. Biotechnol Prog 14:442–449

Riker AW, Banfield W, Wright W, Keitt G, Sagen H (1930) Studies on infectious hairy root of nursery apple trees. J Agric Res 41:887–912

Robins RJ (1998) The application of root cultures to problems of biological chemistry. Nat Prod Rep 549–570

Robins RJ, Bachmann P, Woolley JG (1994) Biosynthesis of hyoscyamine involves an intramolecular rearrangement of littorine. J Chem Soc Perk T I 615–619

Rodriguez S, Compagnon V, Crouch NP, St-Pierre B, De Luca V (2003) Jasmonate-induced epoxidation of tabersonine by a cytochrome P-450 in hairy root cultures of *Catharanthus roseus*. Phytochemistry 64:401–409

Rothe G, Garske U, Dräger B (2001) Calystegines in root cultures of *Atropa belladonna* respond to sucrose, not to elicitation. Plant Sci 160:1043–1053

Saito K, Yamazaki M, Murakoshi I (1992) Transgenic medicinal plants—*Agrobacterium*-mediated foreign gene transfer and production of secondary metabolites. J Nat Prod 55:149–162

Saito K, Sudo H, Yamazaki M, Koseki-Nakamura M, Kitajima M, Takayama H, Aimi N (2001) Feasible production of camptothecin by root culture of *Ophiorrhiza pumila*. Plant Cell Rep 20:267–271

Samuelian S, Kleine M, Ruyter-Spira CP, Klein-Lankhorst RM, Jung C (2004) Cloning and functional analyses of a gene from sugar beet up-regulated upon cyst nematode infection. Plant Mol Biol 54:147–156

Schmülling T, Schell J, Spena A (1988) Single genes from *Agrobacterium rhizogenes* influence plant development. EMBO J 7:2621–2629

Scholl Y, Hoke D, Dräger B (2001) Calystegines in *Calystegia sepium* derive from the tropane alkaloid pathway. Phytochemistry 58:883–889

Scholl Y, Schneider B, Dräger B (2003) Biosynthesis of calystegines: N-15 NMR and kinetics of formation in root cultures of *Calystegia sepium*. Phytochemistry 62:325–332

Schramm DD, German JB (1998) Potential effects of flavonoids on the etiology of vascular disease. J Nutr Biochem 9:560–566

Sevón N, Oksman-Caldentey K-M (2002) *Agrobacterium rhizogenes*-mediated transformation: root cultures as a source of alkaloids. Planta Med 68:859–868

Sevón N, Hiltunen R, Oksman-Caldentey K-M (1992) Chitosan increases hyoscyamine content in hairy root cultures of *Hyoscyamus muticus*. Pharm Pharmacol Lett 2:96–99

Shanks JV, Morgan J (1999) Plant "hairy root" culture. Curr Opin Biotech 10:151–155

Sheludko Y, Gerasimenko I, Kolshorn H, Stöckigt J (2002a) Isolation and structure elucidation of a new indole alkaloid from *Rauvolfia serpentina* hairy root culture: the first naturally occurring alkaloid of the of the raumacline group. Planta Med 68:435–439

Sheludko Y, Gerasimenko I, Kolshorn H, Stöckigt J (2002b) New alkaloids of the sarpagine group from *Rauvolfia serpentina* hairy root culture. J Nat Prod 65:1006–1010

Shepherd T, Davies HV (1993) Carbon loss from the roots of forage rape (*Brassica napus* L.) seedlings following pulse-labeling with (CO_2)-C-14. Ann Bot–Lond 72:155–163

Shiao TL, Doran PM (2000) Root hairiness: effect on fluid flow and oxygen transfer in hairy root cultures. J Biotechnol 83:199–210

Shimomura K, Sudo H, Saga H, Kamada H (1991) Shikonin production and secretion by hairy root cultures of *Lithospermum erythrorhizon*. Plant Cell Rep 10:282–285

Sommer S, Kohle A, Zazaki K, Shimomura K, Bechthold A, Heide L (1999) Genetic engineering of shikonin biosynthesis hairy root cultures of *Lithospermum erythrorhizon* transformed with the bacterial *ubiC* gene. Plant Mol Biol 39:683–693

Stachel SE, Messens E, Van Montague M, Zambryski P (1985) Identification of the signal molecules produced by wounded plant-cells that activate T-DNA transfer in *Agrobacterium tumefaciens*. Nature 318:624–629

Stiller J, Martirani L, Tuppale S, Chian RJ, Chiurazzi M, Gresshoff PM (1997) High frequency transformation and regeneration of transgenic plants in the model legume *Lotus japonicus*. J Exp Bot 48:1357–1365

Stöckigt J, Obitz P, Falkenhagen H, Lutterbach R, Endrea S (1995) Natural products and enzymes from plant cell cultures. Plant Cell Tissue Org 43:97–109

Stockmeyer K, Kempken F (2004) Biotechnology: Production of proteins for biopharmaceutical and industrial uses in transgenic plants. In: Esser K, Lüttge U, Beyschlag W, Murata J (eds) Progress in botany, vol 65. Springer, Berlin, pp 179–192

Stojakowska A, Malarz J (2000) Flavonoid production in transformed root cultures of *Scutellaria baicalensis*. J Plant Physiol 156:121–125

Sudha CG, Reddy BO, Ravishankar GA, Seeni S (2003) Production of ajmalicine and ajmaline in hairy root cultures of *Rauvolfia micrantha* Hook f., a rare and endemic medicinal plant. Biotechnol Lett 25:631–636

Sudo H, Yamakawa T, Yamazaki M, Aimi N, Saito K (2002) Bioreactor production of camptothecin by hairy root cultures of *Ophiorrhiza pumila*. Biotechnol Lett 24:359–363

Suzuki K, Yamashita I, Tanaka N (2002) Tobacco plants were transformed by *Agrobacterium rhizogenes* infection during their evolution. Plant J 32:775–787

Tabata M, FujitaY (1985) Production of shikonin by plant cell cultures. In: Zaitlin M, Day P, Hollaender A (eds) Biotechnology in plant science: relevance to agriculture in the eighties. Academic Press, New York, pp 207–218

Tepfer D (1983) Biology of genetically transformation of plants with *Agrobacterium rhizogenes*. In: Pühler A (ed) Molecular genetics of the bacteria-plant interaction. Springer, Berlin, pp 272–293

Tepfer D (1984) Transformation of several root species of higher plants by *Agrobacterium rhizogenes*: sexual transmission of the transformed genotype and phenotype. Cell 37:959–967

Tepfer D (1990) Genetic transformation using *Agrobacterium rhizogenes*. Physiol Plant 79:140–146

Thompson J, Donkersloot JA (1992) N-(Carboxyalkyl)amino acids—occurrence, synthesis and functions. Annu Rev Biochem 61:517–557

Thurau T, Kifle S, Jung C, Cai D (2003) The promoter of the nematode resistance gene Hs1^{pro-1} activates a nematode-responsive and feeding site-specific gene expression in sugar beet (*Beta vulgaris* L) and *Arabidopsis thaliana*. Plant Mol Biol 52:643–660

Toivonen L (1993) Utilization of hairy root cultures for production of secondary metabolites. Biotechnol Prog 9:12–20

Toivonen L, Ojala M, KauppinenV (1991) Studies on the optimization of growth and indole alkaloid production of by hairy root cultures of *Catharanthus roseus*. Biotechnol Bioeng 37:673–680

Trick HN, Finer JJ (1997) SAAT: sonication-assisted *Agrobacterium*-mediated transformation. Transgenic Res 6:329–336

Uozumi N, Kobayashi T (1995) Artificial seed production through encapsulation of hairy root and shoot tips. In: Bajaj YPS (ed) Biotechnology in agriculture and forestry: somatic embryogenesis and synthetic seed I. Springer, Berlin, pp 170–180

Verpoorte R (2000) Secondary metabolism. In: Verpoorte R, Alfermann AW (eds) Metabolic engineering of plant secondary metabolism. Kluwer, Dordrecht, pp 1–29

Verpoorte R, Memelink J (2002) Engineering secondary metabolite production in plants. Curr Opin Biotech 13:181–187

Wang YM, Wang JB, Luo D, Jia JF (2001) Regeneration of plants from callus tissues of hairy roots induced by *Agrobacterium rhizogenes* on *Alhagi pseudoalhagi*. Cell Res 11:279–284

Washida D, Shimomura K, Takido M, Kitanaka S (2004) Auxins affected ginsenoside production and growth of hairy roots in *Panax* hybrid. Biol Pharm Bull 27:657–660

Weathers PJ, DeJesus-Gonzalez L, Kim YJ, Souret FF, Towler MJ (2004) Alteration of biomass and artemisinin production in *Artemisia annua* hairy roots by media sterilization method and sugars. Plant Cell Rep 23:414–418

Weising K, Kahl G (1996) Natural genetic engineering of plant cells: the molecular biology of crown gall and hairy root disease. World J Microb Biot 12:327–351

White FF, Taylor BH, Huffman GA, Gordon MP, Nester EW (1985) Molecular and genetic analysis of the transferred DNA regions of the root-inducing plasmid of *Agrobacterium rhizogenes*. J Bacteriol 164:33–44

Wielanek M, Urbanek H (1999) Glucotropaeolin and myrosinase production in hairy root cultures of *Tropaeolum majus*. Plant Cell Tissue Org 57:39–45

Willmitzer L, Sanchez-Serrano J, Buschfeld E, Schell J (1982) DNA from *Agrobacterium rhizogenes* is transferred to and expressed in axenic hairy root plant tissues. Mol Gen Genet 186:16–22

Wongsamuth R, Doran PM (1997) Hairy roots as an expression system for the production of antibodies. In: Doran PM (ed) Hairy roots: culture and applications. Harwood, Amsterdam, pp 89–97

Wysokinska H, Chmiel A (1997) Transformed root cultures for biotechnology. Acta Biotechnol 17:131–159

Wysokinska H, Lisowska K, Floryanowicz-Czekalska K (2001) Transformation of *Catalpa ovata* by *Agrobacterium rhizogenes* and phenylethanoid glycosides production in transformed root cultures. Z Naturforsch C 56:375–381

Xu T, Zhang L, Sun X, Zhang H, Tang K (2004) Production and analysis of organic acids in hairy-root cultures of *Isatis indigotica* Fort. (indigo woad). Biotechnol Appl Bioc 39:123–128

Yamanaka M, Shimomura K, Sasaki K, Yoshihira K, Ishimaru K (1995) Glucosylation of phenolics by hairy roots cultures of *Lobelia sessilifolia*. Phytochemistry 39:1149–1150

Yamazaki Y, Urano A, Sudo H, Kitajima M, Takayama H, Yamazaki M, Aimi N, Saito K (2003a), Metabolite profiling of alkaloids and strictosidine synthase activity in camptothecin producing plants. Phytochemistry 62:461–470

Yamazaki Y, Sudo H, Yamazaki M, Aimi N, Saito K (2003b) Camptothecin biosynthetic genes in hairy roots of *Ophiorrhiza pumila*: cloning, characterization and differential expression in tissues and by stress compounds. Plant Cell Physiol 44:395–403

Yamazaki Y, Kitajima M, Arita M, Takayama H, Sudo H, Yamazaki M, Aimi N, Saito K (2004) Biosynthesis of camptothecin. In silico and *in vivo* tracer study from [1–^{13}C]glucose. Plant Physiol 134:161–170

Yaoya S, Kanho H, Mikami Y, Itani T, Umehara K, Kuroyanagi M (2004) Umbelliferone released from hairy root cultures of *Pharbitis nil* treated with copper sulfate and its subsequent glucosylation. Biosci Biotech Biochem 68:1837–1841

Yazaki K, Tanaka S, Matsuoka H, Sato F (1998) Stable transformation of *Lithospermum erythrorhizon* by *Agrobacterium rhizogenes* and shikonin production of the transformants. Plant Cell Rep 18:214–219

Yazaki K, Matsuoka H, Shimomura K, Bechthold A, Sato F (2001) A novel dark-inducible protein, LeDI-2, and its involvement in root-specific secondary metabolism in *Lithospermum erythrorhizon*. Plant Physiol 125:1831–1841

Yeoman MM, Yeoman CL (1996) Manipulating secondary metabolism in cultured plant cells. New Phytol 134:553–569

Yoshikawa T, Furuya T (1987) Saponin production by cultures of *Panax ginseng* transformed with *Agrobacterium rhizogenes*. Plant Cell Rep 6:449–453

Zabetakis I, Edwards R, O'Hagan D (1999) Elicitation of tropane alkaloid biosynthesis in transformed root cultures of *Datura stramonium*. Phytochemistry 50:53–56

Zdravković-Korać S, Muhovski Y, Druart P, Calic D, Radojevic L (2004) *Agrobacterium rhizogenes*-mediated DNA transfer to *Aesculus hippocastanum* L. and the regeneration of transformed plants. Plant Cell Rep 22:698–704

Zenk MH (1988) Biosynthesis of alkaloids using plant cell cultures. Recent Adv Phytochem 23:429–457

Zenk MH (1991) Chasing the enzymes of secondary metabolism—plant cell cultures as a pot of gold. Phytochemistry 30:3861–3863

Zhang C, Yan Q, Cheuk WK, Wu J (2004) Enhancement of tanshinone production in *Salvia miltiorrhiza* hairy root culture by Ag$^+$ elicitation and nutrient feeding. Planta Med 70:147–151

Zhang DY, Wu J, Ye F, Xue L, Jiang SQ, Yi JZ, Zhang WD, Wei HC, Sung M, Wang W, Li XP (2003) Inhibition of cancer cell proliferation and prostaglandin E-2 synthesis by *Scutellaria baicalensis*. Cancer Res 63:4037–4043

Zhang L, Ding R, Chai Y, Bonfill M, Moyano E, Oksman-Caldentey KM, Xu T, Pi Y, Wang Z, Zhang H, Kai G, Liao Z, Sun X, Tang K (2004) Engineering tropane biosynthetic pathway in *Hyoscyamus niger* hairy root cultures. Proc Natl Acad Sci USA 101:6786–6791

Zhou Y, Hirotani M, Yoshikawa T, Furuya T (1997) Flavonoids and phenylethanoids from hairy root cultures of *Scutellaria baicalensis*. Phytochemistry 44:83–87

Zupan J, Muth TR, Draper O, Zambryski P (2000) The transfer of DNA from *Agrobacterium tumefaciens* into plants: a feast of fundamental insights. Plant J 23:11–28

Dr. Inna N. Kuzovkina
Timiryazev Institute of Plant Physiology
Botanicheskaya ul. 35
127276 Moscow, Russia
e-mail: ikuz@ippras.ru

Dr. Bernd Schneider
Max-Planck-Institute for Chemical Ecology
Beutenberg Campus
Hans-Knöll-Str. 8
07745 Jena, Germany
e-mail: schneider@ice.mpg.de

Molecular chaperones—holding and folding

Christoph Forreiter

1 Introduction

Any cellular function is based upon the proper function of native proteins. However, protein function requires a specific 3D structure. Establishing the correct three-dimensional conformation is therefore a premise for the exclusive biochemical function of a given protein. But how can this structure be achieved and maintained throughout the entire lifetime of a protein? In 1972, Christian Anfinsen was awarded with the Nobel Prize for chemistry for showing that the primary amino acid sequence contains all the information required for the native structure and that folding is an autonomous process, which does not require any additional energy. Based on in-vitro experiments with purified proteins, he suggested that the correct three-dimensional structure can form spontaneously *in vivo* once the newly synthesized protein leaves the ribosome. However, this view changed considerably over the last decade. It has become clear that a complex and sophisticated collection of proteins exists, which assists nascent protein folding and allows the functional state of proteins to be maintained under stress conditions in which they would normally unfold and aggregate. These proteins are collectively called "molecular chaperones", because they prevent unwanted interactions between their immature clients, like their human counterparts in Victorian society of the 19th century, who as chaperones had to prevent bourgeoisie female adolescents from unwanted encounters with men. Experimental data indicate that several members of this group of proteins form heterooligomeric complexes, sometimes termed "chaperone machines", interacting with each other to generate a network for maturation, assembly and intracellular targeting of proteins. In this review, the essential information on the structure and function of chaperone and chaperone complexes are summarized and the impact on the cellular stress response in plants will be discussed.

Progress in Botany, Vol. 67
© Springer-Verlag Berlin Heidelberg 2006

2 Molecular chaperones and other elements of the stress response

The prosperous road which led to our current knowledge of what molecular chaperones are and how they work started in 1964, when Ferruccio Ritossa (Ritossa 1964) observed large, uncondensed areas in polytene chromosome preparations of salivary gland cells, so-called "puffs", after elevating the temperature. This was the first demonstration that elevated temperature could result in increased expression of a certain set of genes. Later, it was discovered that these massively transcribed genes mediate the bulk of the organism's response to heat and other stresses in *Drosophila*. Meanwhile, it is clear that all organisms respond to non-permissive temperatures by synthesising a specific set of proteins, which are therefore termed heat stress or heat shock proteins (Hsp). They are needed to protect cells from heat damage and assist in normalisation of functions during recovery (reviewed by Parsell and Lindquist 1993; Morimoto et al. 1994; Forreiter and Nover 1998; Walter and Buchner 2002). The Hsps can be assigned to 11 conserved protein families that can be found in all organisms, including bacteria. Most of them are constitutively present, but markedly induced under stress conditions and/or expressed in certain developmental stages. During the last 2 decades, more details of the biochemical function of Hsps have emerged. Most of them belong to a heterologous group of proteins, for the first time coined as "molecular chaperones" in a remarkable contribution to *Annual Review of Biochemistry* by John Ellis and Saskia van der Vries (1991). By definition, these proteins aid other proteins to maintain or regain their native conformation by stabilising partially unfolded states (Craig et al. 1993; Hendrick and Hartl 1993; Hartl 1996; Walter and Buchner 2002). One very important feature of this class of proteins is that they do not contain specific information for correct folding, but rather prevent non-productive interactions (aggregation) between non-native proteins. Molecular chaperones can therefore be distinguished from other heat-induced proteins acting as direct folding catalysts such as peptidyl-prolyl isomerases or protein disulphide isomerases (Schmid 1993, 1995) and ubiquitin, which is the central element for the degradation of protein debris via the proteasome compartment (Ciechanover 1998).

It is noteworthy that molecular chaperones are sometimes also referred to as heat-stress or heat-shock proteins, since most of the molecular chaperones are strongly induced under heat or other unfavourable conditions. However, this is misleading to a certain extent, since molecular chaperones and heat stress proteins comprise different sets of proteins: Heat stress proteins are not defined by structure or function but by their massive appearance under non-permissive conditions, while molecular chaperones are defined by their mode of action.

3 How do molecular chaperones work?

Information transfer from DNA to mRNA and from mRNA to the polypeptide translates the linear sequence of base triplets into a linear sequence of amino acids. The final step in this process from gene to the functional protein, protein folding, converts this linear information into a three-dimensional structure. This reaction is very complex. Although we have learned much over the past decades about the physical principles underlying the folding process (Creighton 1990; Kim and Baldwin 1992; van Gunsteren et al. 2001), it is still a major challenge for biochemists to predict the structure into which a given polypeptide will fold.

The investigation of the protein-folding problem began in 1960 with the experiments of Anfinsen and co-workers on the reversible folding of ribonuclease A (RNase A; Anfinsen et al. 1961). Incubation with 8 M urea and a reducing agent resulted in an unfolded protein with no disulphide bonds or enzymatic activity. This denatured state of RNase A is thought to resemble the conformation immediately after its synthesis on the ribosome. When the denaturant and the reducing agent were removed by dialysis, the enzyme was found to slowly regain its activity. Apparently, it had refolded *in vitro*. This observation clearly showed that the three-dimensional structure of this protein is encoded in its amino acid sequence, and that no other factors are required for structure acquisition. Thus, Anfinsen deduced, protein folding is an autonomous and, given the proper conditions, spontaneous process (Anfinsen 1973). Following this principle, biochemists studying the folding properties of other small, monomeric proteins were able to confirm his observations (Wetlaufer 1973; Creighton 1975). Eventually, these results led to the notion that, in principle, every protein can be refolded *in vitro* (Jaenicke 1987).

The native state of a protein corresponds to a fairly narrow energy minimum on the conformational energy landscape (Creighton 1990; Dill 1990). On the other hand, the denatured state is represented by a large ensemble of conformations with high internal energy and flexibility. During the folding process, numerous non-covalent interactions are formed that require the exact positioning of the various atoms of the protein. Among these, the hydrophobic interactions seem to play an important role, as already suggested by Kauzmann (1959). Hydrophobic molecules tend to associate with each other in a polar environment for reasons of entropy and enthalpy. Accordingly, hydrophobic amino acids are predominantly found in the core of a folded protein. When biochemists began to study the folding of oligomeric proteins or of larger proteins that consist of multiple domains, it became apparent that the hydrophobic interactions are not only important in stabilizing the folded conformation, but may also have a detrimental effect (Jaenicke and Seckler 1997; Dobson 2004): During early folding stages, many proteins form intermediates that display numerous hydrophobic surfaces. Protein molecules can associate non-specifically through these hydrophobic patches, and ultimately aggregate. Aggregation is a disordered, non-specific association of polypeptide chains that leads to the formation of large heterogeneous protein particles devoid of any biological function. Since aggregation is a second- or higher-order reaction, protein concentration plays an important role in determining whether folding to the native state or non-specific aggregation will predominate (Kiefhaber et. al. 1991). Whereas many proteins can be (re)folded *in vitro* under optimized conditions with good yields, the situation in a living cell is less favourable. In particular, high protein concentration and temperature promote aggregation as an undesired side reaction, competing with productive folding (Jaenicke and Seckler 1997). *In vivo*, all proteins have to fold under the same set of conditions. These conditions seem to be somehow counterproductive for efficient folding, mainly because of the large number of non-native proteins present. Given these circumstances, it seems surprising that cells are usually devoid of aggregated proteins. There are two possible explanations for this observation: (1) aggregation does occur *in vivo*, but its products are rapidly removed by cellular proteases. This would imply that cells waste a lot of energy to produce proteins that never become functional; (2) cells have found a strategy of minimizing the aggregation of newly synthesized proteins in the first place. This has been achieved by complex protein machinery, the chaperones, which influence the spontaneous folding reaction of proteins, thus preventing aggregation. It is important to note that these molecular chaperones do not provide specific steric information for the folding of the target protein, but rather inhibit unproductive interactions and thus allow the protein to fold more efficiently into its native structure.

Molecular chaperones are found in all compartments of a cell where folding or, more generally, conformational rearrangements of proteins occur. Although protein synthesis is the major source of unfolded polypeptide chains, other processes can generate unfolded proteins as well. At non-physiologically high temperatures or in the presence of certain chemicals, proteins can become structurally labile and might even unfold. Eventually, this would result in loss of function of the affected proteins and in the accumulation of protein aggregates.

The term molecular chaperone is used to describe a group of very heterologous but functionally related set of proteins. According to their molecular weight, molecular chaperones are divided into several classes or families. A cell may express multiple members of the same chaperone family. For example,

S. cerevisiae (yeast) produces 14 different versions of the chaperone Hsp70 (Craig et al. 1999). Proteins from the same class of molecular chaperones often show a significant amount of sequence homology and are structurally and functionally related, whereas there are hardly any homologies between chaperones from different families. Despite this diversity, however, most molecular chaperones share some common functional features.

The principal property of any molecular chaperone clearly is its ability to bind unfolded or partially folded polypeptides in a manner generally described in Fig. 1. During the early stages of folding or when misfolding occurs, the hydrophobic residues of a protein are partially solvent accessible and thus render it vulnerable to aggregation. Association of these hydrophobic protein species with molecular chaperones efficiently suppresses aggregation. The low specificity of the hydrophobic interaction and the conformational flexibility of folding intermediates ensure that chaperones act promiscuously: they bind to a wide variety of polypeptides that differ

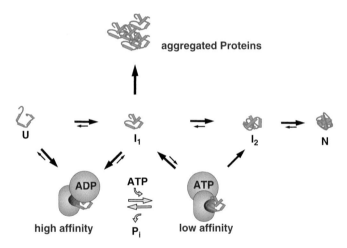

Fig. 1. Chaperone-assisted folding. Unfolded proteins (U) produced either by protein synthesis or as a result of stress-induced unfolding, refold via several folding intermediates (I_1 and I_2), which have an increasing structure complexity. This process leads to the native protein structure (N), usually when the protein has reached its conformational energy minimum. Some of these folding intermediates (I_1) expose hydrophobic elements on the surface that tend to interact randomly with other hydrophobic areas, resulting in misfolding and/or aggregation. Molecular chaperones can interact with these unfolded intermediates via hydrophobic interaction. Some of them, like Hsp70, with different affinity, depending upon ATP/ADP binding. In this case, folding intermediates bind to the high affinity ADP-conformation. After nucleotide exchange (ADP→ATP) the chaperone complex alters its conformation and folding intermediates are released. These may have a structure (I_2), which no longer requires chaperone binding for proper folding or may again result in a folding intermediate prone to aggregation, which would require another round of chaperone-assisted conformational change

widely in amino acid sequence and in conformation. However, since most native proteins and many late folding intermediates do not have hydrophobic patches, they are no longer substrates for molecular chaperones.

Three major aspects in a life cycle of a protein invoke chaperone proteins: (1) they ensure that nascent polypeptides emerging from the ribosome are kept in a folding-competent state until the whole sequence information is available (e.g. Bukau et al. 1997); (2) since fully folded proteins cannot be translocated through membranes, chaperones are needed to maintain a partially unfolded form for proteins destined for transport into mitochondria, plastids or other cellular compartments (e.g. Horst et al. 1997; Lubeck et al. 1997; Neupert 1997; Young et al. 2004); (3) they stabilise damaged proteins generated as a result of chemical or physical stress and thus facilitate renaturation and/or degradation in the recovery period. Many data collected during the last years indicate that members of different Hsp families act together in multi-subunit complexes, so-called chaperone machines (e.g. Hartl 1996; Bukau et al. 1997; Forreiter et al. 1997) and different chaperone complexes may interact to generate a network for protein maturation, assembly and targeting (Frydman and Höhfeld 1997; Johnson and Craig 1997; Young et al. 2004).

4 Structure and function of the different chaperone classes

In the following section, the main chaperone classes will be discussed mainly with respect to their functional properties and their contribution to the chaperone system or chaperone machine in which they are involved. Currently, five basic chaperone systems are known. They are generally named after the apparent molecular weight of the main protein involved, in particular Hsp100, Hsp90, Hsp70, Hsp60 and the low molecular heat shock proteins, the sHsps, with a molecular mass between 15 and 30 kD depending on their host and function. This particular class is most important especially for plants, which as sessile organisms cannot avoid stressfully situations simply by avoidance or flight. Since these classes are common to all organisms and work principally in the same fashion, the prokaryotic counterpart of a given chaperone is given in brackets.

4.1 Hsp70 (DnaK) chaperone machine

The central part of the entire protein folding network is the Hsp70 (DnaK) chaperone machine. A constantly increasing number of seemingly disparate processes are characterised, which depend on the Hsp70 systems. Examples are the processing of newly formed proteins in the cytoplasm, the ER and other cellular organelles, uncoating of clathrin coated vesicles, reorganisation of cytoskeletal systems, translation initiation, nuclear protein import and export, ribosome assembly, protection of nuclear structure and function under stress and protein degradation (Nover and Scharf 1997; Young et al. 2004). The basic function of Hsp70 in all cases is evidently binding and subsequent release of partially unfolded proteins in an ATP-dependent cycle (Fig. 2).

Members of the Hsp70 family consist of a single polypeptide with two functional domains: the N-terminal domain is able to bind and hydrolyse ATP, while the C-terminal peptide binding domain contains at least four

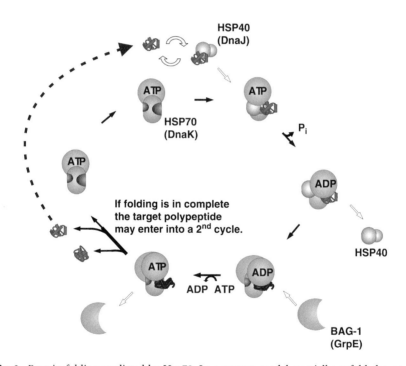

Fig. 2. Protein folding mediated by Hsp70. In a current model, partially unfolded proteins associate loosely with Hsp40 (or DnaJ in prokaryotes) and further with hydrophobic pockets located in the C-terminal domain of Hsp70 (DnaK). Binding of the co-chaperone Hsp40 (DnaJ) results in P_i cleavage of the bound ATP at the N-terminal domain. Conformational change of the ADP bound form results in tighter binding of the substrate polypeptide. After dissociation of Hsp40, the nucleotide exchange factor BAG-1 (GrpE) can bind to initiate an ADP/ATP exchange. Subsequent release of the substrate protein from the binding pocket allows folding. Sometimes several cycles of binding and release to Hsp70 might be required to reach the native conformation of the substrate protein

hydrophobic pockets for binding unfolded target proteins. Crystal structures are available for both subdomains (ATP binding: Flaherty et al. 1991; substrate binding: Zhu et al. 1996). After binding of ATP Hsp70 is able to interact with non-native proteins, characterised by exposition of hydrophobic peptide motifs at their surface. Subsequent ATP hydrolysis generates a Hsp70-ADP complex with tightly bound substrate protein. After ADP→ATP exchange, the target polypeptide is released to undergo folding in solution (Fig. 2). Several cycles of binding, release and folding may be required to obtain the native conformation of a target polypeptide.

Hsp70 needs interaction with other proteins for full chaperone activity (see Table 1). The most important one is Hsp40 (DnaJ). The N-terminal part of the latter harbours the characteristic 70 amino acid residue J domain important for interaction with Hsp70. It is followed by a glycine and phenylalanine rich stretch of 30 residues, a cysteine-rich region probably forming a Zn-finger motif and the C-terminal peptide binding domain. Actually, the highly conserved J-domain is not only found in Hsp40 but also in several other proteins reported to interact with Hsp70 (Cyr et al. 1994). Further examples are auxilin, involved in Hsp70-mediated vesicle uncoating (Ungewickel et al. 1997), the ER membrane bound protein Sec63p (Brodsky 1996), the polyoma virus large T-antigen (Kelly and Georgopoulos 1997) or in Tim44 (MIM44), a protein of the inner mitochondrial membrane, where mitochondrial Hsp70 binds to aid protein import into the organelle (Rassow et al. 1994). Since Hsp70 has only weak ATPase activity, it was proposed that target proteins are first bound by Hsp40 (Bukau and Horwich 1998). In both cases, interaction with the co-chaperone results in a markedly increased ATPase activity of Hsp70. At any rate, Hsp40 is considered as a subunit of the Hsp70 complex essential for coupling ATP hydrolysis to substrate binding.

For effective ADP exchange, an additional factor is required (see Table 1), described as GrpE protein for the *E. coli* DnaK/J-system. As a functionally

Table 1. Core components of the Hsp70 chaperone system

Eukaryotes	Prokaryotes	Function
Hsp/Hsc70	DnaK	Substrate binding; ATP-binding
Hsp40	DnaJ	Hsp70 interaction (J-box) and substrate binding; binds to the peptide binding domain
BAG-1	GrpE	Nucleotide exchange factor; binds to the ATPase domain

related protein for eukaryotes, a BAG-1 like protein was identified (Höhfeld and Jentsch 1997; Zeiner et al. 1997; Sondermann et al. 2001; Alberti et al. 2003). BAG-1 was originally described as component of the BCL-2-system regulating cell cycle and apoptosis in mammals (Takayama et al. 1995). Although BAG-1 and GrpE share no sequence homology, both have similar biochemical functions and bind with a defined sequence to the ATP-binding side of Hsp70 (Sondermann et al. 2002).

Other components connected with Hsp70 are HIP (Hsp70 interacting protein; Höhfeld et al. 1995) and HOP (Hsp70 organising protein), also known as the p60 component of the Hsp90 complex (Smith et al. 1993). HIP binds to the ATPase domain of Hsp70, thus stabilising the ADP-bound form. Subsequently, HOP interacts with the peptide binding domain. Interestingly, HOP/p60 has also a Hsp90 binding site and thus physically links the Hsp70 and Hsp90 chaperone systems (see 4.3 and Fig. 4). This link spotlights a central feature of the Hsp70/Hsp40-chaperone system. Not only Hsp90, but also Hsp101 (see 4.4) and the small Hsps (3.5), require Hsp70 for their full chaperone activity. Thus the Hsp70-system seems to be a central element of a complex chaperone network.

4.2 Hsp60 (GroE) chaperone system—the chaperonins

The prototype for this chaperone-system is the *E. coli* GroEL/GroES complex structurally analysed by Hunt et al. (1996) for GroES and Xu et al. (1997) for GroEL. Data derived from the crystal structure of GroEL confirmed earlier pictures: 14 subunits of the GroEL protein (56 kDa) form a hollow core structure of two heptameric rings associated back to back with each other. Both rings provide an inner cavity of 50 Å in diameter and are separated from each other by the C-terminal 23 amino acids of the GroEL subunits. Each GroEL subunit is composed of three different domains: (1) an equatorial domain interacting with the second ring and harbouring an ATP binding side, (2) a small, flexible, but well-defined intermediate domain required for ATP hydrolysis and (3) an apical domain interacting with the GroES subunit (see Table 2 and Fig. 3). The apical domains of GroEL expose a set of hydrophobic amino acid residues towards the inner surface of the cavity, offering binding sides for partially unfolded proteins. This part of GroEL is the most flexible (Fig. 3). If ATP binds to the ring occupied by the misfolded polypeptide (cis-position), the apical domain is twisted by 90° and lifted from the equatorial domain to increase the size of the inner cavity to more than 80Å (Fig. 3A). Target proteins bound to the hydrophobic surfaces may therefore undergo further mechanical unfolding (Hartl 1996).

Table 2. Core components of the chaperonin (GroE) system

Eukaryotes	Prokaryotes	Function
TRiC/CCT	GroEL	Forms a double ring with 7 (in prokaryotes) or 8 (in eukaryotes) subunits each, providing 2 central cavities for binding misfolded proteins
n.d.	GroES	Forms a lid out of 7 subunits, closing the central cavity of GroEL

Twisting of the equatorial domain buries the hydrophobic sites of the interacting apical domains in two steps: after the first step, the GroES-heptamer is able to bind to the up-shifted apical domain of GroEL (Fig. 3B). The GroES-heptamer forms a dome-like flexible structure. Although it contains a central orifice on top, it covers the inner ring cavity of the GroEL subunit. Therefore the unfolded protein is entrapped and protected in a cage formed by the GroEL ring and covered by the GroES heptamer (Fig. 3C). Binding of GroES is a prerequisite for ATP hydrolysis by the equatorial domains of GroEL. In the second part of conformational changes, remaining hydrophobic residues of the apical GroEL domain disappear. The unfolded protein is released and refolds in the cavity without further interactions with GroEL. In the last part of the cycle, ATP binding to the second ring of the GroEL heptamer (trans) allows removal of (cis) GroES and release of the (re)folded protein (Fig. 3D), which if still incorrectly folded, might enter into a new cycle (Rye et al. 1997). Since there are two identical rings of GroEL, one can easily imagine two simultaneously operating folding events in each ring coupled in their timing by binding, cleavage and release of ATP (Sparrer et al. 1997). However, folding within the GroE-cage is a time consuming process, requiring 15–30 s (Thirumalai and Lorimer 2001).

Two important questions remain to be answered: (1) what type of proteins are folded by the GroEL/S complex and (2) what is the polypeptide flux through this system under normal and stress conditions? Ewalt et al. (1997) found that only a limited number of proteins in *E. coli* interact with the GroEL/S system, i.e. roughly 15% under normal and about 30% under stress conditions. Usually, as mentioned above, these proteins leave the chaperone after 10–30 s, representing one round of GroE binding and release. The authors distinguish between three different groups of polypeptides: (1) a group of small proteins neither requires nor interacts with GroEL/S; (2) a second group, representing the majority of *E. coli* proteins, can bind to the chaperone, but usually folds without GroE contact; (3) only a small set of proteins between 25 and 70 kDa in size really needs interaction

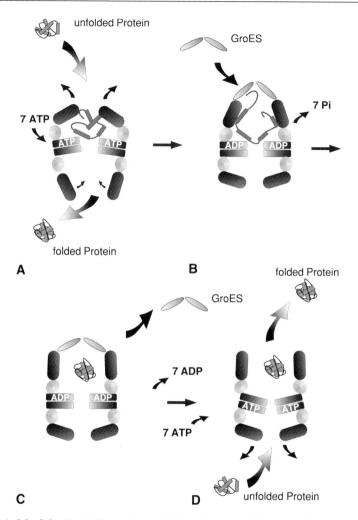

Fig. 3. Model of the GroEL/S complex with bound protein. The GroE-chaperone acts as a two-stroke motor, providing two independent folding cages. Folding occurs in both cavities with an ADP/ATP-exchange dependent delay in time. One complete folding cycle in the upper (cis) cavity is shown. For the trans cavity, only the uptake and release of the substrate protein is given. **A** Unfolded proteins exhibiting hydrophobic amino acids enter the central cavity of the GroEL ring and interact with the inner wall of the ring, formed by hydrophobic amino acid residues of the GroEL apical domain facing the inner GroEL cavity. **B** After binding of seven ATP molecules to the equatorial domains of GroEL, hydrophobic residues disappear from the cavity in two steps with conformational changes in the apical domain. The apical domain is lifted upwards. As a result GroES, providing a lid for the central cavity, can bind to GroEL. In parallel, the apical domain is twisted, resulting in an exchange of the internal hydrophobic residues in favour of more hydrophilic amino acids. **C** As a result, the substrate protein is released from the surface of the inner cavity. Trapped and protected in the cage formed by GroEL/GroES, the polypeptide is now able to (re-)fold. **D** Binding of ATP to the opposite GroEL heptamer (trans) leads to a tilt of the equatorial domains of both GroEL rings. The resulting conformational change releases the GroES lid in cis position allowing the newly folded substrate protein to leave the cavity

with the GroE system. Usually these proteins require several rounds of folding before they reach their native state (Saibil 2000).

Organelles of eukaryotic cells contain chaperone systems similar in structure and function to the prokaryotic GroEL/S, sometimes also referred as "chaperonins". (Viitanen et al. 1995). They belong together with the GroE system to the group I chaperonins. In addition, there is a much more specialised form, referred as group II in the cytosol, called CCT or TRiC (reviewed by Hendrick and Hartl 1993; Hartl 1996). Compared with the group I chaperonins, these complexes have a much lower abundance. They seem to be only needed for a specialised set of proteins such as tubulin (Tina et al. 1996) and actin (Vine and Durban 1994). The CCT/TRiC complex of 970 kDa contains two rings with up to eight different, but structurally related subunits of 52–65 kDa. Though a cofactor similar to GroES is not known, the overall mechanism of folding may be very similar to the GroEL/S complex (Farr et al. 1997). Interestingly, it could be shown that, similar to the situation in prokaryotic cells, Hsp70/Hsp40 and TRiC act together *in vivo*, demonstrated for actin and luciferase as reporter (Frydman and Hartl 1996).

4.3 The Hsp90 (HtpG) chaperone system

Hsp90 belongs to the most abundant constitutively expressed cytosolic chaperones in eukaryotic cells (e.g. Hendrick and Hartl 1993; Buchner 1999). However, its mode of action is far more complex than either GroE or Hsp70. It includes binding to Hsp70 and requires several cofactors. Cytosolic Hsp90 exits in two isoforms (α and β) and its synthesis increases considerably under heat stress conditions. Hsp90 homodimers are able to prevent aggregation of denatured reporter protein *in vitro* (Freeman and Morimoto 1996).

One important peculiarity of Hsp90 is that its chaperone activity can be inhibited by geldanamycin and related ansamycin compounds, described as antitumor antibiotics (Whitesell et al. 1994). This is particular interesting, since Hsp90 is reported to be especially involved in folding of proteins, which have key functions during cell proliferation (Buchner 1999). Hsp90 has therefore become more and more important as a drug target in cancer research (e.g. Beliakoff and Whitesell 2004). The number of proteins known to require Hsp90 for folding is increasing, but a common feature, such as a structural element or sequence-motif targeting proteins to the Hsp90 complex, has not been found yet. It is tempting to speculate that substrate specificity is accomplished by the different co-factor composition of the Hsp90 chaperone complex. The long-lasting discussion about the ATP requirement of Hsp90 (Jakob et al. 1995, 1996) was finally solved (Prodromou et al. 1997; Scheibel et al. 1998). Crystal structure analysis of the Hsp90 N-terminal domain clearly identified the ATP and therefore the geldanamycin binding site: It turns out that ATP binding by Hsp90 is different from other common ATP-binding domains. ATP is incorporated in an unusually kinked formation. Evidently, geldanamycin does not bind to the peptide binding domain as suggested previously (Stebbins et al. 1997), but competes with ATP for this untypical nucleotide binding side. Interestingly, Hsp90 has a second C-terminal chaperone domain, which needs no ATP. Scheibel et al. (1998) present evidence that the N-terminal domain functions in stabilisation of denatured proteins, whereas the C-terminal domain interacts with partially structured substrates, e.g. intermediates of protein de novo synthesis.

Besides its ability to keep target proteins in a folding-competent state, Hsp90 plays an important role in signal transduction pathways. This has been demonstrated in detail for the maturation and activation of Ser/Thr- and Tyr-specific protein kinases and for the maturation and functional cycling of steroid hormone receptors (Cutforth and Rubin 1994; Dey et al. 1996; Nair et al. 1996, 1997; Schulte et al. 1996; Stepanova et al. 1996; Chang et al. 1997). Both processes require interaction with the Hsp70 system as exemplified by the steroid hormone receptors (Fig. 4). The interaction between the Hsp70 and Hsp90 chaperone systems is mediated by two proteins, HIP and HOP (Frydman and Höhfeld 1997; Odununga et al. 2004). Subsequently, Hsp90 complex interacts with at least three other proteins, namely the peptidyl-prolyl isomerases FKBP52 and CYP40 and a small acidic protein, p23. FKBP52 and p23 were reported to act as molecular chaperones themselves (Bose et al. 1996; Davies et al. 2005; see Table 3). After release of Hsp70 the receptor is competent to bind the steroid ligand. Final maturation and acquisition of DNA binding competence is achieved upon dissociation of the hormone receptor from Hsp90.

As mentioned above, the Hsp90 complex might also interact with other co-factors for the sake of substrate specificity, like p50/Cdc37 instead of FKBP52/p23, thus consequently binding other substrate proteins. This type of interaction is predominantly involved in the maturation and functional control of protein kinases (Dey et al. 1996; Owens-Grillo et al. 1996; Hunter and Poon 1997). To date, little is known about the factors regulating the association with Hsp90. Some of them, like HOP, share a docking module, the tetratricopeptide repeat (TRP) motif consisting of an array of helices forming a grove in which extended peptide sequences might bind (Das et al. 1998). HOP consists mainly of TPR motifs, in agreement with its function of bringing Hsp70 and Hsp90 together (Chen and Smith 1998).

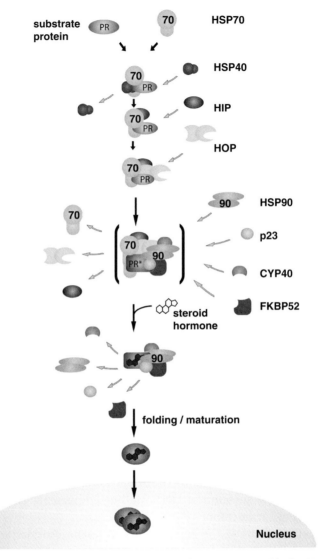

Fig. 4. Maturation of the steroid hormone receptors by interaction with the Hsp70 and Hsp90 chaperone system. Premature steroid receptor protein (PR) binds to the Hsp70/Hsp40 chaperone system. After ATP hydrolysis HIP (Hsp70 interacting protein) binds to the N-terminal domain of Hsp70. The subsequent association of HOP (p60) is evidently required for binding of Hsp90 and subsequently a complex containing CYP40, FKBP52 and a small acidic protein, p23, as additional subunits. At this time point Hsp70, HIP and p60/HOP are released from this complex. The receptor (PR*) is now competent for steroid binding. After release of the Hsp90 components, the steroid receptor complex undergoes dimerisation, enters the nucleus and binds its target promoter sequences of steroid-hormone regulated genes

Table 3. Factors involved in Hsp70/Hsp90 interaction during steroid receptor maturation

Protein	Function
Hsp70/Hsp40	Receptor binding; ATP-hydrolysis
HIP	Stabilising the ADP-bound Hsp70; interacts with the ATPase domain; prerequisite for HOP binding
HOP (p60)	Mediator for Hsp90 interaction; interacts with the peptide binding domain of Hsp70
Hsp90 (HtpG)	Chaperone
p23	Involved in ATP binding of Hsp90; has *in vitro* chaperone activity
CYP40	Proly-peptidyl isomerase
FKBP52	Proly-peptidyl isomerase with chaperone activity

It is worth mentioning that Hsp90 is currently discussed to be a kind of "capacitor of evolution" (Rutherford and Lindquist 1998; Queitsch et al. 2002). In both cases, it could be shown that in *Drosophila* as well as in *Arabidopsis* mutations remain phenotypically silent, since mutational amino acid exchange can be compensated by chaperone activity, unless the population is exposed to a stressful situation or a chaperone malfunctions, as shown for Hsp90. As a consequence, in both cases mutations become manifest. This raises more possibilities for adaptation and thus potentially for overcoming the stressful situation for certain individuals in the population.

4.4 Hsp 100 (Clp) family

Members of the Hsp 100 family were first described as components of the dimeric bacterial Clp protease system (Gottesman et al. 1990). The large Clp-subunit represents an ATP-dependent protein complex, which is able to solubilise aggregated proteins, whereas the small subunit, termed ClpP, acts as a protease. The large Clp-subunit alone has no proteolytic activity. Many Clp-related proteins were characterised in bacteria and eukaryotes as stress-induced proteins and are collectively placed in the Hsp100 family (Schirmer et al. 1996; Gottesman et al. 1997; Dougan et al. 2002; Mogk and Bukau 2004). The proteolytic subunit (ClpP) is only found in bacteria, associated mainly or exclusively with the ClpA protein. The Clp-proteins with "unfoldase" activity belong to a so-called AAA$^+$-ATPase. They form large hexameric rings in the presence of ATP (Beuron et al. 1998; Bochtler et al. 2000; Mogk et al. 2003a; Lee et al. 2003). They can be divided into two subgroups, ClpA and ClpB, also referred as Hsp104 in yeast and Hsp101 in *Arabidopsis* (Sanchez and Lindquist 1990; Squires et al. 1991; Queitsch et al. 2000). Both classes share their capability to promote dissociation of aggregated proteins in an ATP-dependent manner (Fig. 5). Although it was demonstrated that Hsp100 can functionally substitute for the Hsp70/40 system (Wickner et al. 1994; Schmitt et al. 1995), it is very likely that Hsp70 and Hsp100 proteins contribute to stress tolerance

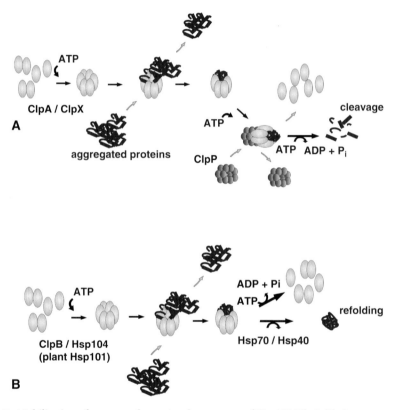

Fig. 5. Mobilisation of aggregated proteins the presence of Hsp100/ClpA. Under severe stress conditions with massive accumulation of aggregated proteins, functionally active Hsp100 (ClpA or ClpB) hexamers are the only chaperone system able to dissolve and remove protein aggregates. At this point, retrieved polypeptides may face two different fates. **A** They can be cleaved by the ClpP protease. However, ClpP proteases were found exclusively in prokaryotes. In an ATP-dependent process polypeptides are resolubilised from protein aggregates by entering the central cavity of the ClpA-ring. After binding of ATP ClpP proteases are recruited and the target proteins undergo cleavage and degradation. **B** In case of ClpB or Hsp101 in plants (Hsp104 in yeast), proteins are also retrieved from pre-existing aggregates but can be refolded in concert with the Hsp70/Hsp40 chaperone system. In contrast to ClpA, proteins of the ClpB-family never recruit proteases

in different ways, with Hsp70 acting primarily by allowing refolding of mis-folded proteins and Hsp100 by removing aggregates already formed (Parsell et al. 1994). Based on the presence of one or two ATP binding domains of about 200 amino acid residues (the AAA⁺-domain), Hsp100 proteins were further divided into the class I-type (ClpA-D), which contain two ATP bind-ing domains, while class II proteins (ClpM, N, X and Y) contain only one ATP binding domain (reviewed by Schirmer et al. 1996). The length of the polypeptide varies between 75 and 100 kDa due to the size of non-conserved spacers between the two domains and some additional sequences at the C- and N-terminus. Functionally active Hsp100 proteins form a hexamer with 12 molecules of ATP bound to it. Structural analysis revealed an image of the complex, which looks somehow similar to the structure of GroE and/or TriC: Hsp100 oligomers form a donut-like structure with a central cavity of 2.5 Å (Lee et al. 2004).

Pak and Wickner (1997) were the first to present a model for refolding of a defined ClpA target protein in *E. coli*, the bacteriophage RepA dimer: the target protein attaches loosely to the ClpA hexamer with bound ATP. In a second step, the protein is tightly bound to the central cavity of the complex. After ATP hydrolysis, the complex dissociates and releases the refolded RepA monomer. At the moment, it is not fully understood at which time point refolding of the target protein occurs, either during ATP independent form-ing of the stable complex or, similar to the Hsp70 system, by ATP-dependent release of the target polypeptide. However, it became evident that in contrast to ClpA, which unfolds aggregates and sentences them to degradation together with ClpP, ClpB/Hsp101 is a bona fide chaperone which interacts with the Hsp70 system (Glover and Lindquist 1998; Goloubinoff et al. 1999; Mogk et al. 1999; Motohashi et al. 1999; Zolkiewski 1999; Weibezahn et al. 2004). Taken together, Hsp100 proteins in concert with the Hsp70-system have the remarkable ability to rescue proteins from an aggregated state.

4.5 Hsp20 (Ibp) family

From the plant physiologist's point of view, the low molecular weight Hsps (also termed the Hsp20- or sHsp-family) are the most intriguing class of molecular chaperons. Members of the Hsp20 family represent the prepon-derant Hsps among plants with more than 20 different members. In other organisms (i.e. insects, mammals), there are only two to four members, and in yeast only one representative of this Hsp family (reviewed by Vierling 1991; van Montfort et al. 2001a). The multiplicity and massive stress-induced abundance (up to 2% of the total cellular protein) of small Hsps led to speculation about their specific role for stress-induced protection of cellular structures, especially in sessile organisms such as plants. Kimpel and Key (1985) postulated that sHsps may serve as a protective matrix for heat or chemically damaged proteins. Meanwhile, it turned out that sHsps bind denatured proteins very efficiently, and current models propose that they function by preventing irreversible protein aggregation, thereby increasing the stress resistance of cells (van Montfort et al. 2001a). sHsps have a molecular mass between 15 and 30 kDa and usually exist in oligomeric complexes of 200–800 kD comprising up to 32 subunits depending on the organism and the sHsp involved. The first high resolution structure data was obtained from of an archaeal sHsp. It revealed a hollow sphere with openings to the inside (Kim et al. 1998). The C-terminal domain of the monomer is predominantly β-structured, while the structure of the N-terminal domain, which is believed to be responsible for substrate binding, is very heterologous among different sHsps and only partly resolved in its structure. The basic building block of the oligomeric structure is a dimer, which associates further to form a sphere. Meanwhile, the three-dimensional structure for a plant sHsp complex is also available, Hsp16,9 from *Triticum aestivum*. In contrast to its prokaryote counterpart, this protein forms a dodecameric disk (van Montfort et al. 2001b).

As mentioned, sHsps have a remarkable protein binding capacity compared to the molecular chaperones discussed above. They are able to bind a large number of non-native proteins, possibly up to one target protein per subunit of the oligomeric sHsp complex (Lee et al. 1997). Furthermore, there seems to be no restriction on the size of proteins that can be bound (Ehrnsperger et al. 1998). Another feature of the sHsps is that they form very large complexes of regular globular shape upon substrate binding (Haslbeck et al. 1999). The simultaneous binding of non-native proteins seems to be a prerequisite for efficient and stable complex formation. The specificity of this reaction is further highlighted by the finding that binding can be saturated at a defined ratio of non-native protein to sHsp. Although the oligomer is the dominant species at optimal temperature for the organism, sHsp oligomers are in rapid equilibrium with dissociated species, as revealed by subunit exchange (Bova et al. 2000, 2002; Sobott et al. 2002; Friedrich et al. 2004). These dynamic properties are likely to be important for sHsp function. The underlying mechanism, however, is still unresolved. Current models suggest that an sHsp dimer or another suboligomeric species is the active substrate-binding unit (Ehrnsperger et al. 1997; van Montfort et al. 2001a). sHsp/substrate interactions are proposed to involve hydrophobic contacts and a few studies have identified regions of sHsps that potentially interact with substrate (Lee et al. 1997; Sharma et al. 1998). The complexes which sHsps form with non-native proteins *in vitro* are remarkably stable (Lee et al. 1997; Ehrnsperger et al. 1998; Haslbeck et al. 1999). Members of the Hsp20 family were shown to prevent aggregation of thermally inactivated reporter proteins *in vitro*. The protective effect does not need any ATP (Horwitz 1992; Jakob et al. 1993; Lee et al. 1995, 1997; Plater et al. 1996; Ehrnsperger et al. 1997; Stromer et al. 2004). The *in-vitro* effects were complemented and extended by in-vivo studies using plant cells and firefly luciferase as reporter (Forreiter et al. 1997).

In contrast to other chaperone families, sequence homology among different sHsps is restricted to a so-called α-crystallin domain, a C-terminal region of about 30 amino acid residues in length found in all members of the Hsp20 family, including α-B-crystallin itself, which is also a major component of the vertebrate eye lens (Horwitz 1992). The latter is a common heat inducible protein of mammalian cells frequently associated with Hsp27 and additionally one major component of the vertebrate eye lens. As mentioned, plants are unusual among eukaryotes in that they express multiple sHsp gene families that appear to have evolved after the divergence of plants and animals. While in other organisms sHsps are found almost exclusively in the cytosol, plants express two cytosolic sHsps-classes and specific isoforms targeted to intracellular organelles. Five separate gene families encode mitochondrion, plastid, peroxisomal, nuclear and endoplasmic reticulum-localized sHsps, each with appropriate organelle targeting signals (Caspers et al. 1995; Waters et al. 1996; Scharf et al. 2001). The two cytoplasmic classes are referred to as sHsp class I and sHsp class II proteins, which share only approximately 50% identity in the α-crystallin domain. Interestingly, their functional difference in the cytosol remains enigmatic: Both classes prevent unfolded proteins from irreversible aggregation in the same manner (Löw et al. 2000). However, there are some lines of evidence indicating that the different cytosolic classes might have different properties in forming large, detergent and salt resistant complexes, termed plant heat stress granules. These stress granules were first described by Nover and co-workers in tomato suspension cultures (Nover et al. 1983, 1989). They appear transiently under elevated temperatures and other stress conditions, i.e. exposure to heavy metals or ethanol. They form granular structures up to 40 nm in size that contain both cytosolic sHsp classes and entrapped partially unfolded target proteins. While class I proteins fail to establish these structures alone, class II sHsps can only form atypically structures *in vivo* (Kirschner et al. 2000; Forreiter et al., unpublished data).

Although the non-native protein is not released spontaneously, these complexes are not dead-end traps for the unfolded protein. It has been demonstrated that during recovery, these aggregates disintegrate and a bound enzyme can be shifted back to the native state by adding a specific ligand that stabilizes the functional conformation of the protein (Ehrnsperger 1998). However, in contrast to other chaperones, no active-release mechanism has been detected so far.

The emerging picture of sHsps function is that they bind non-native proteins once large quantities of unfolded proteins are formed, for example, as a consequence of stress conditions or overexpression of proteins. Binding prevents the formation of large unstructured and non-functional aggregates and makes the subsequent refolding by Hsp70 or other potential ATP-dependent chaperone systems possible (Ehrnsperger et al. 1997; Lee et al. 1997; Veinger et al. 1998; Lee and Vierling 2000; Mogk et al. 2003b). A model explaining these observations is given in Fig. 6. This cooperation of different components of the cellular chaperone machinery allows two key properties of molecular chaperones, binding and folding to be separated in space and time.

4.6 Other proteins with chaperone function

One important common feature of all chaperones is that they are at least in parts structurally disordered to achieve a highly unspecific binding capacity for almost all conceivable denatured target proteins. This feature is common for proteins involved in RNA binding and folding too. Because of this, RNA binding and folding polypeptides are sometimes also referred as "RNA chaperones" (reviewed recently by Tompa and Csermely 2004). RNA chaperones are proposed to interact non-specifically with RNA and to promote RNA folding by either resolving non-native conformations or impeding their formation. *In vitro*, several proteins with diverse functions have been attributed RNA chaperone activity because of their ability to promote RNA strand annealing or displacement and/or stabilising rRNA to prevent the nucleic acid to fall into a kinetic folding trap (e.g. Semrad et al. 2004). However at the moment, the term "RNA chaperone" is not accepted by everybody working in the field. Additionally, a comprehensive presentation of this group of proteins would exceed the scope of this review. Interested readers to this quickly emerging field are therefore referred to Poole et al. (1998), Csermely (2000), Treiber and Williamson (2001) or Christofari and Darlix (2002).

In addition to the evolving RNA chaperone world, several other so far less characterised proteins are believed to act as molecular chaperones, but do not fit into the above-mentioned classes. These proteins have a more or less specialised function, for example, the Skp-protein, recently described as the "jellyfish chaperone" (Walton and Sousa 2004). This protein complex,the structure of which has been recently analysed, consists of a homo-trimer of a 17 kD subunit, which binds proteins destined for membrane association

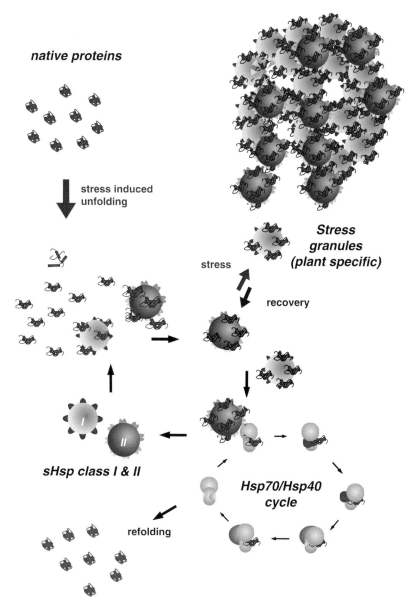

Fig. 6. Protein stabilisation by small stress proteins of the Hsp 20 family. Under stress conditions massive amounts of Hsp20 are synthesised. These form oligomers providing a non-specific surface for binding of partially denatured proteins. This protein complex can interact with Hsp70/40 for immediate refolding, releasing a functionally active protein. Alternatively, at least in plants under prolonged severe heat stress conditions, cellular aggregates were formed which trap partially unfolded proteins holding them in a folding-competent state. These complexes can be detected as large cytosolic complexes with approximately 40 nm in size (plant stress granules). Under recovery conditions, the complex disassembles, releasing Hsp20-oligomers, which subsequently can interact with the Hsp70-chaperone machine for refolding

and keeps them in a folding competent state unless they are integrated properly into the membranous bilayer. This, among others, may serve as an example for the still increasing number of proteins that are necessary to maintain the structural integrity of proteins either during nascency or under denaturing conditions to ensure proper cellular metabolism.

5 Résumé

During the last 2 decades, it has become evident that molecular chaperones are not only an essential component of protein folding, but also a crucial element in maintaining the native structure *in vivo* throughout the lifespan of the protein. Additionally and probably with dramatic impact on human cancer research, chaperones, especially Hsp90, emerge as a key element of post-translational control of protein activity, in particular the regulation of proteins controlling the cell cycle. Although several details in a rapidly increasing network of knowledge are still missing, the benefit in understanding the molecular mechanisms of chaperones in cancer and in stress research becomes more and more evident. It will therefore be exciting to follow this increase in knowledge within this field in the near future.

Acknowledgement. I would like to thank Gottfried Wagner and Jon Hughes for many helpful comments and for carefully reading the manuscript.

References

Alberti S, Esser C, Höhfeld J (2003) BAG-1—a nucleotide exchange factor of Hsc70 with multiple cellular functions. Cell Stress Chaperones 8:225–231
Anfinsen CB (1973) Principles that govern the folding of protein chains. Science 181:223–230
Anfinsen CB, Haber E, Sela M, White FH (1961) The kinetics of formation of native ribonuculease during oxidation of the reduced polypeptide chain. Proc Natl Acad Sci USA 58:1309–1314
Beliakoff J, Whitesell L (2004) Hsp90: an emerging target for breast cancer therapy. Anticancer Drugs 15:651–662
Beuron F, Maurizi MR, Belnap DM, Kocsis E, Booy FP, Kessel M, Steven AC (1998) At sixes and sevens: characterization of the symmetry mismatch of the ClpAP chaperone-assisted protease. J Struct Biol 123:248–259
Bochtler M, Hartmann C, Song HK, Bourenkov GP Bartunik HD, Huber R (2000) The structures of HslU and the ATP-dependent protease HslU–HslV. Nature 403:800–805
Bose S, Weikl T, Bügl H, Buchner J (1996) Chaperone function of Hsp90-associated proteins. Science 247:1715–1717
Bova MP, McHaourab HS, Han Y, Fung BKK (2000) Subunit exchange of small heat shock proteins—analysis of oligomer formation of α-A-crystallin and Hsp27 by fluorescence resonance energy transfer and site-directed truncations. J Biol Chem 275:10351042

Bova MP, Huang QL, Ding LL, Horwitz J (2002) Subunit exchange, conformational stability, and chaperone-like function of the small heat shock protein 16.5 from *Methanococcus jannaschii*. J Biol Chem 277:3846838475

Buchner J (1999) Hsp90 & Co.—a holding for folding. Trends Biochem Sci 24:136–141

Bukau B, Horwich AL (1998) The Hsp70 and Hsp60 chaperone machines. Cell 92:351–366

Bukau B, Hesterkamp T, Luirink J (1997) Growing up in a dangerous environment: a network of multiple targeting and folding pathways of nascent polypeptides in the cytosol. Trends Cell Biol 6:480–485

Caspers GJ, Leunissen JAM, De Jong WW (1995) The expanding small heat shock protein family and structure predictions of the conserved α-crystallin domain. J Mol Evol 40:238–248

Chang HJ, Nathan DF, Lindquist S (1997) *in vivo* analysis of the Hsp90 co-chaperone Sti1 (p60). Mol Cell Biol 17:318–325

Chen S, Smith DF (1998) Hop as an adaptor in the heat shock protein 70 (Hsp70) and hsp90 chaperone machinery. J Biol Chem 273:35194–35200

Ciechanover A (1998) The ubiquitin–proteasome pathway: On protein death and cell life. EMBO J 17:7151–7160

Christofari G, Darlix JL (2002) The ubiquitous nature of RNA chaperone proteins. Prog Nucleic Acid Res Mol Biol 72:223–268

Csermely P (2000) Proteins, RNAs and chaperones in enzyme evolution: a folding perspective. Trends Biochem Sci 22:147–149

Craig EA, Gambill BD, Nelson RJ (1993) Heat-shock proteins: molecular chaperones of protein biogenesis. Microbiol Rev 57:402–414

Craig EA, Yan W, James P (1999) In: Bukau B (ed) Molecular chaperones and folding catalysts. Harwood, Amsterdam, pp 139–162

Creighton TE (1975) The two-disulphide intermediates and the folding pathway of reduced pancreatic trypsin inhibitor. J Mol Biol 25:167–199

Creighton TE (1990) Protein folding. Biochem J 270:1–16

Cutforth T, Rubin G (1994) Mutations in Hsp83 and *CDC37* impair signaling by the sevenless receptor tyrosine kinase in *Drosophila*. Cell 77:1027–1036

Cyr DM, Langer T, Douglas MG (1994) DnaJ-like proteins: molecular chaperones and specific regulators of Hsp70. Trends Biochem Sci 19:176–181

Das AK, Cohen PW, Barford D (1998) The structure of the tetratricopeptide repeats of protein phosphatase 5:implications for TPR-mediated protein-protein interactions. EMBO J 17:1192–1199

Davies TH, Sanchez ER (2005) Fkbp52. Int J Biochem Cell Biol 37:42–47

Dey B, Caplan AJ, Boschelli F (1996) The Ydj1 molecular chaperone facilitates formation of active p60(v-src) in yeast. Mol Biol Cell 7:91–100

Dill KA (1990) Dominant forces in protein folding. Biochemistry 29:7133–715

Dobson CM (2004) Principles of protein folding, misfolding and aggregation. Semin Cell Dev Biol 15:3–16

Dougan DA, Mogk A, Zeth K, Turgay K, Bukau B (2002) AAA+ proteins and substrate recognition, it all depends on their partner in crime. FEBS Lett 529:6–10

Ehrnsperger M, Gräber S, Gaestel M, Buchner J (1997) Binding of non-native protein to Hsp25 during heat shock creates a reservoir of folding intermediates for reactivation. EMBO J 16:221–229

Ehrnsperger M, Hegersberg C, Wienhues U, Nichtl A, Buchner J (1998) Stabilization of proteins and peptides in diagnostic immunological assays by the molecular chaperone Hsp25. Anal Biochem 16:218–225

Ellis RJ, van der Vries SM (1991) Molecular chaperones. Annu Rev Biochem 60:321–347

Ewalt KL, Hendrick JP, Houry WA, Hartl F-U (1997) *In vivo* observation of polypeptide flux through the bacterial chaperonin system. Cell 90:491–500

Farr GW, Scharl EC, Schumacher RJ, Sondek S, Horwich AL (1997) Chaperonin-mediated folding in the eukaryotic cytosol proceeds through rounds of release of native and non-native forms. Cell 89:927–937

Flaherty KM, McKay DB, Kabsch W, Holmes KC (1991) Similarity of the three-dimensional structures of actin and the ATPase fragment of a 70-kDa heat shock cognate protein. Proc Natl Acad Sci USA 88:5041–5045

Forreiter C, Nover L (1998) Heat induced stress proteins and the concept of molecular chaperones. J Bioscience 23:287–302

Forreiter C, Kirschner M, Nover L (1997) Stable transformation of *Arabidopsis* cell suspension culture with firefly luciferase providing a cellular system for analysis of chaperone activity *in vivo*. Plant Cell 7:2171–2181

Freeman BC, Morimoto RI (1996) The human cytosolic molecular chaperones Hsp90 (HSC70) and HDJ-1 have distinct roles in recognition of a non-native protein and protein refolding. EMBO J 15:2969–2979

Friedrich KL, Giese K, Buan NR, Vierling E (2004) Interactions between small heat shock protein subunits and substrate in small heat shock protein/substrate complexes. J Biol Chem 279:1080–1089

Frydman JE, Hartl F-U (1996) Principles of chaperone-assisted protein folding: Differences between *in vitro* and *in vivo* mechanisms. Science 272:1497–1502

Frydman JE, Höhfeld J (1997) Chaperones get in touch: the Hip-Hop connection. Trends Biochem Sci 22:87–92

Glover JR, Lindquist S (1998) Hsp104, Hsp70, and Hsp40:a novel chaperone system that rescues previously aggregated proteins. Cell 94:73–82

Goloubinoff P, Mogk A Zvi, AP Tomoyasu, T, Bukau B (1999) Sequential mechanism of solubilization and refolding of stable protein aggregates by a bichaperone network. Proc Natl Acad Sci USA 96:13732–13737

Gottesman S, Squires C, Pichersky E, Carrington M, Hobbs M, Mattick JS (1990) Conservation of the regulatory subunit for the Clp ATP-dependent protease in prokaryotes and eukaryotes. Proc Natl Acad Sci USA 87:3513–3517

Gottesman S, Wickner S, Maurizi MR (1997) Protein quality control: triage by chaperones and proteases. Genes Develop 11:815–823

Hartl F-U (1996) Molecular chaperones in cellular protein folding. Nature 381:571–580

Haslbeck M, Walke S, Stromer T, Ehrnsperger M, White HE, Chen S, Saibil HR, Buchner J (1999) Hsp26: a temperature-regulated chaperone. EMBO J 18:6744–6751

Hendrick JP, Hartl F-U (1993) Molecular chaperone functions of heat stress proteins. Annu Rev Biochem 62:349–384

Höhfeld J, Jentsch S (1997) GrpE-like regulation of the HSC70 chaperone by the anti-apoptotic protein BAG-1. EMBO J 16:6209–6216

Höhfeld J, Minami Y, Hartl F-U (1995) Hip, a novel co-chaperone involved in the eukaryotic Hsc70/Hsp40 reaction cycle. Cell 83:589–598

Horst M, Azem A, Schatz G, Glick BS (1997) What is the driving force for protein import into mitochondria. Biochim Biophys Acta–Bioenerg 1318:71–78

Horwitz J (1992) α-Crystallin can function as a molecular chaperone. Proc Natl Acad Sci USA 89:10449–10453

Hunt JF, Weaver AJ, Landry SJ, Gierasch L, Deisenhofer J (1996) The crystal structure of the GroES co-chaperonin at 2.8Å resolution. Nature 379:37–45

Hunter T, Poon RYC (1997) Cdc37–a protein kinase chaperone. Trends Cell Biol 7:157–161

Jaenicke R (1987) Folding and association of proteins. Prog Biophys Mol Biol 49:117–237

Jaenicke R, Seckler R (1997) Protein misassembly *in vitro*. Adv Protein Chem 50:1–59

Jakob U, Gaestel M, Engel K, Buchner J (1993) Small heat shock proteins are molecular chaperones. J Biol Chem 268:1517–1520

Jakob U, Lilie H, Meyer I, Buchner J (1995) Transient interactions of Hsp90 with early unfolding intermediates of citrate synthase. Implications for heat shock *in vivo*. J Biol Chem 270:7288–7294

Jakob U, Scheibel T, Bose S, Reinstein J, Buchner J (1996) Assessment of the ATP binding properties of Hsp90. J Biol Chem 271:10035–10041

Johnson JL, Craig EA (1997) Protein folding *in vivo*: unraveling complex pathways. Cell 90:201–204

Kauzmann W (1959) Adv Protein Chem 14:1–63

Kelly WL, Georgopoulos C (1997) The T/t common exon of simian virus 40, JC, and BK polyoma virus T antigen can functionally replace the j-domain of the *Escherichia coli* DnaJ molecular chaperone. Proc Natl Acad Sci USA 94:3679–3684

Kiefhaber T, Rudolph R, Kohler HH, Buchner J (1991) Protein aggregation *in vitro* and *in vivo*: a quantitative model of the kinetic competition between folding and aggregation. Biotechnology 9:825–829

Kim KK, Kim R, Kim SH (1998) Crystal structure of a small heat-shock protein. Nature 394:595–599

Kim PS, Baldwin RL (1992) Specific intermediates in the folding reactions of small proteins and the mechanism of protein folding. Annu Rev Biochem 51:459–489

Kimpel JA, Key J (1985) Heat shock in plants. Trends Biochem Sci 10:353–357

Lee GH, Pokala N, Vierling E (1995) Structure and *in vitro* chaperone activity of cytosolic small heat shock proteins from pea. J Biol Chem 270:10432–10438

Lee GH, Roseman AM, Saibil HR, Vierling E (1997) A small heat shock protein stably binds heat-denatured model substrates and can maintain a substrate in a folding-competent state. EMBO J 16:659–671

Lee GJ, Vierling E (2000) A small heat shock protein cooperates with heat shock protein 70 systems to reactivate a heat-denatured protein. Plant Physiol 122:189197

Lee S, Sowa ME, Watanabe Y, Sigler PB, Chiu W, Yoshida M, Tsai FTF (2003) The structure of ClpB: a molecular chaperone that rescues proteins from an aggregated state. Cell 115:229–240

Lee S, Sowa ME, Choi JM, Tsai FTF (2004) The ClpB/Hsp104 molecular chaperone—a protein disaggregating machine. J Struc Biol 146:99–105

Löw D, Brändle K, Nover L, Forreiter C (2000) Cytosolic heat-stress proteins Hsp17.7 class I and Hsp17.3 class II of tomato act as molecular chaperones *in vivo*. Planta 211:575–82

Lubeck J, Heins L, Soll J (1997) Protein import into chloroplasts. Physiol Plant 100:53–64

Mogk A, Bukau B (2004) Molecular chaperones: structure of a protein disaggregase. Curr Biol 14:78–80

Mogk A, Tomoyasu T, Goloubinoff P, Rüdiger S, Röder D, Langen H, Bukau B (1999) Identification of thermolabile *Escherichia coli* proteins: prevention and reversion of aggregation by DnaK and ClpB. EMBO J 18:6934–6949

Mogk A, Schlieker C, Strub C, Rist W, Weibezahn J, Bukau B (2003a) Roles of individual domains and conserved motifs of the AAA⁺ chaperone ClpB in oligomerization, ATP hydrolysis, and chaperone activity. J Biol Chem 278:17615–17624

Mogk A, Schliecker C, Friedrich KL, Schöfeld HJ, Vierling E, Bukau B (2003b) Refolding of substrates bound to small Hsps relies on a disaggregation reaction mediated most efficiently by ClpB/DnaK. J Biol Chem 278:31033–31042

Morimoto RI, Tissieres A, Georgopoulos C (1994) The biology of heat shock proteins and molecular chaperones. Cold Spring Harbor Laboratory Press, Cold Spring Harbor, N.Y.

Motohashi K, Watanabe Y Yohda M, Yoshida M (1999) Heat-inactivated proteins are rescued by the DnaK/J-GrpE set and ClpB chaperones. Proc Natl Acad Sci USA 96:7184–7189

Nair SC, Toran EJ, Rimerman RA, Hyermstad S, Smithgall TE, Smith DF (1996) A pathway of multi chaperone interactions common to diverse regulatory proteins: estrogen receptor, Fes tyrosine kinase, heat shock transcription factor, and the aryl hydrocarbon receptor. Cell Stress Chaperones 1:237–250

Nair SC, Rimerman RA, Toran EJ, Chen S, Prapapanich V, Butts RN, Smith DF (1997) Molecular cloning of human FKBP51 and comparison of immunophilin interaction with Hsp90 and progesterone receptor. Mol Cell Biol 17:594–603

Neupert W (1997) Protein import into mitochondria. Annu Rev Biochem 66:863–917

Nover L, Scharf K-D (1997) Heat stress proteins and transcription factors. Cell Mol Life Sci 53:80–103

Nover L, Scharf K-D, Neumann D (1983) Formation of cytoplasmatic heat shock granules in tomato cell cultures and leaves. Mol Cell Biol 3:1648–1655

Nover L, Scharf K-D, Neumann D (1989) Cytoplasmatic heat shock granules are formed from precursor particles and are associated with a specific set of mRNAs. Mol Cell Biol 9:1298–1308

Odununga OO, Longshaw VM, Blatch GL (2004) Hop: more than an Hsp70/Hsp90 adaptor protein. Bioessays 26:1058–1068

Owens-Grillo JK, Czar MJ, Hutchinson KA, Hoffman K, Perdew GH, Pratt WB (1996) A model of protein targeting mediated by immunophilins and other proteins that bind to Hsp90 via tetratricopeptide repeat domains J Biol Chem 271:13468–13475

Pak M, Wickner S (1997) Mechanism of protein remodeling by ClpA chaperone. Proc Natl Acad Sci USA 94:4901–4906

Parsell DA, Lindquist S (1993) The function of heat stress proteins in stress tolerance—degradation and reactivation of damaged proteins. Annu Rev Genet 27:437–496

Parsell DA, Kowal AS, Singer MA, Lindquist S (1994) Protein disaggregation mediated by heat shock protein 104. Nature 372:475–478

Plater ML, Goode D, Crabbe MJC (1996) Effects of side-directed mutations on the chaperone-like activity of alpha-B-crystallin. J Biol Chem 271:28558–28566

Poole AM, Jeffares DC, Penny D (1998) The path from the RNA world. J Mol Evol 46:1–17

Prodromou C, Roe SM, O'Brien R, Ladbury JE, Piper PW, Pearl LH (1997) Identification and structural characterization of the ATP/ADP-binding side in the Hsp90 molecular chaperone. Cell 90:65–75

Queitsch C, Hong SW, Vierling E, Lindquist S (2000) Heat shock protein 101 plays a crucial role in thermotolerance in *Arabidopsis*. Plant Cell 12:479–492

Queitsch C, Sangster TA, Lindquist S (2002) Hsp90 as a capacitor of phenotypic variation. Nature 417:618–624

Rassow J, Maarse AC, Krainer E, Kubrich M, Müller H, Meijer M, Craig EA, Pfanner N (1994) Mitochondrial protein import–biochemical and genetic evidence for interaction of matrix Hsp70 and the inner membrane protein MIM44. J Cell Biol 127:1547–1556

Ritossa FM (1964) Chromosome puffs in *Drosophila* induced by ribonuclease. Science 145:513–514

Rutherford SL, Lindquist S (1998) Hsp90 as capacitor of morphological evolution. Nature 396:336–342

Rye HS, Burston SG, Fenton WA, Beechem JM, Xu Z, Siegler PB, Horwich AL (1997) Distinct actions of cis and trans ATP within the double ring of the chaperonin GroEL. Nature 388:792–798

Saibil H (2000) Molecular chaperones: containers and surfaces for folding, stabilising or unfolding proteins. Curr Opin Struct Biol 10:251–258

Sanchez Y, Lindquist S (1990) Hsp104 required for induced thermotolerance. Science 248:1112–1115

Scharf KD, Siddique M, Vierling E (2001) The expanding family of small Hsps and other proteins containing an α-crystallin domain. Cell Stress Chaperones 6:225–237

Scheibel T, Weikl T, Buchner J (1998) Two chaperone sites in Hsp90 differing in substrate specificity and ATP dependence. Proc Natl Acad Sci USA 95:1495–1499

Schirmer EC, Glover JR, Singer MA, Lindquist S (1996) Hsp100/Clp proteins: a common mechanism explains diverse functions. Trends Biochem Sci 21:289–296

Schmid FX (1993) Prolyl isomerases: enzymatic catalysis of slow protein-folding reactions. Annu Rev Biophys Biomol Struc 22:123–1443

Schmid FX (1995) Protein folding: prolyl isomerases join the fold. Curr Biol 5:993–994

Schmitt M, Neupert W, Langer T (1995) Hsp78, a Clp homologue within mitochondria can substitute for chaperone functions of mt-Hsp70. EMBO J 14:3434–3444

Schulte TW, Blagosklonny MV, Rommanova L, Mushinski JF, Monia BP, Johnston JF, Nguyen P, Trepel J, Neckers LM (1996) Destabilization of Raf-1 by geldanamycin leads to disruption of the Raf-1-MEK-mitogene-activated protein kinase signaling pathway. Mol Cell Biol 16:5839–5845

Semrad K, Green R, Schroeder R (2004) RNA chaperone activity of large ribosomal subunit proteins from *Escherichia coli*. RNA 10:1855–1860

Sharma KK, Kumar RS, Kumar GS, Quinn PT (2000) Synthesis and characterization of a peptide identified as a functional element in α-A-crystallin. J Biol Chem 275:3767–3771

Smith DF, Sullivan WP, Marion TN, Zaitsu K, Madden B, McCormick DJ, Toft DO (1993) Identification of a 60kDa stress related protein, p60, which interacts with Hsp90 and Hsp70. Mol Cell Biol 13:869–876

Sobott F, Benesch JLP, Vierling E, Robinson CV (2002) Subunit exchange of multimeric protein complexes—real-time monitoring of subunit exchange between small heat shock proteins by using electrospray mass spectrometry. J Biol Chem 277:38921–38929

Sondermann H, Schneider C, Höhfeld J, Hartl F-U, Moarefi I (2001) Structure of a Bag/Hsc70 complex: convergent functional evolution of Hsp70 nucleotide exchange factors. Science 291:1553–1557

Sondermann H, Ho AK, Listenberger LL, Siegers K, Moarefi I, Wente SR, Hartl F-U, Young JC (2002) Prediction of novel Bag-1 homologs based on structure/function analysis identifies Snl1p as an Hsp70 co-chaperone in *Saccharomyces cerevisiae*. Biol Chem 277:33220–33227

Sparrer H, Rutkat K, Buchner J (1997) Catalysis of protein folding by symmetric chaperone complexes. Proc Natl Acad Sci USA 94:1096–1100

Squires CL, Pedersen S, Ross BM, Squires C (1991) ClpB is the *Escherichia coli* heat shock protein F84.1. J Bacteriol 173:4254–4262

Stebbins CE, Russo AA, Schneider C, Rosen N, Hartl F-U, Pavletich NP (1997) Crystal structure of an Hsp90-geldanamycin complex: targeting of a protein chaperone by an antitumor agent. Cell 89:239–250

Stepanova L, Leng X, Parker SB, Harper JW (1996) Mammalian p50cdc37 is a protein kinase-targeting subunit of Hsp90 that binds and stabilizes CDK4. Genes Dev 10:1491–1502

Stromer T, Fischer E, Richter K, Haslbeck M, Buchner J (2004) Analysis of the regulation of the molecular chaperone Hsp26 by temperature-induced dissociation: the N-terminal domain is important for oligomer assembly and the binding of unfolding proteins. J Biol Chem 279:11222–11228

Takayama S, Sato T, Krajewski S, Kochel K, Irie S, Millan JA, Reed JC (1995) Cloning and functional analysis of BAG-1: a novel Bcl-2 binding protein with anti-cell death activity. Cell 80:279–284

Thirumalai D, Lorimer GH (2001) Chaperonin-mediated protein folding. Annu Rev Biophys Biomol Struct 30:245–269

Tina G, Huang Y, Rommelaere H, Vandekerkhove J, Ampe C, Cowan NJ, (1996) Pathway leading to correct folded β-tubulin. Cell 86:287–296

Tompa P, Csermely P (2004) The role of structural disorder in the function of RNA and protein chaperones. FASEB J 18:1169–1175

Treiber DK, Williamson JR (2001) Beyond kinetic traps in RNA folding. Curr Opin Struct Biol 11:309–314

Ungewickel E, Ungewickel H, Holstein SEH (1997) Functional interaction of the auxilin J domain with the nucleotide and substrate binding molecules of Hsc70. J Biol Chem 272:19594–19600

Van Gunsteren WF, Bürgi R, Peter C, Daura X (2001) The key to solving the protein-folding problem lies in an accurate description of the denatured state. Angew Chem 113:363–367

Van Montfort RLM, Basha E, Friedrich KL, Slingsby C, Vierling E (2001a) Crystal structure and assembly of a eukaryotic small heat shock protein. Nat Struc Biol 8:1025–1030

Van Montfort RLM, Slingsby C, Vierling E (2001b) Structure and function of the small heat shock protein/alpha-crystallin family of molecular chaperones. Adv Protein Chem 59:105–156

Veinger L, Diamant S, Buchner J, Goloubinoff P (1998) The small heat-shock protein IbpB from *Escherichia coli* stabilizes stress-denatured proteins for subsequent refolding by a multichaperone network. J Biol Chem 273:11032–11037

Vierling E (1991) The roles of heat shock proteins in plants. Annu Rev Plant Physiol Plant Mol Biol 42:579–620

Viitanen PV, Schmidt M, Buchner J, Suzuki T, Vierling E, Dickson R, Lorimer GH, Gatenby A, Soll J (1995) Functional characterization of the higher plant chloroplast chaperonins. J Biol Chem 270:18158–18164

Vine DB-N, Durban DG (1994) A yeast TCP-1 like protein is required for actin function *in vivo*. Proc Natl Acad Sci USA 91:9116–9120

Walton TA, Sousa MC (2004) Crystal structure of Skp, a prefoldin-like chaperone that protects soluble and membrane proteins from aggregation. Mol Cell 15:367–374

Walter S, Buchner J (2002) Molecular chaperones—cellular machines for protein folding. Angew Chem Int Ed 41:1098–1113

Waters ER, Lee GJ, Vierling E (1996) Evolution, structure and function of the small heat shock proteins in plants. J Exp Bot 47:325–338

Weibezahn J, Tessarz P, Schlieker C, Zahn R, Maglica Z, Lee S, Zentgraf HW, Weber-Ban EU, Dougan DA, Tsai FTF, Mogk A, Bukau B (2004) Thermotolerance requires refolding of aggregated proteins by substrate translocation through the central pore of ClpB. Cell: 119:653–665

Wetlaufer DB (1973) Nucleation, rapid folding, and globular intrachain regions in proteins. Proc Natl Acad Sci USA 70:697–701

Wickner S, Gottesman S, Skowyra D, Hoskins J, McKenney K, Maurizi MR (1994) A molecular chaperone, ClpA, functions like DnaK and DnaJ. Proc Natl Acad Sci USA 91:12218–12222

Whitesell L, Mimnaugh EG, De Costa B, Myers CE, Neckers LM (1994) Inhibition of heat shock protein 90-p60v-src heteroprotein complex formation by benzoquinone ansamycins: essential role for stress proteins in oncogenic transformation. Proc Natl Acad Sci USA 91:8324–8328

Xu Z, Horwich AL, Siegler PB (1997) The crystal structure of the asymmetric GroEL-GroES-(ATP)$_7$ chaperonin complex. Nature 388:741–750

Young JC, Agashe VR, Siegers K, Hartl F-U (2004) Pathways of chaperone-mediated protein folding in the cytosol. Nat Rev Mol Cell Biol 5:781–791

Zeiner M, Gebauer M, Gehring U (1997) Mammalian protein RAP46: an inter-action partner and modulator of 70 kDa heat shock protein. EMBO J 16:5483–5490

Zhu X, Zhao X, Burkholder WF, Gragerov A, Ogata CM, Gottesman ME, Hendrickson WA (1996) Structural analysis of substrate binding by the molecular chaperone DnaK. Science 272:1606–1614

Zolkiewski M (1999) ClpB cooperates with DnaK, DnaJ, and GrpE in suppressing protein aggregation. J Biol Chem 274:28083–28086

Christoph Forreiter
Pflanzenphysiologie
Justus-Liebig Universität
Senckenbergstr. 3
D-35390 Giessen
Germany
Tel.: +49 641 9935431
Fax: +49 641 9935429
e-mail: christoph.forreiter@bot3.bio.uni-giessen.de

Systematics

Recent progress in floristic research in Korea

Chong-Wook Park

1 Introduction

Korea is a peninsula on the far east coast of continental Asia. Physiographically, it is a mountainous peninsula extending southeast from the Manchurian mainland, and bounded on the north by the two rivers, Ap-rok-kang (Yalu) and Du-man-kang (Tumen). Floristically, it belongs to the eastern Asiatic floristic region, which comprises eastern Himalayas, northeastern India, northern Myanmar, most of continental China, Korea, and Japan (Takhtajan 1986). Within the eastern Asiatic floristic region, the northern part of Korea belongs to the Manchurian and the North Chinese province and the remaining parts to the Japanese-Korean province (Takhtajan 1986).

Although the Korean peninsula is relatively small in size (221,000 km^2), its flora is very rich in species composition because of its topographic and climatic complexities. Extending from 43°1′ N to 33°7′ N, it shows a remarkable variation in climatic conditions as to temperature and precipitation. At its southern extreme, the annual mean temperature is about 14°C, but it decreases progressively northward until it drops to 5°C. Annual precipitation also shows a similar trend, from 1400 mm on the southern coast to 500 mm in the northeastern inland. In addition, the complex mountain system that covers nearly 64% of the total area and ca. 3400 islands along the west and south coast create very diverse habitats throughout the region. The vegetation, therefore, shows a great deal of diversity, ranging from warm temperate in the southern part to cold temperate and alpine in the northern and high mountain regions.

Despite its richness in flora, Korea as a whole received little taxonomic attention in the past, and the progress in floristic work in Korea has been hampered by the lack of support and the inadequacy of field collections and literature. In the present paper, characteristics of the Korean vascular flora and recent progress in floristic research in Korea are discussed.

Progress in Botany, Vol. 67
© Springer-Verlag Berlin Heidelberg 2005

2 Taxonomic diversity and endemism

Korea has a high taxonomic diversity as compared to other countries in temperate regions with similar size. Approximately 3954 species and infraspecific taxa of vascular plants comprising 207 families and 1048 genera are currently distributed in Korea; of these plants, 250 are ferns, 64 are gymnosperms, 2803 are dicots and 837 are monocots (Table 1).

The major families of the Korean vascular plant flora include Asteraceae (261 species), Cyperaceae (211 species), Poaceae (196 species), Rosaceae (152 species), Fabaceae (120 species), Ranunculaceae (109 species), Aspidiaceae (106 species) and Liliaceae (104 species) (Fig. 1); these eight families comprise approximately 40% of the species found in Korea (Fig. 1). On the other hand, 72 families are represented by only one or two species.

At the genus level, the largest is *Carex* L. with 131 species, followed by *Polygonum* L. s. lat. (49 species), *Viola* L. (44 species), *Salix* L. (34 species), *Saussurea* DC. (30 species), *Athyrium* Roth. (29 species), *Artemisia* L. (26 species), and *Prunus* L. (24 species).

Although many of the species distributed in Korea are shared with Japan and northeastern China, including Manchuria, there are a number of unique taxa that are endemic to Korea. Six genera are strictly confined to Korea, which include *Mankyua* Sun et al. (Ophioglossaceae), *Megaleranthis* Ohwi (Ranunculaceae), *Pentactina* Nakai (Rosaceae), *Echinosophora* Nakai (Fabaceae), *Abeliophyllum* Nakai (Oleaceae), and *Hanabusaya* Nakai (Campanulaceae) (Park 1974; Sun et al. 2001). Some authors (Lee 1996a; Paik 1999) also regard *Coreanomecon* Nakai (Papaveraceae) and *Diplolabellum* Maekawa (Orchidaceae) as distinct endemic genera. The number of endemic species and varieties in this region is approximately 467

Table 1. Number of vascular plant taxa distributed in Korea

	Family	Genus	Species and Infraspecific taxon
Ferns	21	64	250
Gymnosperms	8	21	64
Angiosperms	178	963	3640
Dicots	(148)	(734)	(2803)
Monocots	(30)	(229)	(837)
Total	207	1048	3954

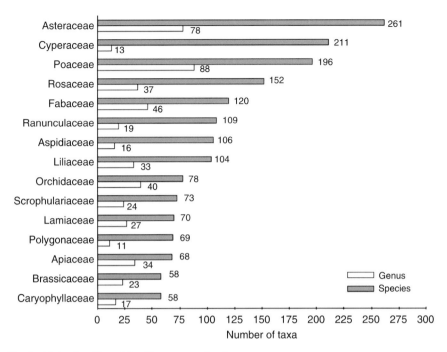

Fig. 1. Major families of the Korean vascular flora

(cf. Paik 1999); 14 in ferns, seven in gymnosperms, 388 in dicots, and 58 in monocots (Table 2). Within Korea, Gangwon and Jeju Provinces show the highest number of endemic taxa. Gyeongbuk Province also has relatively high endemism, but it is mainly due to Ulleung Island; it is a small, volcanic island about 150 km off the east coast and has many unique endemic taxa.

Table 2. Number of species and varieties endemic to Korea

	Species	Subspecies	Variety	Total
Ferns	11	–	3	14
Gymnosperms	4	–	3	7
Angiosperms	271	1	174	446
Dicots	(236)	(1)	(151)	(388)
Monocots	(35)	–	(23)	(58)
Total	86	1	180	467

The number of introduced (alien) vascular plant species in Korea appears to be 271, which constitute approximately 8.6% of the Korean vascular flora; they comprise 38 families and 155 genera, with great concentrations of species in Asteraceae (59 species), Poaceae (45 species), and Brassicaceae (26 species) (C. Park, unpublished data).

3 Floristic affinities

An analysis of the taxa common to Korea and adjacent regions on the basis of the published flora and checklists shows that Korea has close floristic affinities with northeastern China and Japan (C. Park, unpublished data). With northeastern China, Korea shares all of its 207 families, 95% of its genera, and 69% of its species and varieties. It also shares a large number of taxa with Japan; all families, 91% of the genera and 58(−66)% of the species found in Korea are also occur in Japan. As compared to China and Japan, Korea shares fewer taxa with Ussuri and Amur regions of Russia, and approximately 34% of the Korean species occur in this regions. These data indicate that Korea has close floristic relationships with China and Japan as Takhtajan (1986) noted, and occupies a vital position for the analysis and understanding of floristic patterns and relationships in eastern Asia.

4 Historical background of floristic research

Like many other countries in Asia, the earlier investigations on the Korean vascular flora were made by European botanists. The Korean vascular plants were first introduced to the western world by Miquel in his *Prolusio Florae Japonicae* (1865–1867). In this publication, he included some 50 species collected from the east coast of Korea by Schlippenbach in 1854. Since then, Korean plants have been collected from various regions by several European collectors and botanists, including Wilford, Carpenter, Oldham, Bunge, Taquet, and Faurie (cf. Chung 1984). These collections were later identified and published sporadically by European botanists, including Maximowicz (1866–1876), Forbes and Hemsley (1886–1905), Bunge (1893), Palibin (1898, 1900, 1901), and Léveillé (1902, 1903, 1904a,b, 1908, 1910a–e); many new taxa based on these collections were described in these works. However, these works embraced only a few hundred species collected from Korea, and Korea received surprisingly little taxonomic attention from European botanists as compared to Japan and China.

The first comprehensive Korean flora was published in two parts by Japanese botanist, Takenosin Nakai (1882–1952), under the title of *Flora Koreana* (pars 1, 1909; pars 2, 1911). In his *Flora Koreana*, Nakai included 1791 species and 405 infraspecific taxa representing 669 genera and 135 families, and described many new taxa from Korea comprising one genus, 24 species, ten varieties, and four formas. He also published many other important works on Korean plants, including *Flora Sylvatica Koreana* (Parts 1–22; 1915–1939). In 1952, just before he died, Nakai published *A Synoptical Sketch of Korean Flora*, and listed 3176 species and 1015 infraspecific taxa comprising 968 genera and 223 families; he had added more than 1300 species and 600 infraspecific taxa to the Korean flora in about 43 years following the publication of his *Flora Koreana* in 1909. In addition, he described a total of 1118 new taxa from Korea, comprising 642 species, 402 varieties and 74 formas (Paik 1999). However, many new taxa described by Nakai from Korea were mainly based on minor differences in one or two morphological characters, and their taxonomic status, in many cases, is questionable (Park 1974, Paik 1999).

The first illustrated flora of Korea was published in two volumes by Korean botanist, Tae-Hyun Chung (1882–1971), under the title of *Korean Flora*; volume 1. Woody plants (1957), accounted for 1636 species and infraspecific taxa in 324 genera and 95 families, and volume 2. Herbaceous plants (1956), accounted for 2050 species and infraspecific taxa in 666 genera and 154 families. In this work, a brief description and illustration for each species is provided, and it has been the basis of most later floristic works in Korea. The manuscript of this work was finished by him before the Korean War (1950–1953); therefore, he based this work on his extensive collections deposited in the Forest Research Institute at that time. However, many questionable taxa described by Nakai were included without critical reevaluation, and treatments of many major groups also need serious nomenclatural adjustments.

5 Herbarium collections

Collections of plant specimens are the very basis of floristic and systematic research. Unfortunately, however, most of the important collections and vouchers, including Chung's collection, deposited in Korean herbaria were completely destroyed during the Korean War in the early 1950s. Most of the early taxonomic literature in major Korean herbaria and libraries were also destroyed by the war. Furthermore, field trips to North Korea or exchange of specimens with North Korean institutions were no longer possible after the war.

Recent collections have been slowly increasing since then, and new herbaria have been built at several institutions. However, there are only six herbaria housing more than 100,000 specimens, and the biggest herbarium in Korea, the Herbarium of the Department of Biological Sciences of Seoul National University (SNU), currently has only about 300,000 specimens (Table 3). Majority of the other herbaria are much smaller and deal primarily with local plants or specific group(s) of plants (Table 3). Therefore, the unavailability of extensive collections and literature has been a major impediment to the progress in floristic and systematic researches in Korea.

Recognizing the importance of biological collections, including herbarium specimens, the Ministry of Environment recently obtained a budget to establish the "National Biological Resources Center" in Incheon. The center will contain modern research laboratories and biological collections preservation facilities including herbarium. The herbarium space will be approximately 2800 m^2, which can hold up to ca. 3.5 million specimens. The construction will be completed in 2007, and the center will become a core institution for national biodiversity surveys and inventories as well as for other systematic researches in Korea.

6 Current status of floristic research

Several floras with varying quality and completeness have been provided by the Korean botanists since 1960 (Table 4). Chung (1965) published a revised version of his earlier *Korean Flora* (1956, 1957) as volume 5. Tracheophyta of the *Illustrated Encyclopedia of Fauna and Flora of Korea*, a series published by the Ministry of Education of the Korean government.

Tchang Bok Lee (1919–2003) published his *Illustrated Woody Plants of Korea* in 1966 and *Illustrated Flora of Korea* in 1980. In the latter publication, he included 2901 species and 1053 infraspecific taxa within 1048 genera and 190 families (207 families in the Cronquist system) and provided a description and illustration for each taxon. In 2003, Lee issued a new edition of his *Illustrated Flora of Korea* under the title of *Coloured Flora of Korea* in two volumes.

Man-kyu Park (1906–1977) wrote a series of important publications on the Korean ferns, including *Flora of Korean Pteridophyta* (1961) and volume 16. Pteridophyta of the *Illustrated Encyclopedia of Fauna and Flora of Korea* (1975). In the latter publication, he recognized 272 species in Korea and provided keys, descriptions, photographs (or illustrations) and synonymy for the Korean ferns.

Another major work on the specific group of the Korean vascular plants includes Yong No Lee's *Manual of the Korean Grasses* (1966), which contained approximately 240 species and infraspecific taxa in 85 genera,

Table 3. Herbaria in Korea with more than 20,000 specimens

Herbarium	Acronym	Date established	Number of specimens
Dept. of Biological Sciences, Seoul National University	SNU	1953	300,000
Korea University	KUS	1953	120,000
College of Agriculture and Life Sciences, Seoul National University	SNUA	1953	100,000
Ewha Woman's University	EWU	1951	100,000
Chonbuk National University	JNU	1982	100,000
Sungshin Women's University	–	1976	100,000
Kangwon National University	KWNU	1973	80,000
National Institute of Environmental Research	–	2000	53,000
Inha University	IUI	1989	50,000
Kyungpook National University	KNU	1960	50,000
Sungkyunkwan University	SKK	1954	50,000
Yeungnam University	YNUH	1998	50,000
Korea National Arboretum	KH	2003	50,000
Chonnam National University	–	1990	45,000
Chungbuk National University	CBU	1988	40,000
Hannam University	HNHM	1983	30,000
Sunchon National University	–	1994	30,000
Ajou University	AJOU	1993	25,000
Hongnung Arboretum, Forest Research Institute	KFI	1962	25,000
Gyeongsang National University	GNUC	1992	23,000
Andong National University	ANH	1996	20,000
Kyung-Nam University	KNUH	1985	20,000
Daejon University	TUT	2003	20,000
Yong-In University	–	1995	20,000

together with keys to all taxa, descriptions, synonymy, and line drawings of representative floral parts. Recently, Yong No Lee wrote a floristic book entitled *Flora of Korea* that covers gymnosperms and angiosperms (edn 1, 1996; revised edn, 2002). In the revised edition of this work (2002), he included 2763 species and 930 infraspecific taxa in 986 genera and 168 families and

Table 4. Major floristic books on the Korean vascular flora published after 1950. All books except for one published by Y. N. Lee (1966) are in Korean

Publication date	Author	Title
1956	Chung, T. H.	Korean Flora. Vol. 2. Herbaceous Plants
1957	Chung, T. H.	Korean Flora. Vol. 1. Woody Plants
1961	Park, M.-K.	Flora of Korean Pteridophyta
1965	Chung, T. H.	Illustrated Encyclopedia of Fauna and Flora of Korea. Vol. 5. Tracheophyta
1966	Lee, T. B.	Illustrated Woody Plants of Korea
1966	Lee, Y. N.	Manual of the Korean Grasses (in English)
1974	Park, M.-K.	Keys to the Herbaceous Plants in Korea (Dicotyledoneae)
1975	Park, M.-K.	Illustrated Encyclopedia of Fauna and Flora of Korea. Vol. 16. Pteridophyta
1976	Lee, Y. N.	Illustrated Encyclopedia Fauna and Flora of Korea. Vol. 18. Flowering Plants
1980	Lee, T. B.	Illustrated Flora of Korea
1996a	Lee, W. T.	Lineamenta Florae Koreae
1996b	Lee, W. T.	Coloured Standard Illustrations of Korean Plants
1996	Lee, Y. N.	Flora of Korea (1st edn)
2002	Lee, Y. N.	Flora of Korea (Revised edn)
2003	Lee, T. B.	Coloured Flora of Korea (in 2 volumes)

provided a brief description and color photograph for each taxon; some new taxa from Korea mostly described by him were also included.

In 1996, Woo Tchul Lee simultaneously published two floristic books entitled *Lineamenta Florae Koreae* and *Coloured Standard Illustrations of Korean Plants*, respectively. In these works, Lee recognized 3129 species and 942 infraspecific taxa of vascular plants representing 1079 genera and 190 families. In particular, he provided rather extensive synonymy for those Korean taxa included in his *Lineamenta Florae Koreae*.

A number of regional or local floras also have been published recently; these include *Illustrated Flora of Jeju (= Cheju) Island* (Kim 1985), *Indigenous Plants of Gyeongsangbuk-do* (Kim et al. 1998), *Wild Plants of Jeju Island* (Lee et al. 2001), and *Wild Plants of Ulleung Island* (Hong et al. 2002).

7 Recent progress: the new flora of Korea project

The floristic works published so far after the Korean War significantly contributed to the documentation and understanding of the Korean flora. These works were, however, inevitably based on a limited number of specimens, since early collections were entirely lost during the war in the 1950s; in many cases, they contained somewhat limited information on characteristics, morphological variations, relationships, distribution and nomenclature of the Korean vascular plants. Therefore, a need for a new comprehensive vascular flora of Korea based on extensive fieldwork and critical revision of major groups was realized by many Korean botanists.

In 2000, a long-term project to produce a new comprehensive vascular flora of Korea was initiated by the Korean botanists. Fortunately, the 21st Century Frontier R & D Program of the Ministry of Science and Technology supported the initial three-year period of the new flora of Korea project (2000–2003), and the Eco-Technopia-21 project of the Ministry of Environment recently decided to provide funding for another seven years to complete the project. The new flora will be written in English and published in eight volumes with ten parts. It will contain keys for identification, synonymy with bibliographic references, descriptions of all taxonomic groups treated, distribution statements, comments on taxonomic problems, and economic uses. The work will be entirely new and the descriptions and treatments will be specimen-based, rather than extracted from the literature.

Considering the inadequacy of collections at herbaria in Korea (Table 3), extensive fieldwork is essential to the realization of the goal of the new flora project. Therefore, a portion of the funds will be used to support fieldwork by authors to collect throughout Korea, and also short-term visit by authors to major foreign herbaria in which the type specimens and historical vouchers of Korean species are housed.

The new Flora of Korea is on a very tight production schedule with two or three volumes being produced about every three years. Currently 65 Korean plant taxonomists are involved in the project and the three volumes are in preparation. The first book in the series entitled *The Genera of Vascular Plants of Korea* will appear shortly.

8 Conclusion

Plant taxonomy including floristics, in general, has been slowly but steadily expanding over the last 20 years in Korea. Our analysis shows that the total number of papers published by Korean taxonomists each year increases from about 13–23 in the 1980s to 63 in 2003; in particular, about one-third of those papers published in 2003 are carried in major foreign journals. When the Korean Society of Plant Taxonomists was founded in 1971 there were only 28 members, but now there are 235. New herbaria have been built at several institutions, and there is enthusiastic support among all members of the Korean botanical community for the new flora of Korea project. In particular, the new flora, the first to be written in the English language, will fill the last gap in temperate eastern Asia for English speakers, and will contribute to the understanding of the origin and nature of the Korean vascular flora.

Acknowledgments. This work was supported by a grant (no. 052-041-026) from the Core Environmental Technology Development Project for Next Generation funded by the Ministry of Environment of the Korean government. I am also grateful to Min Ha Kim and Tae Kwon Roh for their help during the preparation of this manuscript.

References

Bunge A (1893) Salsolaceae herbarii petropolitani in China, Japonica et Mandshuria collectae. Trudy Imp S-Peterburgsk Bot Sada 13:13–22

Chung TH (1956) Korean flora. Vol. 2. Herbaceous plants. Sinji-sa, Seoul, 1025 pp

Chung TH (1957) Korean flora. Vol. 1. Woody plants. Sinji-sa, Seoul, 507 pp

Chung TH (1965) Illustrated encyclopedia of fauna and flora of Korea. Vol 5. Tracheophyta. Ministry of Education, Korea, 1824 pp

Chung YH (1984) A cairn of taxonomic works in Korea. Seoul, 380 pp

Forbes FB, Hemsley WB (1886–1905) An enumeration of all the plants known from China proper, Formosa, Hainan, Corea, the Luchu Archipelago, and the Island of Hongkong, together with their distribution and synonymy. J Linn Soc Bot 23:1–521, 26:1–592, 36:1–686

Hong SC, Kim YW, Pak JH, Oh SH, Kim JS, Jang BT (2002) Wild plants of Ulleung Island. Ulleung-gun, 404 pp

Kim MH (1985) Illustrated flora of Jeju Island. Jeju-do, 617 pp

Kim YW, Pak JH, Hong SC, Choi K, Yoon CW, Oh SH (1998) Indigenous plants of Gyeongsangbuk-do. Grafica, 476 pp

Lee TB (1966) Illustrated woody plants of Korea. Forest Experiment Station, Seoul, 348 pp

Lee TB (1980) Illustrated flora of Korea. Hyang-moon-sa, Seoul, 990 pp

Lee TB (2003) Coloured flora of Korea (in 2 volumes). Hyang-moon-sa, Seoul, vol. 1: 914 pp, vol. 2: 910 pp

Lee WT (1996a) Lineamenta florae koreae (in 2 parts). Academy, Seoul, 2383 pp

Lee WT (1996b) Coloured standard illustrations of Korean plants. Academy, Seoul, 624 pp

Lee YN (1966) Manual of the Korean grasses. Ewha Womans University Press, Seoul, 300 pp

Lee YN (1996) Flora of Korea (1st edn). Kyo-Hak Publ Co, Seoul, 1237 pp

Lee YN (2002) Flora of Korea (revised edn). Kyo-Hak Publ Co, Seoul, 1265 pp

Lee YN, Lee KS, Sin YM (2001) Wild plants of Jeju Island. Yeomiji Botanical Garden, Jeju, 669 pp

Léveillé AAH (1902) Renonculacées de Corée. Bull Acad Int Géogr Bot 11:297–301

Léveillé AAH (1903) Onothéracées de Corée. Bull Acad Int Géogr Bot 12:17–18

Léveillé AAH (1904a) Cyperaceae (excl. Carices) Japonicae et Coreanae a R. P. Urb. Faurie lectae quas determinavit C. B. Clarke et edidit H. Léveillé. Bull Acad Int Géogr Bot 14:197–203

Léveillé AAH (1904b) Nouveautés Chinoises, Coréennes et Japonaises. Bull Soc Bot France 51:202–206, 217–220, 289–292, 422–424

Léveillé AAH (1908) Carices novae Coreanae. Repert Spec Nov Regni Veg 5:239–241

Léveillé AAH (1910a) Deux nouveaux *Bidens* de Corée. Bull Acad Int Géogr Bot 20:3

Léveillé AAH (1910b) Plantae Taquetianae Coreanae. Bull Acad Int Géogr Bot 20:4–11

Léveillé AAH (1910c) *Vitis* et *Eclipta* de Corée. Bull Acad Int Géogr Bot 20:11

Léveillé AAH (1910d) Clef des *Polygonum* de Chine et de Corée. Bull Soc Bot France 57:443–450

Léveillé AAH (1910e) Clef des *Artemisia* Chinois et Coréens. Bull Soc Bot France 57:456–459

Maximowicz CJ (1866–1876) Diagnoses plantarum novarum japoniae et mandshuriae. Decades 1-20. Bull Acad Imp Sci Saint-Pétersbourg 10:485–490; 11:429–432, 433–439; 12:60–73, 225–231; 15:225–232, 373–381; 16:212–226; 17:142–180, 417–456; 18:35–72, 275–296, 371–402; 19:158–186, 247–287, 475–540; 20:430–472; 22:209–264

Miquel FAW (1865–1867) Prolusio florae japonicae. Ann Mus Bot Lugduno-Batavi 2:69–212, 257–300, 3:1–66, 91–209

Nakai T (1909) Flora koreana. Pars prima. J Coll Sci Imp Univ Tokyo 26:1–304

Nakai T (1911) Flora koreana. Pars secunda. J Coll Sci Imp Univ Tokyo 31:1–573

Nakai T (1915–1939) Flora sylvatica koreana. Parts 1-22. Government of Chosen, Seoul, Korea

Nakai T (1952) A synoptical sketch of Korean flora. Bull Natl Sci Mus 31:1–152

Paik W-K (1999) The status of endemic plants in Korea and our tasks in the 21st century. Korean J Plant Taxon 29:263–274

Palibin IV (1898) Conspectus florae koreae. Pars prima. Trudy Imp S-Peterburgsk Bot Sada 17:1–128

Palibin IV (1900) Conspectus florae koreae. Pars secunda. Trudy Imp S-Peterburgsk Bot Sada 18:147–198

Palibin IV (1901) Conspectus florae koreae. Pars tertia. Trudy Imp S-Peterburgsk Bot Sada 19:101–151

Park M-K (1961) Flora of Korean Pteridophyta. Kyo-Hak Publ Co, Seoul, 353 pp

Park M-K (1974) Keys to the herbaceous plants in Korea (Dicotyledoneae). Chungeum-sa, Seoul, 593 pp

Park M-K (1975) Illustrated encyclopedia of fauna and flora of Korea. Vol. 16. Pteridophyta. Ministry of Education, Korea, 549 pp

Sun B-Y, Kim MH, Kim CH, Park C-W (2001) *Mankyua* (Ophioglossaceae): a new fern genus from Cheju Island, Korea. Taxon 50:1019–1024

Takhtajan AL (1986) Floristic regions of the world. University of California Press, Berkley, 522 pp

List of papers published by Korean taxonomists in 2003 and 2004

An BC, Hong SP (2003) Systematic application of seed morphology in Korean Orobanchaceae. Korean J Plant Taxon 33:411–420

Chang CS, Chang GS (2003) Reexamination on V. L. Komarov's collection sites in North Korea mainly based on Flora Manshuriae. Korean J Plant Taxon 33:203–218

Chang CS, Chang GS, Qin HN (2004) A multivariate morphometric study on *Corylus sieboldiana*. Acta Phytotax Sin 42:222–235

Chang CS, Choi BH (2004) Reexamination on V. L. Komarov's collection sites in North Korea (2) mainly based on Nakai's Flora koreana vol. 2. Korean J Plant Taxon 34:37–41

Chang CS, Choi BH, Kim H, Lee JY (2004) Reexamination on foreign collectors' sites and exploration routes in Korea -with respect to U. Faurie-. Korean J Plant Taxon 34:87–96

Chang CS, Choi H, Chang KS (2004) Reconsideration of *Prunus sargentii* complex in Korea with respect to *P. sargentii* and *P. takesimensis-*. Korean J Plant Taxon 34:221–244

Chang CS, Jeon JI (2003) Leaf flavonoids in *Cotoneaster wilsonii* (Rosaceae) from the island Ulleung-do, Korea. Biochem Syst Ecol 31:171–179

Chang CS, Jeon JI (2004) Foliar flavonoids of the most primitive group, sect. *Distegocarpus* within the genus *Carpinus*. Biochem Syst Ecol 32:35–44

Chang CS, Kim H (2003) Analysis of morphological variation of the *Acer tschonoskii* complex in eastern Asia: implications of inflorescence size and number of flowers within sect. *Macrantha*. Bot J Linn Soc 143:29–42

Chang CS, Kim H (2003) Reconsideration of nomenclature of Korean woody plants. J Korean Forest Soc 92:71–86

Chang CS, Kim H, Jeon JI (2003) Field record of Dr. Tchang-Bok Lee based on herbarium specimens deposited at SNUA. Korean J Plant Taxon 33:455–459

Chang CS, Kim H, Kang HS, Lee DK (2003) Morphometric analysis of *Kalopanax septemlobus* (Thunb.) Koidz. (Araliaceae) in Eastern Asia. Bot Bull Acad Sin 44:337–344

Chang CS, Kim H, Park TY (2003) Patterns of allozyme diversity in several selected rare species in Korea and implications for conservation. Biodiversity and Conservation 12:529–544

Choi BH, Ohashi H (2003) Generic criteria and an infrageneric system for *Hedysarum* and related genera (Papilionoideae-Leguminosae). Taxon 52:567–576

Choi HJ, Jang CG, Ko SC, Oh BU (2004) A taxonomic review of Korean *Allium* (Alliaceae). Korean J Plant Taxon 34:119–152

Choi HJ, Jang CG, Ko SC, Oh BU (2004) Leaf epidermal structure of the *Allium* L. and its taxonomic significance. Korean J Plant Taxon 34:97–118

Choi HJ, Jang CG, Ko SC, Oh BU (2004) Two new taxa of *Allium* (Alliaceae) from Korea: *A. koreanum* H. J. Choi et B. U. Oh and *A. thunbergii* var. *teretifolium* H. J. Choi et B. U. Oh. Korean J Plant Taxon 34:75–85

Choi HJ, Oh BU (2003) A new species of *Allium* sect. *Sacculiferum* (Alliaceae) from Korea: *A. linearifolium* H. J. Choi et B. U. Oh. Korean J Plant Taxon 33:71–78

Choi HJ, Oh BU (2003) Taxonomy of the *Allium* sect. *Sacculiferum* in Korea: with a special reference to the morphology. Korean J Plant Taxon 33:339–357

Choi HJ, Oh BU, Jang CG (2003) An unrecorded species of *Allium* (Alliaceae) in Korea: *A. longistylum* Baker. Korean J Plant Taxon 33:295–301

Chung GY, Oh BU, Park KL, Kim JH, Kim MS, Jang CG (2003) Taxonomic study of Korean *Euphorbia* L. (Euphorbiaceae) by leaf venation. Korean J Plant Taxon 33:135–149

Chung GY, Oh BU, Park KR, Kim JH, Kim MS, Nam GH, Jang CG (2003) Cytotaxonomic study of Korean *Euphorbia* L. (Euphorbiaceae). Korean J Plant Taxon 33:279–293

Chung MG (2003) Genetic structure of age classes in *Camellia japonica* (Theaceae). Evolution 58:62–73

Chung MY, Chung MG (2003) The breeding systems of *Cremastra appendiculata* and *Cymbidium goeringii*: High levels of annual fruit failure in two self-compatible orchids. Ann Bot Fenn 40:81–85

Chung MY, Nason JD, Chung MG (2004) Spatial genetic structure in populations of *Cephalanthera erecta* (Orchidaceae). Amer J Bot 91:92–97

Chung MY, Nason JD, Chung MG (2004) Spatial genetic structure in populations of the terrestrial orchid *Cephalanthera longibracteata* (Orchidaceae). Amer J Bot 91:52–57

Chung MY, Nason JD, Epperson BK, Chung, MG (2003) Temporal aspects of the fine-scale genetic structure in a population of *Cinnamomum insularimontanum* (Lauraceae). Heredity 90:98–106

Epperson BK, Chung MG, Telewski FW (2003) Spatial pattern of allozyme variation in a contact zone of *Pinus ponderosa* and *P. arizonica* (Pinaceae). Amer J Bot 90:25–31

Greimler J, Hermanowski B, Jang CG (2004) A re-evaluation of morphological characters in European *Gentianella* section *Gentianella* (Gentianaceae). Pl Syst Evol 248:143–169

Hong HH, Im HT (2003) *Viburnum japonicum* (Caprifoliaceae): an unrecorded species in Korea. Korean J Plant Taxon 33:271–277

Hong SP, Moon HK (2003) Gynodioecy in *Lycopus maackianus* Makino (Lamiaceae) in Korea: floral dimorphism and nutlet production. Flora (Germany) 198:461–467

Hong SP, Moon HK (2003) Unrecorded and introduced taxon in Korea: *Persicaria wallichii* Greuter & Burdet (*Persicaria*, Polygonaceae). Korean J Plant Taxon 33:219–223

Hong SP, Son JC (2003) Pollination of *Symplocarpus renifolius* Schott ex Miquel (Araceae) in Korea. Korean J Plant Taxon 33:165–179

Hyun JO, Lim Y, Shin H (2003) Validation of *Orobanche filicicola* (Orobanchaceae) from Korea. Novon 13:64–67

Jung EH, Hong SP (2003) Pollen morphology of Thymelaeaceae in Korea. Korean J Plant Taxon 33:255–270

Jung EH, Hong SP (2003) The taxonomic consideration of leaf epidermal microstructure in Korean Thymelaeaceae Adans. Korean J Plant Taxon 33:421–433

Kim C, Choi HK (2003) Genetic diversity and relationship in Korean ginseng (*Panax schinseng*) based on RAPD analysis. Korean J Genet 25:181–188

Kim C, Shin H, Choi HK (2003) A phenetic analysis of *Typha* in Korea and far east Russia. Aquatic Bot 75:33–43

Kim CH (2004) Conservation status of the endemic fern *Mankyua chejuense* (Ophioglossaceae) on Cheju Island, Republic of Korea. Oryx 38:217–219

Kim CH, Sun BY (2004) Infrageneric classification of the genus *Eleutherococcus* Maxim. (Araliaceae) with a new section *Cissifolius*. J Pl Biol 47:282–288

Kim CH, Sun BY, Kim YB (2004) Unrecorded fern species from Korean flora: *Ctenitis maximowicziana*, *Dryopteris lunanensis* and *D. decipiens* var. *diplazioides* (Dryopteridaceae). Korean J Plant Taxon 34:27–35

Kim CH, Sun BY, Park SH (2004) A new species of *Phegopteris* (Thelypteridaceae) from Korea. Novon 14:440–443

Kim DK, Kim JH (2004) Numerical taxonomy of tribe Forsythieae (Oleaceae) in Korea. Korean J Plant Taxon 34:189–203

Kim H, Chang CS (2003) A multivariate morphometric study and revision of *Deutzia* ser. *Parvilforae*. Korean J Plant Taxon 33:47–69

Kim HG, Funk VA, Vlasak A, Zimmer EA (2003) A phylogeny of the Munnoziinae (Asteraceae, Liabeae): circumscription of *Munnozia* and a new placement of *M. perfoliata*. Pl Syst Evol 239:171–185

Kim JS, Lee BC, Chung JM, Pak JH (2004) *Patrinia monandra* (Valerianaceae): an unrecorded species in Korea. Korean J Plant Taxon 34:167–172

Kim JS, Pak JH, Seo BB, Tobe H (2003) Karyotypes of metaphase chromosomes in diploid populations of *Dendranthema zawadskii* and related species (Asteraceae) from Korea: diversity and evolutionary implications. J Pl Res 116:47–55

Kim JS, Pak JH, Tobe H (2004) Chromosome number of *Dendranthema coreana* (Asteraceae). Acta Phytotax Geobot 55:63–64

Kim KJ, Lee HL, Moon YM (2003) Phylogenetic position of *Parasyringa sempervirens* (Franch.) W. W. Smith. Korean J Plant Taxon 33:181–201

Kim M (2004) A taxonomic review of Korean *Lycoris* (Amaryllidaceae). Korean J Plant Taxon 34:9–26

Kim S, Soltis DE, Soltis PS, Suh Y (2004) DNA sequences from Miocene fossils: an *ndh*F sequence of *Magnolia latahensis* (Magnoliaceae) and an *rbc*L sequence of *Persea pseudocarolinensis* (Lauraceae). Amer J Bot 91:615–620

Kim S, Soltis DE, Soltis PS, Zanis MJ, Suh Y (2004) Phylogenetic relationships among earlydiverging eudicots based on four genes: were the eudicots ancestrally woody? Molec Phylogenet Evol 31:16–30

Kim S, Yoo MJ, Albert VA, Farris JS, Soltis PS, Soltis DE (2004) Phylogeny and diversification of B-function MADS-box genes in angiosperms: evolutionary and functional implications of a 260-million-year-old duplication. Amer J Bot 91:2102–2118

Kim SY, Choi BH, Jeon ES (2003) New distribution of *Astragalus sikokianus* Nakai (Leguminosae) in Korea. J Jap Bot 78:167–169

Kim YD, Kim SH, Kim CH, Jansen RK (2004) Phylogeny of Berberidaceae based on sequences of the chloroplast gene *ndh*F. Biochem Syst Ecol 32:291–301

Kim YD, Kim SH, Landrum LR (2004) Taxonomic and phytogeographic implications from ITS phylogeny in *Berberis* (Berberidaceae). J Pl Res 117:175–182

Kim YD, Paik JH, Kim SH, Hong SP (2003) Phylogeny of *Scopolia* Jacq. s. str. based on ITS sequences. Korean J Plant Taxon 33:373–366

Lee B, Kim M (2004) Genome characterization of a Korean endemic species *Lycoris chejuensis* (Amaryllidaceae) by in situ hybridization. Korean J Genet 26:83–89

Lee BW, Park JK, Pak JH (2004) Fruit wall anatomy of the genus *Krigia* (Asteraceae, Lactuceae) and their taxonomic implications. Korean J Plant Taxon 34:321–333

Lee BY, Kwon SG, Oh KH (2004) A taxonomic study on *Daucus* species vegetated in islands off the Korean Peninsula. Korean J Plant Taxon 34: 287–295

Lee C, Wen J (2004) Phylogeny of *Panax* using chloroplast *trn*C-*trn*D intergenic region and the utility of *trn*C-*trn*D in interspecific studies of plants. Molec Phylogenet Evol 31:894–903

Lee CH, Lee S, Suh Y, Yeau SH, Lee NS (2003) A morphological reexamination on the genus *Adonis* L. sensu lato (Ranunculaceae) in Korea. Korean J Plant Taxon 33:435–454

Lee CH, Lee S, Suh Y, Yeau SH, Lee NS (2004) A palynotaxonomic study of Korean *Adonis* (Ranunculaceae). J Pl Biol 47:383–390

Lee CS, Lee NS, Yeau SH (2004) Molecular phylogenetic relationships and speciation of *Ranunculus cantoniensis* (Ranunculaceae). Korean J Plant Taxon 34:335–358

Lee HW, Park CW (2004) New taxa of *Cimicifuga* (Ranunculaceae) from Korea and the United States. Novon 14:180–184

Lee J, Baldwin BG (2004) Subtribes of principally North American genera of Cichorieae (Compositae). Novon 14:309–313

Lee J, Baldwin BG, Gottlieb LD (2003) Phylogenetic relationships among the primarily North American genera of Cichorieae (Compositae) based on analysis of 18S-26S nuclear rDNA ITS and ETS sequences. Syst Bot 28:616–626

Lee KB, Yoo YG, Park KR (2003) Morphological relationships of Korean species of *Sedum* L. subgenus *Aizoon* (Crassulaceae). Korean J Plant Taxon 33:1–15

Lee KH, Yang JY, Morita T, Ito M, Pak JH (2004) Relationship of diploid East Asian *Taraxacum* Wiggers using the capitulum morphological character. Korean J Plant Taxon 34:153–166

Lee KW, Kim HD, Park KR (2003) Numerical taxonomy of Korean *Orostachys* (Crassulaceae). Korean J Plant Taxon 33:359–371

Lee S, Ma S, Lim Y, Choi HK, Shin H (2004) Genetic diversity and its implications in the conservation of endangered *Zostera japonica* in Korea. J Pl Biol 47:275–281

Lee YN (2003) A new taxon of Ranunculaceae. Bull Korea Pl Res 3:2–4

Lee YN (2003) Two new forms of *Pseudolysimachion rotundum* (Nakai) Yamazaki var. *subintegrum* (Nakai) Yamazaki. Bull Korea Pl Res 3:24–25

Lee YN (2003) Two new plants unrecorded in Korea. Bull Korea Pl Res 3:26–27

Lee YN (2003) Two new taxa of Papaveraceae. Bull Korea Pl Res 3:5–6

Lee YN (2003) Two new varieties of *Larix sibirica* Ledb. Bull Korea Pl Res 3:20–23

Lee YN (2003) Two species of the genus *Michelia* in Korea. Bull Korea Pl Res 3:38

Lee YN, Kim YS (2003) A comparative study of Papaveraceae seeds in Korea. Bull Korea Pl Res 3:28–35

Lee YN, Kim YS (2003) *Abelia* in Korea. Bull Korea Pl Res 3:15–19

Lee YN, Kim YS (2003) Two taxa of *Fragaria* in Korea. Bull Korea Pl Res 3:7–14

Lee YN, Lee DH (2003) Additional taxonomic characters. Bull Korea Pl Res 3:36–37

Lim Y, Hyun JO, Shin H (2003) *Aster pseudoglehni* (Asteraceae), a new species from Korea. J Jap Bot 78:203–207

Moon HK, Hong SP (2003) Pollen morphology of the genus *Lycopus* (Lamiaceae). Ann Bot Fenn 40:191–198

Moon HK, Hong SP (2003) The taxonomic consideration of leaf epidermal microstructure in *Lycopus* L. (Mentheae, Lamiaceae). Korean J Plant Taxon 33:151–164

Moon HK, Hong SP (2004) The taxonomic consideration of petal and sepal micromorphology in *Lycopus* L. (Mentheae, Lamiaceae). Korean J Plant Taxon 34:273–285

Moon MO, Kang YJ, Kim CH, Kim CS (2004) An unrecorded species in Korea flora: *Hydrangea luteovenosa* Koidz. (Hydrangeaceae). Korean J Plant Taxon 34:1–7

Oh BU, Jang CG, Yoon CY (2004) *Corydalis namdoensis* B. U. Oh et J. G. Kim: A new species of *Corydalis* sect. *Corydalis* (Fumariaceae) from Korea. Korean J Plant Taxon 34:265–271

Oh IC, Denk T, Friis EM (2003) Evolution of *Illicium* (Illiaceae): mapping morphological characters on the molecular tree. Pl Syst Evol 240:175–209

Oh SH, Potter D (2003) Phylogenetic utility of the second intron of *LEAFY* in *Neillia* and *Stephanandra* (Rosaceae) and implications for the origin of *Stephanandra*. Molec Phylogenet Evol 29:203–215

Oh YC, Lee CS, Heo CY (2004) A taxonomy study on 9 taxa of *Carex* L. (Cyperaceae) in Korea. Korean J Plant Taxon 34:245–264

Oh YC, Lee CS, Yoon JY (2004) A taxonomic study on six sections *Ischnostachyae, Anomalae, Capiteaeae, Debiles, Capillares* and *Molliculae* of *Carex* L. in Korea (Cyperaceae). Korean J Plant Taxon 34:297–319

Oh YJ, Jo MJ (2003) A taxonomic study on six section subgenus *Vigena* Nees of *Carex* L. (Cyperaceae) in Korea. Korean J Plant Taxon 33:227–253

Oh YJ, Lee CS (2003) A taxonomic study on genus *Rhynchospora* Vahl in Korea. Korean J Plant Taxon 33:393–409

Oh YJ, Sung US (2003) A taxonomic study on sections *Montanae, Limosae,* and *Paniceae* of *Carex* L. in Korea (Cyperaceae). Korean J Plant Taxon 33:91–133

Park HJ, Lim CE, Park CW, Cha HC (2004) Analysis of flavonols extracted from leaves of various grapevine cultivars by HPLC. J Korean Soc Hort Sci 45:138–142

Park HJ, Min BM, Cha HC (2003) Flavonoid analysis of *Heloniopsis orientalis* (Thunb.) by high performance liquid chromatography. J Pl Biol 46:250–254

Park HK, Lim Y, Hyun JO, Shin H (2003) Taxonomy of genus *Equisetum* L. (Equisetaceae) in Korea. Korean J Plant Taxon 33:17–46

Park KR (2004) Comparisons of allozyme variation of narrow endemic and widespread species of Far East *Euphorbia* (Euphorbiaceae). Bot Bull Acad Sin 45:221–228

Park KR, Pak JH, Seo BB (2003) Allozyme variation in *Paraixeris*: A test for the diploid hybrid origin of *Paraixeris koidzumiana* (Compositae). Bot Bull Acad Sin 44:113–122

Park SH, Kil JH, Yang YH (2003) Unrecorded and naturalized plants in Korea (18). Korean J Plant Taxon 33:79–90

Park SJ, Kim KJ (2004) Molecular phylogeny of the genus *Hypericum* (Hypericaceae) from Korea and Japan: evidence from nuclear rDNA ITS sequence data. J Pl Biol 47:366–374

Ronse De Craene LP, Hong SP, Smets EF (2004) What is the taxonomic status of *Polygonella*? Evidence of floral morphology. Ann Missouri Bot Gard 91:320–345

Schneeweiss GM, Colwell A, Park JM, Jang CG, Stuessy TF (2004) Phylogeny of holoparasitic *Orobanche* (Orobanchaceae) inferred from nuclear ITS sequences. Molec Phylogenet Evol 30:465–478

Shim HB, Choi BH (2004) RAPD marker variations between and within the species of Korean *Suaeda*. Korean J Plant Taxon 34:63–74

Song GP, Song KM, Hyun HJ, Kim CS, Kim MH (2004) An unrecorded species in Korean flora: *Sedum tosaense* Makino (Crassulaceae). Korean J Plant Taxon 34:359–364

Tae KH, Ko SC (2003) Description of *Lycoris chinensis* Traub var. *sinuolata* K. Tae et S. Ko ex K. Tae et S. Ko (Amaryllidaceae). Korean J Plant Taxon 33:387–392

Tho JH, Kim JH (2004) Numerical taxonomy of *Rhus* sensu lato (Anacardiaceae) in Korea. Korean J Plant Taxon 34:205–220

Yeau SH, Lee CS, Lee NS (2004) A taxonomic review of section *Acris* of *Ranunculus* L. based on ITS sequences. Korean J Plant Taxon 34:173–183

Yoo KO, Lee WT, Kwon OK (2004) Interspecific relationships of Korean *Viola* based on RAPD, ISSR and PCR-RFLP analysis. Korean J Plant Taxon 34:43–61

Chong-wook Park
Department of Biological Sciences
College of Natural Sciences
Seoul National University
Seoul 151-742, Korea
e-mail: parkc@plaza.snu.ac.kr

Recent progress in systematics in China

Jinshuang Ma

1 Introduction

The botanical richness of China is unrivalled in temperate latitudes, with more than 31,000 species of vascular plants, more than half of which are endemic to China. In the past few years of the new century, Chinese botanists, along with their colleagues worldwide, have made a great contribution to Chinese history of botany. In this paper, these research results, as well as their basic facilities (herbaria, libraries, collections and research team), publications (floras, papers, revisions and monographs), and new developments and trends of plant systematic in China are briefly reviewed and discussed.

2 Basic information

Four basic elements in plant taxonomy/systematics are discussed, herbaria, library, collections and research, especially regarding developments in the past few years as well as current information in China.

2.1 Herbaria

The most important and largest herbaria in China are all within the Chinese Academy of Sciences (CAS), which has been supported strongly by the CAS in the past few years. Not only did they get sufficient research funds, but also a solid improvement in basic facilities. They are the backbone of Chinese taxonomic institutes today, and these were represented by Institute of Botany in Beijing (PE; about 2.35 millions), Kunming Institute of Botany in Kunming, Yunnan (KUN; ca. 1.1 million), and South China Botanical Garden in Guangzhou, Guangdong (IBSC; ca. 1 million), as well as others within the Chinese Academy of Sciences (all the above data were collected by the current author when traveling during his latest trip in Beijing, Guangzhou, Kunming and Chengdu in China between November 6 and December 3, 2004). The collections in these herbaria have increased greatly

Progress in Botany, Vol. 67
© Springer-Verlag Berlin Heidelberg 2006

compared with the past 10–20 years (Fu et al. 1993; Ma, and Liu 1998). Research conditions in these institutes, as well as their basic facilities, have been much improved recently, better than at any time in Chinese history. Some conditions for research work in these herbaria are even better than those of western countries, for example, the Herbaria of the University of Tokyo in Japan (Ma, and Cao 2005). However, other herbaria in the universities or in local research institutes are still laboring under difficult conditions, even worse than in the past (Ma, and Liu 1998). The national collections have not been updated in the past 10 years since the first statistics (Fu et al. 1993).

2.2 Library

For various historical reasons, the collections of Chinese botanical libraries have not advanced much, even worse if compared with their collections in the herbaria. All the famous and large botanical libraries in China have been built in the CAS, including those mentioned above (PE, KUN, IBSC). The collections compared with 10 years ago, except for some new publications on molecular biology and conservation, have not greatly changed. There is still a long way to go for online services as well as multi-institutes services, which are common now in western countries. From this point of view, there is certainly a lot of work to be done, not only the 'hardware', which is frequently mentioned, but also 'software' and qualified personnel, both librarians and archivists; these services are currently very weak or non-existent.

2.3 Collections

The collections, an activity of exploration by Chinese botanists, have not been improved much over the past 10 years, especially compared with western collections, such as the Royal Botanical Garden at Edinburgh (Walter 2001), the California Academy of Sciences (Bartholomew 2002) and the Harvard University Herbaria (Boufford 2001) in the past 20 years. Along with the *Flora Reipublicae Popularis Sinicae*, the Chinese edition of *Flora of China* completed (1959–2004), young students are not attracted by traditional taxonomic work. In addition, government policy has been shortsighted in emphasizing and putting SCI papers as well as the influence factors as the only indicators in assessing students' work (Yang 2001). This type of field work (both time and money consuming without any short-term gain) is thus not supported or encouraged in today's research in China. The most recent collections of herbaria are based mainly with those cooperating with western teams in southwest China, such as the expeditions lead by D.E. Boufford from the Harvard University Herbaria and by B. Bartholomew from the California Academy of Sciences; very few researchers have made their own expeditions, either within China or abroad.

2.4 Research

During the early 1990s, there were nearly 1500 registered botanists or plant taxonomic staff registered in the first edition of the Chinese herbarium (Fu et al. 1993). However, most of them are now retired, and the new graduates are mainly focused on new technology such as molecular studies, because they have a better chance for funding and jobs (Ma, and Liu 1998). So far, research on traditional taxonomy is no better than we described previously, around 7 years ago (Ma, and Liu 1998). However, for various reasons, there is no up-to-date figure yet regarding taxonomic staff in national wide within China since the early 1990s

3 Publications

These can be divided into the following areas.

3.1 Flora

These include the following.

3.1.1 Flora Reipublicae Popularis Sinicae (FRPS)

In 1959, Chinese botanists started to document their rich flora within the volumes of the comprehensive *Flora Reipublicae Popularis Sinicae* (FRPS). In 45 years (1959–2004), the massive 80 volumes (published in 126 books), involving more than 300 Chinese botanists throughout China, were finally published in full (Wu and Chen 2004; Ma, and Clemants 2005). Along with detailed descriptions, illustrations, and keys to the genera and species, each family treatment also includes information on economic uses and importance, Chinese names, and much more. FRPS covers about an eighth of the world's flora, and it is clearly of enormous international importance. Unfortunately, as it was written in Chinese, FRPS is not readily accessible to most foreign scientists. Furthermore, due to difficulties in accessing foreign botanical materials during the 1960s and 1970s, some of the volumes published during these periods were out of step with those in surrounding countries and needed to be revised. According to our statistics, there are about 31,000 species, and about 51% are endemic to China only (Ma, and Clemants 2005).

3.1.2 Flora of China

The *Flora of China* project, initiated as a joint Sino–American venture in the mid-1980s, fulfils the need for an English language, condensed revision of *FRPS*. Literally hundreds of scientists worldwide are actively involved in producing revisions for this new Flora, with Chinese and non-Chinese experts working together to produce the manuscripts. Coordination is through a series of Editorial Centers in China, the USA, UK, France, and others, with the managerial center at the Missouri Botanical Garden. In total, 11 volumes of 25 planned (Wu, and Raven 1994, 1995, 1996, 1998, 1999, 2000, 2001a,b, 2003a,b, 2005) have been published, and ten illustrations (Wu, and Raven 1998, 1999, 2000a,b, 2001, 2002, 2003a,b, 2004a,b) have also been published so far. *Flora of China Illustrations* provides companion volumes that accompany the text volumes. Extensive use is made of electronic communication in producing and disseminating the results. The acclaimed *Flora of China* website (http://flora.huh.harvard.edu/china/) has freely available electronic versions of all published volumes and pre-published accounts. There are also images, background information and further technical data. This is a very popular website aimed at both the general public and the scientific community. The *Flora of China* project has attracted worldwide attention and support, and has already become an essential reference for the identification of plants and study of floras of China and adjacent areas. This work has been working for more than 10 years, and most recent Professor D.Y. Hong (PE) and his fellow received solid support from China National Scientific Funds in publishing unfinished parts from Chinese side. This will speed up the project process, and make the whole work easier to complete, as scheduled in the next few years.

3.1.3 Local flora of China

There are two kinds of local flora covered in this section. First are flora at Province Level; and the second are flora above Province Level, or natural region rather than administrative unit. The whole pictures of Chinese local flora can currently be divided into two types, finished and unfinished (Ma, and Gilbert 1998; Ma et al. 2000).

The finished parts are: *Flora of Anhui*, (v. 1–5, 1986–1992), *Flora of Beijing* (ed. 1, v. 1–3, 1962–1975, ed. 2, vol. 1–2, 1984–1987, ed. 2, enlarged/reprinted, 1992). *Flora of Fujian,* (ed. 1, vol. 1–6, 1982–1995, ed. 2, 1–, 1991–). *Flora Hainanica* (vol. 1–4, 1964–1977), *Flora Hebeiensis* (vol. 1–3, 1987–1991), *Henan Flora* (vol. 1–4, 1981–1998), *Flora of Hubei* (ed. 1, vol. 1–2, 1979–, ed. 2, vol. 1–4, 2001–2002), *Flora Bryophytarum Intramongolicarum* (vol. 1, 1997), *Flora Intramongolica* (ed. 1, vol. 1–8, 1977–1985, ed. 2, vol. 1–5, 1989–1998), *Jiangsu Flora* (vol. 1–2, 1977–1982), *Flora Liaoningica* (vol. 1–2, 1988–1991), *Flora Ningxiaensis* (vol. 1–2, 1986–1988), *Flora Qinghaiica* (vol. 1–4, 1996–1999), *Flora Bryophytarum Shandongicorum* (vol. 1, 1998). *Shandong Flora* (vol. 1–2, 1990–1997), *The Plants of Shanghai* (vol. 1–2, 1999), *Flora Shanxiensis* (vol. 1–5, 1992–2004). *Flora of Taiwan (English Edition)* (ed. 1, vol. 1–6, 1975–1979, ed. 2, vol. 1–6 1993–2003). *Bryoflora of Xizang* (vol. 1, 1985), *Flora Xizangica* (vol. 1–5, 1983–1987), *Flora of Zhejiang* (vol. 0–7, 1989–1993), *Flora in Desertis Reipublicae Populorum Sinarum* (vol. 1–3, 1985–1992), *Illustrated Flora of Ligneous Plants of Northeast China,* (vol. 1, 1955), *Flora Muscorum Chinae Boreali-Orientalis,* (vol. 1, 1977), *Flora Hepaticorum Chinae Boreali-Orientalis* (vol. 1, 1977), *Flora Tsinlingensis,* (vol. 1 (part 1–5), 2, 3, 1974–1985)

Unfinished parts covers: *Flora of Guangdong* (vol. 1–4–) 1987–2000–), *Flora of Gaungxi* (vol. 1–, 1991–), *Flora Guizhouensis* (vol. 1–9–, 1982–1989–), *Flora Heilongjiangensis* (vol. 1–,4–11–, 1985–2003–), *Flora of Hunan* (vol. 2–, 2000–), *Flora of Jiangxi* (vol. 1–2–, 1993–2004–), *Flora Sichuanica* (vol. 1–15–, 1981–1999–), *Flora Xinjiangensis* (vol. 1–2–, 5–6–, 1993–1999–), *Flora Yunnanica* (vol. 1–19–, 1977–2003–), *Flora Plantarum Herbacearum Chinae Boreali-Orientalis* (vol. 1–7–,11–12, 1958–, 1998), *Flora Loess-Plateaus Sinicae* (vol. 1–2, 5–, 1989–).

Much more detailed information about Chinese local flora can be found at the author's website www.metasequoia.org/local.htm (Ma et al. 2000). Since the older generations are now retired, and the young generationer is no longer interested in traditional taxonomy, no-one now pays much attention to this type of work in China. If any of the flora is not finished at this time, it will be further delayed, and will remain largely incomprehensible to readers.

4 Journals

About 10 years ago, there were around ten botanical journals in China that covered taxonomic papers (Ma, and Liu 1998). However, for various reasons, the policy of these journals has been largely changed, and their directions have changed to ecology, genetics, conservation, biodiversity, resources or molecular biology. So far, only three are regarded as important taxonomic or systematic journals: *Acta Phytotaxonomia Sinica* (since 1951), *Acta Botanica Yunnanica* (since 1979), plus *Cathaya*, Annals of the Laboratory of Systematic and Evolutionary Botany and Herbarium, Institute of Botany, CAS, at Beijing (since 1989). More and more papers submitted by Chinese authors are being published worldwide (see below for details).

5 Papers, monographs and revisions

This area has grown rapidly in the past few years, especially after the Chinese government focused on the Scientific Citation Index (SCI) and their impact factors, since the promotion or funding chances of staff are directly linked to their publications. More and more papers are being published worldwide in English by Chinese scholars, which have not yet been documented (Cyranoski 2004); and by contrast, fewer and fewer good papers can be expected within China (Yang 2001). In order to give readers more detailed information, we have separated them into the following categories.

5.1 Paleobotany

Research work on this field from China has been largely reported worldwide (Wang Y.F. et al. 2001, 2003), including some important discoveries published in *Science* (Sun et al. 2002) and *Nature* (Zhou, and Zheng 2003)

5.2 Ferns

A series revision on ferns in China as well as in neighboring countries has been made by Dr. X.C. Zhang (PE) and his co-authors (Smith and Zhang 2002; Wang M.L. et al. 2003; Zhang X.C. et al. 2003, Zhang 2004). An interesting species of *Isoetes* (Isoetaceae) has recently been described from southwest China (Wang et al. 2002).

5.3 Gymnosperms

China has more conifers than any other country in the world (Ma, and Cao 2005), and research work has been plentiful in the past few years, such as the new system of gymnosperm classification proposed by Professor D.Z. Fu and colleagues (Fu et al. 2004) based on their longtime works (Fu 1992, Fu, and Yang 1993a,b; Yang, and Fu 2001), with a total of 12 families accepted by this new scheme, as well as many others (Li et al. 2001a,b; Li, and Yang 2002). Another interesting re-discovery of *Thuja sutchuenensis* recently was also a remarkable work in the gymnosperm world (Xiang et al. 2002), since the species has been treated as EW by IUCN-SSC because no wild plant has been found in the past century. One separate species of pine was also described from Taiwan recently (Businsky 2003a,b). A series of reports on *Metasequoia* were published recently, especially on the natural history of the living fossil in the past 60 years (Ma 2003a, b, 2004; Ma, and Shao 2003). An interesting report on the Chinese conifer, *Taxodiomeria*, an intergeneric hybrid genus between *Taxodium* and *Cryptomeria* from China, was published recently, with more than 2000 adult trees raised in Shanghai already (Zhang J.J. et al. 2003). Many other reports exist (Wang W.P. et al. 2003; Wei, and Wang 2003; Yang 2004). Recently, Dr. X.R. Wang, from Sweden, joined the research team in Beijing, and published many papers regarding the conifers (Kormutak et al. 2003; Liu et al. 2003a,b,c; Song et al. 2002, 2003; Yin T et al. 2003).

5.4 Araliaceae

Dr. J. Wen from the Field Museum (now Smithsonian Institution) continued her research work on the family, both in China (Wen 2000, 2002) and worldwide (Wen et al. 2001).

5.5 Asteraceae

Saussurea, one of the largest genera in the family centered in southwest China, has been studied recently based on the chloroplast DNA trnL-F sequences from China (Wang, and Liu 2004), which showed very interesting results.

5.6 Brassicaceae

Dr. Al-Shehbaz (MO) continues his research work on Brassicaceae from China, as well as central Asia (Al-Shehbaz, and Appel 2002; Al-Shehbaz et al. 2002; Al-Shehbaz 2002a,b, 2004; Yue et al. 2004). Within China, there are also some reports about the family (Zhang 2003).

5.7 Burmanniaceae

The phylogenetic reconstruction of *Burmannia* was performed using different outgroups (Zhang 2001). Another genus, *Thismia*, was recorded from Taiwan for the first time (Yang et al. 2002).

5.8 Celastraceae

Revisions to *Euonymus* were made recently, with about 130 species accepted worldwide, about 90 of which are in China (Ma 2001).

5.9 Cornaceae

Dr. Xiang at North Carolina State University and her fellow are still working on her favorite family, with a lot of publications recently (Fan, and Xiang 2001, 2003).

5.10 Corsiaceae

Corsiopsis, a new genus, with the only species from Asia, *Corsiopsis chinensis*, was described from Guangdong, south China. This is also the only record of the family in Asia (Zhang et al. 1999, 2000).

5.11 Cyperaceae

Kobresia from Xizang (Tibet) has been revised from China, based not only on the specimens but also field observation (Zhang 2004). There are 36 species and one subspecies from Xizang, and they occupy about the half of the species in the genus.

5.12 Euphorbiaceae

Gymnanthes Sw. was recorded from Yunnan, China recently (Zhu et al. 2000).

5.13 Gesneriaceae

Two new genera, *Wentsaiboea* (Fang, and Qin 2004) and *Paralagarosolen* (Wei 2004), were described recently from Guangxi, south China. There is a new book entitled "*Plants of Gesneriaceae in China*" just published (Li, and Wang 2004), and a total of 58 genera, 463 species and 421 color photos are presented. This is also the first comprehensive work on the family from China, especially based on long-term collection in the past several decades by the authors.

5.14 Hamamelidaceae

A new superageneric classification of the Hamamelidoideae based on morphology and sequences of nuclear and chloroplast DNA has been proposed recently (Li, and Bogle 2001).

5.15 Labiatae

Changruicaoia, a new genus from Sichuan, southwest China, has been described (Zhu 2001).

5.16 Lauraceae

The family has been studied by Professor X.W. Li (KUN) and his son, Dr. Jie Li, who received his Ph.D. and returned from Australia to Kunming, Yunnan several years ago; several papers have been published by them (Li and Conran 2003; Li, and Li 2004; Li et al. 2004).

5.17 Orchidaceae

A new genus based on previous publications from China was combined and reported (Luo, and Chen 2003), and some new discoveries were made recently from China (Jin et al. 2004a,b).

5.18 Paeoniaceae

The systematics of *Paeonia* is mainly from the laboratory led by Professor D.Y. Hong (Hong et al. 2001a,b, 2004; Hong, and Zhou 2003;, Zhou et al. 2003), and much of his work is based on molecular research (Zhao et al. 2004).

5.19 Phrymataceae

Asian *Phryma* has been revised recently (Li 2000) with one subspecies, *Phryma leptostachya* L. ssp. *asiatica* (Hara) Kitamura, accepted with detailed distribution list.

5.20 Poaceae

The molecular work on *Oryza* has been carried out by Dr. S. Ge (PE) and his fellow (Ge et al. 2002; Bao, and Ge 2003), and another taxon has also been reported recently (Saarela et al. 2003).

5.21 Primulaceae

Molecular phylogeny and biogeography of *Androsace* and the convergent evolution of cushion morphology was reported recently (Wang Y.J. et al. 2004).

5.22 Ranunculaceae

Professor W.T. Wang (PE) continues his research work on *Clematis* and many papers have been published (Wang 2000–2001, 2002–2004). Some other studies on the family have also been reported (Ren et al. 2004).

5.23 Rhamnaceae

Berchemiella wilsonii (Schneid.) Nakai (Rhamnaceae), an endangered species from Hubei, southwest China, was rediscovered for the first time since E.H. Wilson's type collection a century ago (Li et al. 2004).

5.24 Schisandraceae

Worldwide, *Schisandra* was revised with ten species accepted only, but more than 20 synonyms were reduced (Lin 2000), which is much different from the recent monograph (Saunders 2000).

5.25 Scrophulariaceae

The flower biology and pollination biology of *Pedicularis* have been reported recently (Yang et al. 2002, 2003; Wang H. et al. 2003a,b; Yang, and Guo 2004), as well as biogeography and diversity in southwest China (Li et al. 2002; Xu, and Chen 2001).

5.26 Styracaceae

There has been a recent revision on *Styrax* of Asia (Huang et al. 2003).

5.27 Umbelliferae/Apiaceae

Notopterygium, a genus endemic to China, has been revised with five species and one subspecies accepted (Pu et al 2000).

5.28 Zingiberaceae

There are several reports regarding the pollination biology of the family from China, including some in famous journals, including *Nature* (Li et al. 2001, 2002; Zhang L. et al. 2003; Gao et al. 2004; Wang et al. 2004). The phylogenetic analyses of *Amomum* using ITS and *matK* DNA sequence data has also been reported recently (Xia et al. 2004).

6 Floristic work

This has been a very hot topic in the past 20 years in Chinese botany, and many studies have been published, such as the relationships between Taiwan and Mainland of China (Ying, and Hsu 2002), Bryophytes between China and North America (Wu et al. 2001), Southern China (Zhu et al. 2003) and

Indo-Malesian flora (Zhu, and Roos 2004), and Chinese subtropical region and their spatial patterns (Shen, and Zhang 2000). In addition, some excellent papers have come from Chinese scholars who work abroad, for example Dr. Hong Qian at the State Museum of Illinois (Qian 2001a,b, 2002a,b; Qian, and Ricklefs 2000, 2001, 2004; Qian et al. 2003a,b; Ricklefs et al. 2004). Professor Z.Y. Wu and his fellow continue their traditional floristic work and several monographs have been published recently (Wu, and Chen 2004; Wu et al. 1998a,b, 2003a,b)

In Hengduan Mountain and Himalayan Mountain, the unique natural geographical condition in southwest China provides a natural laboratory for the botanists in understanding the origin and evolution of the alpine floras in Hengduan Mountain and the Sino-Himalayan region, an important hotspot in today's Chinese floristics (Sun 2002a,b).

7 Angiosperm system

A famous paper recently in the angiosperm system was the new classification by Professor Z.Y. Wu and his fellow, called "Eight classes system of angiosperms" (Wu et al. 1998a,b). This system was also reviewed in detail in the recent summary work *Flora of China* (Chinese edition, Wu, and Chen 2004). The authors claim that it is a "polyphyletic-polychronic-ploytopic", which divided the "Magnoliidae" into eight classes, 40 subglasses, 202 orders and 572 families (Tang, and Lu 2004). However, this is not the first system proposed by Chinese taxonomists, but the third one after Dr. H.H. Hu (1950) and Professor H.D. Zhang (1986, 2000). Once this new system was published, the critique on it also appeared (Fu 2003), followed by the authors' reply (Tang, and Lu 2003).

8 Higher plants of China

This is a work very similar to the seven volumes of *Iconographia Cormophytorum Sinicorum* (five volumes plus two supplements, 1972–1983, Science Press). *The Higher Plants of China*, edited by Professor L.K. Fu (PE), consisting of 13 volumes, is in Chinese. The work deals with around 17,000 species representing all the currently recognized plant families and genera in China. Representative species for each genus, including those introduced and naturalized ones, are included. Keys to the genera and to species are provided. For each species, its standard and sometimes other commonly used Chinese name, its scientific name with reference, and a basionym with reference when applicable are provided. Scientific names accepted in the ICS and the FRPS are also provided, with page number references in these works when they are treated as synonyms or as misidentified names. Each species is illustrated with brief descriptions of morphology, distribution, habitat, and sometimes uses, and each is accompanied (except for exotic species) by a distribution map. *The Higher Plants of China* is an accumulated work of more than 140 Chinese plant taxonomists from many institutions nationwide. The systematic arrangements for bryophytes, pteridophytes, and gymnosperms follow the *Flora Bryophytarum Sinicorum* and the FRPS (vol. 2–7), and that for angiosperms follows Cronquist's system. Since this is relative new, and most readers are unfamiliar with it, the full contents of each volume are listed as follows:

Vol. 1. Bryophyta (not published yet at the end of 2004);

Vol. 2. Pteridophyta (not published yet at the end of 2004);

Vol. 3 (2000, 757 pages with 1144 figures + 448 color photos) Gymnospermae: 1 Cycadaceae, 2 Ginkgoaceae, 3 Araucariaceae, 4 Pinaceae, 5 Sciadopityaceae, 6 Taxodiaceae, 7 Cupressaceae, 8 Podocarpaceae, 9 Cephalotaxaceae, 10 Taxaceae, 11 Ephedraceae, 12 Gnetaceae, and Angiospermae: 1 Magnoliaceae, 2 Annonaceae, 3 Myristicaceae, 4 Calycanthaceae, 5 Lauraceae, 6 Hernandiaceae, 7 Chloranthaceae, 8 Saururaceae, 9 Piperaceae, 10 Aristolochiaceae, 11 Illiciaceae, 12 Schisandraceae, 13 Nelumbonaceae, 14 Nymphaeaceae, 15 Cabombaceae, 16 Ceratophyllaceae, 17 Ranunculaceae, 18 Circaeasteraceae, 19 Berberidaceae, 20 Sargentodoxaceae, 21 Lardizabalaceae, 22 Menispermaceae, 23 Papaveraceae, 24 Fumariaceae, 25 Tetracentraceae, 26 Trochodendraceae, 27 Cercidiphyllaceae, 28 Eupteleaceae, 29 Platanaceae, 30 Hamamelidaceae, 31 Daphniphyllaceae, 32 Eucommiaceae,

Vol.4 (2001, 745 pages with 1126 figures + 320 color photos), 33 Ulmaceae, 34 Cannabaceae, 35 Moraceae, 36 Urticaceae, 37 Rhoipteleaceae, 38 Juglandaceae, 39 Myricaceae, 40 Fagaceae, 41 Betulaceae, 42 Casuarinaceae, 43 Phytolaccaceae, 44 Nyctaginaceae, 45 Aizoaceae, 46 Cactaceae, 47 Chenopodiaceae, 48 Amaranthaceae, 49 Portulacaceae, 50 Basellaceae, 51 Molluginaceae, 52 Caryophyllaceae, 53 Polygonaceae, 54 Plumbaginaceae, 55 Dilleniaceae, 56 Paeoniaceae, 57 Ochnaceae, 58 Dipterocarpaceae, 59 Theaceae, 60 Actinidiaceae, 61 Pentaphylacaceae, 62 Elatinaceae, 63 Guttiferae.

Vol. 5 (2003, 775 pages + 48 plates), 64 Elaeocarpaceae, 65 Tiliaceae, 66 Sterculiaceae, 67 Bombacaceae, 68 Malvaceae, 69 Lecythidaceae, 70 Nepenthaceae, 71 Droseraceae, 72 Flacourtiaceae, 73 Bixaceae, 74 Cistaceae, 75 Stachyuraceae, 76 Violaceae, 77 Tamaricaceae, 78 Frankeniaceae, 79 Ancistrocladaceae, 80 Passifloraceae, 81 Caricaceae,

82 Cucurbitaceae, 83 Tetramelaceae, 84 Begoniaceae, 85 Salicaceae, 86 Capparaceae, 87 Brassicaceae (Cruciferae), 88 Moringaceae, 89 Resedaceae, 90 Clethraceae, 91 Empetraceae, 92 Ericaceae, 93 Pyrolaceae, 94 Monotropaceae, 95 Diapensiaceae.

Vol. 6 (2003, 833 pages + 32 plates) 96 Sapotaceae, 97 Ebenaceae, 98 Styracaceae, 99 Symplocaceae, 100 Myrsinaceae, 101 Primulaceae, 102 Connaraceae, 103 Pittosporaceae, 104 Hydrangeaceae, 105 Grossulariaceae, 106 Grassulaceae, 107 Saxifragaceae, 108 Rosaceae.

Vol. 7 (2001, 929 pages + 56 plates) 109 Mimosaceae, 110 Caesalpiniaceae, 111 Fabaceae (Papilionaceae), 112 Elaeagnaceae, 113 Proteaceae, 114 Podostemaceae, 115 Haloragaceae, 116 Sonneratiaceae, 117 Lythraceae, 118 Crypteroniaceae, 119 Thymelaeaceae, 120 Trapacae, 121 Myrtaceae, 122 Punicaceae, 123 Onagraceae, 124 Melastomataceae, 125 Combretaceae, 126 Rhizophoraceae, 127 Alangiaceae, 128 Nyssaceae, 129 Cornaceae, 130 Dipentodonetaceae, 131 Olacaceae, 132 Opiliaceae, 133 Santalaceae, 134 Loranthaceae, 135 Viscaceae, 136 Balanophoraceae, 137 Mitrastemonaceae, 138 Rafflesiaceae, 139 Celastraceae, 140 Hippocrateaceae, 141 Salvadoraceae, 142 Aquifoliaceae, 143 Icacinaceae, 144 Cardiopteridaceae, 145 Dichapetalaceae.

Vol. 8 (2001, 748 pages 1151 figures 293) 146 Buxaceae, 147 Pandaceae, 148 Euphorbiaceae, 149 Rhamnaceae, 150 Leeaceae, 151 Vitaceae, 152 Erythroxylaceae, 153 Ixonanthaceae, 154 Linaceae, 155 Malpighiaceae, 156 Polygalaceae, 157 Xanthophyllaceae, 158 Staphyleaceae, 159 Bretschneideraceae, 160 Sapindaceae, 161 Sabiaceae, 162 Hippocastanaceae, 163 Aceraceae, 164 Burseraceae, 165 Anacardiaceae, 166 Simaroubaceae, 167 Coriariaceae, 168 Meliaceae, 169 Rutaceae, 170 Zygophyllaceae, 171 Oxalidaceae, 172 Geraniaceae, 173 Tropaeolaceae, 174 Balsaminaceae, 175 Araliaceae, 176 Umbelliferae (Apiaceae).

Vol. 9 (1999, 627 pages with 921 figures + 195 color photos) 177 Loganiaceae, 178 Gentianaceae, 179 Apocynaceae, 180 Asclepiadaceae, 181 Solanaceae, 182 Convolvulaceae, 183 Cuscutaceae, 184 Menyanthaceae, 185 Polemoniaceae, 186 Hydrophyllaceae, 187 Boraginaceae, 188 Verbenaceae, 189 Lamiaceae.

Vol. 10. Phrymaceae-Theligonaceae (not published yet as the end of 2004).

Vol. 11. Caprifoliaceae-Asteraceae (not published yet as the end of 2004).

Vol. 12. Butomaceae-Poaceae (not published yet as the end of 2004)

Vol. 13. (2002, 806 pages + 104 plates) 244 Sparganiaceae, 245 Typhaceae, 246 Bromeliaceae, 247 Strelitziaceae, 248 Musaceae, 249 Lowiaceae, 250 Zingiberaceae, 251 Costaceae, 252 Cannaceae, 253 Marantaceae, 254 Philydraceae, 255 Pontederiaceae, 256 Liliaceae, 257 Amaryllidaceae,

258 Acanthochlamydaceae, 259 Iridaceae, 260 Agavaceae, 261 Taccaceae, 262 Stemonaceae, 263 Smilacaceae, 264 Dioscoreaceae, 265 Burmanniaceae, 266 Orchidaceae.

9 New developments and trends

In the past few years, molecular work has dominated most of Chinese systematic works, and published either in China (Wu et al. 2000; Ehrendorfer and Samuel 2001; Hong et al. 2001; Li et al. 2002; Bao and Ge 2003; Wang WP et al. 2003) or abroad (Shi et al. 2001; Song et al. 2001; Kong et al. 2002a,b; Sun et al. 2002; Li et al. 2004). More and more Chinese scholars submitted their papers worldwide. For example, there are at least ten papers published in the *Botanical Journal of the Linnean Society* in the past year from China (Hong and Zhou 2003; Li et al. 2003; Lian et al. 2003; Wang WP et al. 2003; Xiang and Farjon 2003; Xu 2003; Liu 2004; Luo 2004; Wang 2004; Xiang et al. 2004; Yue et al. 2004). Another journal is *Novon*, from the Missouri Botanical Garden, where almost every issue contains a paper from China. So does the *Plant Systematics and Evolution* (Liu et al. 2002; Yang et al. 2002, 2003; Wang H et al. 2003a; Wang and Li 2002; Yang 2004).

10 Conclusion

After more than 25 years since its commencement in 1979, botanical research work in the past few years has accumulated largely in China, especially under the CAS organization. Botanists have also won great repute in the academic world for their hard work and excellent contributions. Today, more and more young Chinese botanists have a better chance in pursuing their careers and obtaining good results that at any time in China's history, thanks to solid financial support and higher salaries. The challenges are also increasing, especially when researchers move to the front of their field worldwide. Nevertheless, there is still a long way to go before Chinese taxonomists and systematists even fully understand their own rich floras, not to mention those worldwide.

References

Al-Shehbaz IA (2002a) New species of *Lyussum, Aphragmus, Arabis,* and *Sinosophiopsis* (Brassicaceae) from China and India. Novon 12:309–313
Al-Shehbaz IA (2002b) Six new species of *Draba* (Brassicaceae) from Himalayas. Novon 12:314–318

Al-Shehbaz IA (2004) Novelties and notes on miscellaneous Asian Brassicaceae. Novon 14(2):153–157

Al-Shehbaz IA, Appel O (2002) A synopsis of the central Asian *Rhammatophyllum* (Brassicaceae). Novon 12(1):1–4

Al-Shehbaz I.A, Mummenhoff K, Appel O (2002) *Cardaria, Coronopus,* and *Stroganowia* are united with *Lepidium* (Brassicaceae). Novon 12(1):5–11

Bao Y, Ge S (2003) Phylogenetic relationships among diploid species of *Oryza officinalis* complex revealed by multiple gene sequences. Acta Phytotax Sin 41(6):497–508

Bartholomew B (2002) Checklist of vascular plants of Gaoligong Shan, Yunnan Province, China (http://www.calacademy.org/research/botany/aboutGLGS.html)

Boufford D.E (2001) Harvard University Herbaria—field studies in China (http://www.huh.harvard.edu/research/china/China.html)

Businsky R (2003a) A new hard pine (*Pinus,* Pinaceae) from Taiwan. Novon 13(3):281–288

Businsky R (2003b) Taxonomy and biogeography of Chinese hard pine, *Pinus hwangshanensis* W Y Hsia. Bot Jahrb 125(1):1–17

Cyranoski D (2004) China increases share of global scientific publications. Nature 431:116

Ehrendorfer F, Samuel R (2001) Contributions to a molecular phylogeny and systematics of *Anemone* and related genera (Ranunculaceae—Anemoninae). Acta Phytotax Sin 39(4):293–307

Fan CZ, Xiang QY (2001) Phylogenetic relationships within *Cornus* (Cornaceae) based on 26S rDNA sequences. Am J Bot 88:1131–1138

Fan CZ, Xiang QY (2003) Phylogenetic analyses of Cornales based on 26S rRNA and combined 26S rDNA-MAT*K*-RBC*L* sequence data. Am J Bot 90:1357–1372

Fang D, Qin DH 2004. *Wentsaiboea,* a new genus of the Gesneriaceae from Guangxi, China. Acta Phytotax Sin 42(6):533–536

Fu DZ (1992) Nageiaceae: a new gymnosperm family. Acta Phytota. Sin 30(6):515–528 (Chinese with English abstract)

Fu DZ (2003) A critique on the "eight-class system" of the classification of Angiosperms proposed by Wu et al. (2002). Acta Phytotax Sin 41(1):91–96 (Chinese with English abstract)

Fu DZ, Yang QE (1993a,b) A new morphological interpretation of the female reproductive organs in *Ginkho biloba* L. with a phylogenetic consideration on gymnosperms. Acta Phytotax Sin 31(3):294–296, 31(4):309–317 (Chinese with English abstract)

Fu DZ, Yang Y, Zhu GH (2004) A new scheme of classification of living gymnosperms at family level. *Kew Bull* 59:111–116

Fu LK, Zhang XC, Qin HN, Ma JS (1993) *Index Herbariorum Sinicorum.* Science and Technology Press, Beijing (English and Chinese Edition)

Gao JY, Zhang L, Deng XB, Ren PY, Kong JJ, Li QJ (2004) The floral biology of *Curcumorpha longiflora* (Zingiberaceae): a ginger with two-day flowers. Am J Bot 91:289–293

Ge S, Li A, Lu BR, Zhang, Hong DY (2002) A phylogeny of the rice tribe Oryzeae (Poaceae) based on *matK* sequence data. Am J Bot 89:1967–1972

Hong DY, Pan KY (2004) A taxonomic revision of the *Paeonia anomala* complex (Paeoniaceae). Ann. Missouri Bot Gard 91(1):87–98

Hong DY, Zhou SL (2003) *Paeonia (Paeoniaceae)* in the Caucasus. Bot J Linn Soc 143(2):135–150

Hong DY, Pan KY, Rao GY (2001a) Cytogeography and taxonomy of the *Paeonia obovata* polyploid complex (Paeoniaceae). Plant Syst Evol 227(3–4):123–136

Hong DY, Pan KY, Turland NJ (2001b) *Paeonia anomala* subsp. *veitchii* (Paeoniaceae), a new combination. Novon 11(3):315–318

Hong YP, Chen ZD, Lu AM (2001) Phylogeny of the tribe Menispermeae (Menispermaceae) reconstructed by ITS sequence data. Acta Phytotax Sin 39(2):97–104 (Chinese with English abstract)

Hu HH (1950) A polyphyletic system of classification of angiosperms. Science Record (Peking) 3(2–4):221–230

Huang YL, Fritsch PW, Shi SH 2003. A revision of the imbricate group of *Styrax* Series *Cyrta* (Styracaceae) in Asia. Ann Missouri Bot Gard 90(4):491–553

Jin XH, Chen SC, Qin HN, Zhu GH, Siu GL (2004a) A new species of *Didymoplexiella* (Orchidaceae) from China. Novon 14(2):176–177

Jin XH, Chen SC, Qin HN, Zhu GH, Siu GL 2004b A new species of *Holcoglossum* (Orchidaceae) from China. Novon 14(2):178–179

Kong HZ, Chen ZD, Lu AM 2002a. Phylogeny of *Chloranthus* (Chloranthaceae) based on nuclear ribosomal ITS and plastid TRNL-F sequence data. Am J Bot 89:940–946

Kong HZ, Lu AM, Endress PK (2002b) Floral organogenesis of *Chloranthus sessilifolius*, with special emphasis on the morphological nature of the androecium of *Chloranthus* (Chloranthaceae). Plant Syst Evol 232 (3–4):181–188

Kormutak A, Vookova B, Gomory D, Wang XR, Szmidt AE 2003 Intraspecific variation in chloroplast DNA *psb*AI gene region of silver fir (A*bies alba* Mill.). For Genet 10:19–22

Li CX, Yang Q (2002) Divergence time estimates for major lineages of Cupressaceae (s.l.). Acta Phytotax Sin 40(4):323–333

Li H, Liang HX, Peng H, Lei LG (2003) Sporogenesis and gametogenesis in *Sladenia* and their systematic implication. Bot J Linn Soc 143:305–314

Li J, Conran JG (2003) Phylogenetic relationships in Magnoliaceae subfam. Magnolioides:a morphological cladistic analysis. Plant Syst Evol 242 (1–4):33–47

Li J, Li XW (2004) Advances in *Lauraceae* systematic research on the world scale. *Acta Bot Yunnan* 26(1):1–11

Li J, Christophel DC, Conran JG, Li HW (2004) Phylogenetic relationships within the 'core' Laureae (*Litsea* complex, Lauraceae) inferred from sequences of the chloroplast gene matK and nuclear ribosomal DNA ITS regions. Plant Syst Evol 246(1–2):19–34

Li JH, Bogle AL (2001) A new suprageneric classification system of the Hamamelidoideae based on morphology and sequences of nuclear and chloroplast DNA. Harvard Pap Bot 5(2):499–516

Li JH, Davis CC, Del Tredici P, Donoghue MJ (2001a) Phylogeny and biogeography of *Taxus* (Taxaceae) inferred from sequences of the internal transcribed spacer region of nuclear ribosomal DNA. Harvard Pap Bot 6(1):267–274

Li JH, Davis CC, Donoghue MJ, Kelley S, Del Tredici P (2001b) Phylogenetic relationships of *Torreya* (Taxaceae) inferred from sequences of nuclear ribosomal DNA ITS region. Harvard Pap Bot 6(1):275–282

Li JQ, Huang HW, Sang T (2002) Molecular phylogeny and infrageneric classification of *Actinidia* (Actinidiaceae). Syst Bot 27(2):408–415

Li JQ, Jiang MX, Wang HC, Tian YQ (2004) Rediscovery of *Berchemiella wilsonii* (Schneid.) Nakai (Rhamnaceae), an endangered species from Hubei, China. Acta Phytotax Sin 42(1):86–88

Li QJ, Xu ZF, Kress WJ, Xia YM, Zhang L, Deng XB, Gao JY, Bai ZL (2001) Pollination: flexible style that encourages outcrossing. *Nature* 410:432

Li QJ, Kress WJ, Xu ZF, Xia YM, Zhang L, Deng XB, Gao JY (2002) Mating system and stigmatic behaviour during flowering of *Alpinia kwangsiensis* (Zingiberaceae). Plant Syst Evol 232(1–2):123–132

Li WL, Wang H, Li DZ (2002) Biogeography and species diversity of *Pedicularis* (Scrophulariaceae) of Yunnan. Acta Bot Yunnan 24(5):583–590

Li ZY (2000) Taxonomic notes on the genus *Phryma* L. from Asia. Acta Phytotax Sin 38(4):386–391

Li ZY, Wang YZ 2004 *Plants of Gesneriaceae in* China. Henan Science and Technology Publishing House, Zhengzhou, 1–722 pp (Chinese)

Lian YS, Chen XL, Sun K, Ma RJ 2003 Clarification of the systematic position of *Hippophae goniocarpa* (Elaeagnaceae). Bot J Linn Soc 142:425–430

Lin Q (2000) Taxonomic notes on the genus *Schisandra* Michx. Acta Phytotax Sin 38(6):532–550 (Chinese with English abstract and key)

Liu AZ, Kress WJ, Wang H, Li DZ (2002) Insect pollination of *Musella* (Musaceae), a mono-
typic genus endemic to Yunnan, China. Plant Syst Evol 235(1–4):135–146

Liu JQ (2004) Uniformity of karyotypes in *Ligularia* (Asteraceae:Senecioneae), a highly diversified genus of the eastern Qinghai-Tibet Plateau highlands and adjacent areas. Bot J Linn
Soc 144:329–342

Liu ZL, Zhang DM, Hong DY, Wang XR (2003a) Chromosomal localization of 5S and 18S-
5.8S-25S ribosomal DNA sites in five Asian pines using fluorescence *in situ* hybridization.
Theor Appl Genet 106:198–204

Liu ZL, Zhang DM, Wang XQ, Ma XF, Wang XR (2003b) Intragenomic and interspecific 5S
rDNA sequence variation in five Asian pines (*Pinus*). Am J Bot 90:17–24

Liu ZL, Zhang DM, Wang XR (2003c) Characterization of 5S rRNA gene sequence and secondary structure in gymnosperms. Acta Genet Sin 30:88–96

Luo YB (2004) Cytological studies on some representative species of the tribe Orchideae
(Orchidaceae) from China. Bot J Linn Soc 145:231–238

Luo YB, Chen SC (2003) *Hemipiliopsis*, a new genus of Orchidaceae. Novon 13(4):450–453

Ma JS (2001) A revision of *Euonymus* (Celastraceae). THAISZIA, J Bot 11(1–2):1–264

Ma JS (2003a) The chronology of the "living fossil" *Metasequoia glyptostroboides*
(Taxodiaceae): a review (1943–2003). Harvard Pap Bot 8(1):9–18

Ma JS (2003b) On the unsolved mystery of *Metasequoia*. Acta Bot Yunnan 25(2):155–172
(Chinese with detail summary in English)

Ma JS (2004) The history of the discovery and initial seed dissemination of *Metasequoia glyptostroboides*, A "living fossil". Aliso 21(2):65–75, 2002

Ma JS, Cao W (2005) Review on floras of Mainland Asia – conifers (in press)

Ma JS, Clemants S (2005) The history and review of *Florae Republicae Popularis Sinicae*. Taxon
(in press)

Ma JS, Gilbert MG (1998) The floras of China. In: Boufford DE, Ohba H (eds) Sino-Japanese
flora—its characteristics and diversification, Bull Univ Mus Univ Tokyo 37:37–49

Ma JS, Liu QR (1998) The present situation and prospects of plant taxonomy in China. Taxon
47(1):67–74

Ma JS, Shao GF (2003) Rediscovery of the first collection of the "living fossil", *Metasequoia
glyptostroboides*. Taxon 52(3):585–588

Ma JS, Shao GF, Qian H, Chen JQ (2000) www.metasequoia.org

Pu FT, Wang PL, Zheng ZH, Wang YP (2000) A reclassification of *Notopterygium* Boissieu
(Umbelliferae). Acta Phytotax Sin 38(5):430–436 (Chinese with English abstract)

Qian H (2001a) Floristic analysis of vascular plant genera of North America north of Mexico:
spatial patterning of phytogeography. J Biogeogr 28:525–534

Qian H (2001b) A comparison of generic endemism of vascular plants between East Asia and
North America. Int J Pl Sci 162:191–199

Qian H (2002a) A comparison of the taxonomic richness of temperate plants in East Asia and
North America. Am J Bot 89:1818–1825

Qian H (2002b) Floristic relationships between eastern Asia and North America: test of Gray's
hypothesis. Amer Naturalist 160:317–332

Qian H, Ricklefs RE (2000) Large-scale processes and the Asian bias in temperate plant
species diversity. Nature 407:180–182

Qian H, Ricklefs RE (2001) Diversity of temperate plants in East Asia—reply. Nature
413:130–130

Qian H, Ricklefs RE (2004) Geographic distributions and ecological conservatism of disjunct
genera of vascular plants in eastern Asia and eastern North America. J Ecol 92:253–265

Qian H, Krestov P, Fu P-Y, Wang Q-L, Song J-S, Chourmouzis C (2003a) Phytogeography of
Northeast Asia. In: Kolbek J, Srutek M, Box EO (eds) Forest vegetation of Northeast Asia.
Kluwer Academic Publishers, Dordrecht, pp 51–91

Qian H, Song JS, Krestov P, Guo QF, Wu ZM, Shen XS, Guo XS (2003b) Large-scale phyto-geographical patterns in East Asia in relation to latitudinal and climatic gradients. J Biogeogr 30:129–141

Ren Y, Li ZJ, Chang HL, Lei YJ, Lu AM (2004) Floral Development of *Kindgonia* (Ranunculaceae s.l., Ranunculales). Plant Syst Evol 247(3–4):145–153

Ricklefs RE, Qian H, White PS (2004) The region effect on mesoscale plant species richness between eastern Asia and eastern North America. Ecography 27:129–136

Saarela JM, Peterson PM, Soreng RJ, Chapman RE (2003) A taxonomic revision of the Eastern North American and Eastern Asian Disjunct Genus *Brachyelytrum* (Poaceae): evidence from morphology, phytogeography and AFLPs. Syst Bot 28(4):674–692

Saunders RMK (2000) Monograph of *Schisandra (Schisandraceae)*. American Society of Plant Taxonomists, Ann Arbor, Michigan, pp 1–146 (Systematic Botany Monographs 58)

Shen ZH, Zhang XS (2000) A quantitative analysis on the floristic elements of the Chinese subtropical region and their spatial patterns. Acta Phytotax Sin 38(4):366–380 (Chinese with English abstract)

Shi S, Huang Y, Zhong Y, Du Y, Zhang Q, Chang H, Boufford DE (2001) Phylogeny of the Altingiaceae based on cpDNA matK, PY-IGS and nrDNA ITS sequences. 230 (1–2):13–24

Smith AR, Zhang XC (2002) *Caobangia*, a new genus and species of Polypodiaceae from Vietnam. Novon 12(4):546–550

Song BH, Wang XQ, Li FZ, Hong DY (2001) Further evidence for paraphyly of the Celtidaceae from the chloroplast gene matK. Plant Syst Evol 228(1–2):107–115

Song BH, Wang XQ, Wang XR, Sun LJ, Hong DY, Peng PH (2002) Maternal lineages of *Pinus densata*, a diploid hybrid. Mol Ecol 11:1057–1063

Song BH, Wang XQ, Wang XR, Ding KY, Hong DY (2003) Cytoplasmic composition in *Pinus densata* and population establishment of the diploid hybrid pine. Mol Ecol 12:2995–3001

Sun G, Ji Q, Dilcher DL, Zheng SL, Nixon KC, Wang SF (2002) Archaefructaceae, a new basal angiosperm family. Science 296:899–904

Sun H (2002a) Tethys Retreat and Himalayas–Hengduanshan Mountains Uplift and their significance on the origin and development of the Sino-Himalayan elements and alpine flora. Acta Bot Yunnan 24(3):273–288 (Chinese with English abstract)

Sun H (2002b) Evolution of Arctic–Tertiary flora in Himalayan–Henduan Mountains. Acta Bot Yunnan 24(6):671–688 (Chinese with English Abstract)

Sun K, Chen X, Ma R, Li C, Wang Q, Ge S (2002) Molecular phylogenetics of *Hippophae* L. (Elaeagnaceae) based on the internal transcribed spacer (ITS) sequences of nrDNA. Plant Syst Evol 235(1–4):121–134

Tang YC, Lu AM (2003) Phylogeny of the "polyphyletic-polychronic-polytopic" system of classification of angiosperms—a response to Fu De-Zhi. Acta Phytotax Sin 41(2):199–208 (Chinese with English summary)

Tang YC, Lu AM (2004) A comparison of family circumscription between FRPS and FGAC. Acta Bot Yunnan 26(2):129–138 (Chinese and Latin w/English abstract)

Walter KS (2001) Overview of the living collections at RBGE in Royal Botanic Garden Edinburgh—Catalogue of Plants 2001, pp 9–23, Royal Botanic Garden, Edinburgh

Wang DM, Hao SG, Wang Q (2003) *Hsua deflexa* sp. nov. from the Xujiachong Formation (Lower Devonian) of eastern Yunnan, China. Bot J Linn Soc 142:255–271

Wang H, Mill RR, Blackmore S (2003a) Pollen morphology and infra-generic evolutionary relationships in some Chinese species of *Pedicularis* (Scrophulariaceae). Plant Syst Evol 237(1–2):1–17

Wang H, Li WL, JCai J (2003b) Correlations between floral diversity and pollination patterns in *Pedicularis* (Scrophulariaceae). Acta Bot Yunnan 25(1):63–70

Wang ML, Chen ZD, Zhang XC, Lu SG, Zhao GF (2003) Phylogeny of the Athyriaceae: evidence from chloroplast trnL-F region sequences. Acta Phytotax Sin 41(5):416–426

Wang QF, Taylor WC, He ZR (2002) *Isoetes yunguiensis* (Isoetaceae), a new basic diploid quill-wort from China. Novon 12(4):587–591

Wang WM (2004) On the origin and development of *Artemisia* (Asteraceae) in the geological past. Bot J Linn Soc 145:331–336

Wang WP, Hwang CY, Lin TP, Hwang SY 2003 Historical biogeography and phylogenetic rela-tionships of the genus *Chamaecyparis* (Cupressaceae) inferred from chloroplast DNA polymorphism. Plant Syst Evol 241 (1–2):13–28

Wang WT (2000–2001) Notes on the genus *Clematis* (Ranunculaceae). Acta Phytotax Sin 38(4):305–336, 38(5):401–429, 38(6):497–514, 2000; 39(1):1–19, 39(4):309–336

Wang WT (2002–2004) A revision of *Clematis* (Ranunculaceae). Acta Phytotax Sin 40(3):193–241, 2002; 41(1):1–62, 41(2):97–172, 2003; 42(1):1–72, 42(2):97–135

Wang YF, Li CS, Li ZY, Fu DZ 2001 *Wuyunanthus* gen. nov., a flower of Celastraceae from the Palaeocene of north-east China. Bot J Linn Soc 136:323–327

Wang YF, Li CS, Collinson ME, Lin J, Sun SG (2003) *Eucommia* (Eucommiaceae), a potential biothermometer for the reconstruction of Paleoenvironments. Am J Bot 90(1):1–7

Wang YJ, Li XJ, Hao G, Liu JQ (2004) Molecular phylogeny and biogeography of Androsace (Primulaceae) and the convergent evolution of cushion morphology. Acta Phytotax Sin 42(6):481–499

Wang YJ, Liu JQ (2004) A preliminary investigation on the phylogency of *Saussurea* (Asteraceae:Cardueae) based on chloroplast DNA trn L-F sequences. Acta Phytotax Sin 42(2):136–153 (Chinese with English Abstract)

Wang YQ, Zhang DX, Renner SS, Chen ZY (2004) Botany: a new self-pollination mechanism. *Nature* 431:39–40

Wang YZ, Li ZY (2002) Inflorescence development of *Whytockia* (Epithemateae, Gesneriaceae) and phylogenetic implications within Gesneriaceae. Plant Syst Evol 236(1–2):45–54

Wei YG (2004) *Paralagarosolen*, a new genus of the Gesneriaceae from Guangxi, China. Acta Phytotax Sin 42(6):528–532

Wei XX, Wang XQ (2003) Phylogenetic split of *Larix* evidence from paternally inherited cpDNA trnT-trnF region. Plant Syst Evol 239(1–2):67–77

Wen J (2000) Revision of some problematic taxa of *Aralia* L. (Araliaceae) from China. Acta Phytotax Sin 38(1):1–9 (Chinese with English abstract)

Wen J (2002) Revision of *Aralia* sect. Pentapanax (Seem.) J. Wen (Araliaceae). Cathaya 13–14:1–117

Wen J, Plunkett GM, Mitchell AD, Wagstaff SJ (2001) The evolution of Araliaceae:a phyloge-netic analysis based on ITS sequences of nuclear ribosomal DNA. Syst Bot 26(1):144–167

Wu PC, Jia Y, Wang MZ (2001) Phytogeographical relationships of the bryophytes between China and North America. Acta Phytotax Sin 39(6):526–539 (Chinese with English abstract)

Wu SA, Lu HL, Yang J, Yao GY, You RL, Ge S, Zhong Y (2000) Molecular systematic studies on the tribe Polygonateae (s.l.) in China based on RFLPs data of PCR-amplified chloroplast DNA fragments. Acta Phytotax Sin 38(2):97–110 (Chinese with English abstract)

Wu ZY, Chen SC (2004) Flora Reipublicae Popularis Sinicae, volume 1, Introduction. Science Press, Beijing, p 1–1044 (Chinese)

Wu ZY, Lu AM, Tan YC (1998a) A comprehensive study of "Magnoliidae" sensu lato—with special consideration on the possibility and the necessity a new "polyphyletic-poly-chronic-polytopic" system of angiosperms. In: Zhang AL, Wu SG (eds) Floristic charac-teristics and diversity of East Asian plants. China Higher Education Press and Springer Verlag, Beijijng and Berlin, pp 269–334

Wu ZY, Lu AM, Tang YC, Chen ZD, Li DZ (2003a) The families and genera of Angiosperms in China—a comprehensive analysis. Science Press, Beijing, p 1–1209 (Chinese)

Wu ZY, Raven PH (1994, 1995, 1996, 1998, 1999, 2000, 2001a, 2001b, 2003a, 2003b, 2005) *Flora of China*, vol 17, 16, 15, 18, 4, 24, 8, 6, 9, 5, 14. Science Press, Beijing, Missouri Botanical Gardens, St Louis, Mo.

Wu ZY, Raven PH (1998, 1999, 2000a, 2000b, 2001, 2002, 2003a, 2003b, 2004a, 2004b) Flora of China: illustration, vol. 17, 16, 15, 18, 4, 24, 8, 6, 9, 5. Science Press, Beijing, Missouri Botanical Gardens, St Louis, Mo.

Wu ZY, Tang YC, Lu AM, Chen ZD 1998b On primary subdivisions of the Magnoliophyts–to wards a new scheme for an eight-class system of classification of the angiosperms. Acta Phytotax Sin 36:385–402 (Chinese with English abstract)

Wu ZY, Zhou ZK, Li DZ, Peng H, Sun H (2003b) The areal-types of the world families of seed plants. Acta Bot Yunnan 25(3):245–257 (Chinese with English Abstract)

Xia YM, Kress WJ, Prince LM (2004) Phylogenetic analyses of *Amomum* (Alpinioideae: Zingiberaceae) using ITS and matK DNA sequence data. Syst Bot 29:334–344

Xiang QP, Farjon A (2003) Cuticle morphology of a newly discovered conifer, *Xanthocyparis vietnamensis* (Cupressaceae), and a comparison with some of its nearest relatives. Bot J Linn Soc 143:315–322

Xiang QP, Fajon A, Li ZY, Fu LK, Liu ZY 2002 *Thuja sutchuenensis*: a rediscovered species of the Cpressaceae. Bot J Linn Soc 139:305–310

Xiang XP, Xiang QY, Liston A, Zhang XC (2004) Phylogenetic relationships in *Abies* (Pinaceae): evidence from PCR-RFLP of the nuclear ribosomal DNA internal transcribed spacer region. Bot J Linn Soc 145(4):425–435

Xu FX (2003) Sclerotesta morphology and its systematic implications in magnoliaceous seeds. Bot J Linn Soc 142:407–424

Xu TL, Chen QH (2001) Some new recorded plants and characteristics of ecogeographical distribution of Scrophulariaceae from Guizhou. *Guihaia* 21(1):32–34

Yang CF, Guo YH (2004) Pollen size-number trade-off and pollen-pistil relationships in *Pedicularis* (Orobanchaceae). Plant Syst Evol 247(3–4):177–185

Yang CF, Guo YH, Gituru RW, Sun SG (2002) Variation in stigma morphology—how does it contribute to pollination adaptation in *Pedicularis* (Orobanchaceae)? Plant Syst Evol 236(1–2):89–98

Yang FS, Wang XQ, Hong DY (2003) Unexpected high divergence in nrDNA ITS and extensive parallelism in floral morphology of *Pedicularis* (Orobanchaceae). Plant Syst Evol 240 (1–4):91–105

Yang QE (2001) Over-reliance of SCI damages the research of traditional taxonomy in China – some thoughts after reading two letters in "Nature". Acta Phytotax Sin 283–288 (Chinese)

Yang SZ, Saunders RMK, Hsu CJ (2002) *Thismia taiwanensis* sp. nov. (Burmanniaceae tribe Thismieae):first record of the tribe in China. Syst Bot 27(3):485–488

Yang Y (2004) Ontogeny of triovulate cones of *Ephedra intermedia* and origin of the outer envelope of ovules of Ephedraceae. Am J Bot 91(3):361–368

Yang Y, Fu DZ (2001) Review on the megastrobilus theories of conifers. Acta Phytotax Sin 39(2):169–191 (Chinese with English abstract)

Yin T, Wang XR, B. Andersson B, Lerceteau-Köhler E (2003) Nearly complete genetic maps of *Pinus sylvestris* L. (Scots pine) constructed by AFLP markers in a full-sib family. Theor Appl Genet 106:1075–1083

Ying TS, Hsu KS (2002) An analysis of the floras of seed plants of Taiwan, China: its nature, characteristics, and relations with the flora of the mainland. Acta Phytotax Sin 40(1):1–51 (Chinese with English abstract)

Yue JP, Gu ZJ, Al-Shehbaz IA, Sun H (2004) Cytological studies on the Sino-Himalayan endemic *Solms-laubachia* (Brassicaceae) and two related genera. Bot J Linn Soc 145:77–86

Zhang DX (2001) Phylogeneic reconstruction of *Burmannia* L. (Burmanniaceae) a preliminary study. Acta Phytotax Sin 39(3):203–223

Zhang DX, Saunders RMK, Hu CM (1999) *Corsiopsis chinensis* gen. et sp. nov. (Corsiaceae): first record of the family in Asia. Syst Bot 24:311–314

Zhang DX, Saunders RMK, Hu CM (2000) Addition to the *Florae Reipublicae Popularis Sinicae*:the family Corsiaceae. Acta Phytotax Sin 38(6):578–581 (Chinese with English abstract)

Zhang HD (1986) Outline of spermatophyte classification. Acta Sci Nat Univ Sunyatseni 1:1–13 (Chinese)

Zhang HD (2000) New system of seed plants (Spermatophyta). Chin Bull Bot 17:152–160 (Chinese)

Zhang JJ, Pan SH, Zhu WJ, Niu HJ, Ye ZJ, Zhu JH, Hsu PS (2003) *Taxodiomeria* (Taxodiaceae), an intergeneric hybrid between *Taxodium* and *Cryptomeria* from Shanghai, People's Republic of China. Sida 20(3):999–1006

Zhang L, Li QJ, Deng XB, Ren PY, Gao JY (2003) Reproductive biology of *Alpinia blepharocalyx* (Zingiberaceae): another example of flexistyly. Plant Syst Evol 241(1–2):67–76

Zhang SR (2004) Revision of *Kobresia* (Cyperaceae) in Xizang (Tibet). Acta Phytotax Sin 42(3):194–221 (Chinese with English abstract)

Zhang XC (2004) Miscellaneous notes on Pteridophytes from China and neighgboring regions (IV)—validation of some combinations in *Diplopterygium* (Pteridophyta: Gleicheniaceae). Novon 14(1):149–151

Zhang XC, Liu QR, Xu J (2003) Systematics of *Platygyria* (Polypodiaceae). Acta Phytotax Sin 41(5):401–415

Zhang YH (2003) Delimitation and revision of *Hilliella* and *Yinshania* (Brassicaceae). Acta Phytotax Sin 41(4):305–349

Zhao X, Zhou ZQ, Lin QB, Pan KY, Hong DY (2004) Molecular evidence for the interspecific relationships in *Paeonia* sect. *Moutan*; PCR-RFLP and sequence analysis of glycerol-3-phosphate acyltransferase (GPAT) gene. Acta Phytotax Sin 42(3):236–244 (Chinese with English abstract)

Zhou ZQ, Pan KY, Hong DY (2003) Phylogenetic analyses of *Paeonia* section *Moutan* (tree peonies, *Paeoniaceae*) based on morphological data. Acta Phytotax Sin 41(5):436–446

Zhou ZY, Zheng SL (2003) Palaeobiology: the missing link in *Ginkyo* evolution. Nature 423:821–822

Zhu H, Roos MC (2004) The tropical flora of southern China and its affinity to Indo-Malesian Flora. Telopea 10(2):639–648

Zhu H, Wang H, Li BG (2000) *Gymnanthes* Sw. (Euphorbiaceae), a genus new to China and its biogeographical implication. Acta Phytotax Sin 38(5):462–463

Zhu H, Wang H, Li B, Sirirugsa P (2003) Biogeography and floristic affinities of the Limestone Flora in southern Yunnan, China. Ann Missouri Bot Gard 90(3):444–465

Zhu ZY (2001) *Changruicaoia* Z. Y. Zhu—a new genus of Labiatae from Mount Emei, Sichuan, China. Acta Phytotax Sin 39(6):540–543 (Latin and Chinese)

Jinshuang Ma, Ph.D.
Research Taxonomist,
Brooklyn Botanic Garden,
1000 Washington Avenue,
Brooklyn, NY 11225-1099
Tel.: 718-623-7357
Fax: 718-941-4774
e-mail: jinshuargma@bbg.org
www.bbg.org
www.metasequoia.org

Ecology

Structural determinants of leaf light-harvesting capacity and photosynthetic potentials

Ülo Niinemets and Lawren Sack

1 Introduction

The traits characterizing plant functioning include simple dimensions such as leaf area, leaf thickness (T), display angle, and ratios of simple traits [e.g., leaf dry mass per unit area (M_A)] as well as normalized rates [e.g., net maximum photosynthetic rate per unit mass (A_{mass}) or per unit area (A_{area})], contents [e.g., N per unit dry mass (N_{mass})], and "efficiencies" [gain/cost; e.g., photosynthetic N-use efficiency (PNUE=A_{mass}/N_{mass})]. Current plant science research mostly emphasizes the role of physiological traits in altering plant competitive ability, but the determinants of leaf light-harvesting capacity and foliar photosynthetic potentials also include numerous structural traits. In fact, while chloroplastic metabolism has remained remarkably conserved throughout phylogeny, plant evolution has led to a large diversity in foliar anatomy, morphology and shape that may tremendously modify the resource capture efficiency of leaves with essentially the same metabolic constitution (Smith et al. 2004).

Apart from evolutionary adaptations, all traits have an enormous spatial and temporal variability. The evolutionary, developmental and environmental variations in traits, and the large number of potentially important traits and trait combinations, complicate predictions of relevant plant functions from the collections of traits. However, many traits that alter the same plant function co-vary along environmental gradients and among species. Understanding such coordinated variations among trait assemblages may significantly simplify projections of plant functioning in changing environmental conditions. Analyses of the trait co-variations have identified a series of general correlations among relevant plant structural and functional traits. For instance, across large species sets, A_{mass} correlates positively with N_{mass} and negatively with M_A and leaf life-span (Reich et al. 1997, 1999; Niinemets 2001; Wright et al. 2004b).

We review recent work on inter-coordinated structural traits relevant to the capture and utilization of irradiance and carbon, mainly focusing on leaf tissue and whole-leaf scales. We demonstrate major structural controls on light-harvesting efficiency that result from constraints on leaf size, shape and support investments, and emphasize the important role of leaf venation architecture and internal leaf structure in the supply of water to the leaves and the internal diffusion limitations of photosynthesis. Because in natural environments, N is generally found in concentrations limiting leaf development and photosynthesis, we further examine the coordinated variations in leaf carbon gain potentials with foliar N content. As the cost/benefit ratios of specific leaf constitutions depend on leaf life-span, we also consider the relationships with leaf aging and longevity as part of the fundamental trade-offs.

2 Structural limitations of leaf light-harvesting efficiency

A variety of structural characteristics operative at different hierarchical scales affect light interception per unit leaf area. At a tissue scale, structural modifications alter the amount of light intercepted per unit chlorophyll, while at a leaf scale, structural changes modify the exposure of single leaves. Finally, leaf arrangement and aggregation on the shoot further significantly alter the average irradiance on the leaf surface (Fig. 1).

Although the theoretical light interception efficiencies of specific leaf architectures may be very high, in evaluating the adaptive adjustments, it is

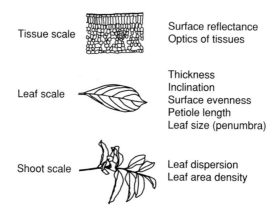

Fig. 1. Summary of the structural determinants of leaf light-interception efficiency at various hierarchical levels. At the tissue scale, the structural characteristics primarily modify the distribution of light within the leaf and the amount of light intercepted per unit chlorophyll. At the leaf and shoot scales, foliage structural characteristics alter the average irradiance and the distribution of light on the leaf surface

necessary to consider that enhanced light interception may be costly. In addition to large N investments in chlorophyll and associated pigment-binding proteins, structural modifications of light interception at every hierarchical level involve costs as well. These may be direct costs such as the biomass required for the construction of extra structures or indirect costs such as the reduction of leaf photosynthetic activity due to enhanced diffusion limitations. Due to large structural costs, the light-harvesting efficiencies of various leaf displays are always lower than the maximum efficiencies even in most light-limited environments (Valladares et al. 2002).

2.1 Tissue-Scale Limitations of Light Interception

The organization of structural elements within the leaf plays a major role in light capture. Because chlorophyll absorbs light very efficiently, most irradiance is often absorbed near the top layers of the leaf (Buckley and Farquhar 2004). To increase the uniformity of the illumination field within the leaves, and the total amount of light intercepted per unit chlorophyll, many plant species possess lens-like epidermal cells that focus the light in the leaf interior (Poulson and Vogelmann 1990; Smith et al. 1997). Leaf transversing bundle-sheath extensions in heterobaric leaves also transfer light into deeper layers (Nikopoulos et al. 2002), implying a potentially important linkage between leaf venation architecture and leaf light interception capacity. In many Mediterranean sclerophylls that possess a sclereid network anchored to the inner surface of epidermis, sclereids appear also to function as optical fibers guiding light into the inner leaf layers (Karabourniotis et al. 1994; Karabourniotis 1998). However, enhancement of light-harvesting capacity by these specific adjustments involves specific trade-offs with leaf photosynthetic capacity. Bundle sheath extensions cover up to 50% of the leaf surface area (McClendon 1992; Nikopoulos et al. 2002), thereby significantly reducing leaf photosynthetic capacity per unit area. Furthermore, bundle sheath extensions also effectively avoid the lateral diffusion among neighboring compartments surrounded by bundle sheath extensions (areoles; Terashima 1992), potentially reducing leaf photosynthesis when stomata are non-uniformly open.

A further important factor modifying leaf light interception efficiency, is the differentiation of leaf mesophyll into palisade and spongy layers. The cells in palisade parenchyma are elongated and parallel to the direction of direct light, facilitating light channeling into the leaf (Vogelmann and Martin 1993). The spongy mesophyll cells are less regularly arranged, leading to greater effective light pathlength and light scattering (Terashima 1989;

DeLucia et al. 1996; Evans et al. 2004). Consequently, the amount of light intercepted per unit chlorophyll is greater in spongy than in palisade mesophyll (Terashima 1989; DeLucia et al. 1996). This implies that the efficiency of leaves with given chlorophyll content and mesophyll thickness varies according to the distribution of mesophyll between palisade and spongy layers and the relative distribution of incident light between direct and diffuse components. Furthermore, due to different light interception efficiencies of palisade and spongy mesophyll as well as differences in distribution of photosynthetic enzymes along the leaf, leaf photosynthesis rates are different for an adaxial and abaxial illumination (Sun and Nishio 2001; Evans and Vogelmann 2003; Buckley and Farquhar 2004).

In many species in stressful environments with potentially high fractions of excess light energy, the leaf surface is covered by trichomes that reduce penetration of visible and UV light (Karabourniotis and Bornman 1999; Karabourniotis et al. 1999). Leaf pubescence is a very effective means of reducing light interception [see Cescatti and Niinemets (2004) for a review], but the overall cost to the plant may be large. Few data of the cost of pubescence are available, but in the tropical evergreen broadleaf species *Metrosideros polymorpha*, at high elevations up to 15% (34 g m^{-2}) of foliar biomass may be invested in trichomes (total M_A=229 g m^{-2}; Geeske et al. 1994).

2.2 Thickness (I) and Inclination Effects on Light Harvesting

Thicker leaves have a greater number of mesophyll cell layers, and a higher light absorption per unit area. However, light capture will not increase linearly with T, as fewer photons penetrate to additional mesophyll layers (Vogelmann and Evans 2002), and the amount of light intercepted per unit chlorophyll invested decreases as more chlorophyll is stacked within the leaf. Due to multiple scattering within the leaves, at a common chlorophyll content, the light interception capacity is larger for thicker leaves (Evans and Poorter 2001). For a series of forest species, it has been demonstrated that an increasing fraction of chloroplasts functions below light saturation with increasing T (Green and Kruger 2001), further demonstrating the trade-offs between high investment of structural and photosynthetic compounds per unit leaf area and efficiency of light interception. Thus, for maximization of whole plant light interception with a certain amount of N and C available for the construction of foliage, plants should optimize leaf chlorophyll content and T simultaneously with total leaf area (Cescatti and Niinemets 2004).

Apart from the internal architecture, thickness and chlorophyll content, the leaf light interception capacity depends on the inclination. Numerous studies have explored the efficiency of light harvesting of leaves with different inclinations [see Valladares (2003) for a review], leading to a general consensus that more horizontal leaves intercept light more efficiently, especially in understory environments where most of the light penetrates at low zenith angles (Valladares 2003; Cescatti and Niinemets 2004). Yet, the adaptive significance of specific leaf inclinations must be assessed within the context of other leaf and shoot structural characteristics. For densely leafed shoots, where the leaf angle is kept constant, foliar light interception efficiency is significantly enhanced by increasing the length of petioles, because this minimizes the self-shading between neighboring leaves (Takenaka 1994; Pearcy and Yang 1998). However, the biomass of the petiole necessary to keep the leaf at a certain angle increases with the cube of petiole length (Niklas 1999), implying that the maintenance of horizontal angles becomes increasingly costly with increasing petiole length.

Overall, there is a strong general relationship between foliar biomass investment in petiole and petiole length (Fig. 2A), further underscoring these biomechanical limitations. As part of this relationship, the fractional biomass investment in the leaf rachis in compound-leaved species increases with the number of leaflets per leaf (Fig. 2B). The latter relationship illustrates the trade-off between the support costs, which are minimized when

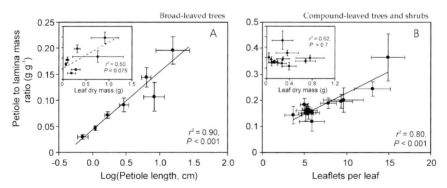

Fig. 2. Petiole dry mass (M_p) to leaf lamina dry mass (M_L) ratio in relation to (**A**) petiole length in seven temperate broadleaved deciduous trees, and (**B**) the number of leaflets in 16 compound-leaved temperate deciduous trees and shrubs. *Insets* demonstrate the relationships between M_p/M_L and total leaf dry mass. *Error bars* indicate ±SD. The data in **A** ($n=18$-156 for each species) were derived from Niinemets and Kull (1999; *Betula pendula, Fraxinus excelsior, Populus tremula, Tilia cordata*) and Niinemets and Fleck (2002b: *Liriodendron tulipifera*, 2002a: *Fagus orientalis, F. sylvatica*). The data in **B** are modified from Niinemets (1998). Data are fitted by linear regressions

the mass is located closer to the axis of rotation, and the light interception efficiency, which is maximized when there is no overlap between neighboring leaves. Thus, for a maximum light interception efficiency, foliar mass should be located farther away from the axis of rotation as more leaflets are added to the leaf rachis.

These relationships collectively demonstrate that effective light interception via maintenance of favorable leaf inclination angles may have a significant cost for the plant. Although the petioles contain chlorophyll and Rubisco, the concentrations of these photosynthetic compounds are four- to ten-fold less than in the leaf lamina (Niinemets 1999a), and the surface area to mass ratio is also several-fold lower, so the photosynthetic activity of petioles is minor compared to that of the leaf lamina (Hibberd and Quick 2001). These data suggest that, depending on the fractional investment of foliar biomass in petiole and rachis, plants investing a constant fraction of biomass in leaves, and with the same M_A of leaf laminas, may have widely differing photosynthetic activities per unit total foliage dry mass.

2.3 Light Capture Dependence on the Flatness of the Leaf Lamina

The inclination of leaf lamina with respect to the horizontal is commonly measured as the angle between the lamina tip and bottom, but leaf laminas are rarely completely flat. More frequently, leaf laminas are curled, folded or rolled to a varying extent, whereas the degree of leaf rolling increases with leaf exposure in the canopy (Fig. 3). Due to complexities in formal analysis

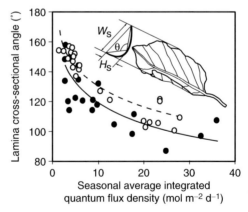

Fig. 3. Lamina cross-sectional angle (θ) in relation to average integrated leaf irradiance in a dominant (*filled symbols*, $r^2=0.64$) and a sub-dominant (*open symbols*, $r^2=0.77$, $P<0.001$ for both) *Fagus sylvatica* tree [modified from Fleck et al. (2003)]. θ as defined in the *inset*. In addition to θ, the degree of leaf rolling can also be characterized by the width (W_S) and height (H_S) of the leaf space.

of uneven lamina light interception efficiency, the influence of such three-dimensional lamina architecture on leaf light harvesting has been characterized in only very few instances (Sassenrath-Cole 1995; Sinoquet et al. 1998; Farque et al. 2001; Fleck et al. 2003). Overall, these studies demonstrate a strong reduction of leaf light-harvesting efficiency in rolled leaves and also shifting contributions of upper and lower leaf surfaces to total leaf light interception (Fleck et al. 2003).

Although the increases in the degree of leaf rolling and lamina total to projected area ratio decrease leaf light-harvesting efficiency, they also increase the photosynthesis per unit projected lamina area due to stacking of photosynthetic biomass. Because enhanced foliage stacking at higher light increases light penetration to deeper canopy layers, such a fundamental trade-off between C gain potential and light-harvesting efficiency may maximize the light interception and C gain of dense canopies at a common biomass investment in leaves.

2.4 Modification of Light Harvesting by Leaf Size

Geometrical models considering the sun as a point light source predict that leaf size per se and leaf lobing do not directly alter leaf light-interception efficiency (Niklas 1989). These models, however, suggest that decreases in leaf size and increases in leaf lobing may increase the occurrence of light-flecks in deeper canopy layers (Niklas 1989). Given that in deeply shaded canopy layers more than half of the total irradiance can penetrate as short-duration lightflecks (Chazdon and Pearcy 1991), the variation in leaf size may modify whole-canopy C gain to a significant extent.

Apart from the modification of the sunlit and shaded leaf area fractions, leaf size can also affect the intensity of direct solar radiation on the leaf surface. This is because, in reality, the solar disk as viewed from the earth has a finite radius of about 0.27 degrees. While in full sun the entire solar disk is visible, and no sun is visible in a complete shade, in intermediate situations that are called penumbra, the solar disk is only partly shaded by phytoelements. Because penumbra makes the light environment more uniform compared with the existence of only sunlit and shaded areas, a large degree of penumbral canopy leaf area increases the whole canopy photosynthetic potential (Cescatti and Niinemets 2004). It is generally thought that the significance of penumbra increases with decreasing leaf size, but the relevant parameter characterizing the role of penumbra is actually leaf size to canopy height ratio rather than leaf size alone (Cescatti and Niinemets 2004). Thus, at a common leaf size, penumbra plays a more prominent role in taller canopies.

Leaf size may further significantly alter the light interception by modifying shoot architecture. In particular, smaller leaves can be more densely packed on the shoot, thereby maximizing the shoot light interception when the irradiance is high (King and Maindonald 1999; Bragg and Westoby 2002, Fig. 1). However, denser leaf packing inevitably results in larger self-shading within the shoot (Valladares and Pearcy 1999), and reduces the average irradiance on the leaf surface (Niinemets et al. 2004b). Thus, less densely leafed shoots become more efficient with decreasing irradiance (Valladares and Pearcy 1999; Niinemets et al. 2004b). As larger leaves require less woody branches to support a common leaf area, and can also be arranged in space relatively independently of stem axes due to lamina translation around the petiole, and changing leaf and petiole curvatures and petiole length, an increase in leaf size may be favored by decreases in light availability.

3 How Structure Determines Leaf Photosynthetic Capacity

Efficient light harvesting must be matched by CO_2 assimilation potentials to ensure optimal conversion of light to chemical energy and finally to growth. A plethora of varying leaf anatomical structures exists across different plant functional types, strongly modifying the photosynthetic efficiency of unit foliar area and mass. Photosynthetic organs differ in the volume fraction of different tissues as well as in the size and packing of cells in specific tissues, and in the thickness and chemical composition of cell walls. To characterize this vast heterogeneity, a general and easily measurable leaf variable, M_A, is employed in studies describing the effect of leaf structure on leaf photosynthesis. However, the relationships between photosynthesis and M_A are often variable. Such variability is partly because M_A is a product of thickness (T) and density (D) that may vary independently, and that affect leaf photosynthetic capacity differently (Witkowski and Lamont 1991; Niinemets 1999b).

3.1 Photosynthetic Capacity in Relation to Leaf Tissue Types and Thickness

The assimilation of CO_2 depends on the diffusion into the leaf intercellular spaces, and absorption by exposed mesophyll cells. The exposed mesophyll surface area per unit lamina area (S_m) ranges from 5 to 50 m^2 m^{-2} among broadleaved species (Hanba et al. 1999, 2001, 2002), and strongly increases with increasing T (Hanba et al. 1999). Because most of the chloroplasts adhere to the inner surface of exposed mesophyll cell walls, scaling of S_m

with T implies also a positive relationship between T and exposed chloroplast to total lamina surface area ratio, directly enhancing the internal diffusion conductance (Syvertsen et al. 1995; Hanba et al. 1999, 2002).

Overall, these data demonstrate a larger surface area for diffusion in thicker leaves (Terashima et al. 2001) such that the increase in T does not necessarily imply enhanced within-leaf diffusion limitations. However, it is further relevant that thicker leaves which contain additional layers of mesophyll also have higher total contents of Rubisco and leaf N per unit area (Nobel 1977; Koike 1988; Niinemets 1999b; Roderick et al. 1999). Therefore, A_{area} increases with increasing T due to stacking of photosynthesis-limiting proteins per unit leaf area (Niinemets 1999b).

Superimposed by these general trends, species- and functional type-specific leaf anatomical characteristics may further modify leaf photosynthetic potentials for any particular T. The correlation between A_{area} and T assumes that changes in thickness are primarily due to changes in mesophyll within the leaves. However, leaf mesophyll is embedded within cuticle and epidermal layers, which contribute differently to total T among species, implying a significant variation in leaf photosynthetic potentials. In addition, plants from high altitudes generally possess a pronounced hypodermis, significantly reducing the fraction of mesophyll in the leaf lamina for a common T (Grubb 1977). Furthermore, at a common T, cells with differing size and shape result in widely varying values of S_m. For instance, more elongated palisade cells relative to spongy mesophyll cells (Nobel 1977; Slaton et al. 2001) or smaller cells with a given geometry (Wilson and Cooper 1970) provide a larger exchange surface area.

3.2 Does the Efficiency of Intraleaf Diffusion Vary for Leaves of Differing Structure? Role of D and M_A

It has been suggested that woody species with inherently larger M_A and D have lower internal diffusion conductances per unit area (g_{area}) than the herbaceous species with less tightly packed mesophyll and thinner cell walls (Lloyd et al. 1992; Epron et al. 1995; Syvertsen et al. 1995). Because the values of g_{area} scale positively with A_{area}, this conclusion of structural determinants of g_{area} has been recently challenged (Evans and Loreto 2000). Evans and Loreto (2000) suggested that because of the coordination of g_{area} with A_{area}, the overall drawdown in the CO_2 mole fraction between sub-stomatal cavities (C_i) and chloroplasts (C_c) due to internal conductance ($C_c = C_i - A_{area}/g_{area}$) is similar among species with differing leaf structure. Indeed, data demonstrating variations in the C_c/C_i ratio in leaves of varying g_{area} are essentially lacking. However, conceptually it is further important that for three-dimensional structures such as leaf laminas the C_c/C_i ratio is a volume-weighted average estimate, and should accordingly scale with g_{mass} (g_{area}/M_A) and A_{mass} (A_{area}/M_A) and not necessarily with the area-based estimates.

We compiled the published data for g_{area} and A_{area}, and g_{mass} and A_{mass} for 22 species, and calculated the C_c/C_i ratios for all measurements (Fig. 4). These data demonstrate a strong coordination between both g_{area} and A_{area} (Fig. 4A) and g_{mass} and A_{mass} (Fig. 4B), but also that the relationships tended

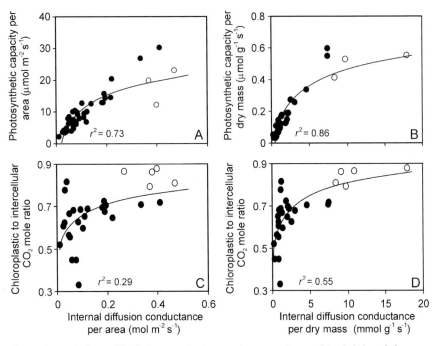

Fig. 4. Co-variations of leaf photosynthetic capacity per unit area (A_{area}) (**A**) and dry mass (A_{mass}) (**B**) with internal diffusion conductance per unit area (g_{area}), and dry mass (g_{mass}), and the relationships between the ratio of chloroplastic to intercellular CO_2 mole ratio and g_{area} (**C**) and g_{mass} (**D**) for a wide range of mature leaves of broad-leaved shrubs and trees (*filled symbols*) and herbs (*open symbols*). Data were fitted by non-linear regressions in the form of $y=\log(ax)+b$ and are all significant at $P<0.001$ ($n=42$). The data were derived from studies of the following species: *Acer mono, Alnus japonica* and *Populus maximowiczii* (Hanba et al. 2001); *Juglans regia* and *J. nigra×regia* (Piel et al. 2002), *Acer mono, A. palmatum* and *A. rufinerve* (Hanba et al. 2002); *Citrus limon, C. paradisi, Macadamia integrifolia* and *Prunus persica* (Lloyd et al. 1992; Syvertsen et al. 1995); *Camellia japonica, Castanopsis sieboldii, Cinnamomum camphora, Ligustrum lucidum, Quercus glauca,* and *Q. phillyraeoides* (Hanba et al. 1999); *Polygonum cuspidatum* (Kogami et al. 2001); *Phaseolus vulgaris, Metrosideros umbellata* and *Weinmannia racemosa* (DeLucia et al. 2003); and *Nicotiana tabacum* (Evans et al. 1994)

to saturate at higher internal conductance (Fig. 4A,B). The C_c/C_i ratios increased with increasing both g_{area} (Fig. 4C) and g_{mass} (Fig. 4D), indicating that the leaves of differing structural characteristics do have varying internal diffusion limitations of photosynthesis. The latter relationship was less scattered with g_{mass}, suggesting that g_{mass} is the true determinant of the volume-weighted average chloroplastic CO_2 concentration.

To further test for the anatomical limitations of diffusion, we examined the dependencies of g_{area} and g_{mass} on M_A for a species set with limited variation in leaf photosynthetic capacity (Fig. 5). In this set of data, g_{area} did not correlate with M_A, but g_{mass} significantly decreased with increasing M_A, conclusively demonstrating the dependence of internal diffusion limitations on leaf structure.

M_A is an integral measure, and it is pertinent to explore which specific structural characteristics are responsible for the variations in internal diffusion conductance and A_{mass}. Differences in cell wall thickness and chemical composition can exert a major control over leaf photosynthetic capacity.

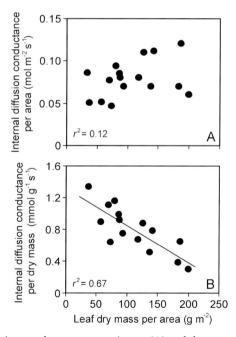

Fig. 5. Internal diffusion conductance per unit area (**A**) and dry mass (**B**) in relation to leaf dry mass per unit area (M_A) for a constrained range of leaf photosynthetic capacities (10 μmol m^{-2} s^{-1}≤A_{area}≥μ5 mol m^{-2} s^{-1}, subset of data in Fig. 4). Because of a strong scaling of mesophyll diffusion conductance with leaf photosynthetic capacity, the relationships between leaf structural variables and diffusion conductance can only be tested conclusively if foliar photosynthetic potentials are kept essentially constant

In particular, thicker and less porous cell walls can strongly impede liquid-phase CO_2 diffusion from the outer surface of cell walls to carboxylation sites in the chloroplasts (Kogami et al. 2001; Terashima et al. 2001; Miyazawa et al. 2003). Thicker and denser cell walls contribute to an overall greater bulk D, and thus, greater volume fraction of non-photosynthetic cell walls and larger liquid-phase diffusion resistance may provide an explanation for the interspecific strong negative relationships between A_{mass} and D (Niinemets 1999b) and A_{mass} and M_A (Wright et al. 2004b).

Increases in D are also associated with decreases in the volume fraction of internal air space within the leaves (Niinemets 1999b; Sack et al. 2003a). A decrease in the air volume fraction is expected to reduce the gas-phase component of the internal diffusion conductance (Parkhurst 1994), thereby potentially further curbing A_{mass}. Such a negative scaling of internal diffusion conductance with the fraction of leaf air space has been observed in some (Syvertsen et al. 1995; Hanba et al. 1999), but not in all studies (Hanba et al. 2001), indicating that the relative significance of various structural traits depends on the overall leaf constitution.

3.3 Structural Determinants of Leaf Water Transport Efficiency

Leaf stomatal conductance is a highly variable quantity, which in well-illuminated plants is primarily determined by plant water status. For the full employment of the structural and biochemical investments in the photosynthetic potentials in non-stressed conditions, A_{area} must be coordinated with the maximum stomatal conductance to water vapor (G_{max}). G_{max} varies with the total number of stomata (stomatal density) and average pore area of single stomata (Sack et al. 2003a), and accordingly, a coordination is expected between the stomatal pore area per unit leaf area, the thickness of mesophyll, and the internal mesophyll surface area (Sack et al. 2003a).

For leaves that operate at a given narrow range of leaf water potential, leaf hydraulic conductance per unit area (K_{leaf}) must further match the transpiration rate ($E=\Delta\psi K_{leaf}$, where $\Delta\psi$ is the water potential gradient between the evaporation sites in the leaf and the point of leaf attachment to the stem). In fact, recent studies have observed strong positive correlations between K_{leaf} and maximum stomatal pore area, and G_{max} and A_{area} (Aasamaa et al. 2001; Sack et al. 2003a; Brodribb and Holbrook 2004).

How does the coordination between G_{max} and K_{leaf} arise? The total leaf hydraulic conductance depends both on the conductances of leaf veins, and on the conductances of the apoplastic and symplasmic pathways of water movement from the xylem to the sites of evaporation in the mesophyll. Both components of the hydraulic pathway are important in determining the overall hydraulic conductance (Cochard et al. 2004; Sack et al. 2004; Sack and Tyree 2005). The total hydraulic conductance of leaf venation, in turn, is mainly determined by the conductances of the higher order major veins and the minor veins (Fig. 6, Sack and Tyree 2005). This may possibly reflect the fact that the midrib and lower order major veins also function as structural support, and leaves tend to invest more biomass than needed for self-support to cope with the dynamic loads experienced during windy conditions. Thus, a large bottleneck in the venation system resides in the transition between vein orders or within the minor veins (Sack et al. 2004). Given the importance of minor veins and apoplastic and symplasmic pathways, total leaf hydraulic conductance is expected to scale with the venation density and with the total number of extravascular pathways, i.e. with internal mesophyll surface area. Currently, only limited data exist on the hydraulic conductance of different venation architectures, the leakiness of vein xylem conduits, membrane permeabilities, and conductances of apoplastic pathways. Nevertheless,

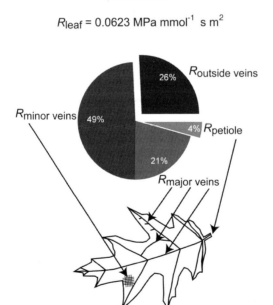

Quercus rubra

$R_{leaf} = 0.0623$ MPa mmol^{-1} s m^2

Fig. 6. Distribution of the total leaf hydraulic resistance (R_{leaf}) among different vascular components in the temperate deciduous tree *Quercus rubra* (modified from Sack et al. 2004). R rather than the inverse, hydraulic conductance (K_{leaf}), has been shown because the total pathway resistance is the sum of the component serial resistances

computer simulations (Roth et al. 1994, 1995) and observations of the occurrence of species with differing leaf venation densities in earth floras (Uhl and Mosbrugger 1999) support the control of K_{leaf} by venation density. In addition, the hypothetical scaling of K_{leaf} by the number of extravascular pathways is confirmed by a strong positive correlation of K_{leaf} with T (Fig. 7).

3.4 Leaf Size and Shape Effects on Photosynthesis: Only a Matter of Leaf Energy Balance?

Gradients in leaf size from tropical rain forests to deserts and temperate forests have been described in many studies (Grubb 1977; Medina 1984), and it is commonly thought that leaf size, and possibly leaf shape will mainly impact on A_{area} via modification of the leaf heat balance. A larger leaf will heat up more at a given air temperature, due to a thicker boundary layer and slower convective cooling. However, with higher temperature, the leaf will potentially show a higher rate of gas exchange at a given value of stomatal conductance (Parkhurst and Loucks 1972), and enhanced latent heat loss buffers against excessive leaf temperatures. As larger leaves are expected to optimize photosynthetic parameters at higher temperature, leaf size-mediated changes in leaf energy balance do not per se facilitate or inhibit photosynthetic potentials.

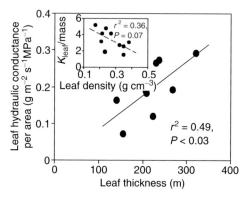

Fig. 7. Leaf hydraulic conductance (K_{leaf}) in relation to leaf thickness in exposed and shaded leaves of five temperate deciduous woody species (*Acer rubrum*, *A. saccharum*, *Betula papyrifera*, *Q. rubra*, *Vitis labrusca*) and in the evergreen vine *Hedera helix* (modified from Sack et al. 2003a). The *inset* demonstrates the correlation between K_{leaf} per unit mass (K_{leaf}/M_A; mg g^{-1} s^{-1} MPa^{-1}) and leaf density. K_{leaf}/M_A characterizes the water supply of average leaf cells. Leaf thickness (T) and density (D) are related through M_A ($M_A=TD$). For other abbreviations, see Figs. 5 and 6

However, in natural conditions, the leaves are often constrained by the availability of water, and if the leaf temperature exceeds the level that can be optimized, water loss relative to C gain becomes uneconomical, and the maximum leaf size will be constrained by water supply.

Whether there is coordination between leaf shape and gas exchange is not yet fully established. While leaf shape affects the capacity to pack the leaves on a shoot for enhanced light capture, leaf shape per se may not play much of a role in modifying leaf energy balance. Classic work using six differently shaped copper model leaves indicated that dissected and lobed leaves may dissipate heat and exchange gases more effectively than entire leaves (Vogel 1970), especially in windless conditions (free convection) or under low wind velocities (mixed convection), which lead to more extensive temperature gradients across the surface of non-lobed leaves (Roth-Nebelsick 2001). The situation may be different under strong wind velocities (forced convection), when a common boundary layer develops above the entire surface of the dissected leaf or a collection of small leaves such as a conifer shoot (Germino and Smith 1999). However, the convection efficiency of different leaf shapes may also strongly depend on the wind direction (Roth-Nebelsick 2001). Lobed and dissected leaves tilted with respect to wind direction are more efficient than their non-lobed counterparts, because wind can freely gush through the interceptions (Vogel 1970; Roth-Nebelsick 2001). Furthermore, lobed leaves are often more jagged and folded, with protruding leaf margins. Such a leaf arrangement significantly increases leaf surface roughness and reduces the critical wind speed for turbulence, thereby potentially increasing leaf heat exchange at a common wind speed (Grace 1978).

Most of the previous experimental work on leaf heat exchange has focused on steady-state conditions. However, leaf light environment in the field strongly fluctuates in time leading to "heatflecks" (Singsaas et al. 1999), which suddenly rise leaf temperature strongly above the ambient temperature. Given that dissected and small leaves generally produce more "flecky" light field, and that lateral heat conductance is larger for dissected structures, coping with temporarily excessive temperatures also favors dissected and lobed leaf shapes.

Recent findings that more lobed leaves possess higher leaf hydraulic conductance, have led to suggestions that leaf lobation is an adaptive adjustment to improve K_{leaf} (Sisó et al. 2001; Zwieniecki et al. 2004). Higher hydraulic conductance of more lobed leaves was explained by a lower amount of mesophyll tissue between the major veins, minimizing the average distance of mesophyll cells from lower-order major veins (Sisó et al. 2001; Zwieniecki et al. 2004). However, this relationship between the complexity of leaf outline and K_{leaf} is not universal (Sack et al. 2003a). In fact, when the conductance of higher-order veins and extravascular pathways limit K_{leaf}, minimization of the distance of mesophyll cells from major veins may have a limited effect on overall leaf hydraulic conductance.

3.5 A Further Linkage Between Leaf Size and A_{mass}

In addition to affecting light interception, leaf energy balance and hydraulic efficiency, differences in leaf size and shape also significantly modify leaf biomass partitioning between support and photosynthetic tissues. Larger leaves may have disproportionately greater biomass investments in vasculature and sclerenchyma to maintain lamina flatness and inclination angles and ensure effective water transport to mesophyll cells (Givnish 1984). Enhanced investment in vasculature is compatible with an efficient water-conducting pathway (Enquist et al. 1999; West et al. 1999), but also with lower photosynthetic rates (Poorter and Evans 1998; Garnier et al. 1999) and lower growth rates (Van Arendonk and Poorter 1994). So far, the biomass partitioning between major veins and the rest of the leaf lamina has been studied in a few broad-leaved species. The results indicate that the fraction of leaf biomass in major veins increases with increasing leaf size (Fig. 8, Niinemets et al. 2004a), which is the optimal strategy to maximize light harvesting and water supply to the mesophyll for a given biomass investment in mechanical support (Niinemets and Fleck 2002b). However, important interspecific differences in investments in major veins apparently also exist (Fig. 8). Given that major leaf veins can contain > 20% of lamina biomass (Fig. 8), scaling of fractional biomass investments in support with leaf size likely provides an important explanation for variation in foliage photosynthetic capacities of co-existing species that possess similar M_A and leaf longevity.

As the previous paragraphs indicate, the relationships between leaf structure and net assimilation primarily arise because leaves of varying structure and chemistry contain different fractions of photosynthetic mesophyll. Aside from modifications of thickness of cuticle, epidermis, and intercellular air space volume of interveinal leaf lamina areas, all leaves contain a significant fraction of vasculature, and also have petioles of varying thickness and length for the attachment to the stem. It is important to note that leaves of the same aggregated structural attributes such as M_A, density and thickness may contain different fractions of support tissues within a unit leaf mass, and accordingly possess contrasting foliar photosynthesis rates.

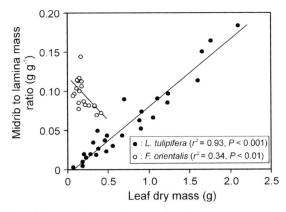

Fig. 8. Leaf midrib relative to the rest of the lamina biomass versus total leaf dry mass in *Liriodendron tulipifera* (data from Niinemets and Fleck 2002b) and *Fagus orientalis* (data from Niinemets and Fleck 2002a). *Each point* corresponds to an individual leaf. Data are fitted by linear regressions

3.6 Variation in PNUE due to Leaf Structure

The availability of limiting elements such as N may be a further factor shaping the structural coordination of leaf photosynthetic capacity. Because roughly half of leaf N is invested in the photosynthetic apparatus (Evans 1989), there is a strong correlation between A_{mass} and N_{mass} (Evans 1989; Wright et al. 2004b). Thus, the relations between leaf architecture and A_{mass} may be interpreted using N as a universal currency that limits maximum leaf photosynthetic capacity. From this perspective, A_{mass} can be expressed as the product of N_{mass} and PNUE ($=A_{mass}/N_{mass}$). Variation in N_{mass} occurs due to concentration or dilution of foliar N, and accordingly, due to variations in the thickness and density of mesophyll cell walls as well as support tissues such as epidermis, hypodermis, sclereids and veins. For instance, increases in foliar biomass investments in vasculature lead to lower foliar N concentrations of the entire lamina (Niinemets and Fleck 2002a, b), while increases in the volume fraction of photosynthetic mesophyll tissue increase N_{mass} (Reich 1998).

The variations in PNUE are associated with differences in the partitioning of leaf N among photosynthetic compounds (Hikosaka and Terashima 1996; Hikosaka et al. 1998), and overall partitioning of leaf N among photosynthetic and non-photosynthetic pools (Hikosaka and Terashima 1996; Hikosaka et al. 1998). The size of the non-photosynthetic pool may depend on leaf structure, as 10–15% of leaf N is associated with cell walls (Takashima et al. 2004), and a significant fraction of N with vasculature

(Niinemets and Fleck 2002b). Thus, as the fraction of support tissues increases, an increasing fraction of total N will be partitioned in non-photosynthetic compartments, reducing leaf PNUE. In addition to N partitioning, variation in internal CO_2 diffusion conductance further modifies the use efficiency of N invested in photosynthetic apparatus, implying a direct link between leaf structure and PNUE. Due to structural controls of leaf N partitioning within the leaf and structural controls on efficiency of diffusion, the slopes of A_{mass} versus N_{mass} may significantly vary for different species assemblages (Wright and Westoby 2002; Niinemets et al. 2004d).

Further, it may be of importance that low nutrient availability itself leads to larger M_A (Beadle 1966; Smith et al. 2004), and that this response enhances leaf nutrient conservation due to a longer leaf life-span (Wright and Westoby 2002, 2003; Wright et al. 2002, 2004a). The hypothesis of nutritional control of sclerophylly has been suggested by several studies, leading to interpretations that sclerophyllous leaf structure in Mediterranean shrublands, tropical heaths and tropical upper montane forests are driven by nutrient-limited soils (Grubb 1977; Turner 1994). However, only a few studies have investigated simultaneously nutrient and water limitations in these communities. Recent studies suggest that low water availability rather than nutrients provide the primary explanation for the occurrence of some sclerophyllous floras (Niinemets 2001; Lamont et al. 2002).

4 Structural Acclimation of Leaf Photosynthesis to Environment

In addition to understanding the broad patterns for species groups, it is important to recognize that a wide range in any leaf trait can be found within every community. The variation in leaf characteristics among species within a site is often as great or greater than the mean difference between the sites, even when the sites are spread out along a dramatic climatic gradient from a desert to a rain forest (Reich et al. 1997, 1999). There is always environmental heterogeneity within each specific site. Thus, scaling of plant photosynthetic productivity from the leaf to the global scale requires an understanding of the fundamental scaling relations as well as a consideration of within-species plasticity in these relations.

4.1 Adjustments to Light

As an acclimation response to the vertical light gradient, M_A varies severalfold in plant canopies (Fig. 9A; Meir et al. 2002). As the result of stacking of the mesophyll with increasing irradiance, A_{area} scales positively with M_A in plant canopies (Fig. 9B), while A_{mass} may be relatively invariable (Fig. 9C).

The within-canopy variation in M_A results from increases in both T and D (Niinemets et al. 1999, 2003; Sack et al. 2003a). Contrary to the interspecific relationships (Niinemets 1999b), increases in density are positively associated with A_{area} within tree canopies (Niinemets et al. 1999), possibly because the light-mediated increase in D is associated with increases in the fraction of palisade tissues in leaf mesophyll more than with increases in cell wall thickness. Such adaptive modifications in leaf anatomy further increase leaf photosynthetic capacity at a common T.

Across habitats, species show similar plastic modifications of M_A and A_{area} to irradiance (Cao 2000; Evans and Poorter 2001; Sack et al. 2003b).

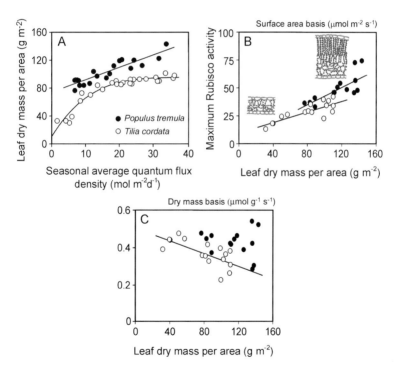

Fig. 9. Increases in M_A with long-term irradiance in the canopies of deciduous temperate trees (**A**), and correlations between M_A and maximum Rubisco carboxylase activity per unit area (**B**) and dry mass (**C**; modified from Niinemets et al. 1998). Analogous relationships were observed for the maximum photosynthetic electron transport rate. In **B**, corresponding changes in half thickness (T) are also shown. The idealized leaf cross-sections are scaled to correspond with the observed three-fold gradient in T in the mixed canopy of *P. tremula* and *T. cordata* (Niinemets et al. 1999). All linear and non-linear regressions are significant at $P<0.005$. For abbreviations, see Figs. 5 and 7

Furthermore, at a global scale, M_A and T increase with increasing global solar radiation (Niinemets 2001), demonstrating evolutionary adjustments in leaf structure to irradiance. In fact, according to experimental studies, structural acclimation is a more effective way of increasing A_{area} in high-light environments than modification of N partitioning among the components of the photosynthetic apparatus (Niinemets et al. 1998; Sack et al. 2003b).

Within the canopy, K_{leaf} scales positively with M_A and its components, D and T (Sack et al. 2003a; Aasamaa et al. 2004), and the values of K_{leaf} also increase across habitats with varying irradiance (K. Aasamaa, A. Sõber and Ü. Niinemets, unpublished data; L. Sack, N. M. Holbrook and M. T. Tyree, unpublished data). These co-variations among leaf structural characteristics and hydraulic conductance along light gradients further support the strong coordination of leaf assimilation and hydraulic characteristics.

4.2 Modifications due to Varying Moisture Supply

Plants adapted to limited moisture supply develop traits that either allow: (1) an early diurnal stomatal closure, or (2) maintenance of xylem function and leaf gas exchange despite falling soil water potentials. An array of structural modifications exists to maintain water flow from drying soil. Commonly, M_A increases in response to drought stress (Abrams et al. 1994; Abrams and Mostoller 1995; Sack 2004). Drought-related changes in M_A are mainly associated with increased cell wall thickness, cell wall lignification and decreased internal air space, collectively leading to larger D [see Niinemets (2001) for a review]. The adaptive significance of larger D in drought-adapted plants is that it rigidifies the plant leaves and accordingly renders them more resistant to pressure-driven changes in leaf volume and water content. The increase in rigidity can be physically measured as a greater bulk modulus of elasticity (ε; pressure change per unit change of symplasmic leaf water content; Niinemets 2001). Thus, plants with higher ε lose less cell water for a common change in leaf water potential, and can in drying soils maintain water extraction with a greater degree of leaf hydration. Adjustment of leaf water extraction capacity through leaf rigidification may further protect the leaves from herbivory and the mechanical injuries that tend to accumulate over time. This suggests that increasing the resistance to mechanical lesions implies a greater leaf life-span in drought-adapted leaves. However, greater diffusion resistance and the resultant lower A_{mass} are the major penalties of more robust leaf design.

The resistance of K_{leaf} to drought-induced decline, or a resilient K_{leaf} via a xylem embolism refilling mechanism, are also the traits of paramount significance in drought tolerance (Trifilo et al. 2003). Ample evidence of a greater safety margin of stem xylem conductance in plants acclimated to drought is available (e.g., Linton et al. 1998; Hacke et al. 2001a), but few data exist on the adjustments of leaf vein conductance to drought (Salleo et al. 2001). Inevitably, a greater safety margin with respect to cavitation is associated with decreased xylem conduit diameters (Linton et al. 1998; Hacke et al. 2001a), implying that the species with water-potential-insensitive K_{leaf} values are characterized by lower potential K_{leaf}. Nevertheless, species with "safe" K_{leaf} can more strongly modify $\Delta\psi$ between the leaf and the point of leaf attachment to the stem, thereby maintaining E ($\Delta\psi K_{leaf}$). Although drought-sensitive species may have a larger potential hydraulic conductance, K_{leaf} decreases with increasing $\Delta\psi$ in these species, typically with a threshold-type response (Salleo et al. 2001), and a certain transpiration rate cannot be maintained after the critical water potential for cavitation has been achieved (Nardini et al. 2001).

There is a trend of a negative scaling of K_{leaf}/M_A with D in six temperate species (Fig. 7, inset). Given that K_{leaf}/M_A is the conductance estimate characterizing the water supply of average leaf cells, this negative trend suggests coordination between leaf hydraulic (hydraulic "safety") and structural variables (extraction of water with lower water loss) that improve leaf-drought tolerance. For stemwood, it has been demonstrated that increases in tissue density effectively prevent xylem implosion at low negative pressures (Hacke et al. 2001a), and an analogous linkage between low water potential tolerance and foliar structure is also feasible for leaves.

5 Age-Dependent Changes in Leaf Photosynthetic Capacity

5.1 Increases in the Functional Activity in Young Leaves: Biochemical Versus Structural Limits

The ontogenetic trajectory of a developing leaf consists of coordinated modifications in all leaf traits. During leaf area expansion and thickening both N_{area} and M_A increase (Hanba et al. 2001; Niinemets et al. 2004c). Although the young leaves have higher N_{mass} than mature leaves (Niinemets et al. 2004c), the N investment in Rubisco, in rate-limiting components of the photosynthetic electron transport chain and in chlorophyll is low, resulting in low C assimilation potentials (Eichelmann et al. 2004; Niinemets et al. 2004c). With advancing leaf ontogeny, N_{area} and the fraction of N in photosynthetic compounds increase, leading to the values of A_{area} typical of mature leaves. There is evidence that these changes are

controlled by photosystem I content, which first reaches the mature-leaf
level (Eichelmann et al. 2004).

Apart from these biochemical modifications, the initial periods of leaf
development are characterized by intensive cell division (Miyazawa and
Terashima 2001; Miyazawa et al. 2003), resulting in tightly packed tissues
with a low volume fraction of internal air space (Miyazawa and Terashima
2001). In the following intensive leaf expansion phase, the volume fraction
of air space increases dramatically, leading to concomitant increases in the
internal CO_2 diffusion conductance from the outer surface of cell walls to
chloroplasts (Hanba et al. 2001; Miyazawa and Terashima 2001; Eichelmann
et al. 2004). Thus, A_{max} in young leaves is co-limited by both the low content
of photosynthetic enzymes as well as by the low efficiency of CO_2 diffusion.

Young leaves with weakly developed vasculature also have low values of
K_{leaf}, suggesting that limited hydraulic conductance may also curb the photosynthetic efficiency of young leaves (Martre et al. 2000; K. Aasamaa, A.
Sõber and Ü. Niinemets, unpublished data). In addition, formation and differentiation of stomata on the leaf surface continues until leaf maturation
(Kouwenberg et al. 2004). These findings suggests that inefficient water-conducting pathways and low stomatal pore area exert a further relevant limitation on C gain in young leaves.

5.2 Time-Dependent Deterioration of Leaf Physiological Activity in Mature Leaves and During Senescence

In non-senescent leaves, there are significant time-dependent decreases in
photosynthetic rates due to decreases in N_{mass} (Wilson et al. 2001; Niinemets
et al. 2004c), as foliar N is diluted by cell wall accretion, and possibly because
of N resorption as a late-season N sink forms in the developing buds.
Mesophyll diffusion conductance also decreases continuously in mature
non-senescent leaves (Miyazawa and Terashima 2001). Decreases in diffusion conductance are associated with increases in the thickness of mesophyll
cell walls (Miyazawa and Terashima 2001), and possibly also with decreases
in the pore volume in cell walls due to accumulation of lignin and cutin
(Suzuki and Itoh 2001; Miyazawa et al. 2003; Niinemets et al. 2004d). At a
"macro-scale", the age-dependent changes reflecting decreased diffusion efficiency are the increase in M_A and leaf C content per unit dry mass
(Miyazawa and Terashima 2001; Wilson et al. 2001; Niinemets et al. 2004d).

The declining biochemical and diffusion efficiencies occur simultaneously with a decline in K_{leaf} (Aasamaa et al. 2002; Salleo et al. 2002; Lo Gullo
et al. 2004, 2005). As leaves age, repeated episodes of dehydration and

embolism in the veins may lead to a progressive loss of conductance (Brodribb and Holbrook 2003), a situation analogous to cavitation fatigue observed in stemwood xylem (Hacke et al. 2001b). The decrease in K_{leaf} is also dependent on the accumulation of tyloses in the leaf (Salleo et al. 2002). In evergreen conifers, the activity of secondary needle growth in non-elastic mature leaves may lead to compression of xylem vessels, thereby further reducing the efficiency of water transport (Gilmore et al. 1995). In addition to vascular limitations, the time-dependent decreases in K_{leaf} are amplified by potential losses of membrane conductances in the bundle sheaths, and in the apoplastic pathways, as cell walls thicken and accumulate hydrophobic constituents (Sack 2005). While decreases in leaf N and diffusion conductance decrease leaf photosynthetic capacity, the hydraulic limitations would result in decreased stomatal conductances, and decrease the realized leaf assimilation rates.

Overall, a series of time-dependent modifications with complex interactions and feedbacks and environmental dependencies occur during leaf aging (Fig. 10). Declining function may interact with a seasonal cue to precipitate leaf senescence (Salleo et al. 2002), finally leading to leaf abscission. In mature non-senescent leaves, changes in leaf functioning occur relatively slowly (Niinemets et al. 2004c), but the decrease in leaf assimilation potentials is fast during senescence due to regulated dismantling of cell organelles and resorption of N (Kitajima et al. 1997, 2002; Wilson et al. 2001; Niinemets et al. 2004c). In evergreens, the time at which the senescence phase commences apparently also depends on the rate of accumulation of mechanical damage, and on the speed of time-dependent changes in leaf diffusion conductance and hydraulics (Fig. 10), which would ultimately determine when the cost/benefit ratios of specific leaves become uneconomical. Furthermore, older leaves of evergreen species are being gradually shaded by new foliage, and the re-acclimation to the new light environment is a critical factor modifying the life-span of the foliage (Niinemets 1997, Fig. 10). Accordingly, leaf longevity depends on the speed and the extent of modifications in leaf structural and physiological traits in a complex manner (Fig. 10).

6 Outlook: a Network of Coordinated Leaf Traits

There are many examples of trait correlations, demonstrating that plant performance is optimized by simultaneous modifications in a series of key traits. We propose that coordinated variation among plant traits may be achieved in at least five ways. The traits can be *allometrically* coordinated,

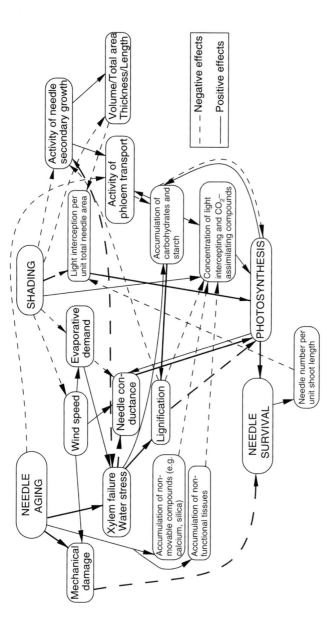

Fig. 10. Dependence of needle survival on the interplay between foliage shading and aging in conifers: a conceptual model. Aging of non-senescent mature needles results in enhanced lignification of cell walls, accumulation of non-structural carbohydrates, and ash minerals (primarily Ca), and deposition of dead tissues (Niinemets 1997). These chemical changes result in dilution of needle chemicals responsible for light interception and C acquisition. As wind speed, and air temperature, are positively correlated with irradiance, and air humidity is negatively correlated with irradiance [see Niinemets and Valladares (2004) for a review], the magnitude of environmental stresses decreases with increasing shading. For simplicity, the influences of tree growth on the foliage carbohydrate pool, and the effects of nutrient requirements for growth on needle life-span have been omitted. Activity of needle secondary growth is assumed to be controlled by the same factors affecting primary growth (Dale 1988). Shade-acclimation state achieved during needle primary growth is assumed to be a central factor controlling the velocity of the decline in C acquisition capacity of the needles during further shading. Most important links are shown with **bold** lines

especially if they are related to plant size, or to developmental age (Sack et al. 2002, 2003b). The traits can be *structurally* coordinated if they share an anatomical and/or a morphological basis. Whether or not the traits are structurally coordinated, they can be *functionally* coordinated. Functionally coordinated traits contribute to a given process in a co-dependent way or independently, and are therefore simultaneously abundant in a given set of conditions. For instance, leaf stomatal and hydraulic conductances are often functionally coordinated (Aasamaa et al. 2001; Brodribb and Holbrook 2003; Sack et al. 2003a). *Genetic* or *developmental* coordination of traits occurs if the expression of the traits during growth and/or during plant plastic responses to environmental stresses proceeds simultaneously. Many trait linkages will fall into multiple categories. Furthermore, observed co-variations among simple and composite traits may reflect the way the composite traits have been defined. For example, M_A is the product of T and D (Witkowski and Lamont 1991; Niinemets 1999b), and accordingly varies if either of these variables changes.

To a certain extent, all traits may be considered as vertices linked by single or double-headed arrows, and path analysis and structural equation modeling can be used to indicate the causality among interrelated traits (Shipley 1995; Shipley and Lechowicz 2000). However, often the causal structure among the traits is less clear. For instance, A_{mass} is given as A_{area}/M_A, and accordingly, these traits are interlinked as in a triangle, with each vertex connected to the other two by double-headed arrows. Taking this further, a super-network of vertices and inter-relationships may be conceived for all possible leaf traits. This would be a purely conceptual framework, and whether the potential trait coordinations are expressed as tight correlations will depend on the set of species considered (Fig. 11). Eventually, when more sets of traits and their coordination are well understood for large species sets, the full potential network can be laid out, and it will allow scaling up from given traits to whole-plant performance.

As in many natural networks, certain vertices will be 'hubs' with a disproportionate number of arrows linking them to other vertices (Barabási 2002). One structural trait that is a notable 'hub' interconnected frequently with other traits is M_A, which connects all traits that are leaf-area based with those that are leaf-mass based. Which relationships of those in the super-network will be realized depends on the species set considered. For instance, there is generally a strong positive interspecific correlation between A_{area} and A_{mass}, while M_A is negatively associated with both estimates of foliar photosynthetic potential (Wright et al. 2004b). However, depending on the variation in M_A in a specific species set, there may be no relationship between A_{area} and A_{mass} (Fig. 11; Niinemets 1999b). In turn, while M_A is for all species the product of T and D,

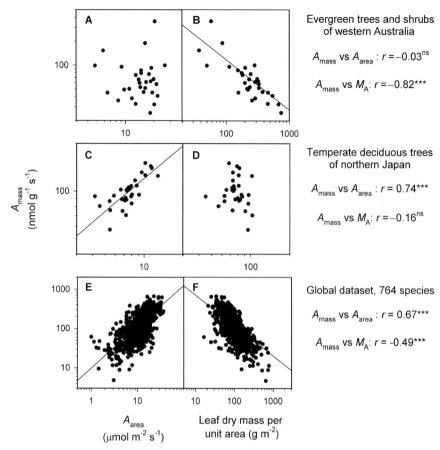

Fig. 11. Correlations among the A_{mass} and A_{area} (**A, C, E**) and A_{mass} and M_A (**B, D, F**) for evergreen trees and shrubs of western Australia (**A, B**; data from E. J. Veneklaas, published in Wright et al. 2004a, b), northern Japan (**C, D**; Koike 1988) and the global dataset (**E, F**; Wright et al. 2004b), which also includes the data presented in **A–D**. M_A links area- and mass-based quantities ($A_{area}=M_A A_{mass}$). Data were fitted by linear regressions. ***$P<0.001$, ns $P>0.05$. For abbreviations, see Figs 4, and 5

T is more variable in some cases, and thus more important in driving the differences in M_A, while for other species sets, the variation in M_A is driven by D (Niinemets 1999b; Shipley 2002). These examples further demonstrate the importance of structural characteristics in modifying the variations in leaf photosynthetic potentials, and underscore the significance of a functional understanding of covariation networks of plant traits.

7 Conclusions

Recently progress has been made in determining the constraints on photosynthesis at tissue, whole-leaf and shoot levels. Our model for a supernetwork of coordinated leaf traits enables a simultaneous consideration of the relationships of disparate traits. Traits show different coordination depending on whether leaves are compared within canopies, among individuals in different habitats, or across species sets, within or across habitats and life forms. In each case, the efficiency of light harvesting varies because of the structural limitations of the exposure of chlorophyll within the leaves and exposure of leaves according to their specific location in the canopy, while the variations in foliar photosynthetic potentials arise from the limitations of either area- or mass-based gas exchange. The area-based gas-exchange involves limitations to surface area for light capture, as well as limitations to gaseous CO_2 diffusion, including the hydraulic constraints that affect stomatal aperture. The mass-based gas-exchange involves constraints due to thick cell walls, as well as low cell wall porosity and a high degree of lignification, and a low fraction of internal leaf air space, which modify the photosynthetic potentials per unit leaf mass or N. Our review also emphasizes the importance of leaf size and shape in altering leaf physiological potentials. Leaf size not only influences leaf energy balance, but also affects the balance between photosynthetic and support biomass within the leaf, thereby changing mass-based leaf photosynthetic activity. Recent work has begun integration of leaf C and N economy traits and water-use traits, and understanding of the coordination of traits across shifting availabilities of irradiance, water, and nutrients is gradually emerging. Due to the large evolutionary and plastic modifications in foliar structure, structural limitations of foliar photosynthesis play at least as significant a role as physiological and biochemical constraints.

Acknowledgements. The authors' work on the integration of plant structural and physiological characteristics has been supported by the Estonian Science Foundation (grant no. 5,702) and by the Estonian Ministry of Education and Science (grant no. 0182468As03). We gratefully thank Ian J. Wright and Peter B. Reich for access to the Glopnet dataset.

References

Aasamaa K, Sõber A, Rahi M (2001) Leaf anatomical characteristics associated with shoot hydraulic conductance, stomatal conductance and stomatal sensitivity to changes of leaf water status in temperate deciduous trees. Aust J Plant Physiol 28:765–774

Aasamaa K, Sõber A, Hartung W, Niinemets Ü (2002) Rate of stomatal opening, shoot hydraulic conductance and photosynthesis characteristics in relation to leaf abscisic acid concentration in six temperate deciduous trees. Tree Physiol 22:267–276

Aasamaa K, Sõber A, Hartung W, Niinemets Ü (2004) Drought acclimation of two deciduous tree species of different layers in a temperate forest canopy. Trees 18:93–101

Abrams MD, Mostoller SA (1995) Gas exchange, leaf structure and nitrogen in contrasting successional tree species growing in open and understory sites during a drought. Tree Physiol 15:361–370

Abrams MD, Kubiske ME, Mostoller SA (1994) Relating wet and dry year ecophysiology to leaf structure in contrasting temperate tree species. Ecology 75:123–133

Barabási AL (2002) Linked: the new science of networks. Perseus, Cambridge, Mass.

Beadle NCW (1966) Soil phosphate and its role in molding segments of the Australian flora and vegetation, with special reference to xeromorphy and sclerophylly. Ecology 47:992–1007

Bragg JG, Westoby M (2002) Leaf size and foraging for light in a sclerophyll woodland. Funct Ecol 16:633–639

Brodribb TJ, Holbrook NM (2003) Changes in leaf hydraulic conductance during leaf shedding in seasonally dry tropical forest. New Phytol 158:295–303

Brodribb TJ, Holbrook NM (2004) Stomatal protection against hydraulic failure: a comparison of coexisting ferns and angiosperms. New Phytol 162:663–670

Buckley TN, Farquhar GD (2004) A new analytical model for whole-leaf potential electron transport rate. Plant Cell Environ 27:1487–1502

Cao KF (2000) Leaf anatomy and chlorophyll content of 12 woody species in contrasting light conditions in a Bornean heath forest. Can J Bot 78:1245–1253

Cescatti A, Niinemets Ü (2004) Sunlight capture. Leaf to landscape. In: Smith WK, Vogelmann TC, Chritchley C (eds) Photosynthetic adaptation. Chloroplast to landscape. (Ecological studies, vol 178) Springer, Berlin Heidelberg New York, pp 42–85

Chazdon RL, Pearcy RW (1991) The importance of sunflecks for forest understory plants. Photosynthetic machinery appears adapted to brief, unpredictable periods of radiation. BioScience 41:760–766

Cochard H, Nardinia A, Coll L (2004) Hydraulic architecture of leaf blades: where is the main resistance? Plant Cell Environ 27:1257–1267

Dale JE (1988) The control of leaf expansion. Annu Rev Plant Physiol Plant Mol Biol 39:267–295

DeLucia EH, Nelson K, Vogelmann TC, Smith WK (1996) Contribution of intercellular reflectance to photosynthesis in shade leaves. Plant Cell Environ 19:159–170

DeLucia EH, Whitehead D, Clearwater MJ (2003) The relative limitation of photosynthesis by mesophyll conductance in co-occurring species in a temperate rainforest dominated by the conifer *Dacrydium cupressinum*. Funct Plant Biol 30:1197–1204

Eichelmann H, Oja B, Rasulov B, Padu E, Bichele I, Pettai H, Vapaavuori E, Niinemets Ü, Laisk A (2004) Development of leaf photosynthetic parameters in *Betula pendula* Roth. leaves: correlations with photosystem I density. Plant Biol 6:307–318

Enquist BJ, West GB, Charnov EL, Brown JH (1999) Allometric scaling of production and life-history variation in vascular plants. Nature 401:907–911

Epron D, Godard D, Cornic G, Genty B (1995) Limitation of net CO_2 assimilation rate by internal resistances to CO_2 transfer in the leaves of two tree species (*Fagus sylvatica* L. and *Castanea sativa* Mill.). Plant Cell Environ 18:43–51

Evans JR (1989) Photosynthesis and nitrogen relationships in leaves of C_3 plants. Oecologia 78:9–19

Evans JR, Loreto F (2000) Acquisition and diffusion of CO_2 in higher plant leaves. In: Leegood RC, Sharkey TD, von Caemmerer S (eds) Photosynthesis: physiology and metabolism. Kluwer, Dordrecht, pp 321–351

Evans JR, Poorter H (2001) Photosynthetic acclimation of plants to growth irradiance: the relative importance of specific leaf area and nitrogen partitioning in maximizing carbon gain. Plant Cell Environ 24:755–767

Evans JR, Vogelmann TC (2003) Profiles of ^{14}C fixation through spinach leaves in relation to light absorption and photosynthetic capacity. Plant Cell Environ 26:547–560

Evans JR, von Caemmerer S, Setchell BA, Hudson GS (1994) The relationship between CO_2 transfer conductance and leaf anatomy in transgenic tobacco with a reduced content of Rubisco. Aust J Plant Physiol 21:475–495

Evans JR, Vogelmann TC, Williams WE, Gorton HL (2004) Sunlight capture. Chloroplast to leaf. In: Smith WK, Vogelmann TC, Chritchley C (eds) Photosynthetic adaptation. Chloroplast to landscape. Springer, Berlin Heidelberg New York, pp 15–41

Farque L, Sinoquet H, Colin F (2001) Canopy structure and light interception in *Quercus petraea* seedlings in relation to light regime and plant density. Tree Physiol 21:1257–1267

Fleck S, Niinemets Ü, Cescatti A, Tenhunen JD (2003) Three-dimensional lamina architecture alters light harvesting efficiency in *Fagus*: a leaf-scale analysis. Tree Physiol 23:577–589

Garnier E, Salager J-L, Laurent G, Sonié L (1999) Relationships between photosynthesis, nitrogen and leaf structure in 14 grass species and their dependence on the basis of expression. New Phytol 143:119–129

Geeske J, Aplet G, Vitousek PM (1994) Leaf morphology along environmental gradients in Hawaiian *Metrosideros polymorpha*. Biotropica 26:17–22

Germino MJ, Smith WK (1999) Sky exposure, crown architecture, and low-temperature photoinhibition in conifer seedlings at alpine treeline. Plant Cell Environ 22:407–415

Gilmore DW, Seymour RS, Halteman WA, Greenwood MS (1995) Canopy dynamics and the morphological development of *Abies balsamea*: effects of foliage age on specific leaf area and secondary vascular development. Tree Physiol 15:47–55

Givnish TJ (1984) Leaf and canopy adaptations in tropical forests. In: Medina E, Mooney HA, Vásquez-Yánes C (eds) Physiological ecology of plants of the wet tropics. Tasks for vegetation science, vol 12. Junk, The Hague, pp 51–84

Grace J (1978) The turbulent boundary layer over a flapping *Populus* leaf. Plant Cell Environ 1:35–38

Green DS, Kruger EL (2001) Light-mediated constraints on leaf function correlate with leaf structure among deciduous and evergreen tree species. Tree Physiol 21:1341–1346

Grubb PJ (1977) Control of forest growth and distribution on wet tropical mountains: with special reference to mineral nutrition. Annu Rev Ecol Syst 8:83–107

Hacke UG, Sperry JS, Pockman WT, Davis SD, McCulloch KA (2001a) Trends in wood density and structure are linked to prevention of xylem implosion by negative pressure. Oecologia 126:457–461

Hacke UG, Stiller V, Sperry JS, Pittermann J, McCulloh KA (2001b) Cavitation fatigue. Embolism and refilling cycles can weaken the cavitation resistance of xylem. Plant Physiol 125:779–786

Hanba YT, Miyazawa S-I, Terashima I (1999) The influence of leaf thickness on the CO_2 transfer conductance and leaf stable carbon isotope ratio for some evergreen tree species in Japanese warm-temperate forests. Funct Ecol 13:632–639

Hanba YT, Miyazawa SI, Kogami H, Terashima I (2001) Effects of leaf age on internal CO_2 transfer conductance and photosynthesis in tree species having different types of shoot phenology. Aust J Plant Physiol 28:1075–1084

Hanba YT, Kogami H, Terashima I (2002) The effect of growth irradiance on leaf anatomy and photosynthesis in *Acer* species differing in light demand. Plant Cell Environ 25:1021–1030

Hibberd JM, Quick WP (2001) Characteristics of C_4 photosynthesis in stems and petioles of C_3 flowering plants. Nature 415:451–454

Hikosaka K, Terashima I (1996) Nitrogen partitioning among photosynthetic components and its consequence in sun and shade plants. Funct Ecol 10:335–343

Hikosaka K, Hanba YT, Hirose T, Terashima I (1998) Photosynthetic nitrogen-use efficiency in leaves of woody and herbaceous species. Funct Ecol 12:896–905

Karabourniotis G (1998) Light-guiding function of foliar sclereids in the evergreen sclerophyll *Phillyrea latifolia*: a quantitative approach. J Exp Bot 49:739–746

Karabourniotis G, Bornman JF (1999) Penetration of UV-A, UV-B and blue light through the leaf trichome layers of two xeromorphic plants, olive and oak, measured by optical fibre microprobes. Physiol Plant 105:655–661

Karabourniotis G, Papastergiou N, Kabanopoulou E, Fasseas C (1994) Foliar sclereids of *Olea europaea* may function as optical fibres. Can J Bot 72:330–336

Karabourniotis G, Bornman JF, Liakoura V (1999) Different leaf surface characteristics of three grape cultivars affect leaf optical properties as measured with fibre optics: possible implication in stress tolerance. Aust J Plant Physiol 26:47–53

King DA, Maindonald JH (1999) Tree architecture in relation to leaf dimensions and tree stature in temperate and tropical rain forests. J Ecol 87:1012–1024

Kitajima K, Mulkey SS, Wright SJ (1997) Decline of photosynthetic capacity with leaf age in relation to leaf longevities for five tropical canopy tree species. Am J Bot 84:702–708

Kitajima K, Mulkey SS, Samaniego M, Wright SJ (2002) Decline of photosynthetic capacity with leaf age and position in two tropical pioneer tree species. Am J Bot 89:1925–1932

Kogami H, Hanba YT, Kibe T, Terashima I, Masuzawa T (2001) CO_2 transfer conductance, leaf structure and carbon isotope composition of *Polygonum cuspidatum* leaves from low and high altitudes. Plant Cell Environ 24:529–538

Koike T (1988) Leaf structure and photosynthetic performance as related to the forest succession of deciduous broadleaved trees. Plant Species Biol 3:77–87

Kouwenberg LLR, Kurschner WM, Visscher H (2004) Changes in stomatal frequency and size during elongation of *Tsuga heterophylla* needles. Ann Bot 94:561–569

Lamont BB, Groom PK, Cowling RM (2002) High leaf mass per area of related species assemblages may reflect low rainfall and carbon isotope discrimination rather than low phosphorus and nitrogen concentrations. Funct Ecol 16:403–412

Linton MJ, Sperry JS, Williams DG (1998) Limits to water transport in *Juniperus osteosperma* and *Pinus edulis*: implications for drought tolerance and regulation of transpiration. Funct Ecol 12:906–911

Lloyd J, Syvertsen JP, Kriedemann PE, Farquhar GD (1992) Low conductances for CO_2 diffusion from stomata to the sites of carboxylation in leaves of woody species. Plant Cell Environ 15:873–899

Lo Gullo MA, Noval LC, Salleo S, Nardini A (2004) Hydraulic architecture of plants of *Helianthus annuus* L. cv. Margot: evidence for plant segmentation in herbs. J Exp Bot 55:1549–1556

Lo Gullo MA, Nardini A, Trifilo P, Salleo S (2005) Diurnal and seasonal variations in leaf hydraulic conductance in deciduous and evergreen trees. Tree Physiol 25:505–512

Martre P, Durand JL, Cochard H (2000) Changes in axial hydraulic conductivity along elongating leaf blades in relation to xylem maturation in tall fescue. New Phytol 246:235–247

McClendon JH (1992) Photographic survey of the occurrence of bundle sheath extensions in deciduous dicots. Plant Physiol 99:1677–1679

Medina E (1984) Nutrient balance and physiological processes at the leaf level. In: Medina E, Mooney HA, Vásquez-Yánes C (eds) Physiological ecology of plants of the wet tropics. Tasks for vegetation science, vol 12. Junk, The Hague, pp 134–154

Meir P, Kruijt B, Broadmeadow M, Barbosa E, Kull O, Carswell F, Nobre A, Jarvis PG (2002) Acclimation of photosynthetic capacity to irradiance in tree canopies in relation to leaf nitrogen concentration and leaf mass per unit area. Plant Cell Environ 25:343–357

Miyazawa S-I, Terashima I (2001) Slow development of leaf photosynthesis in an evergreen broad-leaved tree, *Castanopsis sieboldii*: relationships between leaf anatomical characteristics and photosynthetic rate. Plant Cell Environ 24:279–291

Miyazawa S-I, Makino A, Terashima I (2003) Changes in mesophyll anatomy and sink-source relationships during leaf development in *Quercus glauca*, an evergreen tree showing delayed leaf greening. Plant Cell Environ 26:745–755

Nardini A, Tyree MT, Salleo S (2001) Xylem cavitation in the leaf of *Prunus laurocerasus* and its impact on leaf hydraulics. Plant Physiol 125:1700–1709

Niinemets Ü (1997) Acclimation to low irradiance in *Picea abies*: influences of past and present light climate on foliage structure and function. Tree Physiol 17:723–732

Niinemets Ü (1998) Are compound-leaved woody species inherently shade-intolerant? An analysis of species ecological requirements and foliar support costs. Plant Ecol 134:1–11

Niinemets Ü (1999a) Differences in chemical composition relative to functional differentiation between petioles and laminas of *Fraxinus excelsior*. Tree Physiol 19:39–45

Niinemets Ü (1999b) Research review. Components of leaf dry mass per area – thickness and density – alter leaf photosynthetic capacity in reverse directions in woody plants. New Phytol 144:35–47

Niinemets Ü (2001) Global-scale climatic controls of leaf dry mass per area, density, and thickness in trees and shrubs. Ecology 82:453–469

Niinemets Ü, Fleck S (2002a) Leaf biomechanics and biomass investment in support in relation to long-term irradiance in *Fagus*. Plant Biol 4:523–534

Niinemets Ü, Fleck S (2002b) Petiole mechanics, leaf inclination, morphology, and investment in support in relation to light availability in the canopy of *Liriodendron tulipifera*. Oecologia 132:21–33

Niinemets Ü, Kull O (1999) Biomass investment in leaf lamina versus lamina support in relation to growth irradiance and leaf size in temperate deciduous trees. Tree Physiol 19:349–358

Niinemets Ü, Valladares F (2004) Photosynthetic acclimation to simultaneous and interacting environmental stresses along natural light gradients: optimality and constraints. Plant Biol 6:254–268

Niinemets Ü, Kull O, Tenhunen JD (1998) An analysis of light effects on foliar morphology, physiology, and light interception in temperate deciduous woody species of contrasting shade tolerance. Tree Physiol 18:681–696

Niinemets Ü, Kull O, Tenhunen JD (1999) Variability in leaf morphology and chemical composition as a function of canopy light environment in co-existing trees. Int J Plant Sci 160:837–848

Niinemets Ü, Valladares F, Ceulemans R (2003) Leaf-level phenotypic variability and plasticity of invasive *Rhododendron ponticum* and non-invasive *Ilex aquifolium* co-occurring at two contrasting European sites. Plant Cell Environ 26:941–956

Niinemets Ü, Al Afas N, Cescatti A, Pellis A, Ceulemans R (2004a) Determinants of clonal differences in light-interception efficiency in dense poplar plantations: petiole length and biomass allocation. Tree Physiol 24:141–154

Niinemets Ü, Cescatti A, Christian R (2004b) Constraints on light interception efficiency due to shoot architecture in broad-leaved *Nothofagus* species. Tree Physiol 24:617–630

Niinemets Ü, Kull O, Tenhunen JD (2004c) Within-canopy variation in the rate of development of photosynthetic capacity is proportional to integrated quantum flux density in temperate deciduous trees. Plant Cell Environ 27:293–313

Niinemets Ü, Tenhunen JD, Beyschlag W (2004d) Spatial and age-dependent modifications of photosynthetic capacity in four Mediterranean oak species. Funct Plant Biol 31:1179–1193

Niklas KJ (1989) The effect of leaf-lobing on the interception of direct solar radiation. Oecologia 80:59–64

Niklas KJ (1999) Research review. A mechanical perspective on foliage leaf form and function. New Phytol 143:19–31

Nikopoulos D, Liakopoulos G, Drossopoulos I, Karabourniotis G (2002) The relationship between anatomy and photosynthetic performance of heterobaric leaves. Plant Physiol 129:235–243

Nobel PS (1977) Internal leaf area and cellular CO_2 resistance: photosynthetic implications of variations with growth conditions and plant species. Physiol Plant 40:137–144

Parkhurst DF (1994) Tansley review no. 65. Diffusion of CO_2 and other gases inside leaves. New Phytol 126:449–479

Parkhurst DF, Loucks OL (1972) Optimal leaf size in relation to environment. J Ecol 60:505–537

Pearcy RW, Yang W (1998) The functional morphology of light capture and carbon gain in the redwood forest understorey plant, *Adenocaulon bicolor* Hook. Funct Ecol 12:543–552

Piel C, Frak E, Le Roux X, Genty B (2002) Effect of local irradiance on CO_2 transfer conductance of mesophyll in walnut. J Exp Bot 53:2423–2430

Poorter H, Evans JR (1998) Photosynthetic nitrogen-use efficiency of species that differ inherently in specific leaf area. Oecologia 116:26–37

Poulson ME, Vogelmann TC (1990) Epidermal focussing and effects upon photosynthetic light-harvesting in leaves of *Oxalis*. Plant Cell Environ 13:803–811

Reich PB (1998) Variation among plant species in leaf turnover rates and associated traits: implications for growth at all life stages. In: Lambers H, Poorter H, van Vuuren MMI (eds) Inherent variation in plant growth. Physiological mechanisms and ecological consequences. Backhuys, Leiden, pp 467–487

Reich PB, Walters MB, Ellsworth DS (1997) From tropics to tundra: global convergence in plant functioning. Proc Natl Acad Sci USA 94:13730–13734

Reich PB, Ellsworth DS, Walters MB, Vose JM, Gresham C, Volin JC, Bowman WD (1999) Generality of leaf trait relationships: a test across six biomes. Ecology 80:1955–1969

Roderick ML, Berry SL, Saunders AR, Noble IR (1999) On the relationship between the composition, morphology and function of leaves. Funct Ecol 13:696–710

Roth A, Mosbrugger V, Neugebauer HJ (1994) Efficiency and evolution of water transport systems in higher plants: a modelling approach. 1. The earliest land plants. Philos Trans R Soc Lond Ser B 345:137–152

Roth A, Mosbrugger V, Belz G, Neugebauer HJ (1995) Hydrodynamic modelling study of angiosperm leaf venation types. Bot Acta 108:121–126

Roth-Nebelsick A (2001) Computer-based analysis of steady-state and transient heat transfer of small-sized leaves by free and mixed convection. Plant Cell Environ 24:631–640

Sack L (2004) Responses of temperate woody seedlings to shade and drought: do trade-offs limit potential niche differentiation? Oikos 107:107–127

Sack L, Holbrook NM (2006) Leaf hydraulics. Annu Rev Plant Biol (in press)

Sack L, Tyree MT (2005) Leaf hydraulics and its implications in plant structure and function. In: Holbrook NM, Zwieniecki MA (eds) Vacular transport in plants. Elsevier, Academic Press, Oxford (in press)

Sack L, Marañón T, Grubb PJ (2002) Global allocation rules for patterns of biomass partitioning. Science 296:1923a

Sack L, Cowan PD, Jaikumar N, Holbrook NM (2003a) The 'hydrology' of leaves: co-ordination of structure and function in temperate woody species. Plant Cell Environ 26:1343–1356

Sack L, Grubb PJ, Marañón T (2003b) The functional morphology of juvenile plants tolerant of strong summer drought in shaded forest understories in southern Spain. Plant Ecol 168:139–163

Sack L, Streeter CM, Holbrook NM (2004) Hydraulic analysis of water flow through leaves of sugar maple and red oak. Plant Physiol 134:1824–1833

Salleo S, Gullo MAL, Raimondo F, Nardini A (2001) Vulnerability to cavitation of leaf minor veins: any impact on leaf gas exchange? Plant Cell Environ 24:851–859

Salleo S, Nardini A, Lo Gullo MA, Ghirardelli LA (2002) Changes in stem and leaf hydraulics preceding leaf shedding in *Castanea sativa* L. Biol Plant 45:227–234

Sassenrath-Cole GF (1995) Dependence of canopy light distribution on leaf and canopy structure for two cotton (*Gossypium*) species. Agric For Meteorol 77:55–72

Shipley B (1995) Structured interspecific determinants of specific leaf area in 34 species of herbaceous angiosperms. Funct Ecol 9:312–319

Shipley B (2002) Cause and correlation in biology. Cambridge University Press, Cambridge

Shipley B, Lechowicz MJ (2000) The functional co-ordination of leaf morphology, nitrogen concentration, and gas exchange in 40 wetland species. Ecoscience 7:183–194

Singsaas EL, Laporte MM, Shi J-Z, Monson RK, Bowling DR, Johnson K, Lerdau M, Jasentuliytana A, Sharkey TD (1999) Kinetics of leaf temperature fluctuation affect isoprene emission from red oak (*Quercus rubra*) leaves. Tree Physiol 19:917–924

Sinoquet H, Thanisawanyangkura S, Mabrouk H, Kasemsap P (1998) Characterization of the light environment in canopies using 3D digitising and image processing. Ann Bot 82:203–212

Sisó S, Camarero JJ, Gil-Pelegrín E (2001) Relationship between hydraulic resistance and leaf morphology in broadleaf *Quercus* species: a new interpretation of leaf lobation. Trees 15:341–345

Slaton MR, Hunt ER, Smith WK (2001) Estimating near-infrared leaf reflectance from leaf structural characteristics. Am J Bot 88:278–284

Smith SD, Naumburg E, Niinemets Ü, Germino MJ (2004) Environmental constraints. Leaf to landscape. In: Smith WK, Vogelmann TC, Chritchley C (eds) Photosynthetic adaptation. Chloroplast to landscape. (Ecological studies, vol 178) Springer, Berlin Heidelberg New York, pp 262–294

Smith WK, Vogelmann TC, DeLucia EH, Bell DT, Shepherd KA (1997) Leaf form and photosynthesis. Do leaf structure and orientation interact to regulate internal light and carbon dioxide? BioScience 47:785–793

Smith WK, Vogelmann TC, Critchley C (2004) Background and objectives. In: Smith WK, Vogelmann TC, Chritchley C (eds) Photosynthetic adaptation. Chloroplast to landscape. Springer, Berlin Heidelberg New York, pp 3–11

Sun JD, Nishio JN (2001) Why abaxial illumination limits photosynthetic carbon fixation in spinach leaves. Plant Cell Physiol 42:1–8

Suzuki K, Itoh T (2001) The changes in cell wall architecture during lignification of bamboo, (*Phyllostachys aurea* Carr.). Trees 15:137–147

Syvertsen JP, Lloyd J, McConchie C, Kriedemann PE, Farquhar GD (1995) On the relationship between leaf anatomy and CO_2 diffusion through the mesophyll of hypostomatous leaves. Plant Cell Environ 18:149–157

Takashima T, Hikosaka K, Hirose T (2004) Photosynthesis or persistence: nitrogen allocation in leaves of evergreen and deciduous *Quercus* species. Plant Cell Environ 27:1047–1054

Takenaka A (1994) Effects of leaf blade narrowness and petiole length on the light capture efficiency of a shoot. Ecol Res 9:109–114

Terashima I (1989) Productive structure of a leaf. In: Briggs WR (ed) Photosynthesis. Proceedings of the C.S. French Symposium on Photosynthesis, Stanford, California, 17–23 July 1988. Plant biology, vol 8. Liss, New York, pp 207–226

Terashima I (1992) Anatomy of non-uniform leaf photosynthesis. Photosynth Res 31:195–212

Terashima I, Miyazawa S-I, Hanba YT (2001) Why are sun leaves thicker than shade leaves? Consideration based on analyses of CO_2 diffusion in the leaf. J Plant Res 114:93–105

Trifilo P, Gasco A, Raimondo F, Nardini A, Salleo S (2003) Kinetics of recovery of leaf hydraulic conductance and vein functionality from cavitation-induced embolism in sunflower. J Exp Bot 54:2323–2330

Turner IM (1994) A quantitative analysis of leaf form in woody plants from the world's major broadleaved forest types. J Biogeogr 21:413–419

Uhl D, Mosbrugger V (1999) Leaf venation density as a climate and environmental proxy: a critical review and new data. Palaeogeogr Palaeoclimatol Palaeoecol 149:15–26

Valladares F (2003) Light heterogeneity and plants: from ecophysiology to species coexistence and biodiversity. In: Esser K, Lüttge U, Beyschlag W, Hellwig F (eds) (Progress in botany, vol 64) Springer, Berlin Heidelberg New York, pp 439–471

Valladares F, Pearcy RW (1999) The geometry of light interception by shoots of *Heteromeles arbutifolia*: morphological and physiological consequences for individual leaves. Oecologia 121:171–182

Valladares F, Skillman JB, Pearcy RW (2002) Convergence in light capture efficiencies among tropical forest understory plants with contrasting crown architectures: a case of morphological compensation. Am J Bot 89:1275–1284

Van Arendonk JJCM, Poorter H (1994) The chemical composition and anatomical structure of leaves of grass species differing in relative growth rate. Plant Cell Environ 17:963–970

Vogel S (1970) Convective cooling at low airspeeds and the shapes of broad leaves. J Exp Bot 21:91–101

Vogelmann TC, Evans JR (2002) Profiles of light absorption and chlorophyll within spinach leaves from chlorophyll fluorescence. Plant Cell Environ 25:1313–1323

Vogelmann TC, Martin G (1993) The functional significance of palisade tissue: penetration of directional versus diffuse light. Plant Cell Environ 16:65–72

West GB, Brown JH, Enquist BJ (1999) The fourth dimension of life: fractal geometry and allometric scaling of organisms. Science 284:1677–1679

Wilson D, Cooper JP (1970) Effect of selection for mesophyll cell size on growth and assimilation in *Lolium perenne* L. New Phytol 69:233–245

Wilson KB, Baldocchi DD, Hanson PJ (2001) Leaf age affects the seasonal pattern of photosynthetic capacity and net ecosystem exchange of carbon in a deciduous forest. Plant Cell Environ 24:571–583

Witkowski ETF, Lamont BB (1991) Leaf specific mass confounds leaf density and thickness. Oecologia 88:486–493

Wright IJ, Westoby M (2002) Leaves at low versus high rainfall: coordination of structure, lifespan and physiology. New Phytol 155:403–416

Wright IJ, Westoby M (2003) Nutrient concentration, resorption and lifespan: leaf traits of Australian sclerophyll species. Funct Ecol 17:10–19

Wright IJ, Westoby M, Reich PB (2002) Convergence towards higher leaf mass per area in dry and nutrient-poor habitats has different consequences for leaf life span. J Ecol 90:534–543

Wright IJ, Groom PK, Lamont BB, Poot P, Prior LD, Reich PB, Schulze ED, Veneklaas EJ, Westoby M (2004a) Leaf trait relationships in Australian plant species. Funct Plant Biol 31:551–558

Wright IJ, Reich PB, Westoby M, Ackerly DD, Baruch Z, Bongers F, Cavender-Bares J, Chapin T, Cornelissen JHC, Diemer M, Flexas J, Garnier E, Groom PK, Gulias J, Hikosaka K, Lamont BB, Lee T, Lee W, Lusk C, Midgley JJ, Navas M-L, Niinemets Ü, Oleksyn J, Osada N, Poorter H, Poot P, Prior L, Pyankov VI, Roumet C, Thomas SC, Tjoelker MG, Veneklaas E, Villar R (2004b) The world-wide leaf economics spectrum. Nature 428:821–827

Zwieniecki MA, Boyce CK, Holbrook NM (2004) Hydraulic limitations imposed by crown placement determine final size and shape of *Quercus rubra* L. leaves. Plant Cell Environ 27:357–365

Ülo Niinemets
Department of Plant Physiology,
University of Tartu,
Riia 23,
Tartu 51011,
Estonia
e-mail: ylon@ut.ee
and

Centro di Ecologia Alpina,
38040 Viote del Monte Bondone (TN),
Italy

Lawren Sack
Department of Botany,
University of Hawaii,
3190 Maile Way,
Honolulu,
HI 96822,
Hawaii,
USA
e-mail: LSack@hawaii.edu

Recent trends in plant-ecological modelling: Species dynamics in grassland systems

Eckart Winkler

1 Introduction

A few years ago, Jeltsch and Moloney (2002) published a review on developments in vegetation modelling. They especially stressed the importance of space (in a two-dimensional sense) as a constituent of a dynamic analysis, and of using rules rather than mathematical equations as appropriate building blocks of ecological models. Case studies were communicated where the inclusion of space led to new insights into the functioning of ecological systems.

Since then a large number of papers have been published that present and apply mathematical and simulation models to plant-ecological systems: from single populations in a small area to communities on a landscape scale, from pure data description to prediction and explanation, from matrix and Markov models to individual-based models, from grasslands to forests. This increase in the number and sophistication of model papers follows a trend established in the late 1980s. Due to the stimulating effect of published examples, modelling has become increasingly popular and an integrated part of many ecological projects.

Models, presented in mathematical terms, or simply verbally or graphically, are an inevitable tool for the investigation of complex systems, e.g. ecological ones. They are a means of representing reality, as determined by observation or experiment, to formulate explanatory concepts and to make predictions on the future development of the recorded area of research, possibly under changing conditions. Models must stress features that are considered to be important and must neglect other ones; this is governed by the problem in question and also by subjective judgement. As models arise from an interaction between formulation of hypotheses on the one hand and observations or experiments on the other, i.e. from a loop between brain and reality, they can never be considered as definite.

Reflection on the process of modelling, on kinds and purposes of models, on their role in the discovery process and on techniques in model building started in the 1960s when science considered problems of increasing complexity (see e.g. Mesarović 1968). Some of the actual problems in this debate, focussed on ecology, were presented by Grimm (1999), Wiegand et al. (2003) and Grimm and Railsback (2005). These authors considered "individuals" as the building units of ecological processes and formulated rules for their behaviour. However, the most general concept in ecological modelling is that there is no all-embracing method. Even the concept of individuality will soon come up against its limits: one may simply think of the example of dense grass swards to illustrate this.

This review is intended to continue the Jeltsch–Moloney survey. The material presented is arranged according to two general objectives of modelling:

1. Prediction: Markov and matrix models, cellular automata (CA) and grid-based models, and spatially explicit individual-based (SEIB) models (sections 2 and 3).
2. Explanation: analytical models (section 4).

Due to my field of research, examples are mainly centred on grassland systems: they include population–biological as well as community-dynamic problems. Some aspects are not treated as adequately as they should be (e.g. the impact of herbivory). Forest and agricultural problems are largely excluded, as well as methods of data evaluation, pattern extraction and parameter determination. However, some features of plant dynamics are indicated which, in contrast to their importance, are under-represented in models, such as the interactions of plants with soil resources by consideration of the third dimension or adaptive plasticity in plant behaviour, and they should be tasks for future research (section 5).

2 Markov and Matrix Models

2.1 Markov Models of Vegetation Change

Many ecological models of population growth and vegetation change use transition matrices in the form of Markov models. Such models (Markovian chains) consist of a vector representing the state variables of the system and a matrix containing the (mostly constant) probabilities of possible state transitions. Usually it is assumed that the future dynamics of the system depend only on the present state, but not on its history. It is possible to relax this strict requirement of independence which leads to Markovian chains of higher order (Facelli and Pickett 1990; Balzter 2000) or to Markov chains where additional state variables (e.g. age or size) are introduced which carry the "memory" of the system (Grimm and Wissel 2004). However, such models need a much higher amount of empirical data.

Markov models are ecological models with a highly formalised theoretical background. They are discrete in state, in time and, if included, also in space. Matrix models are special kinds of Markov model designed to predict the development of a population (see below). Markov models in a narrower sense consider vegetation dynamics in space, with transitions given either a species-by-species or a vegetation-type-by-vegetation-type replacement in a spatial array of grid cells, of discrete units of space which are defined by an internal uniformity. If a change is not only governed by the actual stage of a cell itself but also by that of neighbouring cells we generally speak about CA or interacting particle systems (see Czárán 1998; Berec 2002). In strict CA models long-distance interactions between grid cells beyond the neighbourhood, e.g. due to dispersal, are not considered. When such formal restrictions in CA models are relaxed, general grid-based models are the result.

Markov and matrix models are appealing as they can be directly parameterised with respect to field data (e.g. individual transitions, cover, species proportions or species identity); this is aided also by educated guesses and expert opinions. Extrapolation of this knowledge in time, and sometimes in space, is one common objective of such models. They aggregate very complex information so that ecosystems can be examined for which underlying mechanisms are not known in detail. However, usually they assume stationarity, i.e. that the transition probabilities shall be independent of time or the actual values of the state variables. Then the system finally approaches an attractor dynamic, a stable distribution of state variables. Markov models, therefore, may describe succession in a community for some period, but they will fail if features such as colonisation by new species, complex interactions, disturbance, or long-term changes of environmental conditions have to be considered.

Some examples can be used to demonstrate model application. A rule based model for the functional analysis of vegetation change in Australasian forests, designed by Campbell et al. (1999), did not include any neighbourhood interactions. Other models at least included immediate neighbourhood impacts as a constituent component. Soares-Filho et al. (2002) looked for large-scale changes in landscape structures (colonisation-forest frontier in the Amazonian basin), Carmel et al. (2001) for vegetation dynamics in the Galilee mountains, Israel, and Jenerette and Wu (2001) for the transformation of desert landscape in Arizona due to urbanisation. On a smaller scale, Gassmann et al. (2000), relying on long-term observation series, modelled the interaction of five grassland species, each representing different functional types. Lanzer and Pillar (2002) applied earlier data on heathlands in the Netherlands and built a probabilistic CA model. The literature of the last years shows a trend from pure Markov models to CA and general grid-based models.

2.2 Matrix Models of Population Development

Matrix models are the one-species-one-patch version of Markov models. Their practical importance is increasing as they allow workers to include, beyond the formal Markov structure, density dependence, stochastic processes and far-range dispersal by linking several patches. They are the subject of a textbook (Caswell 2001) where a set of methods was presented for formal evaluation. De Kroon et al. (2000) and Grant and Benton (2000) described elasticity (perturbation) analysis as well as loop analysis for models with constant or also density-dependent coefficients, and Caswell (2000) discriminated between projective (e.g. elasticity) and retrospective (life-table response analysis) methods. However, when such tools are applied there is the danger of a very blind, formal approach, promoted by the availability of computer software. Often directed simulations are much better than methods available in computer packages, especially if stochasticities or density dependence have to be considered (see also Fieberg and Ellner 2001).

Matrix models consider discrete age or life-history stages. The integral projection method is an interesting extension of such discrete matrix models: it replaces distinct states by a continuum of states (Easterling et al. 2000). It was applied, for example, to solve evolutionary questions [optimum timing of reproduction in monocarpic perennial plant species (see Metcalf et al. 2003)].

Sensitivity or elasticity analyses in matrix models with constant transition probabilities were done for rather different species: *Coryphantha robbinsorum*, a cactus (Fox and Gurevich 2000); an assembly of species (a forest understorey herb, two tropical forest palms and three tropical forest trees; Zuidema and Franco 2001); *Gentiana pneumonante* (Oostermeijer et al. 2003); *Gentianella campestris* (Lennartsson and Oostermeijer 2001; testing management methods under environmental stochasticity); *Oxalis acetosella*, a clonal herb (Berg 2002); *Hylocomium splendens*, a clonal moss (Rydgen et al. 2001; environmental stochasticity was included in the study); *Plantago media* (Eriksson and Eriksson 2000); *Geum rivale* and *Agrimonia eupatoria* (Kiviniemi 2002; a study that included temporal variability and estimated mean time to extinction); and *Euphorbia clivicola* (Pfab and Witkowski 2000). Ehrlén (2000) succeeded in demonstrating "historical" effects on the life history of a long-lived herb, *Lathyrus vernus*, and hence built a second-order matrix model. Guàrdia et al. (2000) designed a model for a perennial tussock grass (*Achnatherum calamagrostis*) – they did not examine individual tillers but focused on the dynamics of spatial units in order to forecast the spatial growth of tussocks. The model was intermediate between the traditional matrix models of population dynamics and Markovian vegetation models, but it neglected the aspect of interactions between growing tussocks and hence of population density control.

Such a consideration of the effect of density on transition probabilities is a crucial extension of matrix models. Because it is difficult to get a sufficient data base, density-dependent models were mostly used for studying annual species or species with non-overlapping generations. The studies of Buckley et al. (2001) (*Trileurospermum perforatum*), of Watkinson et al. (2000b), of Freckleton et al. (2000), Freckleton and Watkinson (2002a), or of Gustafsson and Ehrlén (2003) (*Sanicula europaea*) are noticeable examples. When density effects are introduced we have a fluid transition to analytical models: such models, with a more conceptual background, are presented in section 4.

Spatially expanding populations can be treated by coupling, via seed dispersal, several matrix models which each describe dynamics in one cell of a grid (with or without density control). Different habitat qualities and different dispersal functions can easily be introduced. Such spatially explicit approaches may also lead to metapopulation models. The importance of this concept for plant populations was discussed in detail by Freckleton and Watkinson (2002b).

This grid-based approach can be illustrated by studies on invasive or weed species. Wadsworth et al. (2000) examined control programmes for two weeds of riparian habitats, *Impatiens glandulifera* and *Heracleum mantegazzianum*. Watkinson et al. (2000a) investigated the population dynamics of *Vulpia ciliata* including density control and dispersal. However, as they only considered restricted, neighbourhood dispersal, the impact of fat-tailed dispersal functions on species spread was neglected (see below). Directed dispersal was included by Levine (2003) in his model of downstream transport. An example of a model that neglects grid structure but considers continuous space is the simulation of the spread of four annual weed species by integro-difference models (Woolcok and Cousens 2000).

The impact of aspects of the environment on genotype diversity of the annual species *Arabidopsis thaliana* and of those with a more complicated life history were studied by Damgaard (2003).

An important application of matrix models is the assessment of species extinction risk and the estimation of the size of viable populations. Recent surveys (Menges 2000; Brigham and Schwartz 2003) cover the relevant literature.

3 SEIB Models

The different kinds of spatially explicit modelling given in the last section included rather large areas and sometimes complex phenomena in a crude manner. Researchers go into more detail when they look at the behaviour of individuals, not only of vegetation or of populations. However, in this case the number of questions a modeller must ask a field ecologist in order to fix rules and parameterise the model will inevitably increase.

SEIB models are, in general, very flexible, and are not fixed to any formal set of rules (such as Markov models or CA). They are usually, but not inevitably, discrete in space and in the possible developmental stages of the individuals. What remains is the discreteness of time: time units are often years or seasons, governed by the life cycle of the model species, or also months or even days in growth models where the discreteness is only limited by computer time. The behaviour of individuals is governed by rules, describing demography, competition, dispersal, resource acquisition, stochasticity, disturbance, or landscape dynamics.

On a grid basis, SEIB models very often assume that one grid cell can host at most one (adult) individual, and additionally a seed bank, but deviations from this basic assumption can be found in the literature. Higgins and Cain (2002) considered grid cells that could be occupied by K plants of the competing species (where K is the capacity of the cell): local competition occurred only within the grid cells. The filling of cells with populations instead of individuals corresponds to the coupled matrix models of the last section. However, matrix models usually pretend that a population of a given species lives alone in its habitat. The individual-oriented method can now consider the multitude of biotic and abiotic interactions. If the growth of individuals, e.g. of grass tufts, and their mutual competition for space are included in the model it may also be assumed that one individual covers more than one grid cell, with rules for assigning grid cells to growing individuals (Winkler and Klotz 1997; Winkler and Fischer 2002). A combination of grid-based local competition and continuous growth and dispersal is also possible (Winkler et al. 1999).

The grid-like structure of space can even completely be relaxed leading to models that define interaction neighbourhoods of individuals in some way [i.e. zone of influence or field of neighbourhood (Czárán 1998; Berger and Hildenbrandt 2000; Stoll et al. 2002; Bauer et al. 2004; Berger et al. 2004)].

The tremendous increase in computer power will enable modellers to overcome any technical problems in dealing with the richness of details that can be included in SEIB models. Forest models especially can handle dozens of species (Grimm and Railsback, 2005). Grassland models are, in practice, much more limited in terms of species numbers due to the difficulty of obtaining all the parameters needed.

The power and limits of rule-based SEIB models can best be presented by some examples together with their basic features and questions. We start with a one-patch one-species model. Buckley et al. (2003a, b) treated the dynamics of an invasive plant species, *Hypericum perforatum*, in an individual-based model that lacks a spatial aspect (thus it covers only partially the property of being invasive). The behaviour of one species in discrete space, whilst stressing the aspects of dispersal or of landscape dynamics, was modelled by Higgins et al. (2001) (invasion of *Acacia cyclops* and *Pinus pinaster* in Fynbos), or by Wiegand et al. (2000) (an *Acacia* species in the Negev desert). Density effects were limited to one grid cell. In contrast, Uriarte et al. (2002) using *Solidago altissima*, applied a model with continuous spatial coordinates and individual competition neighbourhoods. Other multi-species models looked for mechanisms underlying species persistence and coexistence in a community. The model of Favier et al. (2004), treating vegetation succession in a savanna by a model which was a mixture of a species- and vegetation-oriented approach, stressed key processes such as fire or succession cycle, but the deciding mechanisms for species replacement was hidden in the complexity of the model. Geertsema et al. (2002) proposed a model that relates plant population dynamics of several species to habitat quality, configuration and dynamics. The habitat was arranged as a dynamic network of suitable and unsuitable patches, each patch including at most K individuals of different species (a means of including intraspecific density control). Whereas the work stressed the importance of dispersal the authors treated interspecific competition, a core element of community dynamics, in insufficient detail. Much more explicit in terms of decisive mechanisms was a model by Higgins et al. (2000) that investigated the coexistence of grass and trees in a savanna; they emphasized a storage effect, and temporal variations in recruitment opportunities, as being key mechanisms. Groeneveld et al. (2002) published a spatially explicit rule-based model for three co-occurring *Banksia* species which investigated coexistence-mediating processes in a fire-prone shrubland in Western Australia. They arrived at a spatial establishment gradient as a key factor – a mechanism that works in the model for three species but which obviously lacks some generality (as species coexistence in small, homogeneous areas will still remain unclear). In the model a seedling lottery alone (see below) was not sufficient to explain coexistence. We will show that the interpretation of such findings on mechanisms by means of simulations can be aided by simple, analytical models.

In SEIB models, in contrast to strict Markov models or CA, dispersal over different distances by distribution of seeds or other propagules or by clonal spread is a key process. Seed-dispersion patterns vary among plant species, populations and individuals. Field methods for evaluating phenomenological and mechanistic dispersal models have to be designed in order to obtain distance distributions from dispersion patterns. Models try to extract mechanisms behind these patterns. Also, consequences of dispersal for species spread and migration have been the subject of a great deal of attention.

The state of the art in modelling dispersal processes and in field work to gather relevant data is summarised by Nathan and Muller-Landau (2000), Cain et al. (2000), Levin et al. (2003) and Nathan et al. (2003). Bullock and Clarke (2000) evaluated dispersal field data for *Calluna vulgaris* and *Erica cinerea* and fitted them to a stratified dispersal model consisting of two exponential functions that represent two different dispersal mechanisms. Higgins et al. (2003b) linked theory to data for transport of seeds of *Xanthium strumarium* by the Iberian lynx, *Lynx pardinus*. Problems of species migration were thoroughly discussed by Higgins et al. (2003a, c). They argued that traditional diffusion models of dispersal cannot cope with the decisive mechanism, i.e. rare far-distance dispersal and recruitment events. Thus, fat-tailed dispersal functions are necessary, and at least a combination of two exponential functions. A quite different approach to the dispersal problem was presented by Tackenberg (2003). He did not propose a phenomenological model as just described but tried to develop models that rely on physical factors such as landscape topography, wind velocity, turbulence, and weather conditions on the one hand, and seed weight and properties on the other. Admittedly, this is a very appealing and promising approach, but it exhibits problems in parameterisation and adaptation to real landscapes.

Clonal plants place their vegetatively formed offspring modules mostly in the immediate neighbourhood of the mother modules, but with high variability according to environmental conditions. Sexual offspring are only rarely successful. Winkler and Fischer (2002), Winkler and Stöcklin (2002), and Stöcklin and Winkler (2004) studied the relationship between both forms of reproduction whereas Birch (2002) looked for the intermingling or separation of genets in stands of a clonal bracken species. Oborny and her group (Oborny et al. 2000, 2001; Oborny and Kun 2002; Kun and Oborny 2003) as well as Herben and Suzuki (2002) looked at an important feature of clonal plants, their ability to exchange resources between modules and to divide labour via integrating modules over large areas as contrasted to an early splitting of the genet. Needless to say, all such questions which also include the patchiness of soil resources demand analysis by SEIB models.

4 Analytical Models: Explanation of Coexistence as an Example

4.1 Coexistence: The Basic Problem

The multitude of models presented up to now served primarily for predictive purposes under conditions that were real or hypothetically assumed: "what will happen if ...". Such predictions can directly lead to practical results, such as the assessment of extinction risk or recommendations for management. But the aim of scientific investigation is also explanation: the reduction of the main features of observed phenomena (of their patterns) to some basic mechanisms. Model types differ in their ability to uncover processes that govern development of populations or communities. Here, analytical (mathematical) models will serve as a complement to simulation-oriented models such as Markov or SEIB models.

Coexistence of species in a community is a classic, however abstract, pattern to be explained by models. Observations will only lead to an assessment of co-occurrence. The question arising from observations is therefore the probability of the continuation of such co-occurrence over a given period. Coexistence now means co-occurrence over an infinite period with a probability of 1 – this is the theoretical background for practical management problems which can be treated by analytical methods.

A multitude of mechanisms was proposed to explain species coexistence. For plants this question is especially striking as they mostly share only a few common resources. Simple niche separation therefore cannot be the dominant mechanism when multispecies coexistence, e.g. in homogeneous grassland, has yet to be explained. Barot (2004) gave an overview (including references) of existing proposals and suggested how to reconcile them. Without repeating his considerations I would like to discuss three basic aspects that appear in statements on coexistence: competition vs. interaction, lottery, and trade-off.

4.2 Competition and Coexistence

Competition is an interaction between individuals that leads to a reduction in net reproductive rate (R_0) by affecting survivorship or reproduction of the interacting individuals. It determines and limits coexistence. In textbooks, also in those for plant population ecology (Silvertown and Charlesworth 2001), we find the basic statement that stable coexistence implies that each species is more sensitive to its own density than to that of its competitors. This was derived from the Lotka–Volterra equations which are based on the parameters growth rate (r) and carrying capacity (K) and competition coefficients (α and β) (see Wissel 1989). Figure 1a transforms the coexistence conditions as derived from Lotka–Volterra equations into conditions that include demographic parameters such as birth and death rates as well as coefficients for individual interactions in order to bridge the gap between the classic equations and models based on life history. The system of equations of this figure links competition with fecundity in a simple, linear manner; it could easily be adapted for density-dependent mortality or mortality by interspecific "overgrowth" (the impact of facilitation will need further consideration). Such simple demographic models approximate to individual-based models but can also be considered as an expansion of matrix models which in their strict form are not suited to the investigation of phenomena such as coexistence, competitive exclusion, or succession. The figure shows that we must formulate coexistence or exclusion conditions by certain expressions that can be denoted as "competition functions". These expressions indicate overall population behaviour (the intercepts of the isolines) and are quite different from those coefficients that describe interactions of individuals (coefficients β_{ij}). Thus, competition as a process determining coexistence has two aspects: an individual and a population-dynamic one. This differentiation has consequences for field work (see below).

In the model of Fig. 1a stable coexistence must always include differences in the slopes of the isolines. These slopes are only determined by individual competitive abilities, by combinations of the β_{ij}-values (β_{11}/β_{12} and β_{21}/β_{22}, respectively). Demographic parameters affect the intercepts of isolines with the axes and thus determine the regions of coexistence. As a consequence, both demography and individual interactions together give rise to coexistence or exclusion. This is not only valid for different species but also for genetically different types of the same species: for coexistence they must differ not only in demography but also in their interactions in order to ensure long-term genetic diversity (as long as we neglect genetic mechanisms and consider a homogeneous habitat).

4.3 Seedling Lottery

The simple model shown in Fig. 1a only considers interactions between adult individuals. For a systematic investigation of coexistence and to build SEIB models of communities it is useful to differentiate interactions according to developmental stages (e.g. seedlings, juveniles, and adults). Some consequences of such differentiations are discussed by Levins and Culver (1971), Tilman (1994), Winkler et al. (1999), Higgins and Cain (2002) and Bruno et al. (2003).

The term lottery competition denotes an interaction between seedlings or emerged juveniles. It assumes that the space is divided up into elementary units (cells), where each unit can finally bear, after competition between juveniles, only one adult of any of the species in question. It is often a constituent of grid-based models (see the Groeneveld model discussed above), but can also implicitly be covered by analytical models. The outcome of lottery competition is a chance event, depending on the number of juveniles of the different species and their species-dependent relative competition strength. This type of competition was introduced by Skellam (1951) and then promoted by Chesson and Warner (1981) and mainly by Chesson in many subsequent papers (for relevant literature, see Chesson 2000; Muko and Iwasa 2000).

Sometimes, this "seedling lottery" is viewed as being the decisive component of a complete coexistence model. However, its contribution to coexistence depends on two factors: (1) the number of competing seedlings (not only of seeds!) relative to the number of available microsites (grid cells) – low and high percentages of occupied microsites may give a very different outcome of species competition (see the Skellam and Chesson cases in Fig. 1b); and (2) its incorporation into the whole demography of a species, including adult interactions, stochastic features and dispersal. Lottery competition for microsites alone is insufficient to explain coexistence.

4.4 Plant Performance under Trade-offs

During evolution a species tends to maximize its fitness, in competition with other species and between different phenotypes of the given species. This fitness maximisation concerns the evolution of different life history components. However, components may be linked to each other by restricting trade-offs. If these trade-offs also include interactions between individuals, each species maximising its fitness may nevertheless coexist. The model shown in Fig. 1a, b allows us to discuss some consequences of such trade-offs, irrespective of their underlying morphological or physiological mechanisms.

Following Skellam (1951), a combination of colonisation and competition may enable coexistence. However, both notions are somewhat vague. Colonisation can be defined as (net) fecundity (seed production plus germination plus seedling establishment) as well as "ability to reach new sites" and "ability to establish there", both indirect impacts on net reproductive rate. In small, homogeneous habitats the fecundity and establishment aspects are more relevant (as in Skellam's model) whereas in large, heterogeneous areas dispersal abilities become more and more important. Also competition has

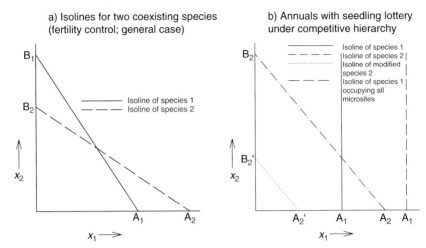

Fig. 1a, b. Isoline analysis of species coexistence. **a** Conditions for coexistence of two species in a homogeneous landscape without dispersal limitations. A simple demographic model of population growth of polycarpic perennial plants with mutual (linear) density control of fecundity as an example of density effects is given by $x_{1t+1}=x_{1t}-d_1 x_{1t}+b_{10} x_{1t}(1-\beta_{11}x_{1t}-\beta_{12}x_{2t})$ and $x_{2t+1}=x_{2t}-d_2 x_{2t}+b_{20} x_{2t} (1-\beta_{21} x_{1t}-\beta_{22}x_{2t})$, where x_1 and x_2 are the density of species 1 and 2, respectively [number of individuals relative to the total number (n) of available microsites], d_1 and d_2 the mortalities, and b_{10} and b_{20} the uncontrolled fecundities. Density control is expressed by factors β_{ij}. An isoline analysis (calculation of intercepts A_1, A_2, B_1, and B_2 with axes x_1 and x_2) gives the conditions for species coexistence: $A_2>A_1$ and $B_1>B_2$, with $A_1=(1-d_1/b_{10})/\beta_{11}$, $A_2=(1-d_2/b_{20})/\beta_{21}$, $B_1=(1-d_1/b_{10})/\beta_{12}$, and $B_2=(1-d_2/b_{20})/\beta_{22}$. For other conditions governing mutual species exclusion see Wissel (1989). These conditions include coefficients β_{ij} indicating the impact of individual interactions plus demographic rates (excluding density control). Thus, the outcome of population interaction depends on both sets of parameters. **b** Outcome of seedling-lottery competition in populations of two species with non-overlapping generations where within a microsite seedlings of species 1 are absolutely superior to those of species 2. *Solid lines* and *dashed-dotted lines* show the emergence of coexistence in the Skellam case (Skellam 1951), where species 1 is unable to occupy all available microsites; species 2 can coexist in a stable manner if $A_2>A_1$, leading to a fecundity condition $b_2>1/(1-x_1^*)$, where x_1^* is the stationary density of species 1 (Crawley and May 1987; competition–colonisation trade-off). *Dotted line* gives a less fertile species 2 that cannot fulfil this condition. When species 1 is able to occupy all n available microsites (*dashed line*; intercept A_1) we have density $x_1=1$, and species 2 has no chance in the presence of species 1, irrespective of its fecundity (Chesson case; Chesson and Warner 1981)

several aspects: the underlying interactions will concern different stages. These interactions may counteract each other. Species 1 can be totally superior to species 2 in a local seedling lottery: for annuals this feature can ensure coexistence (Fig. 1b). But in adult–juvenile competition species 2 may be superior to species 1 if we adopt a general hypothesis that adults cannot be replaced by juveniles, irrespective of species identity. The complicated situation arising from such conflicting assumptions (an "incomplete colonisation–competition trade-off") was first investigated by Fagerström and Agren (1979) and recently by Kisdi and Geritz (2003). The situation is clearer if species 1 is superior to species 2 in all interactions, i.e. if their offspring also can replace adults. Such total superiority of one species to another can theoretically lead to the coexistence even of many species in a community (Levins and Culver 1971; Tilman 1994). Because of such complications (Yu and Wilson 2001) the colonisation vs. competition model was predominantly applied to communities of annuals (Levine and Rees 2002).

In addition, a trade-off between competitive ability and stress tolerance links individual interactions to demography and may lead to species coexistence (Groeneveld 2003; Winkler et al. 2003).

Other trade-offs directly include only demographic parameters and not interactions. The model in Fig. 1a gives rise to the hypothesis that species only developing according to such trade-offs will not coexist. Indeed, the aim of modelling studies that included such demographic trade-offs was fitness maximisation of single species, not interactions in communities. There are discussions about a trade-off between seed number and seed mass (Leishman 2001; Levine and Rees 2002; Coomes and Grubb 2003). Larger dispersal distances may reduce the degree of interaction with other species, and seed mass may finally affect competitive ability of seedlings – in this way this trade-off also indirectly affects the degree of interactions. In clonal plants there may be trade-offs (linear or nonlinear) between sexual and vegetative reproduction – both with different "dispersal" distances and consequences for species behaviour in disturbed habitats (Winkler and Fischer 2002; Winkler and Stöcklin 2002; Stöcklin and Winkler, in press). Finally, trade-offs between reproductive capacity (i.e. fecundity and thus colonisation ability) and mortality are discussed (Crowley and McLetchie 2002; Metcalf et al. 2003). Also in this context the inclusion of interactions and hence of density effects turns out to be decisive in order to cope with species coexistence.

4.5 Limits of Analytical Models

Simple analytical equation models reach their limits if space is a key factor in community dynamics. One may introduce sophisticated mathematical methods to cope with ranges of interaction and dispersal (e.g. Bolker and Pacala 1999; Bolker et al. 2003). However, the explanatory value of such models sharply decreases, and systematic evaluation of well-designed SEIB models is now the better way to analyse species behaviour in a community.

Nevertheless, some simple a posteriori interpretation of simulation results in the light of equation models is possible through the calculation of different correction factors from SEIB results (Winkler et al. 1999).

Finally equation models must capitulate in the face of nonequilibrium phenomena of large spatial or temporal fluctuations, of unique events or catastrophes. Jeltsch and Moloney (2002) strongly recommend the application of SEIB models in such cases.

4.6 Theory and Empirical Work

Field ecologists often argue that models do not meet the problems they are really interested in. However, from the viewpoint of modelling there are questions for field workers: do their experiments always meet the complete problem that they are designed for? As an example I list some problems that arise in population-dynamic investigations.

There is the danger of looking only for parts of a life cycle, not even including all demographic events. Some examples of this are:

1. What is the fate of seeds in the soil: reliable data on seed-bank dynamics are often difficult to obtain.
2. Are dispersal and subsequent establishment always sufficiently distinguished between? Even long-distance dispersal may be more frequent than often thought, but difficulties in establishment far from the parent location give low colonisation rates (Nathan et al. 2003).
3. Is seedling and juvenile establishment studied under relevant site conditions over the entire vegetation period? Will germination experiments in the greenhouse or the experimental garden always be a sufficient substitute? Does birth rate always include the whole story from seed production until the onset of reproduction of the developing adults?
4. Is death rate sufficiently studied?

The distinction between individual interaction and population competition (see section 4.2.) has a consequence for experiments and field observations. Artificial settings that allow for the investigation of individual interactions in an isolated manner will clearly not reveal an overall competition effect as defined by population-dynamic equations, whereas long-term observation in nature on species replacement will demonstrate just this. The net effect has, according to the theory, different components which may, at a first glance, not be considered as "competition" at all. But all aspects have to be studied together (Aarssen and Keogh 2002).

The example model given in Fig. 1a shows that interactions have to be studied in such a way that one can see how they really affect demographic events (birth rate in Fig. 1a). However, many experiments on competition observe such a short period, often only some weeks, that the demographic effect will not be revealed. Weigelt and Jolliffe (2003) listed many indices, formed by combining several primary measures that summarize results from plant competition experiments. These competition indices are based on plant performance: yield, biomass, relative growth rate, cover. Such measures are, at best, indirectly linked with reproduction or mortality. The number of vegetative offspring may sometimes be included, certainly the probability of one species' individuals to be overgrown by other species. Seed number is not considered – there may be a relationship between size of an individual and its fertility, but that is not directly addressed in most competition experiments. What do such experiments really tell us about the development of populations in the field?

To summarize, many competition experiments do not aim to study demography, and the demographic matrix models mostly used by plant ecologists neglect competition. Plant population control in the field and the long-term observation of community development are obviously underrepresented fields in empirical research.

Another problem in plant population research is the sufficient separation of causes and effects. Stoll and Prati (2001) studied, in a broadly discussed paper, the effect of species aggregation. Experiments over (only) 6 weeks showed that the spatial distribution of plants profoundly affects competition in such a way that weaker competitors increased their fitness. It was concluded that the spatial arrangement of plants in a community can be an important determinant of species coexistence and biodiversity.

Can a 6-week experiment really tell us something about coexistence? What is the cause of aggregation? Are there some external conditions, or is the latter due to limited growth or dispersal ability as an internal feature? The question must be to determine the effect of such features on population competition; aggregation will be only a mediator.

Field ecologists may respond to such a criticism of their work that modellers also neglect a lot of important factors. Indeed, most models do not distinguish between above-ground and below-ground competition. Modellers argue that for the problems actually in question it is sufficient to summarize both features in one coefficient. Future research will show if this is always sufficient. This holds also for other features. If experimental findings demand their inclusion, there will be many ways to achieve this. In short: no model component, no model features without questions derived from real situations!

5 Outlook

Different modelling approaches and a lot of studies have shown the importance of (local) interactions between individuals and of dispersal when analysing species coexistence, exclusion, spread and succession. As space turns out to be indispensable (Jeltsch and Moloney 2002) SEIB models must play the leading role at the forefront of plant-ecological modelling, with analytical models as their complement in explaining basic patterns and mechanisms. Markov and matrix models are important for practical work, as they can be tightly linked with data, and Markov models (including neighbourhood interactions as in CA) will even increase in importance when large and heterogeneous landscapes are the focus of research.

Looking at SEIB models, some problems can be stressed which should be investigated in the near future in more detail. The habitat, mainly the soil, is not a fixed, invariable template. It acts on plants, but it is equally influenced by the individuals growing in it. Gassmann et al. (2000) talks about "circular" causality in the interaction between local plant cover and soil properties. Therefore we have to ask: what are the general impacts of soil–plant circularity on community composition?

A well-known property of plants is their high degree of plasticity. We should distinguish between plasticity in the broad sense of a reaction norm and adaptive phenotypic plasticity, a phenotypic response to local environmental conditions that enhances fitness, as far as trade-offs will allow for it (Sultan 1995; Callaway et al. 2003). In the plant kingdom, clonal species are the best candidates to exhibit such situation-oriented fitness-enhancing behaviour, provided the costs of plasticity are limited. What are the consequences for the community if several co-occurring species react in such a way?

Last but not least: long-term succession is important in many communities; there are many concepts related to this and also Markov models, but this process deserves more attention by SEIB modellers (e.g. Groeneveld 2003).

A modeller needs data and patterns, but a field ecologist needs concepts. We must conclude that we need more research that integrates field work and modelling into a loop, if possible, even into a multiple loop. There is especially a lack of long-term studies which are planned or qualified according to a theoretical concept which should create testable hypotheses. Thus, the integration of long-lasting field work and modelling is still in its infancy!

Acknowledgements. I am grateful to V. Grimm and H. Auge for their suggestions and criticism of earlier drafts of the manuscript.

References

Aarssen LW, Keogh T (2002) Conundrums of competitive ability in plants: what to measure? Oikos 96:531–542

Balzter H (2000) Markov chain models for vegetation dynamics. Ecol Model 126:139–154

Barot S (2004) Mechanisms promoting plant coexistence: can all the proposed processes be reconciled? Oikos 106:185–192

Bauer S, Wyszomirski T, Berger U, Hildenbrandt H, Grimm V (2004) Asymmetric competition as a natural outcome of neighbour interactions among plants: results from the field-of-neighbourhood modelling approach. Plant Ecol 70:135–145

Berec L (2002) Techniques of spatially explicit individual-based models: construction, simulation, and mean-field analysis. Ecol Model 150:55–81

Berg H (2002) Population dynamics in *Oxalis acetosella*: the significance of sexual reproduction in a clonal, cleistogamous forest herb. Ecography 25:233–243

Berger U, Hildenbrandt H (2000) A new approach to spatially explicit modelling of forest dynamics: spacing, ageing and neighbourhood competition of mangrove tress. Ecol Model 132:287–302

Berger U, Hildenbrandt H, Grimm V (2004) Age-related decline in forest production: modelling the effects of growth limitation, neighbourhood competition and self thinning. J Ecol 92:846–853

Birch CPD (2002) The influence of position on genet growth: a simulation of a population of bracken [*Pteridium aquilinum* (L.) Kuhn] genets under grazing. Evol Ecol 15:463–483

Bolker BM, Pacala SW (1999) Spatial moment equations for plant competition: understanding spatial strategies and the advantages of short dispersal. Am Nat 153: 575–602

Bolker BM, Pacala SW, Neuhauser C (2003) Spatial dynamics in model plant communities: what do we really know? Am Nat 162:135–148

Brigham CA, Schwartz MW (eds) (2003) Population viability in plants. Conservation, management, and modeling of rare plants. (Ecological studies, no. 165) Springer, Berlin Heidelberg New York

Bruno JF, Stachowicz JJ, Bertness MD (2003) Inclusion of facilitation into ecological theory. Trends Ecol Evol 18:119–125

Buckley YM, Hinz HL, Matthies D, Rees M (2001) Interactions between density-dependent processes, population dynamics and control of an invasive plant species, *Trileurospermum perforatum* (scentless chamomile). Ecol Lett 4:551–558

Buckley Y, Briese DT, Rees M (2003a) Demography and management of the invasive plant species *Hypericum perforatum*. I. Using multi-level mixed-effects models for characterizing growth, survival and fecundity in a long-term data set. J Appl Ecol 40: 481–493

Buckley Y, Briese DT, Rees M (2003b) Demography and management of the invasive plant species *Hypericum perforatum*. II. Construction and use of an individual-based model to predict population dynamics and the effects of management strategies. J Appl Ecol 40:494–507

Bullock JM, Clarke RT (2000) Long distance seed dispersal by wind: measuring and modelling the tail of the curve. Oecologia 124:506–521

Cain M, Milligan BG, Strand AE (2000) Long-distance seed dispersal in plant populations. Am J Bot 87:1217–1227

Callaway RM, Pennings SC, Richards CL (2003) Phenotypic plasticity and interactions among plants. Ecology 84:1115–1128

Campbell BD, Stafford Smith DM, Ash AJ (1999) A rule-based model for the functional analysis of vegetation change in Australasian grasslands. J Veg Sci 10:723–730

Carmel Y, Kadmon R, Nirel R (2001) Spatiotemporal predictive models of Mediterranean vegetation dynamics. Ecol Appl 11:268–280

Caswell H (2000) Prospective and retrospective perturbation analysis: their role in conservation biology. Ecology 81:619–627

Caswell H (2001) Matrix population models. Construction, analysis, and interpretation, 2nd edn. Sinauer, Sunderland, Mass.

Chesson P (2000) Mechanisms of maintenance of species diversity. Annu Rev Ecol Syst 31: 343–366

Chesson P, Warner RR (1981) Environmental variability promotes coexistence in lottery competitive systems. Am Nat 117:923–943

Coomes DA, Grubb PJ (2003) Colonization, tolerance, competition and seed-size variation within functional groups. Trends Ecol Evol 18:283–291

Crawley MJ, May RM (1987) Population dynamics and plant community structure: competition between annuals and perennials. J Theor Biol 125:457–489

Crowley PF, McLetchie DN (2002) Trade-offs and spatial life-history strategies in classical metapopulations. Am Nat 159:190–208

Czárán T (1998) Spatiotemporal models of population and community dynamics. Chapman & Hall, London

Damgaard C (2003) Modeling plant competition along an environmental gradient. Ecol Model 170:45–53

De Kroon H, Van Groenendael J, Ehrlén J (2000) Elasticities: a review of methods and model limitations. Ecology 81:607–618

Easterling MR, Ellner SP, Dixon PM (2000) Size-specific sensitivity: applying a new structured population model. Ecology 81:694–708

Ehrlén J (2000) The dynamics of plant populations: does the history of individuals matter? Ecology 81:1675–1684

Eriksson A, Eriksson O (2000) Population dynamics of the perennial *Plantago media* in semi-natural grasslands. J Veg Sci 11:245–252

Facelli JM, Pickett STA (1990) Markovian chains and the role of history in succession. Trends Ecol Evol 5:27–29

Fagerström T, Agren GI (1979) Theory for coexistence of species differing in regeneration properties. Oikos 33:1–10

Favier, C, Chave J, Fabing A, Schwartz D Duvois MA (2004) Modelling forest–savanna mosaic dynamics in man-influenced environments: effects of fire, climate and soil heterogeneity. Ecol Model 171:85–102

Fieberg J, Ellner SP (2001) Stochastic matrix models for conservation and management: a comparative review of methods. Ecol Lett 4:244–266

Fox GA, Gurevich J (2000) Population numbers count: tools for near-term demographic analysis. Am Nat 156:242–256

Freckleton RP, Watkinson AR (2002a) Are weed population dynamics chaotic? J Appl Ecol 39:699–707

Freckleton RP, Watkinson AR (2002b) Large-scale spatial dynamics of plants: metapopulations, regional ensembles and patchy populations. J Ecol 90:419–434

Freckleton RP, Watkinson AR, Dowling PM, Leys AR (2000) Determinants of the abundance of invasive annual weeds: community structure and non-equilibrium dynamics. Proc R Soc Lond Ser B 267:1153–1161

Gassmann F, Klötzli F, Walther GR (2000) Simulation of observed types of dynamics of plants and plant communities. J Veg Sci 11:397–408

Geertsema W, Opdam P, Kropff MJ (2002) Plant strategies and agricultural landscapes: survival in spatially and temporally fragmented habitat. Landsc Ecol 17:263–279

Grant A, Benton TG (2000) Elasticity analysis for density-dependent populations in stochastic environments. Ecology 81:680–693

Grimm V (1999) Ten years of individual-based modelling in ecology: what have we learned, and what could we learn in the future? Ecol Model 115:129–148

Grimm V, Wissel C (2004) The intrinsic mean time to extinction: a unifying approach to analysing persistence and viability of populations. Oikos 105:501–511

Grimm V, Railsback SF (2005) Individual-based modeling and ecology. Princeton University Press, Princeton, N.J.

Groeneveld J (2003) Raum-zeitliche Artenmuster entlang von Stressgradienten – ein Simulationsmodell in Anlehnung an die Pflanzensukzession auf den trockengefallenen Aralseeböden. PhD thesis. University of Marburg, Marburg

Groeneveld J, Enright NJ, Lamont BB, Wissel C (2002) A spatial model of coexistence among three *Banksia* species along a topographic gradient in fire-prone shrublands. J Ecol 90:762–774

Guàrdia R, Raventós J, Caswell H (2000) Spatial growth and population dynamics of a perennial tussock grass (*Achnatherum calamagrostis*) in a badland area. J Ecol 88: 950–963

Gustafsson C, Ehrlén J (2003) Effects of intraspecific and interspecific density on the demography of a perennial herb, *Sanicula europaea*. Oikos 100:317–324

Herben T, Suzuki J (2002) A simulation study of the effects of architectural constraints and resource translocation on population structure and competition in clonal plants. Evol Ecol 15:403–423

Higgins SI, Cain ML (2002) Spatially realistic plant metapopulation models and the colonization-competition trade-off. J Ecol 90:616–626

Higgins SI, Bond WJ, Trollope WSW (2000) Fire, resprouting and variability: a recipe for grass-tree coexistence in savanna. J Ecol 88:213–229

Higgins SI, Richardson DM, Cowling RM (2001) Validation of a spatial simulation model of a spreading alien plant population. J Appl Ecol 38:571–584

Higgins SI, Clark JS, Nathan R, Hovestadt T, Schurr F, Fragoso JMV, Aguiar MR, Ribbens E, Lavorel S (2003a) Forecasting plant migration rates: managing uncertainty for risk assessment. J Ecol 91:341–347

Higgins SI, Lavorel S, Revilla E (2003b) Estimating plant migration rates under habitat loss and fragmentation. Oikos 101:354–366

Higgins SI, Nathan R, Cain ML (2003c) Are long-distance dispersal events in plants usually caused by nonstandard means of dispersal? Ecology 84:1945–1956

Jeltsch F, Moloney KA (2002) Spatially explicit vegetation models: what have we learned? Prog Bot 63:326–343

Jenerette GD, Wu JG (2001) Simulation of land-use change in the central Arizona–Phoenix region, USA. Landsc Ecol 16:611–626

Kisdi E, Geritz SAH (2003) On the coexistence of perennial plants by the competition colonization trade-off. Am Nat 161:350–354

Kiviniemi K (2002) Population dynamics of *Agrimonia eupatoria* and *Geum rivale*, two perennial grassland species. Plant Ecol 159:153–169

Kun Á, Oborny B (2003) Survival and competition of clonal plant populations in spatially and temporally heterogeneous habitats. Commun Ecol 4:1–20

Lanzer ATS, Pillar VD (2002) Probabilistic cellular automaton: model and application to vegetation dynamics. Commun Ecol 3:159–167

Leishman MR (2001) Does the seed size/number trade-off model determine plant community structure? An assessment of the model mechanisms and their generality. Oikos 93: 294–302

Lennartsson T, Oostermeijer GB (2001) Demographic variation and population viability in *Gentianella campestris*: effects of grassland management and environmental stochasticity. J Ecol 89:451–463

Levin SA, Muller-Landau HC, Nathan R, Chave J (2003) The ecology and evolution of seed dispersal: a theoretical perspective. Annu Rev Ecol Evol Syst 34:575–604

Levine JM (2003) A patch modeling approach to the community-level consequences of directional dispersal. Ecology 84:1215–1224

Levine JM, Rees M (2002) Coexistence and relative abundance in annual plant assemblages: the role of competition and colonization. Am Nat 160:452–467

Levins R, Culver D (1971) Regional coexistence of species and competition between rare species. Proc Natl Acad Sci USA 68:1246–1248

Menges ES (2000) Population viability analyses in plants: challenges and opportunities. Trends Ecol Evol 15:51–56

Mesarović MD (1968) Systems theory and biology – view of a theoretician. In: Mesarovic MD (ed) Systems theory and biology. Springer, Berlin Heidelberg New York, pp 59–87

Metcalf JC, Rose KE, Rees M (2003) Evolutionary demography of monocarpic perennials. Trends Ecol Evol 18:471–480

Muko S, Iwasa Y (2000) Species coexistence by permanent spatial heterogeneity in a lottery model. Theor Popul Biol 57:273–284

Nathan R, Muller-Landau HC (2000) Spatial patterns of seed dispersal, their determinants and consequences for recruitment. Trends Ecol Evol 15:278–285

Nathan R, Perry G, Cronin JT, Strand AE, Cain ML (2003) Methods for estimating long-distance dispersal. Oikos 103:261–273

Oborny B, Kun A (2002) Fragmentation of clones: how does it influence dispersal and competitive ability? Evol Ecol 15:319–346

Oborny B, Kun Á, Czárán T, Bokros S (2000) The effect of clonal integration on plant competition for mosaic habitat space. Ecology 81:3291–3304

Oborny B, Czárán T, Kun Á (2001) Exploration and exploitation of resource patches by clonal growth: a spatial model on the effect of transport between modules. Ecol Model 141:151–169

Oostermeijer JGB, Suijten SH, den Nijs JCM (2003) Integrating demographic and genetic approaches in plant conservation. Biol Conserv 113:389–398

Pfab MF, Witkowski ETF (2000) A simple population viability analysis of the critically endangered *Euphorbia clivicola* R.A. Dyer under four management scenarios. Biol Conserv 96:263–270

Rydgren K, De Kroon H, Økland, RH, Van Groenendael J (2001) Effects of fine-scale disturbances on the demography and population dynamics of the clonal moss *Hylocomium splendens*. J Ecol 89:395–405

Silvertown J, Charlesworth D (2001) Introduction to plant population biology, 4th edn. Blackwell, Oxford

Skellam JG (1951) Random dispersal in theoretical populations. Biometrika 38:196–218

Soares-Filho BS, Coutinho Cerqueira G, Lopes Pennachin C (2002) DINAMICA – a stochastic cellular automata model designed to simulate the landscape dynamics in an Amazonian colonization frontier. Ecol Model 154:217–235

Stöcklin J, Winkler E (2004) Optimum reproduction and dispersal strategies of a clonal plant in a metapopulation: a simulation study with *Hieracium pilosella*. Evol Ecol 18:563–584

Stoll P, Prati D (2001) Intraspecific aggregation alters competitive interactions in experimental plant communities. Ecology 82:319–327

Stoll P, Weiner J, Muller-Landau H, Muller E, Hara T (2002) Size symmetry of competition alters biomass-density relationships. Size symmetry of competition alters biomass density relationships. Proc R Soc Lond Ser B 269:2191–2195

Sultan SE (1995) Phenotypic plasticity and plant adaptation. Acta Bot Neerl 44:363–383

Tackenberg O (2003) Modeling long-distance dispersal of plant diaspores by wind. Ecol Monogr 73:173–189

Tilman D (1994) Competition and biodiversity in spatially structured habitats. Ecology 75:2 16

Uriarte M, Canham CD, Root R (2002) A model for simultaneous evolution of competitive ability and herbivore resistance in a perennial plant. Ecology 83:2649–2663

Wadsworth RA, Collingham YC, Willis SG, Huntley B, Hulme PE (2000) Simulating the spread and management of alien riparian weeds: are they out of control? J Appl Ecol 37[Suppl 1]:28–38

Watkinson AR, Freckleton RP, Forrester L (2000a) Population dynamics of *Vulpia ciliata*: regional, patch and local dynamics. J Ecol 88:1012–1029

Watkinson AR, Freckleton RP, Robinson RA, Sutherland WJ (2000b) Predictions of biodiversity response to genetically modified herbicide-tolerant crops. Science 289: 1554–1557

Weigelt A, Jolliffe P (2003) Indices of plant competition. J Ecol 91:707–720

Wiegand K, Schmidt H, Jeltsch F, Ward D (2000) Linking a spatially-explicit model of acacias to GIS and remotely sensed data. Folia Geobot 35:211–230

Wiegand T, Jeltsch F, Hanski I, Grimm V (2003) Using pattern-oriented modeling for revealing hidden information: a key for reconciling ecological theory and application. Oikos
100:209–222

Winkler E, Fischer M (1999) Two fitness measures of clonal plants and the importance of spatial aspects. Plant Ecol 141:191–199

Winkler E, Fischer M (2002) The role of vegetative spread and seed dispersal for optimal life
histories of clonal plants: a simulation study. Evol Ecol 15:281–301

Winkler E, Klotz S (1997) Long-term control of species abundances in a dry grassland: a spatially explicit model. J Veg Sci 8:189–198

Winkler E, Stöcklin J (2002) Sexual and vegetative reproduction of *Hieracium pilosella* L.
under competition and disturbance: a grid-based simulation model. Ann Bot 89:525 536

Winkler E, Fischer M, Schmid B (1999) Modelling the competitiveness of clonal plants by
complementary analytical and simulation approaches. Oikos 85:217–233

Winkler E, Prati D, Peintinger M (2003) Clonal plants in lake-shore grasslands: a 15-year
study at Lake Constance plus simulation modelling. Verh Ges Okol 33:185

Wissel C (1989) Theoretische Ökologie. Springer, Berlin Heidelberg New York

Woolcock JL, Cousens R (2000) A mathematical analysis of factors affecting the rate of spread
of patches of annual weeds in an arable field. Weed Sci 48:27–34

Yu DW, Wilson HB (2001) The competition-colonization trade-off is dead; long live the competition-colonization trade-off. Am Nat 158:49–63

Zuidema PA, Franco M (2001) Integrating vital rate variability into perturbation analysis: an
evaluation for matrix population models of six plant species. J Ecol 89:995–1005

Eckart Winkler
UFZ Centre for Environmental Research
Department of Ecological Modelling
POB 500 135
04301 Leipzig
Germany
e-mail: eckart.winkler@ufz.de

Atmospheric carbon dioxide enrichment effects on ecosystems – experiments and the real world

Martin Erbs and Andreas Fangmeier

1 Introduction

The beneficial effect of atmospheric CO_2 on the growth of plants has been known since 1804 (De Sassure 1804; cited in Kimball et al. 1993), and its role as a C source for vegetation was proven by Justus von Liebig 125 years ago. Atmospheric CO_2 enrichment has been used to promote the growth of legumes in greenhouse cultures for >50 years. As early as 1961, greenhouses covering >1,600 ha were under enriched CO_2 in the Netherlands alone.

However, only after atmospheric CO_2 concentrations had been recorded from continuous monitoring sites such as the Mauna Loa Observatory at Hawaii, where $[CO_2]$ has been recorded since 1958, was there a growing awareness that CO_2 enrichment occurs on a global scale and that it affects ecosystems throughout the world. In 2005, the concentration of atmospheric CO_2 will be around 380 $\mu mol\ mol^{-1}$; thus already exceeding by ca. 35% the background concentration of ca. 280 $\mu mol\ mol^{-1}$ before the beginning of industrialization. A further increase to at least 550 $\mu mol\ mol^{-1}$ will have occurred by the end of this century (IPCC 2001). Consequently, numerous studies have been performed to test the response of vegetation and ecosystems to CO_2 enrichment, and great progress has been made in developing experimental facilities and in our understanding of biosphere–atmosphere interactions with respect to CO_2 by means of both experimentation and modelling.

CO_2 enrichment effects on crops were reviewed as early as the mid 1980s by Cure and Acock (1986) who reported an average increase in C_3 crop yield due to CO_2 doubling of approximately 41%. A mechanistic understanding of the physiological background for this CO_2 fertilization effect in C_3 plants that has been widely accepted was provided by von Caemmerer and Farquhar (1981). This CO_2 gas exchange model for C_3 plants was later extended and modified by Sage (1994) and other authors to explain photosynthetic acclimation to CO_2 enrichment. Down-regulation of photosynthesis in C_3 plants

Progress in Botany, Vol. 67
© Springer-Verlag Berlin Heidelberg 2006

may largely be attributed to sink limitation due to experimental shortcomings during exposure, such as a soil volume which is too low (Arp 1991), thus triggering feedback inhibition of photosynthesis (Stitt 1996).

Meta-analyses of existing data revealed that there is a relationship between plant functional type and growth response to CO_2 enrichment (e.g. Poorter 1993; Poorter et al. 1996). Data compiled in the latter study comprised 500 observations of CO_2 exposure of individually grown plants without competition and without flowering or fruiting. Poorter et al. (1996) found an average plant weight increase of 58% in C_3 crops, 44% in wild herbaceous C_3 species, 40% in woody species (which are all C_3), 14% in C_4 species, and 15% in CAM species, due to CO_2 doubling. However, it is also clear from these meta-analyses that variation between the results of different studies involving the same species is huge and that species behaviour in the real world may not be predicted from short-term experiments under controlled conditions.

An overall result from the numerous studies performed to date is that atmospheric CO_2 enrichment will lead to more "efficient" plants (Drake et al. 1997). This not only concerns the increase in biomass production and yield which is most pronounced in C_3 species, but also water use efficiency which is positively affected by CO_2 enrichment in all species irrespective of their photosynthetic pathway (Amthor 1995; Bowes 1996). In addition, plants with a C_3 photosynthetic pathway generally exhibit widened C/N ratios (Cotrufo et al. 1998) and have a potentially higher N-use efficiency because of acclimation to CO_2 enrichment involving lower N allocation to Rubisco and enzymes of the photorespiratory pathway (Webber et al. 1994; Fangmeier et al. 2000; Ainsworth and Long 2005).

As the physiological and chemical background of the response of plant species to CO_2 enrichment has been largely clarified in earlier work, in several more recent studies the response of different types of ecosystems to CO_2 enrichment has been investigated in experimental field trials. A summary of such approaches published in 1996 states that crop systems show the clearest correspondence between photosynthetic and growth responses since resources other than CO_2 are not strongly limiting, whereas in unmanaged ecosystems there is often little or no significant measurable increase in biomass (Koch and Mooney 1996).

Most of the latest studies involve free air carbon dioxide enrichment (FACE) exposure systems which nearly completely overcome the problem of microclimatic artefacts during exposure. At present, > 30 FACE experiments are being performed worldwide (see http://cdiac.esd.ornl.gov/programs/FACE/face.html for the most recent information). One of these studies at Rhinelander, Wisconsin, also includes the experimental manipulation of both

CO_2 and O_3 concentrations (Isebrands et al. 2001). A meta-analytic review of the knowledge gained from FACE experiments has been presented recently (Ainsworth and Long 2005).

One common focus of most of the recent experiments is not only to test the response of the ecosystem under investigation with regard to shifts in species composition, productivity, and other ecosystem traits, but also to understand the C cycling in these ecosystems, and their potential role as C sinks or sources under atmospheric CO_2 enrichment. This also has to be interpreted in the context of the role of CO_2 as the major anthropogenic greenhouse gas (IPCC 2001) and international agreements of greenhouse gas emission reductions such as the Kyoto Protocol which came into force on 16 February 2005.

In this contribution we will be not able to cover even a minority of all these aspects. We will not deal with the role of CO_2 as a greenhouse gas, without disregarding the importance of global climate change driven by CO_2 enrichment. We will also not deal with modelling approaches employed to understand CO_2 effects at different scales. Instead, we will focus on experimental field work on the direct effects of increasing atmospheric CO_2 concentrations on vegetation. In this context, we will report the progress of the experimental approaches currently used in CO_2 research, and we will critically evaluate the potential shortcomings of those approaches.

2 Experimental Manipulation of Atmospheric CO_2

2.1 The History of CO_2 Enrichment Experimentation

The vast number of CO_2 experiments that have been conducted in the last decades are documented by several reviews and meta-analyses of experimental results and technical features (e.g. Kimball 1983; Bazzaz 1990; Allen et al. 1992; Lee and Barton 1993; Drake et al. 1997; Curtis and Wang 1998; McLeod and Long 1999; Wand et al. 1999; Körner 2000; Amthor 2001; Fangmeier and Jäger 2001; Norby et al. 2001; Jablonski et al. 2002; Nowak et al. 2004; Long et al. 2004; Ainsworth and Long 2005). The history of CO_2 fertilization as used in glasshouses and of the first CO_2 enrichment studies has been compiled by Wittwer (1986). Early examinations of plant responses to elevated CO_2 exposure were mainly conducted on single plants under laboratory conditions and revealed the basic understanding of physiological effects. However, as early as 1922 a first field exposure experiment on crop growth under CO_2 enrichment involving a chamberless tube release system was performed in Sweden (Lundegårdh 1927).

Knowledge gained with laboratory experiments was not adequate to be applied to the real-world situation since the CO_2 response of plants is highly interactive with other environmental factors such as water and nutrient availability, temperature and light (Mooney et al. 1999). Particularly these microclimatic conditions are typically found to be altered in any kind of CO_2 enrichment system relying on chambers or enclosures. In order to overcome this problem different fumigation systems were developed in the 1970s and 1980s (Heagle et al. 1973; Mandl et al. 1973; Lockyer et al. 1976; Shinn et al. 1977; Heck et al. 1978; Rogers et al. 1983). Among these approaches, open-top chambers became the exposure system most commonly used because up until then they provided the best approximation of a natural microenvironment and were easy to run.

Nevertheless, it was shown that also in open-top chambers growth conditions are not comparable to the open field and that the technical apparatus affects the plants. This chamber effect is characterized by increased ambient air temperatures of up to 3 K (Heagle et al. 1973, 1989; Fangmeier et al. 1992; Lee and Barton 1993), reduced air humidity of up to 5% (Fangmeier et al. 1992; Lee and Barton 1993), and a light retention of the plastic foils depending on material and age of 12–25% (Heagle et al. 1989; Van Oijen et al. 1999). Additionally, the quality of solar radiation is substantially altered by the foils used for most open-top chambers since UV radiation is absorbed by them (Heagle et al. 1989). Recent meta-analyses demonstrated that reactions of single plants grown under artificial and constant environmental conditions are not suitable for appropriately predicting their response in the field (Poorter and Navas 2003; Long et al. 2004). Therefore, there is a clear necessity to examine the effects of CO_2 enrichment on plants growing under real-world conditions (Drake et al. 1997; Bazzaz and Catovsky 2002).

2.2 FACE Technology

Great progress was made due to the development of FACE systems in the early 1990s. FACE systems typically consist of several units, each of them comprising a circular plenum connecting vertical pipes through which CO_2 is released onto the test plots. A computer-controlled feedback system controlled by wind speed, wind direction and CO_2 concentration achieved maintains the CO_2 level at the FACE plot at the target value. Since there is no enclosure of the plot by walls or panels, microclimatic alterations at the test plots are much smaller than in any other previously used exposure system.

The first FACE system reported and the prototype of this new category of experimental setup was the Brookhaven National Laboratory (BNL)-type FACE system (Brookhaven National Laboratory, N.Y.) operated in Maricopa, Arizona (Hendrey and Kimball 1990). Since then, further FACE systems have been established at different locations all around the world (http://cdiac.esd.ornl.gov/programs/FACE/face.html). The plot size of the current FACE systems is up to 30 m in diameter, and the height of the area receiving CO_2 enrichment is up to 20 m.

The large plot size as typically realized in BNL-type FACE systems is useful since the area exposed to CO_2 enrichment is large enough to facilitate interdisciplinary research involving several research groups. On the other hand, fumigation of large-scale field plots is costly in terms of CO_2 consumption even for short vegetation (Leadley et al. 1997; Nowak et al. 2004). For example, the CO_2 consumption of the BNL-type FACE system at Maricopa, Arizona, which is 23 m in diameter and is used to assess effects on short vegetation, approached ca. 1.2 t CO_2 per day per FACE unit. Hence, it is not surprising that up until 1999 no crop, semi-natural or natural vegetation type had been examined at more than one location under FACE (McLeod and Long 1999). Furthermore, in order to save costs the number of replicates in FACE experiments is often low, at least in the large BNL-type FACE systems. This may cause problems, particularly when plant communities are examined, since variability between plots may be so pronounced that there is a failure to detect CO_2 effects (Chiariello and Field 1996). Consequently, in their meta-analysis of FACE experiments performed on semi-natural or natural plant communities Nowak et al. (2004) pointed out the need for further free-air CO_2 research on species mixtures. The need for FACE studies on natural vegetation is also underlined by other authors (Field et al. 1992; Mooney et al. 1999; Bazzaz and Catovsky 2002; Weltzin et al. 2003; Long et al. 2004).

Besides the BNL type, other types of FACE systems with smaller plot sizes have been developed. Miglietta et al. (1996) developed a simple midi-FACE system enveloping plots 8 m in diameter. Another midi-FACE system was described by Jäger et al. (2003). This system also encloses plots 8 m in diameter but is equipped with a circular plenum which recirculates part of the CO_2-enriched air after passage across the plot. In addition, several other mini-FACE systems have been developed, most of them supplying plots 1–2 m in diameter. Many of the mini-FACE systems lack the involvement of wind direction to control the CO_2 supply (e.g. Miglietta et al. 2001a). Nevertheless, their operation costs with respect to CO_2 consumption are sufficiently low to run a satisfactory number of replicates.

2.3 Deviations From the Real World Despite FACE Technology

A potential shortcoming of many of the recent FACE experiments is that CO_2 fumigation is discontinued during the night in order to save costs. Thus, the daily course of $[CO_2]$ achieved in these exposure systems does not reflect the time course of $[CO_2]$ expected in a future CO_2-rich world (Fig. 1B) since a future CO_2-rich world will see even higher night-time concentrations. Around-the-clock CO_2 enrichment would represent a much more realistic scenario of anticipated future $[CO_2]$. There is some debate whether additional night-time CO_2 enrichment would render significantly different experimental results than daylight-hour fumigation alone (e.g. Bunce 2003; Holtum and Winter 2003; Ainsworth and Long 2005).

Unfortunately, there are only few experiments that have tested day- and night-time versus day-time CO_2 enrichment, and none of these studies have been performed in the field with FACE systems. In a recent study the responses of seedlings of five species (two C_3 herbs, one C_4 herb, and two tree species) were tested at day-time/night-time CO_2 concentrations of 700/350 and 700/700 μmol mol^{-1}, respectively; significant differences in dry mass were found between the CO_2 treatments as well as interactions with day-time/night-time temperature regimes (Bunce 2003). Since results from climate chamber experiments with seedlings may not be predictive for ecosystem responses under field conditions, we cannot state the potential error due to missing night-time CO_2 enrichment in many of the current studies. The potential mode of action of night-time CO_2 enrichment on the growth response is not clear. Increased $[CO_2]$ at night will affect water relations (Wullschleger et al. 2002), though only to a small degree, and thus lead to growth responses. It has been shown that CO_2 enrichment inhibits mitochondrial respiration (Drake et al. 1999). However, these observations might be attributed to artefacts of earlier measurement systems (Ainsworth and Long 2005).

Another technical feature of the BNL-type FACE systems is the involvement of powerful blowers that carry CO_2 diluted with ambient air into the FACE rings (Nagy et al. 1992). The use of blowers means a two-step dilution

---

Fig. 1A–C. Course of diurnal average $[CO_2]$ for ambient and CO_2-enriched plots (1-h means±SE). **A** Data from the 2003 exposure period (12 May–27 July) in Hohenheim from five ambient and five free air carbon dioxide enrichment (*FACE*) plots, respectively. **B, C** Data from a potato growth experiment performed in 1999 with an exposure period from 15 May to 15 September in Giessen. **B** FACE data from only one plot (no SE; data kindly provided by S. W. Schmidt, L. Grünhage and H.-J. Jäger, Giessen). **C** Data from four replicate CO_2 treatments in open-top chambers (*OTC*) [for more information see De Temmerman et al. (2002)]

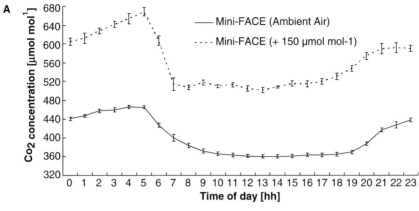

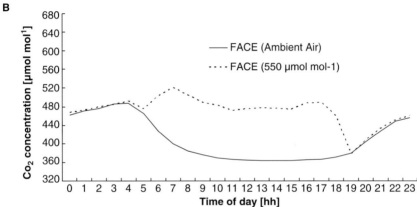

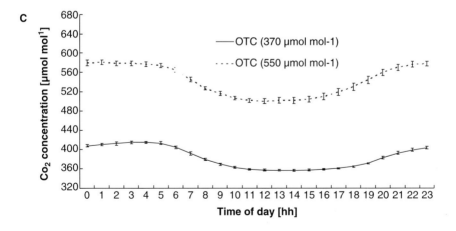

of pure CO_2 from the storage tank down to the desired concentration at the FACE plot and is a prerequisite for, as far as is possible, homogeneous mixing and an even concentration profile across the FACE plots. On the other hand the use of blowers creates microclimatic artefacts since they carry additional energy into the plots and produce a specific wind profile in the vicinity of the vents of the pipes. This microclimatic alteration is most significant under stable night-time wind conditions. Pinter et al. (2000) demonstrated that night-time foliage and air temperatures were on average 0.6–1.0 °C warmer than controls without blowers and that dew duration was significantly lower, leading to alterations in the development of the investigated wheat canopies. Pinter et al. (2000) therefore concluded that the desirability of night-time CO_2 enrichment might be questionable–assuming that blowers are used.

2.4 The Hohenheim Mini-FACE System

A new mini-FACE system was recently developed at the University of Hohenheim. This system was designed for minimum CO_2 consumption and to avoid the use of a blower.

The system consists of 15 field plots of 2 m in diameter designed to expose a spring wheat–weed mixture to CO_2 enrichment. Five plots each are assigned to one of the following three treatments: (1) FACE plots; (2) ambient air plots with the same technical equipment as the FACE plots but without an elevated $[CO_2]$ treatment; (3) control plots without any technical installations. Elevated CO_2 levels (ambient $[CO_2]$+150 µmol mol^{-1}) are achieved by using a computer-based control system. According to wind speed and direction, undiluted labelled CO_2 ($\delta^{13}C$; −48 %) is dispensed into the FACE plots. Each plot is divided into four sectors with a separate CO_2 supply according to the prevailing wind direction, thus ensuring that the CO_2 is dispensed only from the two upwind directions. Two transparent windscreens made of UV-stabilized PVC shields, 1 mm thick and 0.15 m high surround each FACE and ambient plot (Fig. 2) and accelerate the mixing of CO_2 and ambient air. Perforated PTFE tubes are used for the release of CO_2 and are attached to the middle of the inner surfaces of the two windscreens. The windscreens create turbulence and thus ensure good mixing of the CO_2 released from the tubes. This functional principle has been applied earlier (Walklate et al. 1996; Leadley et al. 1997; Volk et al. 2003; Erbs and Fangmeier 2005) and proven to be reliable and efficient. No blowers have to be used, thus largely avoiding their influence, described above, on microclimatic conditions. Costs for CO_2 consumption are comparatively low (<5 t

Fig. 2. Photograph of a Hohenheim mini-FACE plot during the early growing season

liquid CO_2 month-1 at an average wind speed of 0.77 m s^{-1}) and allow for full-time CO_2 enrichment without a cut-off at night.

The system showed high performance in terms of reliability, achievement and homogeneity of the target $[CO_2]$, and microclimatic disturbance. A comparison of the Hohenheim FACE system with other systems is given in Table 1. In 2003 the system was operated 24 h day-1 at 96.9% reliability. Similar reliability values have been reported by Hendrey et al. (1999) and Miglietta et al. (2001b) (96.7% and 93%, respectively). In the remaining 3.1% of the time fumigation was interrupted due to the calibration routine of the CO_2 analysers and fine adjustments of control parameters. The season-long increment of $[CO_2]$ at the FACE plots in 2003 was +160.2 µmol mol^{-1} (SE ±1.56), i.e. about 10 µmol mol^{-1} higher than intended (Fig. 1A). For comparison, the diurnal course of $[CO_2]$ from a FACE system in operation only during day-time (Fig. 1B) and from an open-top chamber system involving the dispensing of CO_2 for 24 h (Fig. 1C) is also shown.

A main criterion used to assess the performance of a fumigation system is the minimum duration of oscillations in $[CO_2]$ that cause a measurable effect on the C uptake of the plants. According to Hendrey et al. (1997) oscillations of approximately ±20% in amplitude from the average $[CO_2]$ lasting longer than 1 min may influence photosynthesis. At the Hohenheim mini-FACE deviations from the target $[CO_2]$ of <±10% and ±20% were achieved for 63.2% and 89.9% of the time, based on the 1-min readings at the individual plots, respectively, despite not using a blower to dispense CO_2. Figure 3 shows

Table 1. Technical details and performance of different free air-carbon dioxide enrichment (*FACE*) systems

Type of vegetation	Plot size (diameter; m)	Target $[CO_2]$ (μmol mol^{-1})	Operation time	Method of exposure	Percentage of 1-min $[CO_2]$ data within deviation from the set point $[CO_2]$ of $\pm10\%$	Percentage of 1-min $[CO_2]$ data within deviation from the set point $[CO_2]$ of $\pm20\%$	Fluctuation of $[CO_2]$ in the central area	Reference
Cotton field	22.0	550 (Fixed)	Daytime only	Blower/diluted CO_2	88%	98%	$\pm5\%$	Nagy et al. (1992)
Wheat field (Dwarf cultivar)	1.0	600 (Fixed)	24 h Continuous	Blower/diluted CO_2	>90%	–	$\pm5\%$	Miglietta et al. (1996)
Grassland on limestone	0.2	700 (Fixed)	Daytime only	Blower/diluted CO_2	64%	–	–	Spring et al. (1996)
Potato field	8.0	460, 560, 660 (Fixed)	Daytime only	Blower/diluted CO_2	83%, 74%, 66% (for three different $[CO_2]$)	–	<$\pm10\%$	Miglietta et al. (1997)

Loblolly pine forest	26.0	550 (Fixed)	Daytime only	Blower/diluted CO_2	69%	92%	±10% –±20%	Hendrey et al. (1999)
Bog ecosystem	1.0	560 (Fixed)	24 h Continuous	Blower/diluted CO_2	–	>95%	±20%	Miglietta et al. (2001a)
Poplar plantation	22.2	550 (Fixed)	24 h Continuous	No blower/pure CO_2	75%	91%	–	Miglietta et al. (2001b)
Rice paddy	12	+ 200 Above ambient	24 h Continuous	No blower/pure CO_2	60%	90%	±15%	Okada et al. (2001)
Small-stature (treeline)	1.3	550 (Fixed)	Daytime only	No blower/pure CO_2	63%	90%	±10%	Hättenschwiler et al (2002)
Semi-natural grassland	8.0	+ 100 Above ambient	Dayti only	Blower/pure CO_2	–	–	±6%	Jäger et al. (2003)
Wheat and weeds of cultivation	2.0	+ 150 Above ambient	24 h Continuous	No blower/pure CO_2	63.2%	89.9%	<±10%	Hohenheim mini-FACE

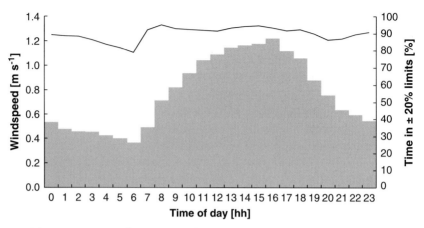

Fig. 3. Mini-FACE system performance in relation to wind speed throughout the 2003 exposure period. Data represent averaged daily time-course of 1-min means of wind speed (*grey
bars, left-hand y-axis*) and the percentages of 1-min [CO$_2$] measured on the FACE plots
within a ±20% range of the target [CO$_2$] (*solid line, right-hand y-axis*)

the relation between the 1-min averages of wind speed and the percentage of
time with 1-min mean [CO$_2$] within ±20% of the target concentration.
Obviously, the ±20% criterion was achieved less often only between 6 a.m.
and 7 a.m. when the wind speeds were lowest. Throughout the day-time
(7 a.m.–7 p.m.) the ±20% criterion was met >90% of the time.

Due to rapid changes in wind regime, fluctuations in [CO$_2$] are inevitable
at the plot borders in the proximity of the fumigation tubes (Karnosky et al.
2001). Taking this into consideration, only the central 1 m^2 was used for biological surveys at the Hohenheim FACE system. Comparable restrictions of
the usable area were imposed by Spring et al. (1996) and Miglietta et al.
(2001a) in their studies. The spatial distribution of [CO$_2$] for this central
1 m^2 is shown in Fig. 4, based on 12 different monitoring locations at five
different heights above ground level: 0.1, 0.2, 0.3, 0.4 and 0.5 m. With the
exception of the points at the outer corners (SE<±4%) differences in [CO$_2$]
obtained at the particular heights did not vary considerably compared to the
average (SE<±1%). In the central area fluctuations in [CO$_2$] did not exceed
10%. Largest differences were found at the outer corners, especially those
facing north-east and north-west. A somewhat better performance was
found for the BNL-type FACE systems involving blowers. However, compared with several other FACE systems the spatial CO$_2$ homogeneity of the
Hohenheim mini-FACE system is satisfactory (Table 1).

The technical installation of the Hohenheim system led to some microclimate alterations at the FACE plots. The temperature was increased during

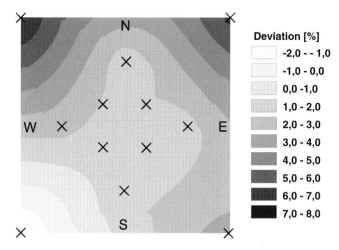

Fig. 4. Distribution of [CO_2] at the central 1 m^2 of the Hohenheim Mini-FACE plots. A comparison is given between [CO_2] measured simultaneously at the standard measuring point in the plot centre and 12 adjacent sample points (*crosses*). The results shown were calculated as percentages of deviation based on the [CO_2] at the plot centre (=100%)

the day-time (7 a.m.–8 p.m.) by 0.73 K on average, with a maximum increase of 1.24 K between 12 a.m. and 1 p.m. At the same time, air humidity was found to be reduced by 2.39% on average with a maximum difference around noon (−3.49%). These differences were slightly above the accuracy range of the sensors (±0.3 K and ±1.5% relative humidity at 296.15 K, respectively). Comparable data on air temperature are available for the screen-aided CO_2 control system described by Leadley et al. (1997) where mean daytime temperatures were elevated by 1 K.

The Hohenheim mini-FACE system represents a reliable and cost-effective exposure system with which to apply predicted future [CO_2] to short vegetation under a realistic exposure regime. Assembling the system at the beginning of the vegetative period can be performed within a few days. The running costs are low enough to run a sufficient number of replicates. It is also easy to extend the system for the fumigation with other trace gases, such as O_3, in order to test interactive effects at the ecosystem level.

3 Conclusions

Obviously, the ideal exposure system with which to test the effects of varying atmospheric CO_2 concentrations on vegetation and ecosystems does not yet exist. What are the main shortcomings of current systems?

First, it must be stated that FACE systems represent the most sophisticated exposure systems which presently exist for assessments at the ecosystem level. Nevertheless, they still do not completely represent the predicted real world of the future. In spite of fast feedback systems for the dispensing of CO_2, it is physically impossible to exactly maintain the desired $[CO_2]$, since there is a time lag between (fast) changes in wind speed and the response of the dispensing system. In addition, deviations from average air movement due to small eddies will always cause spatially different mixing ratios. Holtum and Winter (2003) exposed tropical trees to oscillating CO_2 enrichment with half-cycles of 20 s and found significantly reduced photosynthesis stimulation compared to under continuous CO_2 enrichment. However, FACE operators typically negate such effects for CO_2 fluctuations observed in these systems (Hendrey et al. 1997).

As long as blowers are used, there is additional thermal energy provided to the test plots which may affect canopy temperatures and dew formation, which in turn may alter phenological development but also pest infestation. Not using blowers avoids these problems, but on the other hand, when only one CO_2 dilution step is involved, dispensing CO_2 uniformly is technically more crucial and requires even more sophisticated CO_2 monitoring and control and other measures, such as the windscreens used in the Hohenheim Mini-FACE, to provide the most uniform mixing.

The BNL-type FACE systems facilitate CO_2 enrichment at plots large in size compared to midi- or mini-FACE systems. In turn, running costs per plot are high and limit the number of true replicates in large FACE systems to mostly to two or three, and fumigation is typically stopped at night. Experiments in midi-and, in particular, in mini-FACE systems may be run with a higher number of replicates. Nevertheless, their efficiency in terms of costs per area enriched in CO_2 is lower than that of large FACE systems. This is because the area subjected to edge effects, which must not be used for biological assessments, becomes larger relative to the total area with shrinking plot size. Since most of the mini-FACE systems lack a CO_2 dispensing unit driven by wind direction, the cost efficiency per area of these systems is even worse. In the Hohenheim mini-FACE, this disadvantage is overcome by dividing the plenum into four sections with two of them in operation according to wind direction, respectively.

Despite all technical progress achieved, technical constraints will still prevent the ideal exposure system for CO_2 enrichment studies in the near future, although we are getting closer. Furthermore, we should reflect on the fact that FACE systems can only be used to answer certain questions. The starting point is always the present day $[CO_2]$ which is already 35% above the natural background concentration. It should not be expected that natural ecosystems are already in a new steady state equilibrium as ambient $[CO_2]$ has been increasing for the past 150 years at a pace which has never been experienced for as long as we can track back atmospheric $[CO_2]$. Since CO_2 enrichment experimentation always means a sudden change in $[CO_2]$, it is also questionable whether responses at the ecosystem level observed after a few years of abrupt CO_2 enrichment will really reflect the situation which will be seen by the end of this century after a steady increase in $[CO_2]$. After sudden CO_2 enrichment, ecosystems may instead exhibit an overshoot response for some time, and therefore longer-lasting experiments are required. How can we convince policy makers and the public that on the one hand research on CO_2 enrichment effects requires some time to be considered reliable, and that on the other emission reductions must commence immediately?

References

Ainsworth EA, Long SP (2005) What have we learned from 15 years of free-air CO_2 enrichment? A meta-analytic review of the responses of photosynthesis, canopy properties and plant production to rising CO_2. New Phytol 165:351–372

Allen LH, Drake BG, Rogers HH, Shinn JH (1992) Field techniques for exposure of plants and ecosystems to elevated CO_2 and other trace gases. In: Hendrey GR (ed) FACE: free-air CO_2 enrichment for plant research in the field. CRC Press, Boca Raton, Fla., pp 85–119

Amthor JS (1995) Terrestrial higher-plant response to increasing atmospheric $[CO_2]$ in relation to the global carbon cycle. Global Change Biol 1:243–274

Amthor JS (2001) Effects of atmospheric CO_2 concentration on wheat yield: review of results from experiments using various approaches to control CO_2 concentration. Field Crops Res 73:1–34

Arp WJ (1991) Effects of source-sink relations on photosynthetic acclimation to elevated CO_2. Plant Cell Environ 14:869–875

Bazzaz FA (1990) The response of natural ecosystems to the rising global CO_2 levels. Annu Rev Ecol Syst 21:167–196

Bazzaz FA, Catovsky S (2002) Plant Competition in an elevated CO_2 world. In: Mooney HA, Canadell J (eds) Encyclopaedia of global environmental change. Wiley, Chichester, pp 471–481

Bowes G (1996) Photosynthetic responses to changing atmospheric carbon dioxide concentrations. In: Baker NR (ed) Photosynthesis and the Environment. Kluwer, Dordrecht, pp 387–407

Bunce J (2003) Responses of seedling growth to daytime or continuous elevation of carbon dioxide. Int J Plant Sci 164:377–382

Caemmerer S von, Farquhar GD (1981) Some relationships between the biochemistry of photosynthesis and the gas exchange of leaves. Planta 153:376–387

Chiariello NR, Field CB (1996) Annual grassland responses to elevated CO_2 in multiyear community microcosms. In: Körner C, Bazzaz FA (eds) Carbon dioxide, populations, and communities. Academic Press, San Diego, Calif., pp 139–157

Cotrufo MF, Ineson P, Scott A. (1998) Elevated CO_2 reduces the nitrogen concentration of plant tissues. Global Change Biol 4:43–54

Cure JD, Acock B (1986) Crop response to carbon dioxide doubling: a literature survey. Agric For Meteorol 38:127–145

Curtis PS, Wang XZ (1998) A meta-analysis of elevated CO_2 effects on woody plant mass, form, and physiology. Oecologia 113:299–313

De Temmerman L, Fangmeier A., Craigon J (2002) EU project: Changing climate and potential impacts on potato yield and quality (CHIP). Eur J Agron 17:231–381

Drake BG, Gonzàlez-Meler MA, Long SP (1997) More efficient plants: a consequence of rising atmospheric CO_2? Annu Rev Plant Physiol 48:609–639

Drake BG, Azcon-Bieto J, Berry J, Bunce J, Dijkstra P, Farrar J, Gifford RM, Gonzalez-Meler MA, Koch G, Lambers H, Siedow J, Wullschleger S (1999) Does elevated atmospheric CO_2 concentration inhibit mitochondrial respiration in green plants? Plant Cell Environ 22:649–657

Erbs M, Fangmeier A (2005) A chamberless field exposure system for ozone enrichment of short vegetation. Environ Pollut 133:91–102

Fangmeier A, Jäger H-J (2001) Wirkungen erhöhter CO_2-Konzentrationen. In: Guderian R (ed) Handbuch der Umweltveränderungen und Ökotoxikologie, vol 2A: Terrestrische Ökosysteme. Springer, Berlin Heidelberg New York, pp 382–433

Fangmeier A, Stein W, Jäger H-J (1992) Advantages of an open-top chamber plant exposure system to assess the impact of atmospheric trace gases on vegetation. Angew Bot 66:97–105

Fangmeier A, Chrost B, Högy P, Krupinska K (2000) CO_2 enrichment enhances flag leaf senescence in barley due to greater grain nitrogen sink capacity. Environ Exp Bot 44:151–164

Field CB, Chapin FS, Matson PA, Mooney HA (1992) Responses of terrestrial ecosystems to the changing atmosphere: A resource-based approach. Annu Rev Ecol Syst 23:201–235

Hättenschwiler S, Handa IT, Egli L, Asshoff R, Ammann W, Körner C (2002) Atmospheric CO_2 enrichment of alpine treeline conifers. New Phytol 156:363–375

Heagle AS, Body DE, Heck WW (1973) An open-top field chamber to asses the impact of air pollutant on plants. J Environ Qual 2:365–368

Heagle AS, Philbeck RB, Ferrell RE, Heck WW (1989) Design and performance of a large, field exposure chamber to measure effects of air quality on plants. J Environ Qual 18:361–368

Heck WW, Philbeck RB, Dunning JA (1978) A continuous stirred tank reactor (CSTR) system for exposing plants to gaseous air contaminants: principles, specifications, construction, and operation. US Department of Agriculture, New Orleans, La., pp 1–32

Hendrey GR, Ellsworth DS, Lewin KF, Nagy J (1999) A free-air enrichment system for exposing tall forest vegetation to elevated atmospheric CO_2. Global Change Biol 5:293–309

Hendrey GR, Kimball B (1990) FACE: Free-air carbon dioxide enrichment. In: DOE FACE project brochure, application to field-grown cotton. National Technical Information Service, U.S. Department of Commerce, Springfield, Mass., pp 1–17

Hendrey GR, Long SP, McKee IF, Baker NR (1997) Can photosynthesis respond to short-term fluctuations in atmospheric carbon dioxide? Photosynth Res 51:179–184

Holtum JAM, Winter K (2003) Photosynthetic CO_2 uptake in seedlings of two tropical tree species exposed to oscillating elevated concentrations of CO_2. Planta 218:152–158

IPCC (2001) Houghton JT, Ding Y, Griggs DJ, Noguer M, van der Linden PJ, Dai X, Maskell K, Johnson CA (eds) Climate change 2001: the scientific basis. Cambridge University Press, Cambridge

Isebrands JG, McDonald EP, Kruger E, Hendrey G, Percy K, Pregitzer K, Sober J, Karnosky DF (2001) Growth responses of *Populus tremuloides* clones to interacting elevated carbon dioxide and tropospheric ozone. Environ Pollut 115:359–371

Jablonski LM, Wang X, Curtis PS (2002) Plant reproduction under elevated CO_2 conditions: a meta-analysis of reports on 79 crop and wild species. New Phytol 156:9–26

Jäger H-J, Schmidt SW, Kammann C, Grünhage L, Müller C, Hanewald K (2003) The University of Giessen free-air carbon dioxide enrichment study: description of the experimental site and of a new enrichment system. J Appl BotAngew Bot 77:117–127

Karnosky DF, Gielen B, Ceulemans R, Schlesinger WH, Norby RJ, Oksanen E, Matyssek R, Hendrey GR (2001) FACE Systems for studying the impacts of greenhouse gases on forest ecosystems. In: Karnosky DF, Ceulemans R, Scarascia-Mugnozza G, Innes JL (eds) The impact of carbon dioxide and other greenhouse gases on forest ecosystems. CABI, Oxford, pp 297–324

Kimball BA (1983) Carbon dioxide and agricultural yield: an assemblage and analysis of 430 prior observations. Agron J 75:779–788

Kimball BA, Mauney JR, Nakayama FS, Idso SB (1993) Effects of increasing atmospheric CO_2 on vegetation. In: Rozema J, Lambers H, Van de Geijn SC, Cambridge ML (eds) CO_2 and biosphere. Kluwer, Dordrecht, pp 65–75

Koch GW, Mooney HA (1996) Response of terrestrial ecosystems to elevated CO_2: A synthesis and summary, In: Koch GW, Mooney HA (eds) Carbon dioxide and terrestrial ecosystems. Academic Press, San Diego, Calif., pp 415–429

Körner C (2000) Biosphere response to CO_2 enrichment. Ecol Appl 10:1590–1619

Leadley PW, Niklaus P, Stocker R, Körner C (1997) Screen-aided CO_2 control (SACC): a middle ground between FACE and open-top chambers. Acta Oecol Oecol Plant 18:207–219

Lee HSJ, Barton CVM (1993) Comparative studies on elevated CO_2 using open-top chambers, tree chambers and branch bags. In: Schulze ED, Mooney HA (eds) Design and execution of experiments on CO_2 enrichment. Commission of the European Communities, Brussels, pp 239–259

Lockyer DR, Cowling DW, Jones LHP (1976) A system for exposing plants to atmospheres containing low concentrations of sulphur dioxide. J Exp Bot 27:397–409

Long SP, Ainsworth EA, Rogers A, Ort DR (2004) Rising atmospheric carbon dioxide: plants FACE the future. Annu Rev Plant Biol 55:591–628

Lundegårdh H (1927) Carbon dioxide evolution of soil and crop growth. Soil Sci 23:417–453

Mandl RH, Weinstein LH, McCune DC, Keveny M (1973) A cylindrical, open-top chamber for the exposure of plants to air pollutants in the field. J Environ Qual 2:371–376

McLeod AR, Long SP (1999) Free-air carbon dioxide enrichment (FACE) in global change research: a review. Adv Ecol Res 28:1–56

Miglietta F, Giuntoli A, Bindi M (1996) The effect of free air carbon dioxide enrichment (FACE) and soil nitrogen availability on the photosynthetic capacity of wheat. Photosyn Res 47:281–290

Miglietta F, Lanini M, Bindi M, Magliulo V (1997) Free air CO_2 enrichment of potato (*Solanu m tuberosum*, L.): design and performance of the CO_2-fumigation system. Global Change Biol 3:417–427

Miglietta F, Hoosbeek MR, Foot J, Gigon F, Hassinen A, Heijmans M, Peressotti A, Saarinen T, Van Breemen N, Wallen B (2001a) Spatial and temporal performance of the miniFACE (free air CO_2 enrichment) system on bog ecosystems in northern and central Europe. Environ Monit Assess 66:107–127

Miglietta F, Peressotti A, Vaccari FP, Zaldei A, DeAngelis P, Scarascia-Mugnozza G (2001b) Free-air CO_2 enrichment (FACE) of a poplar plantation: the POPFACE fumigation system. New Phytol 150:465–476

Mooney HA, Canadell J, Chapin JR, Ehleringer JR, Körner C, McMurtrie RE, Parton WJ, Schulze ED (1999) Ecosystem physiology responses to global change. In: Walker B, Steffen W, Canadell J, Ingram J (eds) The terrestrial biosphere and global change. Cambridge University Press, Cambridge, pp 141–189

Nagy J, Lewin KF, Hendrey GR, Lipfert FW, Daum ML (1992) FACE facility engineering performance in 1989. Crit Rev Plant Sci 11:165–185

Norby RJ, Kobayashi K, Kimball BA (2001) Rising CO_2 – future ecosystems -commentary. New Phytol 150:215–221

Nowak RS, Ellsworth DS, Smith SD (2004) Functional responses of plants to elevated atmospheric CO_2 – do photosynthetic and productivity data from FACE experiments support early predictions? New Phytol 162:253–280

Okada M, Lieffering M, Nakamura H, Yoshimoto M, Kim HY, Kobayashi K (2001) Free-air CO_2 enrichment (FACE) using pure CO_2 injection: system description. New Phytol 150:251–260

Pinter PJ, Kimball BA, Wall GW, Lamorte RL, Hunsaker DJ, Adamsen FJ, Frumau KFA, Vugts HF, Hendrey GR, Lewin KF, Nagy J, Johnson HB, Wechsunge F, Leavitt SW, Thompson TL, Matthias AD, Brooks TJ (2000) Free-air CO_2 enrichment (FACE): blower effects on wheat canopy microclimate and plant development. Agric For Meteorol 103:319–333

Poorter H (1993) Interspecific variation in the growth response of plants to an elevated ambient CO_2 concentration. In: Rozema J, Lambers H, Van de Geijn SC, Cambridge ML (eds) CO_2 and biosphere. Kluwer, Dordrecht, pp 77–97

Poorter H, Navas ML (2003) Plant growth and competition at elevated CO_2: on winners, losers and functional groups. New Phytol 157:175–198

Poorter H, Roumet C, Campbell BD (1996) Interspecific variation in the growth response of plants to elevated CO_2: a search for functional types. In: Körner C, Bazzaz FA (eds) Carbon dioxide, populations, and communities. Academic Press, San Diego, Calif. pp 375–412

Rogers HH, Heck WW, Heagle AS (1983) A field technique for the study of plant responses to elevated carbon dioxide concentration. J Air Pollut Contr Assoc 33:42–44

Sage RF (1994) Acclimation of photosynthesis to increasing atmospheric CO_2: the gas exchange perspective. Photosynth Res 39:351–368

Shinn JH, Clegg BR, Stuart ML (1977) A linear gradient chamber for exposing field plants to controlled levels of air pollutants. UCRL reprint no. 80411. Lawrence Livermore Laboratory, University of California, Calif.

Spring GM, Priestman GH, Grime JP (1996) A new field technique for elevating carbon dioxide levels in climate change experiments. Funct Ecol 10:541–545

Stitt M (1996) Metabolic regulation of photosynthesis. In: Baker NR (ed) Photosynthesis and the environment. Kluwer, Dordrecht, pp 151–190

Van Oijen M, Schapendonk AH, Jansen MJ, Pot CS, Maciorowski R (1999) Do open-top chambers overestimate the effects of rising CO_2 on plants? An analysis using spring wheat. Global Change Biol 5:411–421

Volk M, Geissmann M, Blatter A, Contat F, Fuhrer J (2003) Design and performance of a free-air exposure system to study long-term effects of ozone on grasslands. Atmos Environ 37:1341–1350

Walklate PJ, Xu ZG, McLeod AR (1996) A new gas injection method to enhance spatial utilization within a free-air CO_2 enrichment (FACE) system. Global Change Biol 2:75–78

Wand SJE, Midgley GF, Jones MH, Curtis PS (1999) Responses of wild C_4 and C_3 grass (Poaceae) species to elevated atmospheric CO_2 concentration: a meta-analytic test of current theories and perceptions. Global Change Biol 5:723–741

Webber AN, Nie GY, Long SP (1994) Acclimation of photosynthetic proteins to rising atmospheric CO_2. Photosynth Res 39:413–425

Weltzin JF, Belote RT, Sanders NJ (2003) Biological invaders in a greenhouse world: will elevated CO_2 fuel plant invasions? Front Ecol Environ 1:146–153

Wittwer SH (1986) Worldwide status and history of CO_2 enrichment – an overview. In: Enoch HZ, Kimball BA (eds) Carbon dioxide enrichment of greenhouse crops. CRC, Boca Raton, Fla., pp 3–15

Wullschleger SD, Tschaplinski TJ, Norby RJ (2002) Plant water relations at elevated CO_2 – implications for water-limited environments. Plant Cell Environ 25:319–331

Martin Erbs
Andreas Fangmeier
Universität Hohenheim
Fakultät Agrarwissenschaften
Institut für Landschafts-und Pflanzenökologie (320)
Fachgebiet Pflanzenökologie und Ökotoxikologie
August-von-Hartmann-Strasse 3
70599 Stuttgart
Germany
Tel.: +49-711-4592189
Fax: +49-711-4593044
e-mail: afangm@uni-hohenheim.de

Quaternary Palaeoecology:
Central and South America, Antarctica and the Pacific Ocean Region

Burkhard Frenzel

Huntley (1996) quite correctly stated that "palaeoecology is today more important than ever before because of the insights that it provides into the response of species to a changing environment." Regrettably this statement very often seems to be ignored. Instead, nowadays palaeoecological changes of the past are repeatedly interpreted in terms of climate changes, only. But can other ecological influences, like competition between the involved taxa, pedogenesis, or human influence simply be neglected?

1 The Problem of Exactly Dating Palaeoecological Processes

In the former Soviet Union thermoluminescence dating of fossil soils and of aeolian sediments, which are older than the last interglacial, is widely used. Yet Zhou et al. (1995) showed at the Orkutsay profile, Usbekistan, that age data older than ca. 130,000 years are highly unrealistic and must be rejected. This agrees well with observations already made in other regions of the globe. Moreover, Olley et al. (1996) stress that even datings by the U/Th series very often produce inaccurate values, and Pillans et al. (1996) used several independent dating techniques when trying to date the Rangitawa tephra, New Zealand, which can be widely used as a marker horizon of Mid-Pleistocene sediments there. They showed that out of 51 datings only 15 seem to be more or less reliable. As for the aminoacid-racemisation technique, the situation is even worse for littoral sediments, caused by redeposition of older fossils (Murray-Wallace et al. 1996).

Palaeomagnetic reversals found in various sediment types have been repeatedly used as marker horizons. Lyons (1996), studying the Blake event in New Zealand for the first time, which dates from the beginning of the last glaciation, stresses that this method works reliably well only if longer vertical sequences in geological profiles are investigated.

In palaeoecology the ^{14}C dating technique is widely used. Yet it is a well-known fact that, for most phases of the past, ^{14}C ages strongly differ from

Progress in Botany, Vol. 67
© Springer-Verlag Berlin Heidelberg 2006

absolute age data. Shore et al. (1995) convincingly demonstrated that even the ^{14}C age data obtained from fulvic acids, humic acids and from humin of the same sample from a northern English peat bog can differ from one another by about 630–1,210 years, and Nilsson et al. (2001) stress that the ^{14}C ages of various microrelief elements of a certain peat bog can differ by ca. 365–1,000 years (even by 2,125 years). Alkali-treated samples always gave higher age data than untreated samples. It is recommended that for peat bogs only *Sphagnum* samples should be used for dating.

Furthermore, ^{14}C datings can be strongly influenced in lakes and seas by the so-called reservoir effect, i.e. by older carbon. Examples of this will be dealt with later. Age data of Lake Aricota, southern Peru (Placzek et al. 2001) may serve as an example only. Although here in general the reservoir effect typically amounts to <350 years, even values of up to 12,000 years can be detected. The opposite was found by Martinez et al. (2001), investigating the geochronology of Pleistocene beach sediments on the coast of Uruguay. Here, sediments evidently dating from the last interglacial constantly gave ^{14}C age data of about 29,500–35,500 years before present (BP), only.

As far as younger sediments are concerned, even charcoal layers can frequently reveal unreliable age data (Carcaillet 2001). In alpine regions this is caused by bioturbation, frost action, etc. Due to these unfavourable factors Lowell (1995) recommends only relying on a wealth of ^{14}C age data for a specific region (here the Miami sublobe, Ohio, of the last glaciation), which together should be interpreted based on geographical and geological positions. Of course, the situation as regards dendrochronology is much better as has been shown by Roig et al. (1996) for the south coast of Tierra del Fuego, Patagonia, using subfossil *Nothofagus* wood.

Widely spread volcanic ash layers can be used as stratigraphical marker horizons in different regions of the globe. A good example for this is the Toba tephra dating from about 73,000 years BP (Westgate et al. 1998; Schulz et al. 2002), provided that a certain tephra layer [as to its identification in geological profiles, see Rose et al. (1996)] can be geochemically and petrographically reliably attributed to the dated eruption. If a geochemical/petrographical control is either impossible or has not been used, palaeoecological misinterpretations are often the consequence (Shane et al. 1998).

2 Upper Pleistocene and Holocene Palaeoecology of Central America and the Surrounding Islands

To understand the palaeoecological development in perimarine areas it is important to know whether these regions have been tectonically stable or not. The Bahamian Archipelago and Bermuda have been stable at least since the last interglacial (Vacher and Hearty 1989; Carew and Mylroie 1995; Muhs 2002), when the sea level was about 6 m above its present day position. Interestingly more or less the same value holds for the 5a interstadial. This contradicts results of deep-sea isotope investigations (Vacher and Hearty 1989; Gascoyne 1992). On the other hand there are comparable contrasts between deep-sea records and geological observations on tectonically stable coastlines concerning the transition from the penultimate glaciation to the last interglacial (Muhs 2002). When the sea level fell at the end of the last interglacial (5e/5d transition), strong north-easterly winds and waves are assumed to have piled up more or less regularly spaced sediment ridges on the Bahamas, which even contain huge blocks (Hearty et al. 1998). I wonder whether this situation was caused by tsunamis? According to physical datings critically done for the Lesser Antilles these repeatedly occurring sea level changes seem to have negatively influenced the living conditions of the big rodent *Amblyrhiza*, finally causing its extinction there at about 51,000 BP; up until now this was attributed to human activities (McFarlane et al. 1998).

In Central America it is often quite difficult to correctly interpret palaeoecological processes, because climatic, tectonic and volcanic processes have strongly interacted. In the Chihuahuan Desert, North Mexico (Metcalfe et al. 2002) the climate between 38,000 and approximately 29,000 years BP was obviously appreciably moister than before and since. At that time the amount of arboreal pollen (AP) was about 90%, mostly consisting of *Quercus*, *Picea*, *Pinus*, *Juniperus* and *Tamarix* pollen. Yet between 29,000 and 21,000 years BP *Quercus* had virtually disappeared, the share of *Picea* had strongly decreased, but *Pinus* had become the dominating taxon. There is no pollen flora left dating from the last glacial maximum (LGM), but Islebe et al. (1996) state that at that time the climate in northernmost Guatemala had become extremely arid and some 6.5–8.0°C colder than it is at present. Most of the lakes had disappeared at that time, and during late glacial times the frequency of fire was quite high. On the other hand, in the vicinity of Mexico City (Lake Chalco Basin), the hydrological situation during long parts of the last glaciation was obviously dictated by volcanic activities of the surrounding mountains (Caballero and Ortega-Guerrero 1998). For the synchronisation of geological profiles by volcanic tephra layers within the Basin of Mexico see Ortega-Guerrero and Newton (1998).

It is interesting to note that in lowland Guatemala tropical rain forests at lower altitudes existed only from about 9,000 years BP onwards (Islebe et al. 1996). However, on Haiti (Higuera-Gundy et al. 1999) moist forests, which are at present limited to altitudes above 800 m in the mountains, existed in the lowlands since at least 10,300 BP. The Holocene vegetation history of this region has been very carefully investigated, too. Moisture increased since 10,000 years BP with maximum values between 3,950 and 2,490 BP, when deciduous taxa of a secondary vegetation spread remarkably due to human activities. About 93% of the rodent fauna of the Dominican Republic and Haiti became extinct due to these activities (or climatic effects?).

In Mexico, the Holocene history of vegetation and climate has been intensively studied by Goman and Byrne (1988) and Leyden et al. (1998), on the Yucatan Peninsula by Whitmore et al. (1996) and by McLaren and Gardner (2000), in Guatemala by Goman and Byrne (1998) and Leyden et al. (1998) and in Costa Rica by Northrop and Horn (1996) and by Clement and Horn (2001). Most of these papers show the appreciably early cultivation of maize in this area [e.g. in the Tuxtla region, Veracruz, from at least 4,830 calendar years (cal) BP; Goman and Byrne 1998], paralleled by strong human impact on the vegetation. These human activities have been so intensive, at least regionally, that Islebe et al. (1996) and Clement and Horn (2001) stress that it is frequently impossible to correctly distinguish between climatic and human influences on the vegetation. In view of these difficulties it is astonishing to see that Haug et al. (2003) try to explain former changes in human activities and population densities in Central America only by climate change, whereas Rosenmeier et al. (2002) point to the fact that forest clearances of the Mayas seemed to have caused an increase in lake levels.

3 Quaternary Glaciation History of South America and the Problem of the Younger Dryas Equivalent in the Southern Hemisphere

Andriessen et al. (1993) and Hooghiemstra et al. (1993) reinvestigated the famous Funza pollen and sediment sequences of the Bogotá basin, Colombia, to achieve better dating of the wealth of Quaternary palaeoecological changes observed there. The investigation was based on fission track datings, done on volcanic zircon minerals, and on a comparison of the pollen curves with the $\delta^{18}O$ curves obtained from deep-sea boring in the Ocean Drilling Progamme (ODP) site 677 of the eastern Pacific. In doing so, earlier published age data could be disproved. It could be shown that a strong tectonic uplift of the Eastern Cordillera had happened between 5,000,000 and 3,000,000 years ago and also at about 1,000,000 years ago. During the last 2,700,000 years these processes were superimposed by strong climatic changes (Andriessen et al. 1993), and for the last 800,000 years, Hooghiemstra et al. (1993) found oscillations in the contents of arboreal pollen at a frequency of approximately 100,000, 23,000, 40,000, 30,000, and 15,000 years BP, etc. However, the question is, are these values reliable or not? Since the pollen diagrams have been adjusted to the $\delta^{18}O$ curve of ODP site 677, it is not astonishing to find the same "periodicities" in the pollen diagram as in the deep-sea diagram. According to Ortlieb et al. (1996) deep-sea stage MIS-11 experienced a remarkably warmth-loving marine fauna at the north Chilean coast. From this the authors state that MIS-11 "was the warmest interglaciation, at least in the southern hemisphere." Comparable results concerning the northern hemisphere were recorded much earlier (Frenzel 1973).

The timing and extent of the Upper Pleistocene and Holocene phases of glacier advance and retreat have been intensively investigated in southern Chile (including Patagonia and Tierra del Fuego) by Gordillo et al. (1993), Marden and Clapperton (1995), Aniya (1996), Harrison and Winchester (1998), Denton et al. (1999), Moreno et al. (1999) and Schlüchter et al. (1999). The same was done in the Atacama and in adjacent regions by Grosjean et al. (1991, 1998), Kuylenstierna et al. (1996), Geyh et al. (1999) and Cross et al. (2001), but here extremely high reservoir effects of the lakes studied sometimes caused severe problems (Geyh et al. 1999; Placzek et al. 2001). Rodbell et al. (2002) dealt with the stratigraphically important problem of characterizing tephra layers in Ecuador.

Nowadays, there is a vivid discussion about the southern hemisphere equivalent of the northern hemisphere Younger Dryas period (Jüngere Tundrenzeit). It is a well-known fact that this cold spell in North America or in Europe was not equally experienced everywhere (Frenzel 1983; Alley et al. 1993; Peteet 1995; Thomas and Thorp 1995; Wohlfarth 1996). One reason for this finding seems to be a remarkable lack of reliable physical age data. To overcome these difficulties Wohlfarth (1996) recommended using widely spread marker horizons like volcanic tephra layers or the beginning and end of ^{14}C plateaus.

There are data which indicate the occurrence of this Younger Dryas cold spell in Venezuela (Curtis et al. 1999), in the Cariaco Basin to the north of Venezuela (Hughen et al. 1996, 1998), within the Llanos Orientales, Colombia (Behling and Hooghiemstra 1998), in Patagonia (Marden and Clapperton 1995; Pendall et al. 2001), within the Chilean lake district [Denton et al. 1999; Moreno et al. 1999; Moreno 2000; although here other factors like human impact may also have played a role (Moreno 2000)], in Tierra del Fuego (Heusser 1993a, b; Heusser et al. 1999a) on the South Island of New Zealand (Ivy-Ochs et al. 1999), and possibly in the South China Sea (Steinke et al. 2001). Other observations are contradictory, like those made in some parts of the Cordillera system of South America (Osborn et al. 1995), the Central Andes (Lynch 1990), and even in eastern Colombia where Kuhry et al. (1993) described the El Abra stadial as equivalent to the Younger

Dryas period. This has been highly criticized by Heine (1993), yet van der Hammen and Hooghiemstra (1995) contributed more details in favour of the observations of Kuhry et al. (1993). On the other hand, Hansen (1995) showed that at the Laguna Junín, Peru, this cold spell had happened between 12,000 and 11,000 ^{14}C BP, i.e. already one millennium earlier than the classic Younger Dryas period. Finally, this cold spell does not seem to have influenced the climate in various regions of the southern hemisphere: south-eastern Asia (Maloney 1995); Cordillera Blanca, Peru (Rodbell and Seltzer 2000); southernmost South America (Markgraf 1993); New Zealand (McGlone 1995); north-western New Zealand (Hellstrom et al. 1998); Tasmania (Colhoun 1996). It may be that on the Auckland Island to the south of Tasmania this cold spell had already begun at about 12,500 ^{14}C years BP, i.e. much earlier than in Europe (McGlone et al. 2000). Evidently the situation is equivocal, but is it correct, as has been stated repeatedly, to think that phases of climate change, particularly when reconstructed with respect to vegetation history, always begin and end abruptly according to well-defined age data? Would it not be better to think about more or less long-lasting processes, whose amplitude will only be felt if certain threshold values are exceeded, even if this is always complicated by local and regional conditions (for Upper Quaternary changes of climate in Beringia and adjacent regions see e.g. Mann and Hamilton 1995; and Elias 2000).

An equally interesting question is, when might the climatologically and ecologically important El Niño/Southern Oscillation (ENSO) have begun to act? According to palaeoecological records for coastal Peru (Fontugne et al. 1999; Sandweiss et al. 2001) faint traces of the action of ENSO seem to have been felt there at about 8,900 cal BP and, after a long period of interruption, since about 5,800 (i.e. ca. 6,700 cal BP), culminating at between 3,400 and 1,000 cal BP (Sandweiss et al. 2001). The Mid-Holocene period of faint ENSO activities seems to have been favourable for human settlements in this region, because freshwater was constantly available. For reconstruction of timing and intensity of ENSO phases since 1525 see Enfield (1989). For frequency and intensity of ENSO events during the last decades see Yarnal (1985), and Yarnal and Kiladis (1985). Interestingly these ENSO events did not have a strong influence on the radial increments of tree growth along the Mendoza River, Argentina (del Prieto et al. 1999).

4 South American Upper Quaternary Vegetation History

The problem of last-glacial forest-refuge areas has been investigated repeatedly in southern Chile. By about 27,000–14,000 years BP very open *Nothofagus dombeyi* forest–grassland communities existed in the Chilean Lake District, which were comparable to those thriving there at present above the alpine timberline, yet the fossil and modern plant communities differ from each other (Moreno et al. 1999). Comparable *Nothofagus*–Gramineae–Ericaceae–Tubuliflorae communities were found by Heusser et al. (1999b) for the middle part of the last glaciation on the Isle of Chiloë. These refuge areas seem to have had only a relatively open vegetation, which varied spatially and temporally. It had to retreat when the ice masses invaded this region. But by about 14,000 BP, when glaciers had strongly retreated, the open Subantarctic parkland had already been replaced by forests composed of *Nothofagus* and evergreen taxa (Heusser 1993b; Heusser et al. 1999b; Moreno 2000). The climate of these periods was reconstructed by Heusser et al. (1999a, b) by comparing past pollen floras with those of modern surface samples, and by Pendall et al. (2001) by relying on δD and δ¹³C values of *Sphagnum* peat. The late-glacial pollen flora on Isla Grande de Tierra del Fuego at about 13,000–12,000 years BP was characterized by astonishingly high values of *Nothofagus* pollen, together with high amounts of pollen of *Empetrum*, Gramineae and Caryophyllaceae. Long-distance transport or redeposition of older material are discussed as possible sources of these high *Nothofagus* values (Heusser 1993a). Up until to now, it has not been possible to better analyse the positioning and biological composition of possible last-glacial forest-refuge areas localised in protected areas of the South-Chilean Andes. Sayago (1995) mapped the neotropical loesses of Argentina, which should have been situated beyond former forests. Unfortunately, they still cannot be dated correctly. Thus these loesses do not contribute to a better knowledge of last-glacial forest-refuge areas. Between about 11,000 and 10,000 years BP Palaeo-Indians immigrated into southernmost South America. These tribes are believed to have been already endangered at that time, together with horses and the ground sloth found there (Lynch 1990). During these late-glacial times fire seems to have been an important factor in the development of vegetation (Heusser 1993a). Because of this, Markgraf (1993) thought that the previously discussed late-glacial changes in the South American vegetation had not been caused by climate change but by fire only. It may also be that both of these factors affected the forest vegetation there, as has been long known for European vegetation.

A major problem regarding the study of the history of South American vegetation is the last-glacial distribution pattern of forests within the Amazonas catchment (for older publications, see Frenzel 2000). Cowling et al. (2001) have made interesting computations using models based on different atmospheric CO_2 contents (220–180 p.p.m.), mean annual temperatures (2 or 6°C lower than present day values) and mean annual precipitation rates (−20 or −60% of present day values). Using these values, the results are 80% or only 18% of last glacial forest cover in the Amazonian lowlands. Colinvaux (1996) dealt at length with these problems, together with that of forest biodiversity at that time, by means of pollen analysis, concentrating his efforts most of all on the interstadial periods of the last glaciation. Ledru et al. (1998) pointed to the fact that, in general, sediments of the last glacial maximum (LGM) are lacking in the South American tropical forest regions. This is interpreted as an indicator of stronger soil erosion when the climate was appreciably much drier there than today. On the other hand, Ledru et al. (2001) investigated the vegetation history of the late glacial and Holocene in the eastern Amazon lowland (2°58′S, 43°25′W). Here, during early late-glacial times (14,000–13,500 cal BP) the formerly strongly heliophytic *Podocarpus* forest (about 5% AP), presently found at much higher altitudes, was replaced by *Byrsomina* and *Didymopanax* Cerrado communities. During the LGM grass savannah thrived in southern Brazil. Tropical rain forest trees were very rare at that time, or had even completely disappeared. Only during late-glacial times did forest vegetation began to immigrate into these areas (Behling and Negrelle 2001).

During late-glacial times the Colombian Llanos Orientales were covered by various types of savannah, into which *Alchornea* and *Mauritia* had immigrated along the rivers (Behling and Hooghiemstra 1998). Comparable observations were made by the same authors (Behling and Hooghiemstra 1999) at the lakes of El Pinal and Carimagua within the Colombian Llanos Orientales, and Behling et al. (1999) described a relatively open late-glacial forest vegetation from the middle Caquetá river, and in Central Colombian Amazonia at about 160 m a.s.l. *Podocarpus* was clearly a characteristic element of this vegetation, which is presently thriving in the upper montane forests. At that time, the mean annual temperature seems to have been about 5°–7°C lower than at present, together with less precipitation. On the other hand, Auler and Smart (2001), dating travertines and speleothems of northeastern Brazil, stated that the climate of the LGM there has not been dry. Yet according to Arz et al. (1998) during the LGM a strong terrestrial sedimentation took place via the rivers into the Atlantic Ocean off the coast of northeastern Brazil. The climatological background of these observations was confirmed by Rodriguez-Filho et al. (2002) when studying the palaeoecological situation of Lake Silvana, mid Rio Dolce region, Minas Gerais. Here a very open late-glacial grassland vegetation was replaced at the beginning of the Holocene by some tropical trees and shrubs (see also Meunier et al. 1999). The Holocene vegetation history in the south-western Brazilian Amazonas region was studied by Pessenda et al. (1998, 2001), using $\delta^{13}C$ and ^{14}C values of organic soil material.

Quite a contrasting reconstruction of the late-Quaternary vegetation history of the southern Amazon Basin was communicated by de Freitas et al. (2001). However, this study only used the $\delta^{13}C$ values of soil organic matter. It is said that the area was covered by forests from ca. 17,000– 9,000, while afterwards savannah vegetation began to dominate. Yet the $\delta^{13}C$ values used for the reconstruction of forests lie between −26 and −19%, and those of the savannah between −27 and −14%. According to our own work on $\delta^{13}C$ values of forest trees in the German Black Forest, I think that the $\delta^{13}C$ values found in Amazonia have been misinterpreted.

For reasons of space, the most interesting observations on the Holocene vegetation history of the eastern Amazon Basin (Behling and da Costa 2000), from the Pacific lowlands of Colombia (Behling et al. 1998; Berrio et al. 2000), from Laguna Loma Linda in eastern Colombia (Behling and Hooghiemstra 2000), from the surroundings of Arequipa, Peru (Holmgren et al. 2001; rodent middens), and from Rio Limay, Neuquen Province (Markgraf et al. 1997; rodent middens), cannot be discussed here. Bouchard (1988) gave an interesting summary of our knowledge of human Prehispanic activities within the north Ecuadorian lowlands, and MacFadden (2000) has shown that the Middle Pleistocene Ensenadan mammal fauna (*Equus* and the mastodon *Cuvieronius*) had grazed at an elevation of ca. 1,800 m a.s.l. during periods of a cold climate on C_3 plants, but in periods of a warm climate on C_4 plants. This may serve as a warning against an overly simple extrapolation from animal diets to climates or to plants preferred.

The above-mentioned papers already contain a wealth of palaeoclimatic data. However the Andes of Peru and Chile show some special problems. These are caused by Holocene changes in the atmospheric circulation, by neotectonics and by volcanic activities. Grosjean et al. (1997) stated that Holocene climate at an elevation of 4,125 m a.s.l. was characterized by aridity between 6,000 and 3,800 BP, followed by oscillating moister climates, but Placzek et al. (2001) point to regionally strongly changing moisture conditions in southern Peru, i.e. the Lake Aricota and Lake Titicaca region. When discussing these discrepancies these authors point to the strong difficulties in exactly dating these various episodes, due to frequent extremely high reservoir effects. In this respect observations by Hirschmann (1973) on changes in the present day solar radiation in Chile are welcome, and Barnett et al. (1996) after a comparison with palaeoclimate data, very explicitly warn of making climate predictions too carelessly based on climate modelling: "it will be hard to say, with confidence, that an anthropogenic climate signal has or has not been detected." Much less cautious is the argument given by Scott and Collins (1996) who discuss sea level changes along the east coast of North America, where eustatic, isostatic and perhaps neotectonic movements are mixed together; the authors use these results to explain the end of pyramid construction in Egypt.

5 Upper Quaternary Palaeoecology of Antarctica

The history of the Antarctic ice shield, which is so important for understanding Quaternary palaeoclimatology and palaeooceanology, is still a matter of debate. The difficulties already arise with the question of whether this ice shield has existed for >40,000,000 years, reacting dynamically to climatic change during the first ca. 38,000,000 years, or whether it has been quite stable in its present form during the last 14,000,000 years (Wilson 1995). For the Upper Quaternary, Grootes et al. (2001) show that in different regions of Antarctica quite remarkable differences in climate and repeated asynchronous changes in the ice volume existed. According to Domack et al. (2001) the LGM occurred on the Antarctic peninsula before 13,180 cal BP [for the thickness and extent of the ice masses during the last glaciation see Goodwin and Zweck (2000) and ÓCofaigh et al. (2001)], but the region seems to have been free of ice between 13,180 and 11,460 cal BP. This contradicts the reports on an equivalent of the Younger Dryas cold spell in southernmost South America discussed above. The Hypsithermal period seems to have lasted in Antarctica – with regional differences – from about 9,100 cal BP to 3,400 cal BP (Roberts and McMinn 1999; Masson et al. 2000; Domack et al. 2001; Hodell et al. 2001; Taylor and McMinn 2001, 2002). It is interesting to note that in Ace Lake, Vestfold Hills (68°S, 77°E) several fossil rotifers of the genus *Notholea*, together with *Phormidium* ssp., Chrysophyceae and Dinoflagellates could be found in Early Holocene sediments. They are either held to be part of an endemic Antarctic flora and fauna or their occurrence points to a very effective long-distance transport (Swadling et al. 2001). For Upper Holocene traces of a deteriorating climate since about 3,000 years BP see Cunningham et al. (1999), Domack and Mayewski (1999), Roberts et al. (1999, 2001a), Fabrés et al. (2000), and McMinn et al. (2001). Again, it should be taken into consideration that, due to the large and at least regionally changing reservoir effects, the ^{14}C age data given are somewhat problematic. Morgan and van Ommen (1997) described a palaeoclimatologically interesting observation: analysing seasonal δ^{18}O values of the last 700 years, it could be shown that, in contrast to stronger changes in the winter temperatures, the summer temperatures seem to have changed only relatively insignificantly. Thus, the calculation of mean annual temperatures using these proxy data can give quite misleading results.

6 Palaeoecology of the East Asian Monsoon Region

The long sequence of loesses and of fossil soils, which are so characteristic for northern China, have stimulated much geological and palaeoecological research work. According to Pye (1995) it is not necessary to think about a special "loessification" process as has been done in the past, mostly by Russian and Hungarian scientists, but to better differentiate between various genetic loess types. It is often difficult to exactly evaluate the intensity of chemical weathering of loesses. To do so, Chen et al. (1999, 2000) used the Rb/Sr ratios in loesses and fossil soils of the Luochuan and Huanxian loess profiles, northern China. The data are compared to those of magnetic susceptibility. It is said that these parameters correspond with each other. Finally, these values have been compared to the $\delta^{18}O$ data of SPECMAP. After this, the observations were extended to the whole of the Luochuan loess profile, but nothing has been said about the wealth of hiatuses in this profile. Derbyshire et al. (1995) studied changes in micromorphology, magnetic susceptibility and grain sizes of loesses and the palaeosols of the North Chinese loess plateau in a west–north-west to east–south-east direction. Evidently this procedure can help to better understand regional differences of an earlier climate, and Ding et al. (1999) analysed the sand contents of nine loess profiles, situated between the southern border of the Mu Us Desert and the surroundings of Xian. This was done for loesses and fossil soils between the last interglacial and the Holocene. This helps to better understand the changing influence of strong winds. Sun and Liu (2000) stress that in northern China loess sedimentation had strongly increased at about 1.1 to ca. 0.9 million years ago, due to a strong uplift of the Tibetan Plateau of that time. Indeed traces of this tectonic phase can easily be observed in the deep valleys of western Sichuan and eastern Xizang, where traces of the North Chinese S_5 soil complex can be repeatedly found near the base of the younger sediment masses. According to Han et al. (1998) this soil complex is held to date from deep-sea isotope stages MIS 13–15, though till now this correlation could not be proven exactly. From the $\delta^{13}C$ values of organic material in the former A-horizons ($\delta^{13}C$ between −23% and −24%) it is concluded that when this soil was formed, no typical forest could have existed there, but that a large fraction of C_4 plants seems to have been typical. If this is correct, according to our own investigations the mountains of the German Black Forest would be covered at present by a forest–steppe-like vegetation! Ji et al. (2001) investigated the 135-m-high Luochuan loess profile, North China, by means of diffuse reflectance spectrometry. The brightness of the photos is taken as an indicator of magnetic susceptibility and of mineral contents. From these it is hypothesized that during the formation of the Lower Pleistocene Wucheng loess warm climate periods were characterized by "dry summer monsoons", as compared to the "wet summer monsoons" of later times. Repeatedly one gets the impression that much more critical methodological research is necessary before publication of such data. According to Lu et al. (2000) the intensity of loess deposition due to winter monsoons had oscillated there remarkably, when the Early to Middle Pleistocene loess layers L_9 and L_{15} were formed. Yet in view of the impossibility of dating these profiles correctly, I wonder whether the calculated frequencies of these changes in wind strengths (ca. 1,000 years, 2,770 years and maximum 1,450 years) are reliable. Rutter and Ding (1993) visually divided the 37 fossil soils of the Baoji profile (to the west of Xian) into four different classes of pedogenetic activities. These "soil types" are ascribed to various types of formerly prevailing vegetation (without any palaeobotanical analyses). From this it is concluded that the strength of the summer monsoon had repeatedly changed during the last 2,500,000 years. This is also stressed by Rousseau and Wu (1999), who studied the mollusc faunas of the upper Mid-Pleistocene S_2–L_2 loesses and fossil soils. The general geographic tendency of modern climate change influencing the deposition of loess dust and the formation of Holocene soils within the North Chinese Loess Plateau was studied by Porter et al. (2001) along a profile from the Mu Us to the south-east. It is stated that the magnetic susceptibility and mean grain sizes reflect quite well the governing climatic conditions. Two loess profiles at the southeastern border of the Mu Us with the North Chinese loess plateau each exhibit three well-developed fossil soils, dating from the beginning of the Last Interglacial to that of the last glaciation, i.e. they were formed between <130,000±10,000 and 75,000±9,000 years BP. These profiles were compared to other loess soil sequences of the same span of time and to those of SPECMAP climate-change and insolation curves (Sun and Ding 1998) [(as to the changing activity of the summer monsoon during the last interglacial and at the beginning of the last glaciation of Japan see Morley and Heusser (1989)]. Moreover these profiles at the southern border of Mu Us are characterized by repeatedly occurring sand layers, which point to phases of increased wind strengths (see also Sun 2000) and it is stated "that the occurrence of modern Mu Us desert is the result of reactivation of the ancient sand dunes buried within the loess deposits, mainly due to agriculture since the Tang Dynasty 1,300 years ago". These anthropogene phases of sand accumulation in loess profiles were studied intensively by Roberts et al. (2001b), who also concentrated their efforts on the loess-covered mountains to the south of the old town of Xining. Pan and Chao (2003) have already shown the high vulnerability to human activities of the transition zone between oases and the surrounding desert. Thus it is somewhat difficult to follow the argumentation of Liu et al. (2002) who interpret Holocene palaeoecological changes along the western slope of the Ta Xing An mountains, north-western Mandzhuria, in terms of climate change only, although it is known that tribes which repeatedly endangered China came from here in the past. These anthropogenic influences on the East Asian ecological situation – it is said that during the last 3,000 years >60% of vegetation in China has been changed by humans – were studied by Fu (2003) in view of anthropogenic changes of climate. It is concluded that aridity was increased by these activities (though the data base seems to be scarce). According to Wang et al. (2001) between 8,000 and 3,000 years BP temperatures in China were higher by about 1.5–2.0°C than at present. This agrees quite well with observations in other parts of the Northern Hemisphere at least. Moreover the "medieval climatic optimum" can be traced in Eastern China, too. Also, Feng et al. (1999) tried to investigate the Holocene's climate history by using the δD contents of tree-ring cellulose. However, the database seems to be very poor; only data of three tree trunks and various tree species are mixed together. According to our own experience in southern Germany this causes remarkable differences, which cannot be interpreted in terms of former climate.

According to Jiang and Piperno (1999) wild rice was collected along the Yang tze river (surroundings of Po yang hu) at about 12,000 BP, and has been cultivated there since between 10,000 and 9,000 BP. By about 12,830–10,500 years BP the vegetation was obviously still very open with ≤50% AP, only. Unfortunately neither here nor at Tai hu (vicinity of Shanghai; Wang and Liu 2000) can the Early Holocene changes in vegetation and bioproduction be traced, due to a long hiatus in the sediment sequences. Only periods after ca. 6,500 years BP can be investigated in both of these lakes. This is regrettable, because Xue et al. (1995) studied the history of delta formation of the Hoang Ho. Formerly it was thought that four Holocene deltas of the Hoang Ho had existed, situated between its present day position and somewhat to the north of the mouth of Yang tse djiang. It can be shown now that the former "deltas" of Hoang Ho at the east coast of Shan tung and to the south of Qing dao (Chíng tao) had probably never existed. Thus only the modern (since 1855 AD) and the southernmost delta (1128–1855 AD) seem to be proven. Regrettably the reasons for the build-up of these deltas were not investigated: were they related to the extremely long-lasting and intensive agriculture within the catchment area of this river?

The Upper Holocene climate history in the surroundings of Beijing was analysed by Qin et al. (1999) and by Qian and Zhu (2002), using stalagmites. The thickness of the "annual layers" was regrettably interpreted only in terms of available moisture, which is assumed to be solely governed by climate. But what are the effects of anthropogenic changes of the surrounding vegetation?

On at least Hokkaido these Upper Holocene changes of "climate" cannot be traced (Igarashi 1996), although most interesting observations on the last-glacial and Holocene vegetation history have been made, which show that the Younger Dryas cold spell was felt here intensely, as is shown by the strong expansion of *Larix* forests at that time.

During the middle of the last glaciation, i.e. during the younger part of deep-sea stage MIS 3, the bioproduction in two karstic lakes of Central Yunnan, surroundings of K'un ming, was high, though not as high as it was during the Holocene (before 2,200 cal BP). Yet during the LGM it had strongly decreased, having been replaced by an intensive influx of quartz sand (Hodell et al. 1999). This points to the remarkable influence of the winter monsoon at that time, just as has been described earlier, concerning North China. Zheng and Li (2000) showed that in the coastal plain of south-eastern China (mouths of the Han and Rong jiang) an appreciably rich forest vegetation, composed of *Castanopsis*, Celastraceae, *Distylium*, *Dacrydium*, *Liquidambar*, *Pinus* and *Tsuga* accompanied by mangrove species, had characterized the landscape during these interstadial times. Yet almost nothing is known about the vegetation of full glacial times, which seems to have been very open, accompanied by strong weathering and sediment transport. Only beginning at about 11,000 BP can vegetation be traced analytically according to pollen here. It seems that at first grasses, Cyperaceae and ferns were important, while in the forests *Castanopsis* and *Cyclobalanopsis* were characteristic genera. Since about 8,300 years BP mangrove has been a new, very characteristic element. Although climate history and sea level changes are discussed quite comprehensively, nothing is said about human impact, which has been strong here.

For the history of the summer and winter monsoon during the Upper Quaternary as revealed by deep-sea boring, see Jian et al. (2001) and Steinke et al. (2001).

7 Upper Quaternary Palaeoecology of Australia and of the Pacific Islands

It is most interesting to learn that on O'ahu, Hawai'i, 460 m a.s.l., a mixture of dry and mesic forests with extremely high biodiversity had existed just before ca. 21,000 years BP. At about 20,000 years BP forests seem to have remarkably given way to a much more open vegetation (Hotchkiss and Juvik 1999). Already by about 14,500 years BP (or so) very rich forests had formed a dense canopy. A modern equivalent of this vegetation could not be found, due to the maximum values of *Pritchardia*. On other islands of this archipelago dunes and loesses had replaced the formerly existing forests during full-glacial times (Hotchkiss and Juvik 1999). Climate history was also discussed at some length in this paper, but I got the impression that the geological boring doubled the sequence twice. O'ahu seems to have risen due to tectonic activity during the Pleistocene (Jones 1993), yet, at least during the Holocene, this was not the case for this island, Funafuti and at least some of the Fiji Islands which were tectonically stable (Dickinson 1999; Nunn and Peltier 2001). It seems to be difficult to determine from which time on human impact affected the spontaneous vegetation on O'ahu [according to Hotchkiss and Juvik (1999), from as early as about 11,400 and 9,700 years BP?]. Haberle (1996) presents convincing arguments that human activities, perhaps together with ENSO events, caused the differentiation between forests on the south-facing slopes and of grasslands on the plains in the northern parts of Guadalcanal, Solomon Islands. This view is strengthened by archaeological findings from 28,000 years BP and according to Maloney (1995), based on older investigations, forests of *Dacrycarpus, Podocarpus, Engelhardia, Myrsine* and Ericaceae seem to have been diminished in Central Sumatra and Java by about 2,000 m during the LGM. These changes, together with their palaeoclimatological background, were intensively discussed for the Sunda–Sahul region by Flenley (1996). As far as the marine situation off the north-west coast of New Guinea is concerned, the pollen flora in core C 4402, which is 1,108 cm long, seems to have originated during full glacial times in the vegetation of the montane belt of the island mentioned above (Kawahata et al. 2002). In view of the recent catastrophe on western Sumatra and on the coasts of the Indian Ocean it is remarkable to learn from Chappell et al. (1996) that the collision of the West Pacific Plate with the Australian Plate had repeatedly caused strong tectonic uplifts in south-east New Guinea by, on average, ca. 3 m, occurring at an interval of 1,000–1,300 years. For the youngest vertical movements of 2 m in 1855 AD and their ecological consequences for the North Island, New Zealand see Dunbar et al. (1997).

Denton et al. (1999) mapped the full-glacial vegetation on New Zealand. At that time only the Northland Peninsula of the North Island was covered by forests (McGlone 1995), but most parts of this island were covered by forests again between 14,500 and 11,500 BP, and by about 11,000 years BP bushes and tree-ferns had replaced the last-glacial grasslands of the eastern part of the South Island of New Zealand. Positive $\delta^{18}O$ excursions in speleothems of two deep caves in the very north-western corner of southern New Zealand agree well with seven known phases of glacier advances between 27,200 and 3,000 cal BP. Since the speleothems seem to have constantly grown, it is thought – though not yet proven – that the alpine tree-line was never >600–700 m below its present position. Comparable palaeoclimatic interpretations were made by

Williams et al. (1999) regarding speleothems of the Waitomo district, North Island, New Zealand. In this connection it needs to be mentioned that the results from the analysis of various speleothems frequently differ from each other. So it is recommended to always base palaeoclimatic conclusions on investigations of more than one speleothem. Further, the changes in the $\delta^{18}O$ values may have been caused by quite different processes so that the palaeoclimatic interpretation of such values becomes much more difficult than it is normally thought. Based on the elevation at which they were located and the ages of fossil tree trunks found on mount Hauhunqatahi, south-eastern North Island, New Zealand, Ogden et al. (1997) drew the conclusion that the alpine tree-line moved from about 1,150–1,400 m a.s.l. between 12,000 and 9,000 years BP, but that during the Holocene its climate never changed. Moar and Suggate (1996) investigated the Upper Pleistocene vegetation history on the west coast (42°30′S) of the South Island, New Zealand. Here, the *Podocarpus–Nothofagus* forests of the last interglacial were replaced for about 10,000 years during the LGM by a grassland vegetation, dominated by grasses, Asteraceae, *Coprosma*, and *Bulbinella*, when tree pollen was mainly absent. It is thought that annual precipitation at that time had decreased by about 2,400–1,800 mm. During these full-glacial times the surface area of active dunes in Australia is thought to have covered about 40% of the continent, strongly contributing to the sand accumulation of that time in the Tasman Sea (Hesse and McTainsh 1999). For the Upper Quaternary history of North Australian dune fields see Nott et al. (1999). This arid phase can also be traced in the history of the Australian lakes (Harrison 1993). At that time a cryocratic vegetation also seems to have covered Tasmania at elevations of >500 m a.s.l. The development of the flora of Tasmania (differing in time and character) is comprehensively described by Colhoun (1996), who shows very interesting diagrams of the spatial and temporal occurrence of various taxa. It is quite convincingly stated that these major changes were not only governed by climate, but by biological and physical processes as well. For well-dated vegetation history in various parts of Tasmania see Harle et al. (1999) and Dodson (2001a).

The history of glacier advances and retreats in south-eastern Australia was intensively studied by Barrows et al. (2001). Evidently here, like in other parts of the globe, the maximal extent of ice seems to have been during MIS 4, i.e. during the earlier part of the last glaciation. Yet even at that time humans seems to have already lived in south-eastern Australia (Oyston 1996), so this palaeoecological factor has to be taken into consideration, too.

The Mediterranean-type Australian vegetation shows a very high biodiversity (Dodson 2001b; Thomas et al. 2001). Evidently this rich flora had survived full-glacial climates in the vicinity of its present-day distribution areas.

The regionally somewhat warmer and moister climate of Early and Middle Holocene times seems to have been favourable for this flora and type of vegetation. Since about 4,000 years BP it experienced a higher fire frequency than before, causing to some extent changes in vegetation composition. However, the effects of fire were never completely deleterious. The cause of higher fire frequency may have been climatic or anthropogenic, or both (Thomas et al. 2001). Comparable changes in the history of prevailing vegetation types of South Australia are evidenced by macrofossils in *Leporillus* middens (McCarthy and Head 2001) which also reveal a clear increase in the extent of vegetation typical of more arid climates by about 4,000 BP, which is thought to have been caused by a change in climate towards more arid conditions. This also negatively affected lake levels (Cann et al. 2000). Nevertheless, a fine-resolution palaeoclimatic analysis of lake sediments of south-eastern Australia (analysis of pollen, charcoal, magnetic susceptibility, sedimentology and total carotenoids) showed that a mere palaeoclimatic interpretation of these changes may be misleading (Mooney 1997). At least the strong ecological consequences of agriculture practised by the Europeans has to be taken into consideration, too (Gale and Pisanu 2001; Taffs 2001).

Some 300–499 km to the south of New Zealand lie the Auckland and Campbell Islands. On Campbell Island the oldest traces of the former vegetation (ca. 13,000–11,750 BP) point to widely spread macrophyllous herbs, while bushes were still completely absent. It is thought that mean annual temperatures were approximately 5°C lower than at present, creating a tundra-like vegetation. However at about 11,000 years BP *Myrsine* and *Coprosma* bushes seem to have immigrated into the island (McGlone et al. 1997). Comparable sequences in vegetation development were observed on Auckland Island (McGlone et al. 2000). Here the macrophyllous vegetation (12,500–10,580 years BP) was replaced by a vegetation rich in *Myrsine* bushes by about 10,580 years BP. At about 10,000 years BP long-distance transport of tree-pollen affected Auckland Island, but only from 8,000 years BP onwards do *Metrosideros–Myrsine–Dracophyllum* forests seem to have thrived here, though it is thought that most of the modern woody taxa (without *Metrosideros*) were able to survive the last glaciation on this island.

References

Alley R, Bond G, Chapellaz J, Clapperton C, Del Genio A, Keigwin L, Peteet D (1993) Global Younger Dryas? Eos 74(50)

Andriessen PAM, Helmens KF, Hooghiemstra H, Riezebos PA, van der Hammen (1993) Absolute chronology of the Pliocene – Quaternary sediment sequence of the Bogotá area, Colombia. Quat Sci Rev 12:483–501

Aniya M (1996) Holocene variations of Ameghino Glacier, southern Patagonia. Holocene 6:247–252

Arz HW, Pätzold J, Wefer G (1998) Correlated millennial-scale changes in surface hydrography and terrigenous sediment yield inferred from last-glacial marine deposits off northeastern Brazil. Quat Res 50:157–166

Auler AS, Smart PL (2001) Late Quaternary paleoclimate in semiarid northeastern Brazil from U-series dating of travertine and water-table speleothems. Quat Res 55:159–167

Barnett TP, Santer BD, Jones PD, Bradley RS, Briffa KR (1996) Estimates of low frequency natural variability in near surface air temperature. Holocene 6:255–263

Barrows TT, Stone JO, Fifield LK, Cresswell RG (2001) Late Pleistocene glaciation of the Kosciuszko Massif, Snowy Mountains, Australia. Quat Res 55:179–189

Behling H, da Costa ML (2000) Holocene environmental changes from the Rio Curuá record in the Caxiuaña region, Eastern Amazon Basin. Quat Res 53:369-377

Behling H, Hooghiemstra, H (1998) Late Quaternary palaeoecology and palaeoclimatology from pollen records of the savannas of the Llanos Orientales in Colombia. Palaeogeogr Palaeoclim Palaeoecol 139:251–267

Behling H, Hooghiemstra H (1999) Environmental history of the Colombian savannas of the Llanos Orientales since the last glacial maximum from lake records El Pinal and Carimagua. J Paleolimnol 21:461–476

Behling H, Hooghiemstra H (2000) Holocene Amazon rainforest–savanna dynamics and climatic implications: high-resolution pollen record from Laguna Loma Linda in eastern Colombia. J Quat Sci 15:687–695

Behling H, Negrelle RRB (2001) Tropical rain forest and climate dynamics of the Atlantic lowland, Southern Brazil, during the Late Quaternary. Quat Res 56:383–389

Behling H, Hooghiemstra H, Negret AJ (1998) Holocene history of the Chocó rain forest from Laguna Piusbi, southern Pacific lowlands of Colombia. Quat Res 50:300–308

Behling H, Berrio JC, Hooghiemstra H (1999) Late Quaternary pollen records from the middle Caquetá river basin in Central Colombian Amazon. Palaeogeogr Palaeoclim Palaeoecol 145:193–213

Berrio JC, Behling H, Hooghiemstra H (2000) Tropical rain forest history from the Colombian Pacific area: a 4,200 year pollen record from Laguna Jotaordó. Holocene 10:749–756

Bouchard JF (1988) Culturas prehispánicas del Lítoral Pacifico nor-ecuatorial; el hombre y el medio ambiente. Arte de la Tierra – Cultura Tumaco. Fondo de la promoción de la cultura del Banco Popular, Bogotá, pp 8–11

Caballero M, Ortega Guerrero B (1998) Lake levels since about 40,000 years ago at Lake Chalco, near Mexico City. Quat Res 50:69 -79

Cann JH, Bourman RP, Barnett EJ (2000) Holocene foraminifera as indicators of relative estuarine lagoonal and oceanic influences in estuarine sediments of the River Murray, South Australia. Quat Res 53:378–391

Carcaillet C (2001) Are Holocene wood-charcoal fragments stratified in alpine and subalpine soils? Evidence from the Alps based on AMS ^{14}C dates. Holocene 11:231–242

Carew JL, Mylroie JE (1995) Quaternary tectonic stability of the Bahamian Archipelago – evidence from fossil coral reefs and flank margin caves. Quat Sci Rev 14:145–153

Chappell J, Ota Y, Berryman K (1996) Late Quaternary coseismic uplift history of Huon Peninsula, Papua New Guinea. Quat Sci Rev 15:7–22

Chen J, An ZhSh, Head J (1999) Variation of Rb/Sr ratios in the loess-sequences of Central China during the last 130,000 years and their implications for monsoon paleoclimatology. Quat Res 51:215–219

Chen J, Ji JF, Chen Y, An ZhSh, Dearing JA, Wang YJ (2000) Use of rubidium to date loess and paleosols of the Louchan sequence, Central China. Quat Res 54:198–205

Clement RM, Horn SP (2001) Pre-Columbian land-use history in Costa Rica: a 3000-year record of forest clearances, agriculture and fires from Laguna Zoncho. Holocene 11:419–426

Colhoun EA (1996) Application of Iversen's glacial-interglacial cycle to interpretation of the late last glacial and Holocene vegetation history of western Tasmania. Quat Sci Rev 15:557–580

Colinvaux PA (1996) Quaternary environmental history and forest diversity in the Neotropics. In: Jackson JBC, Budd AF, Coates AG (eds) Evolution and environment in Tropical America. University of Chicago Press, Chicago, Ill., pp 359–405

Cowling SA, Maslin MA, Sykes MT (2001) Paleovegetation simulations of lowland Amazonia and implications for neotropical allopatry and speciation. Quat Res 55:140–149

Cross SL, Baker PA, Seltzer GO, Fritz SC, Dunbar RB (2001) Late Quaternary climate and hydrology of Tropical South America inferred from an isotopic and chemical model of Lake Titicaca, Bolivia and Peru. Quat Res 56:1–9

Cunningham WL, Leventer A, Andrews JT, Jennings AE, Licht KJ (1999) Late Pleistocene–Holocene marine conditions in the Ross Sea, Antarctica: evidence from the diatom record. Holocene 9:129–139

Curtis JH, Brenner M, Hodell DA (1999) Climate change in the Lake Valencia Basin; Venezuela, ~12,600 years BP to present. Holocene 9:609–619

Denton GH, Heusser CJ, Lowell TV, Moreno PI, Andersen BG, Heusser LE, Schlüchter C, Marchant DR (1999) Interhemispheric linkage of paleoclimate during the last glaciation. Geogr Ann 81A:107–153

Derbyshire E, Kemp R, Meng XM (1995) Variations in loess and palaeosol properties as indicators of palaeoclimatic gradients across the loess plateau of North China. Quat Sci Rev 14:681–697

Dickinson WR (1999) Holocene sea level record on Funafuti and potential impact of global warming on Central Pacific atolls. Quat Res 51:124–132

Ding ZhL, Sun JM, Rutter NW, Rokosh D, Liu TSh (1999) Changes in sand content of loess deposits along a north–south transect of the Chinese loess plateau and its implications for desert variations. Quat Res 52:56–62

Dodson JR (2001a) A vegetation and fire history in a subalpine woodland and rain forest region, Solomons Jewel Lake, Tasmania. Holocene 11:111–116

Dodson JR (2001b) Holocene vegetation change in the Mediterranean-type climate region of Australia. Holocene 11:673–680

Domack EW, Mayewski PA (1999) Bi-polar ocean linkages: evidence from Late-Holocene Antarctic marine and Greenland ice-core records. Holocene 9:247–251

Domack E, Leventer A, Dunbar R, Taylor F, Brachfeld S, Sjunneskog C, ODP Leg 178 Scientific Party (2001) Chronology of the Palmer Deep site, Antarctic Peninsula: a Holocene palaeoenvironmental reference for the circum-Arctic. Holocene 11:1–9

Dunbar GB, Barrett PJ, Goff JR, Harper MA, Irwin SL (1997) Estimating vertical tectonic movements using sediment texture. Holocene 7:213–221

Elias SA (2000) Late Pleistocene climates of Beringia, based on analysis of fossil beetles. Quat Res 53:229–235

Enfield DB (1989) El Niño, past and present. Rev Geophys 27:159–187

Fabrés J, Calafat A, Canals M, Bárcena MA, Flores JA (2000) Bransfield Basin fine-grained sediments: Late-Holocene sedimentary processes and Antarctic oceanographic conditions. Holocene 10:703–718

Feng XH, Cui HT, Tang KL, Conkey LE (1999) Tree-ring δD as an indicator of Asian monsoon intensity. Quat Res 51:262–266

Flenley JR (1996) Problems of the Quaternary on mountains of the Sunda–Sahul region. Quat Sci Rev 15:549–555

Fontugne M, Usselmann P, Lavallée D, Julien M, Hatté Ch (1999) El Niño variability in the coastal desert of Southern Peru during the Mid-Holocene. Quat Res 52:171–179

Freitas HA de, Pessenda LCR, Aravena R, Gouveia SEM, Souza Ribeiro A de, Boulet R (2001) Late Quaternary vegetation dynamics in the southern Amazon Basin inferred from carbon isotopes in soil organic matter. Quat Res 55:39–46

Frenzel B (1973) Climatic fluctuations of the Ice Age. Western Reserve University, Cleveland, Ohio

Frenzel B (1983) Die Vegetationsgeschichte Süddeutschlands im Eiszeitalter. In:Müller, Beck HJ (ed) Urgeschichte in Baden-Württemberg. Theiss, Stuttgart, pp 91–165

Frenzel B (2000) History of flora and vegetation during the Quaternary. Progr Bot 61:303–334

Fu CB (2003) Potential impacts of human-induced land cover change on East Asia monsoon. Global Planet Change 37:219–229

Gale SJ, Pisanu PC (2001) The Late-Holocene decline of Casuarinaceae in southeast Australia. Holocene 11:485–490

Gascoyne M (1992) Palaeoclimate determination from cave calcite deposits. Quat Sci Rev 11:609–632

Geyh MA, Grosjean M, Núñez L, Schotterer U (1999) Radiocarbon reservoir effect and the timing of the Late Glacial/Early Holocene humid phase in the Atacama Desert (Northern Chile). Quat Res 52:143–153

Goman M, Byrne R (1998) A 5000-year record of agriculture and tropical forest clearance in the Tuxtlas, Veracruz, Mexico. Holocene 8:83–89

Goodwin ID, Zweck C (2000) Glacio-isostasy and glacial ice load at Law Dome, Wilkes Land, East Antarctica. Quat Res 53:285–293

Gordillo S, Coronato AMJ, Rabassa JO (1993) Late Quaternary evolution of a Subantarctic palaeofjord, Tierra del Fuego. Quat Sci Rev 12:889–897

Grootes PM, Steig EJ, Stuiver M, Waddington ED, Morse DL (2001) The Taylor Dome Antarctic ^{18}O record and globally synchronous changes in climate. Quat Res 56:289–298

Grosjean M, Messerli B, Schreier H (1991) Seenhochstände, Bodenbildung und Vergletscherung im Altiplano Nordchiles: Ein interdisziplinärer Beitrag zur Klimageschichte der Atacama. Erste Resultate. Bamberger Geogr Schr 11:99–108

Grosjean M, Valero-Garcés BL, Geyh MA, Messerli B, Schotterer U, Schreier H, Kelts K (1997) Mid- and Late-Holocene limnogeology of Laguna del Negro Francisco, northern Chile, and its palaeoclimatic implications. Holocene 7:151–159

Grosjean M, Geyh MA, Messerli B, Schreier H, Veit,H (1998) A Late-Holocene (<2,600 BP) glacial advance in the south-central Andes (29°S), northern Chile. Holocene 8:473–479

Haberle S (1996) Explanations for palaeoecological changes on the northern plains of Guadalcanal, Solomon Islands: the last 3,200 years. Holocene 6:333–338

Hammen T van der, Hooghiemstra H (1995) The El Abra stadial, a Younger Dryas equivalent in Colombia. Quat Sci Rev 14:841–851

Han JT, Fyfe WS, Longstaffe FJ (1998) Climatic implications of the S5 paleosol complex on the southernmost Chinese Loess Plateau. Quat Res 50:21–33

Hansen BCS (1995) A review of lateglacial pollen records from Ecuador and Peru with reference to the Younger Dryas event. Quat Sci Rev 14:853–865

Harle KJ, Hodgson DA, Tyler PA (1999) Palynological evidence for Holocene palaeoenvironments from the lower Gordon River valley, in the World Heritage Area of southwest Tasmania. Holocene 9:149–162

Harrison SP (1993) Late Quaternary lake-level changes and climates of Australia. Quat Sci Rev 12:211–231

Harrison S, Winchester V (1998) Historical fluctuations of the Gualas and Reicher Glaciers, North Patagonian Icefield. Holocene 8:481–485

Haug GH, Günther D, Peterson LC, Sigman DM, Hughen KA, Aeschlimann B (2003) Climate and the Maya. PAGES News 11(2, 3):28–30

Hearty PJ, Neumann AC, Kaufman DS (1998) Chevron ridges and runup deposits in the Bahamas from storms late in oxygen-isotope substage 5e. Quat Res 50:309–322

Heine JT (1993) A reevaluation of the evidence for a Younger Dryas climatic reversal in the Tropical Andes. Quat Sci Rev 12:769–799

Hellstrom J, McCulloch M, Stone J (1998) A detailed 31,000-year record of climate and vegetation changes, from the isotope geochemistry of two New Zealand speleothems. Quat Res 50:167–178

Hesse PP, McTainsh GH (1999) Late glacial maximum to Early Holocene wind strengths in the mid-latitudes of the Southern Hemisphere from aeolian dust in the Tasman Sea. Quat Res 52:343–349

Heusser CJ (1993a) Late-Glacial of southern South America. Quat Sci Rev 12:345–350

Heusser CJ (1993b) Late Quaternary forest–steppe contact zone, Isla Grande de Tierra del Fuego, subantarctic South America. Quat Sci Rev 12:169–177

Heusser CJ, Heusser LE, Lowell TV (1999a) Palaeoecology of the southern Chilean Lake District – Isla Grande de Chiloë during Middle–Late Llanquihue Glaciation and deglaciation. Geogr Ann 81A:231–284

Heusser L, Heusser C, Kleczkowski A, Crowhurst (1999b) A 50,000-year pollen record from Chile of South American millennial-scale climate instability during the last glaciation. Quat Res 52:154–158

Higuera-Gundy A, Brenner M, Hodd DA, Curtis JH, Leyden BW, Binford MW (1999) A 10,300 ^{14}C record of climate and vegetation change from Haiti. Quat Res 52:159–170

Hirschmann J (1973) Records of solar radiation in Chile. Solar Energy 14:129–138

Hodell DA, Brenner M, Kanfoush SL, Curtis JH, Steuer JS, Song XL, Yuan W, Whitmore TJ (1999) Paleoclimate of southwestern China for the past 50,000 years inferred from lake sediment records. Quat Res 52:369–380

Hodell DA, Kanfoush SL Shemesh A, Crosta X, Charles CD, Guilderson TP (2001) Abrupt cooling of Antarctic surface waters and sea ice expansion in the South Atlantic sector of the Southern Ocean at 5,000 cal years BP. Quat Res 56:191–198

Holmgren CA, Betancourt JL, Rylander KA, Roque J, Tovar O, Zeballos H, Linares E, Quade J (2001) Holocene vegetation history from fossil rodent middens near Arequipa, Peru. Quat Res 56:242–251

Hooghiemstra H, Melice JL, Berger A, Shackleton NJ (1993) Frequency spectra and paleoclimatic variability of the high-resolution 30–1,450 ka Funza I pollen record (Eastern Cordillera, Colombia). Quat Sci Rev 12:141–156

Hotchkiss S, Juvik JO (1999) A Late-Quaternary pollen record from Ka'au Crater, O'ahu, Hawai'i. Quat Res 52:115–128

Hughen KA, Overpeck JT, Peterson LC, Trumbore S (1996) Rapid climate changes in the Tropical Atlantic region during the last deglaciation. Nature 380:51–54

Hughen KA, Overpeck JT, Lehman SJ, Kashgarian M, Southon J, Peterson LC, Alley R, Sigman DM (1998) Deglacial changes in ocean circulation from an extended radiocarbon calibration. Nature 391:65–68

Huntley B (1996) Quaternary palaeoecology and ecology. Quat Sci Rev 15:591–606

Igarashi Y (1996) A lateglacial climatic reversion in Hokkaido, Northeast Asia, inferred from the *Larix* pollen record. Quat Sci Rev 15:989–995

Islebe GA, Hooghiemstra H, Brenner M, Curtis JH, Hodell DA (1996) A Holocene vegetation history from lowland Guatemala. Holocene 6:265–271

Ivy-Ochs S, Schlüchter C, Kubik PW, Denton GH (1999) Moraine exposure dates imply synchronous Younger Dryas glacier advances in the European Alps and in the Southern Alps of New Zealand. Geogr Ann 81A:313–323

Ji JF, Balsam W, Chen J (2001) Mineralogic and climatic interpretations of the Luochuan loess section (China) based on diffuse reflectance spectrophotometry. Quat Res 56:23–30

Jian ZhM, Huang BQ, Kuhnt W, Lin HL (2001) Late Quaternary upwelling intensity and East Asian monsoon forcing in the South China Sea. Quat Res 55:363–370

Jiang QH, Piperno DR (1999) Environmental and archaeological implications of a Late Quaternary palynological sequence, Poyang Lake, Southern China. Quat Res 52:250–258

Jones AT (1993) Review of the chronology of marine terraces in the Hawaiian Archipelago. Quat Sci Rev 12:811–823

Kawahata H, Maeda R, Ohshima H (2002) Fluctuations in terrestrial-marine environments in the western Equatorial Pacific during the Late Pleistocene. Quat Res 57:71 -81

Kuhry P, Hooghiemstra H, van Geel B, van der Hammen T (1993) The El Abra stadial in the Eastern Cordillera of Colombia (South America). Quat Sci Rev 12:333 -343

Kuylenstierna JL, Rosqvist GC, Holmlund P (1996) Late-Holocene glacier variations in the Cordillera Darwin, Tierra del Fuego, Chile. Holocene 6:353–358

Ledru MP, Bertaux J, Siffedine A (1998) Absence of Last Glacial maximum records in lowland Tropical forests. Quat Res 49:233–237

Ledru MP, Cordeiro RC, Dominguez JML, Martin L, Mourguiart P, Siffedine A, Turcq B (2001) Late-glacial cooling in Amazonia inferred from pollen at Lagoo de Caçó, Northern Brazil. Quat Res 55:47–56

Leyden BW, Brenner M, Dahlin B (1998) Cultural and climatic history of Cobá, a lowland Maya city in Quintana Roo, Mexico. Quat Res 49:111–122

Liu HY, Xu LH, Cui HT (2002) Holocene history of desertification along the woodland–steppe border in Northern China. Quat Res 57:259–270

Lowell TV (1995) The application of radiocarbon age estimates to the dating of glacial sequences: an example from the Miami Sublobe, Ohio, USA. Quat Sci Rev 14:85–99

Lu HY, van Huissteden K, Zhou J, Vandenberghe J, Liu XD, An ZhSh (2000) Variabiliy of East Asian winter monsoon in Quaternary climatic extremes in North China. Quat Res 54:321–327

Lynch TF (1990) Quaternary climates, environment and human occupation of the South-Central Andes. Geoarchaeology 5:199–228

Lyons RG (1996) The elusive Blake: a record of paleomagnetically reversed deposits in New Zealand post 120 ka. Quat Sci Rev 15:719–726

MacFadden BJ (2000) Middle Pleistocene climate change recorded in fossil mammal teeth from Taríja, Bolivia, and upper limit of the Ensenadan land-mammal age. Quat Res 54:121–131

Maloney BK (1995) Evidence for the Younger Dryas climatic event in Southeast Asia. Quat Sci Rev 14:949–958

Mann DH, Hamilton TD (1995) Late Pleistocene and Holocene paleoenvironments of the North Pacific coast. Quat Sci Rev 14:449–471

Marden CJ, Clapperton CM (1995) Fluctuations of the South Patagonian ice-field during the last glaciation and the Holocene. J Quat Sci 10:197–210

Markgraf V (1993) Younger Dryas in southernmost South America – an update. Quat Sci Rev 12:351–355

Markgraf V, Betancourt J, Rylander KA (1997) Late-Holocene rodent middens from Rio Limay, Neuquen Province, Argentina. Holocene 7:325–329

Martínez S, Ubilla M, Verde M, Perea D, Rojas A, Guérèquiz R, Piñeiro G (2001) Paleoecology and geochronology of Uruguayan coastal marine Pleistocene deposits. Quat Res 55:246–254

Masson V, Vimeux F, Jouzel J, Morgan V, Delmotte M, Ciais P, Hammer C, Johnsen S, Lipenkov VYa, Mosley-Thompson E, Petit JR, Steig E, Stievenard M, Vaikmae R (2000) Holocene climate variability in Antarctica based on 11 ice-core isotopic records. Quat Res 54:348–358

McCarthy L, Head L (2001) Holocene variability in semi-arid vegetation: new evidence from *Leporillus* middens from the Flinders Ramps, South Australia. Holocene 11:681–689

McFarlane DA, MacPhee RDE, Ford DC (1998) Body size variability and a Sangamonian extinction model for *Amblyrhiza*, a West Indian megafaunal rodent. Quat Res 50:80–89

McGlone MS (1995) Lateglacial landscape and vegetation change and the Younger Dryas climatic oscillation in New Zealand. Quat Sci Rev 14:867–881

McGlone MS, Moar NT, Wardle P, Meurk CD (1997) Late-glacial and Holocene vegetation and environment of Campbell Island, far southern New Zealand. Holocene 7:1–12

McGlone MS, Wilmhurst JM, Wiser SK (2000) Lateglacial and Holocene vegetation and climatic change on Auckland Island, Subantarctic New Zealand. Holocene 10:719–728

McLaren S, Gardner R (2000) New radiocarbon dates from a Holocene aeolianite, Isla Cancun, Quintana Roo, Mexico. Holocene 10:757–761

McMinn A, Heijnis H, Harle K, McOrist G (2001) Late-Holocene climatic change in sediment cores from Ellis Fjord, eastern Antarctica. Holocene 11:291–300

Metcalfe S, Say A, Black S, McCulloch R, O'Hara S (2002) Wet conditions during the last glaciation in the Chihuahuan Desert, Alta Babicora Basin, Mexico. Quat Res 57:91–101

Meunier AA, Meunier JO, Mariotti A, Soubies F (1999) Late Holocene phytolith and carbon-isotope record from a latosol at Salitre, South-Central Brasil. Quat Res 51:187–194

Moar NT, Suggate RP (1996) Vegetation history from the Kaihinu (last) Interglacial to the present, west coast, South Island, New Zealand. Quat Sci Rev 15:521–547

Mooney S (1997) A fine-resolution palaeoclimatic reconstruction of the last 2,000 years, from Lake Keilambete, southeastern Australia. Holocene 7:139–149

Moreno PI (2000) Climate, fire, and vegetation between about 13,000 and 9,200 ^{14}C years BP in the Chilean Lake District. Quat Res 54:81 89

Moreno PI, Lowell TV, Jacobson GL jr, Denton GH (1999) Abrupt vegetation and climate changes during the Last Glacial maximum and last termination in the Chilean Lake District: a case study from Canal de la Puntilla (41°S). Geogr Ann 81A:285–311

Morgan V, Ommen TD van (1997) Seasonality in Late-Holocene climate from ice-core records. Holocene 7:351–354

Morley JJ, Heusser LE (1989) Late Quaternary atmospheric and oceanographic variations in the Western Pacific inferred from pollen and radiocarbon analyses. Quat Sci Rev 8:263–276

Muhs DR (2002) Evidence for the timing and duration of the last interglacial period from high-precision uranium-series ages of corals on tectonically stable coastlines.Quat Res 58:36–40

Murray–Wallace CV, Ferland MA, Roy PS, Sollar A (1996) Unravelling patterns of reworking in lowstand shelf deposits using amino acid racemisation and radiocarbon dating. Quat Sci Rev 15:685–697

Nilsson M, Klarqvist M, Bohlin E, Possnert G (2001) Variation in ^{14}C age of macrofossils and different fractions of minute peat samples dated by AMS. Holocene 11:579–586

Northrop LA, Horn SP (1996) Pre Columbian agriculture and forest disturbance in Costa Rica: palaeoecological evidence from two lowland rainforest lakes. Holocene 6:289–299

Nott J, Bryant E, Price D (1999) Early-Holocene aridity in tropical northern Australia. Holocene 9:231–236

Nunn PD, Peltier WR (2001) Far-field test of the ICE–4G model of global isostatic response to deglaciation using empirical and theoretical Holocene sea level reconstructions for the Fiji Islands,Southwestern Pacific. Quat Res 55:203–214

ÓCofaigh C, Dowdeswell JA, Pudsey CJ (2001) Late Quaternary iceberg rafting along the Antarctic Peninsula continental rise and in the Weddell and Scotia Seas. Quat Res 56:308–321

Ogden J, Horrocks M, Palmer JG, Fordham RA (1997) Structure and composition of the sub-alpine forest on Mount Hauhungatahi, New Zealand, during the Holocene. Holocene 7:13–23

Olley JM, Murray A, Roberts RG (1996) The effects of disequilibria in the uranium and thorium decay chains on burial dose rates in fluvial sediments. Quat Sci Rev 15:751–760

Ortega-Guerrero B, Newton AJ (1998) Geochemical characterization of Late Pleistocene and Holocene tephra layers from the Basin of Mexico, Central Mexico. Quat Res 50:90–106

Ortlieb L, Diaz A, Guzman N (1996) A warm interglacial episode during oxygen isotope stage 11 in northern Chile. Quat Sci Rev 15:857–871

Osborn G, Clapperton C, Davis PT, Reasoner M, Rodbell DT, Seltzer GO, Zielinski G (1995) Potential glacial evidence for the Younger Dryas event in the Cordillera of North and South America. Quat Sci Rev 14:823–832

Oyston B (1996) Thermoluminescence age determinations for the Mungo III human burial, Lake Mungo, southeastern Australia. Quat Sci Rev 15:739–749

Pan XL, Chao JP (2003) Theory of stability, and regulation and control of ecological systems in oasis. Global Planet Change 37:287–295

Pendall E, Markgraf V, White JWC, Dreier M, Kenny R (2001) Multiproxy record of Late Pleistocene–Holocene climate and vegetation changes from a peat bog in Patagonia. Quat Res 55:168–178

Pessenda LCR, Gomes BM, Aravena R, Ribeiro AS, Boulet R, Gouveia SEM (1998) The carbon isotope record in soils along a forest–Cerrado ecosystem transect: implications for vegetation changes in the Rondonia State, southwestern Brazilian Amazon region. Holocene 8:599–603

Pessenda LCR, Boulet R, Aravena R, Rosolen V, Gouveia SEM, Ribeiro, Lamotte M (2001) Origin and dynamics of soil organic matter and vegetation changes during the Holocene in a forest–savanna transition zone, Brazilian Amazon region. Holocene 11:250–254

Peteet D (1995) Global Younger Dryas? Quat Internat 28:93–104

Pillans B, Kohn BP, Berger G, Froggatt P, Duller G, Alloway B, Hesse P (1996) Multi-method dating comparison for Mid-Pleistocene Rangitawa tephra, New Zealand. Quat Sci Rev 15:641–653

Placzek C, Quade I, Betancourt JL (2001) Holocene lake-level fluctuations of Lake Aricota, southern Peru. Quat Res 56:181–190

Porter SC, Hallet B, Wu XH, An ZhSh (2001) Dependence of near-surface magnetic susceptibility on dust accumulation rate and precipitation on the Chinese Loess Plateau. Quat Res 55:271–283

Prieto MR del, Herrera R, Dussel P(1999) Historical evidence of streamflow fluctuations in the Mendoza River, Argentina, and their relationship with ENSO. Holocene 9:473–481

Pye K (1995) The nature, origin and accumulation of loess. Quat Sci Rev 14:653–667

Qian WH, Zhu YF (2002) Little Ice Age climate near Beijing, China, inferred from historical and stalagmite records. Quat Res 57:109–119

Qin XG, Tan M, Liu TSh, Wang XF, Li TY, Li JP (1999) Spectral analysis of a 1,000-year stalagmite lamina-thickness record from Shihua Cavern, Beijing, and its climatic significance. Holocene 9:689–694

Roberts D, McMinn A (1999) A diatom-based palaeosalinity history of Ace Lake, Vestfold Hills Antarctica. Holocene 9:401–408

Roberts D, Roberts JL, Gibson AE, McMinn A, Heijnis H (1999) Palaeohydrological modelling of Ace Lake, Vestfold Hills, Antarctica. Holocene 9:515–520

Roberts D, Ommen TD van, McMinn A, Morgan U, Roberts JL (2001a) Late-Holocene East Antarctic climate trends from ice-core and lake-sediment proxies. Holocene 11:117–120

Roberts HM, Wintle AG, Maher BA, Hu MY (2001b) Holocene sediment-accumulation rates in the western Loess Plateau, China, and a 2,500-year record of agricultural activity, revealed by OSL dating. Holocene 11:477–483

Rodbell DT, Seltzer GO (2000) Rapid ice margin fluctuations during the Younger Dryas in the Tropical Andes. Quat Res 54:328–338

Rodbell DT, Bagnato S, Nebolini JC, Seltzer GO, Abbott MB (2002) A Late Glacial–Holocene tephrochronology for glacial lakes in Southern Ecuador. Quat Res 57:343–354

Rodriguez-Filho S, Behling H, Irion G, Müller G (2002) Evidence for lake formation as a response to an inferred Holocene climatic transition in Brazil. Quat Res 57:131–137

Roig F jr, Roig C, Rabassa J, Boninsegna J (1996) Fuegian floating tree-ring chronology from subfossil *Nothofagus* wood. Holocene 6:469–476

Rose NL, Golding PNE, Batterbee RW (1996) Selective concentration and enumeration of tephra shards from lake sediment cores. Holocene 6:243–246

Rosenmeier MF, Hodell DA, Brenner M, Curtis JH, Guilderson TP (2002) A 4,000-year lacustrine record of environmental change in the southern Maya Lowlands, Petén, Guatemala. Quat Res 57:183–190

Rousseau DD, Wu NQ (1999) Mollusk records of monsoon variability during the L2-S2 cycle in the Luochuan sequence, China. Quat Res 56:286–292

Rutter N, Ding Z (1993) Paleoclimates and monsoon variations interpreted from micro-morphogenic features of the Baoji paleosols, China. Quat Sci Rev 12:853–862

Sandweiss DH, Maasch KA, Burger RL, Richardson III JB, Rolling HB, Clement A (2001) Variation in Holocene El Niño frequencies: climate records and cultural consequences in ancient Peru. Geology 29:603–606

Sayago JM (1995) The Argentine neotropical loess : an overview. Quat Sci Rev 14:755 -766

Schlüchter C, Gander P, Lowell TV, Denton GH (1999) Glacially folded outwash near Lago Llanquihue, southern Lake District, Chile. Geogr Ann 81A:347–358

Schulz H, Emeis K-C, Erlenkeuser H, Rad U von, Rolf C (2002) The Toba volcanic event and interstadial/stadial climates at the marine isotope stage 5 to 4 transition in the northern Indian Ocean. Quat Res 57:22–31

Scott DB, Collins ES (1996) Late Mid-Holocene sea level oscillation: a possible cause. Quat Sci Rev 15:851–856

Shane P, Froggatt P, Smith I, Gregory M (1998) Multiple sources for sea-rafted Loisels Pumice, New Zealand. Quat Res 49:271–279

Shore JS, Bartley DD, Harkness DD (1995) Problems encountered with the ^{14}C dating of peat. Quat Sci Rev 14:373–383

Sun JM (2000) Origin of eolian sand mobilization during the past 2,300 years in the Mu Us Desert, China. Quat Res 53:78–88

Sun JM, Ding ZhL (1998) Deposits and soils of the past 130,000 years at the desert–loess transition in Northern China. Quat Res 50:148–156

Sun JM, Liu TSh (2000) Stratigraphic evidence for the uplift of the Tibetan Plateau between ~1,1 and ~0,9 million years ago. Quat Res 54:309–320

Steinke S, Kienast M, Pflaumann U, Weinelt M, Stattegger K (2001) A high resolution sea-surface temperature record from the Tropical South China Sea (16,500–3,000 years BP). Quat Res 55:352–362

Swadling KM, Dartnall HJG, Gibson JAE, Saulnier-Talbot É, Vincent WF (2001) Fossil rotifers and the early colonization of an Antarctic lake. Quat Res 55:380–384

Taffs KH (2001) Diatoms as indicators of wetland salinity in the upper southeast of South Australia. Holocene 11:281–290

Taylor F, McMinn A (2001) Evidence from diatoms for Holocene climate fluctuations along the East Antarctic margin. Holocene 11:455–466

Taylor F, McMinn A (2002) Late Quaternary diatom assemblages from Prydz Bay, Eastern Antarctica. Quat Res 57:151–161

Thomas I, Enright NJ, Kenyon CE (2001) The Holocene history of Mediterranean-type plant communities, Little Desert National Park, Victoria, Australia. Holocene 11:691–697

Thomas MF, Thorp MB (1995) Geomorphic response to rapid climatic and hydrologic change during the Late Pleistocene and Early Holocene in the humid and subhumid Tropics. Quat Sci Rev 14:193–207

Vacher HL, Hearty P (1989) History of stage 5 sea level in Bermuda: review with new evidence of a brief rise to present sea level during substage 5a. Quat Sci Rev 8:159–168

Wang JQ, Liu JL (2000) Amino acids and stable carbon isotope distributions in Taihu Lake, China, over the last 15,000 years, and their paleoecological implications. Quat Res 53:223–228

Wang ShW, Gong DY, Zhu JH (2001) Twentieth-century climatic warming in China in the context of the Holocene. Holocene 11:313–321

Westgate JA, Shane PAR, Pearce NJG, Perkins WT, Korisettar R, Chesner CA, Williams MAJ, Acharyya SK (1998) All Toba tephra occurrences across Peninsular India belong to the 75,000 years BP eruption. Quat Res 50:107–112

Whitmore TJ, Brenner M, Curtis JH, Dahlin BH, Leyden BW (1996) Holocene climatic and human influences on lakes of the Yucatan Peninsula, Mexico: an interdisciplinary, palaeolimnological approach. Holocene 6:273–287

Williams PW, Marshall A, Ford DC, Jenkinson AV (1999) Palaeoclimatic interpretation of stable isotope data from Holocene speleothems of the Waitomo District, North Island, New Zealand. Holocene 9:649–657

Wilson GS (1995) The Neogene East Antarctic ice sheet: a dynamic or stable feature? Quat Sci Rev 14:101–123

Wohlfarth B (1996) The chronology of the last termination: a review of radiocarbon-dated, high-resolution terrestrial stratigraphies. Quat Sci Rev 15:267–284

Xue ChT, Zhu XH, Lin HM (1995) Holocene sedimentary sequence, foraminifera and ostracoda in west coastal lowlands of Bohai Sea, China. Quat Sci Rev 14:521–530

Yarnal B (1985) Extratropical teleconnections with El Niño/Southern Oscillation (ENSO) events. Progr Phys Geogr 9:315–345

Yarnal B, Kiladis G (1985) Tropical teleconnections associated with El Nino/Southern Oscillation (ENSO) events. Progr Phys Geogr 9:524–558

Zheng Zh, Li QY (2000) Vegetation, climate and sea level in the past 55,000 years, Hanjiang Delta, Southeastern China. Quat Res 53:330–340

Zhou LP, Dodonov AE, Shackleton NJ (1995) Thermoluminescence dating of the Orkutsay loess section in Tashkent region, Uzbekistan, Central Asia. Quat Sci Rev 14:721–730

Burkhard Frenzel,
Institut für Botanik – 210
Universität Hohenheim,
70593 Stuttgart,
Germany

Biodiversity experiments – artificial constructions or heuristic tools?

Carl Beierkuhnlein and Carsten Nesshöver

1 Introduction

Biodiversity experiments are important tools for basic research. Here, we discuss their scientific basis, some major hypotheses and review the most prominent results according to the role of biodiversity in productivity, invasion and nutrient cycling. Progress in methods and insights is documented. Shortcomings and restrictions of recent approaches are identified. Finally, we point out research perspectives.

At the global level, the loss of rare species is recorded. Regional extinctions of populations are also documented. Communities are becoming more and more uniform due to standardized management techniques and increased dispersal of ruderal species. Alien and invasive species are contributing at the regional scale to an increase in species diversity whereas rare species may be lost from the same area. Changes of biodiversity have many facets at different scales. The mechanisms that are responsible for the ongoing changes are generally being identified, with rapid land use changes being the most prominent (Sala et al. 2000). Early biodiversity research concentrated on the detection and description of taxa and their diversity within certain ecosystems or regions and on their loss or decline (e.g. Ehrlich and Ehrlich 1981). Due to the complexity of natural and anthropogenic ecosystems and landscapes, monitoring of such changes is rather time consuming and will deliver sound results only for selected case studies and areas.

Science has to analyse the mechanisms of change and to contribute to coping strategies through predictive conclusions. The long-term preservation of sound ecosystems is needed. Strategies that consider biodiversity and ecosystem functioning are requested (Schwartz et al. 2000; Hector et al. 2001b). One category that adds complexity to the system is the interaction between biotic compartments, and between biotic and abiotic compartments differing from stand to stand. As early as in the 1920s the recognition of this fact lead to the development of the individualistic concept by Gleason

(1926). Even if this variability can be reduced and categorized, it is an obstacle to the identification of general rules in ecology. What holds true for one community does not necessarily do so for another one. The role of diversity in the functioning of ecosystems was not recognized as a highly important scientific issue until the end of the twentieth century. However, recently the relationship between (plant) species diversity and ecosystem functions has been highlighted in many studies. Observational case studies yielded varying results depending on the respective system, spatial scale and time frame. Thus, an urgent need for experimental studies became obvious to close the gap of knowledge when comparing theory and reality (Schmid and Hector 2004).

During the last decade, there was a controversial debate about the relevance of biological diversity for ecosystem functioning (e.g. Schulze and Mooney 1993; Naeem et al. 1994, 1995; Mooney et al. 1996; Tilman et al. 1996; Grime 1997; Naeem and Li 1997; Chapin et al. 1998; Hector et al. 1999; Wall 1999; Chapin et al. 2000; Ghilarov 2000; Hall et al. 2000; Huston et al. 2000; Kaiser 2000; Loreau 2000; Wardle et al. 2000b; Fridley 2001; Loreau et al. 2001; Huston and McBride 2002; Loreau et al. 2002a; Mooney 2002; Naeem et al. 2002; Schmid 2002; Schmid et al. 2002b; Vandermeer et al. 2002; Symstad et al. 2003; Beierkuhnlein and Jentsch 2005; Hooper et al. 2005; Schläpfer et al. 2005). Some aspects of the discussion had already been raised decades ago, the diversity–stability debate initiated by MacArthur (1955) being the most prominent one (e.g. Leigh 1965; May 1972). The loss of interest in the diversity–stability relationship in the 1970s was due to difficulties in defining stability, diversity and – closely connected to these issues – complexity. Global environmental changes and the uncertainty about their consequences ignited this discussion again (Tilman and Downing 1994; Johnson et al. 1996; Tilman 1996; McGrady-Steed et al. 1997; Naeem and Li 1997; Doak et al. 1998; Tilman et al. 1998; Ives et al. 2000; Lehman and Tilman 2000; McCann 2000; Wardle et al. 2000a; Fonseca and Ganade 2001; Loreau et al. 2002b; Pfisterer and Schmid 2002; Tilman et al. 2002b; Wardle and Grime 2003). Stability in a strict sense does not exist in ecosystems. It is ecologically neither positive nor realistic and there are no standard ways to measure or calculate it (Grimm and Wissek 1997). Recent approaches aim to produce as clear results as possible on the basis of specific and measurable parameters. This is one reason why much emphasis is put on aboveground biomass production. The net biomass that is produced per area and in a defined time is a prominent trait of ecosystems, and comparable data can be achieved across different ecosystems.

The perception of environmental problems as well as the lack of explanatory and predictive scientific capacity stimulated the debate. Wrong conclusions could result in serious consequences (Tilman 1999;

Chapin et al. 2000). Direct connections to global change ecology have been pointed out (Reich et al. 2001; Lloret et al. 2004; Thomas et al. 2004; Hooper et al. 2005). Effects of increasing nutrient inputs and global climate change on diversity and on ecosystem processes are expected (see Hartley and Jones 2003), respectively. For example, experimental results support predicted interactions between the loss of species and an increased CO_2 concentration in the atmosphere (Stocker et al. 1999; Reich et al. 2001; He et al. 2002). Further, there is an indication that not only trends in average values will occur, but extreme climate events will become more frequent (Meehl et al. 2000). It is assumed that species-poor communities will be less resilient when disturbance frequency increases (White and Jentsch 2001) or when single events of extreme magnitude occur (White and Jentsch 2004). Stochastic short-term events are important historical factors that contributed to invasion and determine present-day species composition (Davis and Pelsor 2001).

Biodiversity research is not only a fascinating scientific discipline but is also of high socio-economic importance and political relevance (Gowdy and Daniel 1995; Perrings 1995; Montgomery and Pollack 1996; Jentsch et al. 2003). It has been the subject of a great deal of attention in society at large and political bodies not only because of the ethical responsibility of mankind or of aesthetic needs. More than that, the fear of losing ecosystem functions, and especially those which are of societal importance, are the reasons for political concern and societal awareness. Most biodiversity experiments focus on element cycling, nutrient use and biomass production, because these ecological parameters are of socio-economic relevance (e.g. Spehn et al. 2000; Tilman et al. 2002b; Roscher et al. 2004).

From agricultural systems it is well known that the cultivation of species mixtures (intercropping or multiple cropping) may increase productivity, in comparison with monocultures, via positive interactions (Trenbath 1974, 1976; Mead and Willey 1980; Chetty and Reddy 1984; Willey 1985; Vandermeer 1989). Indirect effects occur if suboptimal site conditions are improved, resources are made available or microclimatic conditions are modified in a way that prevents plant pathogens or restricts herbivores (Altieri 1994; Bertness and Callaway 1994; Hooper and Vitousek 1997). Mainly on resource-limited sites or under low input conditions, at least one of the components in a mixture may benefit as a result of facilitation that does not restrict the productivity or performance of the dominant species. Combinations of trees and annuals in agro-forestry systems have increased stand productivity remarkably (Schroth 1999). Nevertheless, there are obvious limits to extrapolating results from agronomy to biodiversity research. Plants in land use systems are planted and supported (by weeding, fertilizing and specific management techniques). In many cases, the plants are extremely different (e.g. trees and annuals). But above all, only a few species are included, even in intercropping systems, compared to the diversity of natural ecosystems or biodiversity experiments (Swift and Anderson 1993; Vandermeer et al. 2002). In addition, it can be assumed that there are even more species in these systems (weeds) which are ignored because they are not of any economic interest.

The basic thought behind biodiversity experiments is that ecosystems have evolved a characteristic diversity of species and that this diversity contributes to their functioning. In managed systems, the influence of humans on environmental conditions and the disturbance regime via management of diversity and functioning will be strong and overwhelm the potential influence of diversity (Fig. 1). But if human impacts are kept constant, a change in diversity would thus be accompanied by changes in ecosystem functions (central arrow, Fig. 1). In most experiments, establishing communities with varying diversity tests this relationship. The main aim of biodiversity experiments is to understand the nature of this relationship and identify general trends. Experiments are considered to be of value as examples, and their results may be applied to various other ecosystems. In this article, we will discuss the findings of the types of experiments outlined and try to identify important results, shortcomings and future applications.

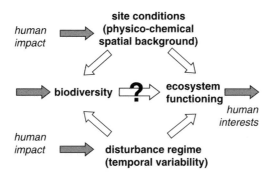

Fig. 1. Humans are altering the environment and nature in many ways. The effects of such activities can hardly be differentiated. Pollution and environmental change may directly affect ecosystem functioning but also biodiversity. Land use change modifies the temporal variabiity of ecosystems and this will have repercussions for biodiversity and ecosystem functioning. Changes in biodiversity can result from direct and indirect human impacts. This web of cause and effects implies that it is necessary to separate the various processes experimentally. As a consequence, it is almost impossible to detect and measure the impact of biodiversity loss on ecosystem functions in natural ecosystems

2 Some Philosophical and Basic Aspects of Ecology and Diversity Research

Ecology deals with complex systems, especially when communities and ecosystems are studied. Ecology is considered a precise natural science (or "hard" science), but workers recognise that ecological units cannot be characterized completely due to the complexity of the subject, and this creates tension. Decisions have to be made by ecologists on how to proceed with research, which aspect should be focused on, and what has to be ignored. The complexity and the spatio-temporal heterogeneity of ecological objects provoke simplifications, such as reductionism, or generalisations, such as holism. In any case, there is a necessity to concentrate or filter the scientific topic. Either parts are taken out of their context and investigated in detail, or the whole system is considered to be one unit. It is impossible to cope comprehensively with an entire ecosystem. The variety of paradigms and methodological approaches that are partly complementary, partly competing, reflects the range of the theoretical and philosophical background of ecologists. Research aims of ecology have shifted from descriptive studies, including sampling and monitoring of field data (which is still urgently needed), i.e. from "what, when, where" questions, to analytical investigations of cause and effect ("why" questions) (Fridley 2001). Thus ecology has become progressively theory-oriented.

The quality of pure science depends on the standard of theory. Theory is used to search for general rules or even laws. Such rules have to be empirically proven and the aim is to understand the mechanisms that give rise to certain phenomena. When a new gap in knowledge has been identified, when patterns or processes cannot be explained or predicted, and when an urgent need to find an answer to a question is felt, the appropriate approach is to formulate hypotheses which are reasonable as well as innovative. Hypotheses comprise testable predictions about cause and effect. In ecology this is not an easy task, even if regular structures are evident. The complexity of interactions between organisms (see Connell 1983), but also between biotic and abiotic processes, inevitably leads to a high variance in data. This "noise" is an integral part of the subject. As a matter of fact, it creates restrictions to statistical analyses and tests. Simberloff (1980) illustrates this ambiguity of ecology in a comparison with other fields of natural science "physicists may view it as noise, but it is music to the ecologist".

Does this vagueness and uncertainty exclude the application of criteria of hard sciences, such as the search for theories and laws with predictive and/or explanatory power (Loehle 1987)? Certainly not! However, restrictions have to be considered. The hypothetico-deductive approach requires an elaborated theoretical framework. Its appropriateness varies according to the maturity of a theory. Biodiversity research is a young discipline and, as we will show, many divergent theories and hypothesis have been worked out. This can be taken as a indication of immaturity, and if this were true, it is difficult to conduct conclusive tests (Loehle 1988). Data and results are then related to a certain approach or even paradigm (see also Naeem 2002b). Predictions will be vague or even contradictory.

According to Popper (1959, 1963), only the falsification of a hypothesis can be regarded as a sound result. However, if a hypothesis is based on previous knowledge, if it is intelligent, innovative and reasonable, it is hard to accept that it is wrong. Scientists tend to defend their ideas. If the hypothesis has to be rejected, this means that it is based on false assumptions or theory or that it is valid only under certain circumstances, which were not given in this specific case. Results that confirm a theory do so only under certain constraints and are never able to support a general truth of it or, in the worst case scenario, may be just a result of artefacts and autocorrelations but not of any causal importance at all. Coming back to the ecological complexity of nature and to the point that this goes hand in hand with a certain individuality of ecological communities, we are thus faced with essential logical, if not philosophical, frontiers. When can we be sure that a hypothesis is falsified or verified? When can an explanation be accepted as adequate? To which degree are we allowed to or even have to generalize, classify and sweep aside differences between ecological units?

One way to find a solution is to accept the fact that observations in natural systems, as exactly measured as they may be, "only" serve as a source for generating theories and hypotheses but are not adequate for their approval. Observations and descriptions are necessary and essential to identify problems, regular patterns or remarkable exceptions. Field data will stimulate scientists to think about reasons for their structure. Such thoughts can be compared with existing knowledge and if there are conflicts, this is an indication that there is a need for more research. Nonetheless, the falsification of theories can only be successful with the support of models and experiments (Fig. 2). In addition, there is no absolute truth in theory, especially not in ecological theory. Theoretical considerations, as necessary as they are, are nothing more than idealized approaches. They serve as tools for a better understanding of complex phenomena. Like ideal gases they simulate the exclusion of interfering factors. This is helpful when considering isolated interactions, but it often fails to explain the variation of data in reality.

In biodiversity research we are confronted with a mixture of assumptions (that are not tested but taken for granted) and hypotheses. Science requires that researchers agree about their objects. Terminological conventions have

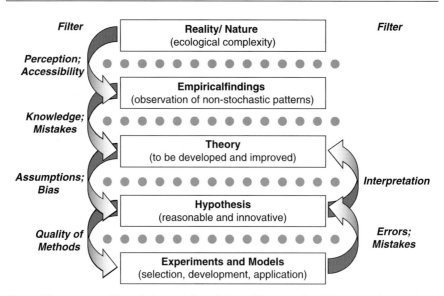

Fig. 2. The progress of knowledge and the solving of theoretical problems are always subject to various filters. For example, we can detect only a narrow range of radiation and sound waves, and many natural phenomena are simply not within our range of detection. In addition we are limited in terms of temporal and spatial scales. Our lifetime sets the limit for our individual access to resources and space, and important habitats are simply out of our reach, either being too high (forest canopies), too deep (oceans) or too hidden (soils). The creation and the further development of theoretical concepts, definitions and models for a better understanding of nature are at the heart of any scientific approach. When we aim to prove their validity, hypotheses have to be formulated which are both logical on the basis of previous findings and novel in terms of their focus. Technical and scientific progress constantly lead to improvement in methods. The interpretation of results is influenced by paradigms, knowledge and creativity. The progress which is achieved will always be a relative and transient one. Note that empirical findings merely stimulate the development of theories but can never be used to prove hypotheses!

to be created, such as the concept of species and their nomenclature. However, taxa differ a lot according to their within-species divergence or variability. Species are not discreet units that integrate the same amount of variance. A basic assumption, which is implied if species are stochastically mixed, is that species are functionally different due to niche complementarity (see also Grime 2002; Fargione et al. 2003), and even more that they are more or less equally different. However, between-species dissimilarity also differs. Species vary gradually according to their morphological similarity and phylogenetic proximity, longevity, life cycles and reproductive turnover, but also according to the importance of sexual reproduction within certain taxa (e.g. Weiher et al. 1999; Poschlod et al. 2000; Thompson et al. 2001; Craine et al. 2002; Lavorel and Garnier 2002). Clonal or apomictic species underlie very different operating mechanisms than species for which the focus is sexual reproduction and selection (Dietz and Steinlein 2001). Distances between the units (species) are not regular and will differ in magnitude according to the criteria that are applied. Ignoring functional dissimilarity between organisms creates problems, but so does a priori classification when selecting species for an experiment. Many dimensions and parameters would have to be considered. However, very presumably, positive interactions and facilitation are more effective if species differ noticeably. And, competition is likely to be strongest when species are quite similar. Assuming that the evolution of species is mainly triggered by selection and niche occupation, taxonomic diversity has to be related closely to functional divergence. Biodiversity will strongly reflect biocomplexity (Reich et al. 2003). This view ignores the fact that evolution is related to other temporal scales than present-day ecological processes.

As operational units species are related to human perception and the definition of taxa. This is indicated by the fact that taxa undergo permanent modification due to increasing knowledge and access to molecular methods. Although using species is a widely used approach when working with biota, they may be not the best tool for biodiversity investigations when ecosystem functioning is the subject under investigation. If we accept species as a concept and not as a reality there is no problem with this, but if we integrate this construct into units of a higher hierarchy, as is done in diversity experiments, an awareness of the concept of species is required, it being merely a very successful and pragmatic approach, which levels out a large part of the diversity within and between species.

Hypotheses about community performance have to consider site-dependent limits in the supply of energy, water and nutrients, too. Site-specific constraints, which cannot be exceeded, have to be taken into consideration. Thus, knowledge, for instance on the carrying capacity of a stand, is needed as an extensive variable, which is not dependent on fixed species traits (Fig. 3). If this value can be exhausted by one single species, e.g. during competitive-exclusion experiments, then mixtures and biodiversity are irrelevant. If intrinsic mechanisms within the community are able to slightly modify the site and increase its capacity, this is a question of temporal scale and no longer within the scope of the investigated system.

The shortcomings of various approaches used to identify repercussions of changes of species diversity can only be superficially touched upon here. However, the arguments needed to develop reductionistic models and experiments with defined conditions and environmental interactions are clear, as are the restrictions of such approaches. Theoretical considerations and experiments are mostly based on unrealistic simplifications. Nevertheless,

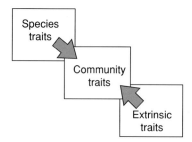

Fig. 3. Assembly experiments in biodiversity research focus on the community level. When aiming to explain the community performance at different levels of diversity, the discussion concentrates on the species level (composition, complementarity, dominance, selection, functional attributes, niches). However, some inconsistent results can be put down to external variables such as plot size, nutrient availability or climatic site conditions

they are crucial for the development of knowledge as they concentrate on specific arguments, assumptions or aspects of interest and thus will provoke a willingness to identify the limits of their validity. This will deepen and extend our fundamental theories.

3 Hypotheses and Concepts Addressing Biodiversity and Ecosystem Functioning

The purest hypothesis addressing the relation between species diversity and ecosystem functioning is based on theoretical niche occupation. It states that each species plays a specific additive functional role and no two species share exactly the same function within a community. High species diversity then would be associated with efficient use of resources and high performance of the whole system. However, this assumption mixes evolutionary and current processes. Ehrlich and Ehrlich (1981) compare the species of an ecosystem with the rivets of an airplane. Each one has a function, and the security of the plane will be reduced if one after the other is lost. Even if this does not directly lead to a crash, however, the risk of severe damage increases with the loss of units (rivet hypothesis). The focus of the rivet hypothesis is on saturated systems with large populations. In contrast to this, many examples of communities exist where several ruderal species occupy the same niche with few individuals without any interference. In addition, ecological niches are multidimensional. When looking at ecosystem functions, mostly only one parameter, such as biomass production, is selected. Coexisting species are in most cases to some degree complementary (Hector 1998) but not necessarily so according to the parameter that is selected as a dependent variable. Just recording the number of species ignores the individual importance of different species. The taxonomic relation between species (e.g. their affiliation to a genus or family) does not necessarily provide information about their functional relatedness or resemblance. However, in many cases, traits are restricted to a limited set of phylogenetically related taxa (Reich et al. 2003).

When classifying types according to specific ecosystem functions, this leads to the concept of functional redundancy. Here, complementarity is related mainly to functional groups and not to single species. The occurrence of various groups then is important and the diversity within these groups contributes less or minimally to ecosystem performance. Species within one group are redundant (Walker 1992; Gitay et al. 1996). They are able to replace each other without a decline of ecosystem functioning. Reactions of the system occur only if one group is lost or strongly reduced. A high diversity of functional groups or types is considered to be decisive (Diaz and Cabido 2001).

The next approach is to look more closely at single species, which may be well adapted, powerful, competitive or highly efficient in their use of resources. In the long term, this could be the successful K-selected species, but such important species can also be identified in a short time period. Many communities are regularly dominated by certain species, which control overall functioning (Smith and Knapp 2003). In the first biodiversity experiments, such observations were ignored to some degree. Bias was avoided and all species were regarded as equally important. However, it is just a question of statistics that the probability of occurrence of certain powerful (e.g. highly productive) species will increase with species richness in stochastic mixtures of species (Aarssen 1997; Huston 1997; Loreau 1998b; Wardle 1999; Deutschman 2001). Huston (1997) calls this the "sampling effect". It corresponds to the "selection probability effect" (see Van der Heijden et al. 1998; Wardle 1999). If this was true, a tendency for an increase in productivity with species richness could be related to the occurrence of single species which tend to be lacking in species-poor mixtures. They represent only one of many monocultures. Mean values obtained from various monocultures will not make the potential of single species obvious. After the first debate on this topic, statistical methods were developed to separate complementarity effects from sampling effects (Spaèková and Lepš 2001; Loreau and Hector 2001).

Up until now, all hypotheses concentrate on contributions made by species either via complementarity, redundancy or species-specific characteristics. When ecosystem functioning is related to ecological complexity, we additionally have to integrate direct interactions between species such as competition or facilitation (Cardinale et al. 2002; Bruno et al. 2003). If such interactions take place, and this is highly probable, we can expect effects which are related to the occurrence of certain pairs or combinations of species that cannot be explained just by species numbers or by the presence of specific traits. If system behaviour is related to species number, it may give the impression of being chaotic ("idiosyncratic response effect") (Lawton and Brown 1993). Such communities are sensitive to and react to the addition or loss of certain interacting groups of organisms.

Finally, biodiversity can be considered to have just a potential value. According to findings on certain ecosystem functions, many species may behave as if redundant today. They can be rare and quantitatively unimportant. Nevertheless, the possibility exists that the dominant and successful species in recent communities could be challenged by novel environmental conditions. In the case of climate change, there are the first hints that temperature increase might favour insect outbreaks in monodominant forests. If there are no other species that could replace the formerly dominant tree species, this leads inevitably to a crash of the system. Redundancy can be considered to contribute to stability or "reliability" when environmental conditions are changing (Naeem and Li 1997; Naeem 1998). The theoretical concept is called the "insurance hypothesis" (Yachi and Loreau 1999). It integrates aspects of redundancy and sampling effects. The higher the number of species with comparable requirements and functional traits, the higher is the probability that environmental hazards will not necessarily cause a loss of the community and its contribution to ecological functioning (Chapin et al. 1998). Higher resilience of diverse communities is predicted (Naeem and Li 1997; Peterson et al. 1998; White and Jentsch 2004). In the face of global climate change, some authors expect that diverse ecosystems will have the capacity to shift dominances within certain communities and to better adapt to novel conditions (Peters 1994).

Nevertheless, the effects of rapid changes in ecosystems are ambiguous (Van der Maarel 1993; Mackey and Currie 2000). Nonlinear interactions are highly probable. Such disruptions (e.g. drought) may be buffered by diversity but they will impact diversity and functioning themselves. For many systems it is not clear which intensity or magnitude of environmental change will cause irreversible effects (Sprugel 1991; Fay et al. 2000). The debate remains controversial in this respect. Models seem to confirm the theory (Yachi and Loreau 1999; Fonseca and Ganade 2001; Norberg et al. 2001), but are strongly simplified in comparison with nature. However, it is evident that biodiversity or other biotic aspects can influence the behaviour of a system to a certain degree (Hooper et al. 2005). There are thresholds and strong physico-chemical influences that render biotic processes marginal. Without considering the abiotic compartments and fluxes, biodiversity experiments will remain as reduced artificial constructs and perhaps irrelevant.

4 Experiments in Community Ecology and Biodiversity Research

4.1 The Role and the Value of Experiments

Methodological concepts of natural science differentiate between theoretical considerations, hypotheses and models on the one hand and field work, measurements and observations on the other hand. Experiments are placed somewhere in between, as they closely relate to theoretical considerations but apply and test these with the support of real objects. The gap between statistical requirements, such as replication or randomisation, and reality is a fundamental one. Experience, recognition and observation are important scientific skills. However, if potential problems are identified that might become important in the future, field work and the monitoring of developments is not a very helpful approach. It has only limited predictive power. Processes and effects will not be detected until they occur. Models are able to make significant contributions, but they share the restriction that they simply cannot be better than the knowledge that is entered into them. The problem with ongoing global changes and with biodiversity loss is that no comparable development has ever been recorded.

Experiments may contribute to a simulation of future conditions of ecological communities. As we do not know how systems will behave under novel circumstances (e.g. at higher temperatures, with less species), experiments are always not only reductionistic but also actualistic. A decisive advantage of the experimental approach is that abiotic site conditions are controlled. This limits the "noise" of other processes. The value of experiments and their explanatory power depends strongly on the rigour of their experimental design. It has to integrate controls, replicates, random distributions and interspersion (Hurlbert 1984). Theoretical and conceptual problems and suggestions regarding the design of biodiversity experiments have been extensively discussed (e.g. Allison 1999; Naeem 2002a; Schmid et al. 2002a). Non-adequate design can easily lead to systematic errors that cannot be erased later on. Efficient experiments aim at minimizing the variance of the SE in the statistical analyses, but this requires good level of knowledge about plant traits and their contribution to heterogeneity via specific spatial and temporal performance. This knowledge is mostly implicit and based on experience, and only rarely on hard data. On the other hand, special care has to be taken to avoid effects that are attributed to certain species (Benedetti-Cecchi 2004).

However, with increasing complexity of the experiment, the number of plots becomes very large (Table 1) if the experiment is based on a random draw of species in a mixture from a given pool (Huston and McBride 2002). The required surface area becomes large and the workload to maintain it becomes high.

Many ecological experiments concentrate on rather limited research questions, such as autecological or ecophysiological questions or on competition, and investigate them on the basis of simple and species-poor experimental setups (Scheiner and Gurevitch 1993; Underwood 1997). Experience from competition experiments is helpful for diversity studies, but limits are obvious according to methodological transfer. The quality of an experimental analysis depends on whether the effective variables are really considered. In complex experiments, the key variables are often superimposed by other factors that have not been considered in the experimental design. Huston (1997) criticises experimental approaches, which ignore "hidden treatments" in species assemblages. Being confronted with this problem in biodiversity experiments contributed enormously to improving experimental methodology.

4.2 Historical Experiments

The first documented experiment which considered the importance of species diversity in ecological communities was set up at Woburn Abbey in Bedfordshire, England, at the beginning of the nineteenth century. This

Table 1. Number of possible treatments in biodiversity experiments [five types and a maximum (*Max.*) of six diversity levels]. Statistical analyses would require three replicates and thus increase the numbers threefold. *Level 1* Monocultures

Max.	Level of species diversity					
	1	2	4	8	16	32
32	32	496	35,960	10,518,300	601,080,390	1
16	16	120	1,820	12,870	1	–
8	8	28	70	1	–	–
4	4	12	1	–	–	–
2	2	1	–	–	–	–

experimental garden was documented for the first time in 1816 (Hector and Hooper 2002). In addition to a varying diversity of species, different substrates were also applied. In total, 242 plots, each 4 square feet, were set up. Even though this experiment had several obvious methodological shortcomings (e.g. species-poor stands and monocultures were sown and diverse plots were transplanted from nature), it remains impressive. Obviously, the aim was to find out the role of diversity and under which circumstances diversity is important. This question, which was apparently new to ecologists at the end of the twentieth century, had clearly already been addressed almost 200 years earlier.

This historic experiment is indirectly referred to by Charles Darwin in his work The Origin of Species, when he says "it has been experimentally proven that, if a plot of ground be sown with one species of grass, and a similar plot be sown with several distinct genera of grasses, a greater number of plants and a greater weight of dry herbage can thus be raised" (after Hector and Hooper 2002). The first part of Darwin's conclusion can hardly be proven, as in these communities of mostly clonal grasses apparently only the number of tillers and not the number of plants was meant by "number". The second part anticipates the recent discussion on the relationship between species diversity and productivity of a stand. Once again, the discovery of this old treatment reveals the importance of bibliographic sources. During the twentieth century, diversity experiments were rarely applied and when they were, focussed on the effects *on* diversity and not *of* diversity.

4.3 Modern Experiments

Since the formulation of essential research needs regarding the relationship between diversity and ecosystem functioning (Schulze and Mooney 1993), numerous experiments have been designed, performed and analysed in different systems, using different designs, on different spatial and temporal scales and regarding different ecosystem functions, during the last decade. Such concerted activity is unprecedented in ecology. Beginning in the early 1990s, research groups started with mesocosms, glasshouses and chamber experiments to exclude external influences due to small-scale differences in site conditions (Naeem et al. 1996; Symstad et al. 1998). The ECOTRON experiment (Naeem et al. 1994; Lawton et al. 1998) at the Centre for Population Biology in Silwood close to London aimed to investigate the importance of ecological complexity and not only species diversity. In isolated chambers, where the water, light, nutrient, wind and temperature regime was controlled, mixtures of two, five and 16 plant species were established. In addition, a diversity gradient of invertebrates simulated diversity at higher trophic levels. In total, 31 species (among them 15 invertebrates) were cultivated in 16 chambers. The numbers of replicates per level of diversity differed. The fact that the replicates were composed of exactly the same species combinations was criticized as producing pseudoreplicates (Hurlbert 1984, 2004). In such cases, diversity effects and effects of certain combinations of species or of species identity cannot be separated (but see Oksanen 2001). Field experiments with model ecosystems are an efficient tool with which to simulate biodiversity loss under semi-natural conditions (Lawton 1995; Tilman et al. 1997b; Hooper and Vitousek 1997; Schläpfer and Schmid 1999). Results from these experiments have already been subject to extensive reviews (e.g. Schläpfer and Schmid 1999; Schmid et al. 2002a; Tilman et al. 2002b; Spehn et al. 2005).

However, such experiments in the field are associated with many restrictions (Lamont 1995; Grime 1997; Huston 1997; Allison 1999). They are more representative of nature than microcosms or mesocosms in the laboratory or greenhouse. Nevertheless, they can hardly be compared with natural communities even if their diversity resembles one in nature, because their composition of species has to ignore natural combinations due to statistical requirements. Although their environment (water and nutrient availability, climate) is strongly standardized and small-scale differences are not to be expected, it remains difficult to link the processes and data that are recorded to the diversity gradient. Uncontrolled influences, such as small-scale differences in soil biota and nutrient status, cannot be avoided completely. If the soil remains undisturbed, it is difficult to measure fluxes of gases and liquids. In any case, the site conditions have to be manipulated and this creates deviations from conditions in natural grasslands. Pre-experimental site preparation includes the application of herbicides (e.g. methyl bromide) in some cases, in order to reduce unintentional contributions by the soil seed bank. Herbicides can disrupt microbial activities in the upper soil horizons and even eliminate soil biota.

Technical problems have to be understood. In contrast to greenhouse research, field experiments are exposed to unplanned extreme events, such as heavy rainfall, drought, or to unintended biotic impacts, such as herbivory. If the whole experiment is affected in the same way, non-linearity may be integrated as part of the work; if this is not the case, the entire experiment may be lost. The direct simulation of the loss of diversity by the removal of species (e.g. Hobbie et al. 1999; Symstad and Tilman 2001; Berlow et al. 2003; Diaz et al. 2003; Bret-Harte et al. 2004) is accompanied by direct effects of the manipulation itself. Root decomposition and nutrient fluxes are affected (McLellan et al. 1995; Wardle et al. 1999). In this case, the artefact of increased short-term production would respond a reduced functioning of the system.

The removal is accompanied by either mechanical disturbances or by the application of toxic compounds. We can better differentiate between controlled diversity experiments (with installed communities under controlled and comparable site conditions) and biodiversity manipulation experiments (e.g. exclusion of species from natural communities or artificial increase of diversity). The terminology here is not always clear (Hector et al. 2002b). The quality of seeds gave rise to problems in many biodiversity experiments. Limited availability of large numbers of seeds from natural varieties is a consequence of the low demand for and limited longevity of seeds. They are economically unimportant. On the other hand, cultivars which can be obtained from commercial traders, cannot be used in experiments which are aimed at simulating natural grasslands. Harvesting of seeds is time consuming and need not be carried out by highly skilled scientists. However, distinguishing sub-species, identifying adequate ecotypes and even distinguishing species (e.g. in the case of many grasses) is not a trivial matter. Some experiments (Cedar Creek, BIODEPTH) partly suffered due to single seed portions that were contaminated with plants from other origins.

The largest portion of field experiments focuses on herbaceous plant communities on a plot scale ranging from 0.5 m×0.5 m up to 20 m×20 m and with a regular duration of 1–3 years. In some cases, this timeframe is exceeded, e.g. at the Cedar Creek field experiment in Minnesota (Tilman et al. 2001), or at some of the BIODEPTH sites (Pfisterer et al. 2004; Nesshöver et al., submitted). Some experiments are undertaken for even shorter time scales (Garnier et al. 1997; Fridley 2002, 2003). Only recently have experiments been designed and established that manipulate tree species diversity and thus cover a longer time span (Scherer-Lorenzen et al. 2005a). In this article, we focus on results from the most comprehensive grasslands experiments, Cedar Creek (e.g. Knops et al. 1999, 2001; Tilman et al. 1996, 1997a, 2001, 2002a) and BIODEPTH (e.g. Hector et al. 1999, 2002b; Spehn et al. 2005).

The Cedar Creek biodiversity experiment in Minnesota started in 1993. It had a strong impact on the scientific debate. It was set up within the Cedar Creek Natural History Area on a nutrient-poor sandy soil. The topsoil was removed from a surface area of 9 ha and a strong herbicide applied to avoid germination of the existing soil seed bank. The soil was superficially mixed and seeds were sown at a standardized number of seeds per plot in a stochastic design in 168 out of 342 available plots, 9×9 m each. The diversity gradient ranged from one (39 plots), to two (35), to four (29), to eight (30), to 16 (35) species. All species were cultivated in monocultures as well. Species mixtures were defined on the basis of a random draw from a pool of 18 selected species. They reflect the composition of natural grasslands nearby and represent various functional groups such as C3 grasses, C4 grasses, legumes, herbs and woody plants. The plots were continuously weeded to make sure that no additional species could invade the plots. The fulfilling of basic statistical requirements and the high number of plots and replicates were the major innovative contributions of this experiment.

In 1996, the BIODEPTH project was launched (Biodiversity and Ecosystem Processes in Terrestrial Herbaceous Ecosystems) (e.g. Diemer et al. 1997; Hector et al. 1999; Minns et al. 2001; Hector 2002; Spehn et al. 2005). The project was run for a 3-year period. It made reference to the Cedar Creek experiment. For instance, the same diversity gradient was installed. However, at some sites, the highest level of 16 species could not be reached. On the other hand, an additional 32-species level experiment was installed in Lupsingen, Switzerland. Its novel contribution was that this multi-site experiment applied the same experimental design to eight sites following biogeographical gradients across Europe, from Portugal to northern Sweden and from Ireland to Greece (Hector 2002; Hector et al. 2002b). So, the experiment was carried out with the same design in each country, but under very different environmental conditions across Europe. However, regional species pools had to be considered. Of course, it was impossible to use the same species everywhere. The plots (2×2 m) were weeded during the 3 years of the experiment. The questions which can be answered with this approach, are whether diversity effects are site-specific and arise only when certain environmental conditions are met or if they are a common phenomenon. On the BIODEPTH sites in Switzerland and Germany, the investigation of the field plots was continued after the end of the project in 1999, leading to additional insights into the relationship between diversity and productivity. At both sites, the maintenance of the original diversity gradient was stopped and species not present in the plots were able to invade giving some idea about the stability of species compositions and diversity patterns (Pfisterer et al. 2004; Nesshöver 2004).

In 2002, a new experiment was installed in Jena, Germany, which also focuses on grasslands (Roscher et al. 2004). Ninety plots (20×20 m) were established. Diversity levels followed the Cedar Creek and BIODEPTH experiment (one, two, four, eight, and 16 species) with additional, extremely diverse 60-species plots being composed by stochastic mixtures. The species pool of 60 species adapted to moist meadows was classified according to morphological, phenological and physiological traits and four functional groups (grasses, small herbs, tall herbs, and legumes) were distinguished. The experimental design is based on four blocks to eliminate soil heterogeneity within this rather large area. First results are available, but the experiment is designed to continue for 10 years.

Many diversity experiments are based on rather simple communities with short-lived (or at least assumed to be) species. Results are obtained within several months (e.g. Garnier et al. 1997) or within the first years. However, for many herbaceous species, their optimal performance in time is not known. The experiments assume that this is achieved when the plants are established. Experiments are set up in pots or on artificial substrate (e.g. Sheffield in the BIODEPTH experiment). Other sites used widely undisturbed developed soil. To reduce weeding effort either steam sterilization or the application of methyl bromide is carried out. The latter may have effects on soil organisms such as fungi or bacteria. These artificial environments have often been criticized (Grime 1997; Huston 1997; Fridley 2002), but their artificial nature goes along with the advantage of delivering well-known and, first of all, comparable site conditions. This allows for replicates and comparisons.

Many experiments are based on grassland communities. Grasslands can easily be established in a short time and their structure is simple compared to other perennial communities such as forests. Grasslands have a rather constant performance in time (compared with fields and ruderal communities). According to the disturbance regime, most experiments simulate mowing, because this allows a more precise experimental approach and aboveground biomass can be harvested at defined stages. Pastures and grazing would afford much more spatial and temporal heterogeneity in biomass, trampling and local nutrient hotspots. Grasslands share the advantage of being spatially as well as economically important. Under continental climates, they represent the natural vegetation. In more oceanic climates grasslands are a major land use type. Besides being a food supply for ruminants, grasslands are also relevant for water cycles and atmospheric gas fluxes. There are many indices showing that the diversity of European grasslands has declined during the twentieth century. Technical advances – especially physical (drainage) and chemical (fertilization) improvements of the sites as well as efficient mowing facilities – are responsible for this development.

5 Advances and Frontiers – Insights into Mechanisms and Processes

5.1 The Biodiversity–Productivity Relationship

One major focus of modern diversity experiments has been on the relationship between species richness and measurable abiotic ecosystem functions, especially aboveground productivity. The BIODEPTH plots show in general a positive logarithmic relationship between species richness as well as functional richness (number of plant functional types) and aboveground productivity (Hector et al. 1999; Spehn et al. 2005). The strength of this relationship varied between sites as well as between the years of observation within the 3-year period of measurements (Spehn et al. 2005). Partitioning this effect into a complementarity and a sampling component showed that species richness complementarity was significant and positive at most sites (Loreau and Hector 2001) and increased over time (Spehn et al. 2005). The sampling effect was more variable between sites and within time. One important factor for the species richness–productivity relationship at some sites was the presence of legumes in the seed mixture and their positive influence on productivity (Spehn et al. 2002). Besides high productivity of the legume species themselves, a complementarity effect between legumes and other species was also detected (Scherer-Lorenzen 1999; Spehn et al. 2002). At the German BIODEPTH site, this effect was strongest and even continued for 5 years after weeding had been stopped (Fig. 4a; Nesshöver et al., in review). Since legume abundance strongly decreased over time (see the graph of *Trifolium pratense* data in Fig. 4b), this effect was driven by the facilitation of other species due to the early legume development. For other ecosystem functions, such as light use and aboveground space use, similar effects were found indicating greater resource use by niche differentiation (Spehn et al. 2000, 2005). The experiments at Cedar Creek showed similar patterns in the relationship between diversity and productivity (Tilman et al. 1996, 1997a, 2001, 2002a). There, it was shown that the diversity effect increased over time. Complementarity effects were believed to be responsible for this (Tilman et al. 2001). It was found that few species or functional groups play a major role in this process, although an additional influence of richness remained (Lambers et al. 2004).

In contrast, various publications support the low importance of facilitation and complementarity or even no effect of diversity on productivity at all (Kenkel et al. 2000). An experiment carried out at different sites across Europe, where different seed mixtures were added to the natural secondary succession, showed no consistent effect of richness on productivity during 3 years (Van der Putten et al. 2000; Lepš et al. 2001). The response was mainly triggered by single species and their traits. A short-term experiment by Fridley (2002, 2003) suggests that resource availability may influence productivity more than species richness. This is not surprising, as the experimental communities were mainly composed of annual grasses and herbs. The species composition of communities and the presence of single species were more important than richness per se. This experiment was carried out with very small raised-bed plots (50×50cm). Edge effects make facilitative

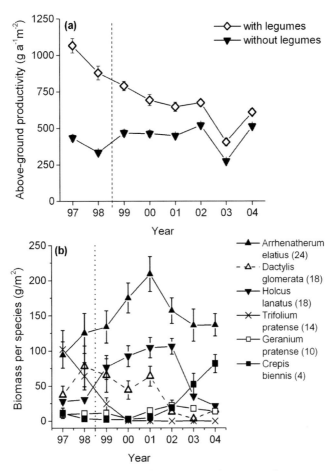

Fig. 4a, b. Development of above ground biomass production in the BIODEPTH experimental field site in Bayreuth, Germany. The *vertical dotted line* indicates cessation of weeding after the BIODEPTH experiment terminated. Species were then able to immigrate into plots in which they had not been sown. **a** Total above ground productivity per year separated by plots with and without legumes in the seed mixtures, **b** mean biomass production of single species in summer harvest. Numbers *in parentheses* indicate the number of plots in which the species were present in 1997 and 1998; after that, the species shown also immigrated into other plots. See text for further explanation and also Nesshöver (2004), Nesshöver et al. (in review). Means +/−SE are shown

interactions between species less probable. Only ten species were used. Polley et al. (2003) found negative complementarity and selection effects in an experiment with annual grasses and herbs on small plots of 1×1 m. However, the short-term development of annual species on small plots can hardly be taken as a proof for general mechanisms in vegetation. It is strongly influenced by resource availability and founder effects.

When multi-species assemblages are compared to species-poor communities or even monocultures, it has to be noted that it makes a difference whether the mean values for monocultures or the maximum yield of the most successful monoculture are taken (Garnier et al. 1997; Huston et al. 2000). To test the second aspect would require monocultures of all the species that are used in the mixtures available. This is not the case in many experiments (e.g. BIODEPTH). Garnier et al. (1997) conclude that there are only rare cases where superior productivity of multi-species assemblages as compared to monocultures has been clearly shown.

If mixtures show higher productivity than species-poor stands or monocultures, this is called "overyielding" (Hector 1998; Loreau 1998b). Pretzsch (2005) analyses forest experiments with tree species mixtures and finds that overyielding can occur if early and late successional species are combined or if the trees are considerably different in their light use strategy or seasonal culmination (e.g. *Larix decidua* and *Picea abies*). Mixtures of such species can yield higher biomass than pure stands. However, most mixtures do not reach the yield of the best performing species. When species were competing for the same resources, total yield was reduced by up to 30%. Several analyses on overyielding in grassland experiments were performed (e.g. Hector et al. 2002a; Hooper and Dukes 2004; Roscher et al. 2005). Hector et al. (2002a), analysing the BIODEPTH results, found a low explanatory power of the overyielding effect based on the monocultures of the most productive species (but see Drake 2003). Hooper and Dukes (2004) analysed the results from an 8-year experiment in serpentine grassland and find that overyielding effects got stronger with time and varied with the functional characteristics of the included species and the abiotic conditions. Combining two different sets of experimental plots on different spatial scales in the Jena experiment, Roscher et al. (2005) also found an overyielding effect mainly caused by complementary effects. Lambers et al. (2004) point out that both positive and negative interspecific interactions determine whether species-specific overyielding occurs.

To summarize the discussion on the diversity–productivity relationship, apparently two sets of mechanisms, sampling and competition on the one hand, and niche complementarity and facilitation on the other hand, are effective at varying levels of importance in most ecosystems. Their relative importance may vary depending on the life stages of individuals and populations (Walker and Vitousek 1991; Tilman and Lehman 2002), physiological attributes (Callaway et al. 1996; Holmgren et al. 1997) and abiotic stressors (Bertness and Callaway 1994). It has further become apparent that species richness per se is not the main driver behind the diversity effects, but rather a combination of different diversity parameters, including the composition of the communities and the presence of single functionally important species (Hooper et al. 2005). Ecosystem engineers (Jones et al. 1994) deliver improved conditions to other species such as shading, higher nitrogen supply or favourable soil structures. Various studies indicate that the productivity of diverse grasslands depends more on the occurrence (not their abundance or quantitative share) of *certain* functional types than on their diversity (Hooper and Vitousek 1997; Hooper 1998; Scherer-Lorenzen 1999; Nesshöver 2004). Our results on the lasting effect of originally sown legumes on productivity are a good example of this (Fig. 4a; Nesshöver et al., submitted) So, the question is, whether comparisons with stochastic combinations of functional types or species are adequate, or if the best-adapted type should be used as a standard? The degree of overyielding is strongly dependent on the specific traits of the species used in mixtures, which is in contrast to the Grime's "mass-ratio" hypothesis (1998).

5.2 Plant Species Diversity, Invasibility and Community Dynamics

In the context of biodiversity loss, the question that is raised is: does enhanced diversity stabilize community structure and thus hinder the invasion of alien species (e.g. Levine and D'Antonio 1999)? The resistance against invasion becomes increasingly important due to efficient dispersal vectors. This topic is even more complex than the biodiversity–productivity discussion. Not only the qualities of the invaded communities (of different diversity) but also those of the invading organisms or populations have to be considered. Generally, invasion, or more general immigration, is an indication of species turnover. This mechanism is also important for native species within the community, but there it is less obvious. Mouquet et al. (2004) have experimentally investigated the contribution of seed rain to biodiversity and found that it affects communities differently depending on the competitive abilities of the present species. The authors suggest that the mechanisms which control species dynamics in grasslands are a mixture of colonization–extinction processes and a competitive weighted lottery.

It has been documented by various authors that species diversity seems to control the invasibility of communities. Most experimental approaches found a negative relationship between species and/or functional diversity and invasion (e.g. Tilman 1997; Knops et al. 1999; Naeem et al. 2000b; Prieur-Richard et al. 2000; Symstad 2000; Kennedy et al. 2002; Fargione et al. 2003; Pfisterer et al. 2004). Species-rich communities are supposed to possess a higher resistance against invasion and to be more stable in their composition (Sankaran and McNaughton 1999; Hector et al. 2001a; Kennedy et al. 2002). But results of other studies diverged from these findings (e.g. Robinson et al. 1995; Palmer and Maurer 1997; Crawley et al. 1999; Lavorel et al. 1999). In general, experimental studies tend to find negative correlations, while observational studies more often find no significant or positive correlations, between diversity and invasibility (reviews in Prieur-Richard and Lavorel 2000; see also Hector et al. 2001a; Meiners et al. 2004). Fox (1987) formulated the assembly rule, that a decrease or lack of certain functional groups will facilitate the invasion of species which belong to this specific group. Given a certain pool of species and functional types there should be a correlation between functional diversity and invasibility. Symstad (2000) and Dukes (2001a) found negative relationships between the functional diversity of experimental grassland plots and the success of an invasive annual weed.

Some experimental findings indicate that species composition (certain species-specific traits or interactions) determines invasibility more strongly than richness per se (Crawley et al. 1999). There, species identity matters more than richness in determining the number and the biomass of invaders. Other case studies point at the role of dominance patterns (here of C4 grasses) for the control of invasion by exotic plants rather than at the diversity of the stands (Smith et al. 2004). The latter would indicate that sampling effects might conceal diversity effects. With the experiment's duration, with increasing invasion and decreasing dissimilarity between plots, the diversity effects (complementarity) on productivity tend to become less significant (Pfisterer et al. 2004).

Hector et al. (2001a) report the influence of diversity on invasion by analysing the number of species and individuals weeded out during the experiment. They found a clear, positive relationship between number of sown species and invasion resistance. Confounding effects appear due to the fact that the weeding of species-poor communities causes higher disturbances than weeding of species-rich ones. During long-term observation on the Swiss BIODEPTH site, the number of species invading after the end of the weeding period showed a negative log-linear relationship with the numbers of sown species (Pfisterer et al. 2004). The results from Bayreuth showed the same pattern (Nesshöver 2004). We additionally analysed whether this effect was triggered by species richness or rather by compositional effects. Since the most productive species (*Arrhenatherum elatius* and *Holcus lanatus*) had a strong influence on the biomass of invaders, selection effects have to be considered (Fig. 4b). Similar trends were found by Crawley et al. (1999), Van der Putten et al. (2000) and Van Ruijven et al. (2003). In this respect, Wardle (2001a) criticises studies which find a positive relationship between species richness and invasion resistance. He states that no study was able to separate the richness effect from effects triggered by single species. In addition to this, individual immigrating species might react differently towards species richness of communities (Meiners et al. 2004).

The importance of richness versus compositional effects in this context may further be altered by the influence of environment and management. The German BIODEPTH site was relatively nutrient-rich, although the plots were not fertilized (Scherer-Lorenzen 1999). According to the hump-shaped model (Grime 1973), the dominance of single species was relatively strong (Nesshöver 2004) and thus influenced invasion. Davis and Pelsor (2001) also find that resource availability affects competition intensity and that this restricts invasibility. These authors point at the role of history when discussing invasion processes. Short-term stochastic disturbances may locally reduce competition and allow for invasion (White and Jentsch 2001). Van der Putten et al. (2000) detected in their multi-site comparison across Europe a dependence of invasion success on the overall productivity.

Invasion and invasibility have to be considered as source and sink models. But again, experiments which integrate differences in diversity at the multi-patch scale have not yet been set up. Up until now, only a few experiments have been carried out that manipulate the genetic diversity within species of recipient plant communities. Weltzin et al. (2003) could not find an effect of genotypic richness of *Arabidopsis thaliana* on invasion success, emergence and survivorship of *Arabidopsis suecica*.

Some studies have been conducted which analyse the influence of plant species diversity on the diversity, performance and biomass of other trophic levels (Mulder et al. 1999; Knops et al. 1999; Koricheva et al. 2000; Naeem 2002b; Holt and Loreau 2002). Andow (1991) finds a lower probability that specialised herbivores are successful in diverse stands. In addition, predators of herbivores are supported. In contrast to this, Hanley (2004) could not identify effects of plant species diversity on the herbivory of seedlings by molluscs in experimental grasslands which compared three, six, and 12 species mixtures of seedlings. There seemed to be even a slight increase in seedling mortality with increasing diversity. Obviously, facilitating effects are unimportant in the seedling phase. Using field experiments in moorlands and the ECOTRON facility, Hartley and Jones (2003) conclude that the effects of climate change on diversity and indirect effects on herbivores are likely to be strongly site specific. Initial community structures and dominant species are important. It is most likely that belowground factors such as the composition of soil biota are more important than species diversity. On the other hand, structures that provide microclimatic shelter can reduce insolation and deliver wind protection and this can improve the water use efficiency during photosynthesis (Vandermeer 1989). The better regulation of fungal plant diseases with increasing diversity was documented by Mitchell et al. (2002). However, again, species richness correlates with biomass, and thus, a direct influence is hard to prove.

5.3 Influences of Species Diversity on Element Cycling

Plant species diversity is also proclaimed to contribute to a higher efficiency in resource use in the rhizosphere (Tilman et al. 1997b; Loreau 1998a). Tilman et al. (1996) report elevated nitrogen uptake from soil in diverse stands via complementary resource use. This was confirmed in various investigations (see Schläpfer and Schmid 1999). Hooper and Vitousek (1998) found a better use of resources and increased nitrogen and phosphorous uptake at higher levels of species diversity but no effect on nitrate leaching. Only when nitrogen fixers were present did nitrate leaching occur. Scherer-Lorenzen (1999) and Scherer-Lorenzen et al. (2003) show, based on the experimental plots of the BIODEPTH experiment in Bayreuth, Germany, that stands of low diversity may intensify nitrate leaching without being fertilized. Enhanced values were mainly found in soil depths below 15 cm, hinting at the possible impacts on ground water. Even in the absence of legumes, at certain times critical concentrations of nitrate occurred in species-poor communities.

A decrease in unconsumed soil nitrate with increasing richness was found in Cedar Creek (Tilman et al. 1996, 2002b). However, the latter effect could be attributed to the presence of certain functional groups (legumes and C4 grasses) and not to species richness per se. Knops et al. (2002) conclude in a review of plant species' impacts on nitrogen cycling that in the short-term no effect of plant species richness on nitrogen cycling can be expected, since the main driver is the microbial community within the system. Impacts of single species can be found, if they alter nitrogen inputs and losses directly (e.g. legumes).

However, the image we have of the diversity–function relationship is rather hazy, when looking at belowground processes (Naeem et al. 2000a; Wardle et al. 2000a; Van der Heijden and Cornelissen 2002; Wardle and Van der Putten 2002). The influence of legumes on the nitrogen pool may differ between sites and alter the relationship between diversity and soil nitrogen. At some BIODEPTH sites, fixed nitrogen was transferred to non-legume species (Mulder et al. 2002; Spehn et al. 2002), and thus, diverse communities could profit from increased nitrogen availability. Plots without legumes showed no coherent reaction of diversity to soil nitrogen (Scherer-Lorenzen et al. 2003). Decomposition of cotton strips and wood was found to be generally unaffected by species richness and only slightly affected by functional group richness (Spehn et al. 2005). In addition to plant diversity, small-scale site conditions, such as microclimate, are decisive for leaf litter decomposition (Hector et al. 2000a).

Zaller and Arnone (1999) report a linear decrease in lumbricid populations together with decreasing plant species richness. In contrast, Wardle et al. (1997, 1999) could not find a positive effect of species diversity such as enhanced activity of decomposers (Collembola, lumbricids). Stephan et al. (2000) detected positive effects of plant diversity on bacterial diversity. Zak et al. (2003) could experimentally prove significant increases in microbial community biomass, respiration and fungal abundances with increasing plant species diversity, but could not separate biomass from diversity effects.

It is assumed that niche complementarity will support the efficient use of resources and thus enhance productivity in diverse communities (Tilman and Downing 1994; Hooper and Vitousek 1997; Hector 1998; Dukes 2001b). Excluding legumes, Van Ruijven and Berendse (2003, 2005) experimentally demonstrated complementarity effects. They attribute this effect to increased nutrient use efficiency in species-rich stands. This can be explained by species-specific root morphology (Kutschera et al. 1992a, 1992b), rooting depth (Sullivan and Zedler 1999) and turn-over as well as by physiological complementarity in the use of the variety of chemical molecules and linkages. Structural heterogeneity produces micro-hotspots (nutrient patches) in soils, which – depending on symbiotic partners – some plant species are able to utilize and others are not. Plants differ in their capability to release protons, chelates or other compounds. This may explain coexistence and complementary resource use. Nutrient uptake by plants is correlated with the biodiversity of mycorrhizal fungi in grasslands (Van der Heijden et al. 1998). Hyphae of mycorrhizal fungi have access to smaller soil pores and the nutrients therein. The diversity of mycorrhizal fungi is in turn correlated with plant species diversity (Van der Heijden et al. 1998). Thus, indirect effects in carbon and nitrogen turnover in the soil can be expected (Hooper and Vitousek 1998). Bardgett and Shine (1999) point to the role of plant litter diversity in microbial activity and biomass in temperate grasslands. The significance of tree species diversity for litter quality and decomposition is reviewed by Hättenschwiler (2005). Only few studies focus on diversity effects, and here results are inconsistent. There is some indication of specific effects of chemical compounds, which are more likely to occur in mixtures, but research is still in its infancy in this area.

6 Developments in Methods and Approaches

6.1 Functional Types, Traits and Attributes

A major point that is raised in the context of biodiversity–ecosystem functioning experiments is, whether diversity should be treated at the species level, or on the basis of functional characteristics (Lavorel and Garnier 2002; Hooper et al. 2002). The common application of functional characteristics is to identify plant functional types or functional groups (see Gitay and Noble 1997; Westoby and Leishman 1997; Woodward and Kelly 1997) as groups of species which have a similar response to disturbance or a similar effect on ecosystem properties. Functional types are considered to be a promising approach to cope with functional diversity and ecological processes and to assess the consequences of biodiversity loss. When using plant functional types, we have to keep in mind that this reduces the information according to a continuous gradient of functional diversity (set by the functional traits) to a few groups. Furthermore, the individual decision on which functional traits are used as criteria for the classification of groups, is in most cases a subjective one (Smith et al. 1997; Lavorel and Garnier 2002). It can lead to circular reasoning when using functional groups to characterize functionality (Beierkuhnlein and Schulte 2001). Especially in grassland experiments, the categorization is simple. Growth type (grasses vs. herbs), nitrogen-fixing ability (legumes vs. non-legumes) and carbon-metabolism (C3 vs. C4 grasses) may serve as examples. The consistency of the approach is lost when groups are mixed that have been classified on the basis of different criteria (e.g. herbs, grasses and legumes).

Besides functional redundancy, one source of redundancy may be rareness. There are only some hints that the loss of rare species might interfere with ecosystem functioning (Lyons and Schwartz 2001). If organisms are rare or afforded little space, they normally will not interact strongly. Then, the same ecological niche can be occupied by different species within one community. Even if edge effects and transient emergence are excluded, dynamics of species turnover in microsites, non-uniform seed dispersal and spatio-temporal site heterogeneity will support such co-occurrences (Shmida and Ellner 1984). They will contribute to species diversity in the sense of "functional analogues" (Barbault et al. 1991) but not enhance the complexity of the system.

Functional traits are derived from direct functional properties of organs or from specific forms of metabolism. "Response traits" are attributes which determine how individuals (or species) will react to environmental changes (Walker et al. 1999). "Effect traits", in contrast, influence certain ecosystem properties (Lavorel and Garnier 2002). Especially the latter ones are interesting for biodiversity–ecosystem functioning studies. With regard to biodiversity loss scenarios, response traits also become important (Hooper et al. 2002). The trait approach is more flexible and also more specific than the functional type approach. It can be directly linked to specific processes, such as fluxes of carbon or water. For some ecosystem properties, effect and response traits may be tightly linked (Weiher et al. 1999). Functional traits are not always strictly based on certain mechanisms but on merely structural or temporal attributes. In such cases, it is better to stay with the established terminology of morphological traits, and then use the classification of growth forms or life forms or – when time is the scale – life-history traits (Poschlod et al. 2000), but not functional types. Phenological complementarity may contribute to the functional performance of ecosystems (Loreau and Behara 1999; Stevens and Carson 2001). Depending on the ecosystem function under consideration, the relevant functional traits are specific too. Thus, on a functional level, it is difficult to design experiments with proper gradients of functional diversity if various traits are investigated. Furthermore, we are often not provided with satisfying information about the relevance of functional traits for a certain function. For many species, important functional traits are not documented (see Weiher et al. 1999).

A data-defined analysis of functional diversity and species diversity measures based on data from the BIODEPTH experiment demonstrated that only few functional characteristics – in this case the presence of legumes–explain the biodiversity–ecosystem functioning relationship best, and that simple functional groups, i.e. only grasses, herbs and legumes, also serve this purpose well (Petchey et al. 2004). Thus, it may be concluded that simple measures of functional diversity, such as functional types, are good surrogates for the description of the biodiversity–ecosystem functioning relationship, at least in grasslands, where the functional diversity gradient of species is not very broad (Nesshöver 2004). However, the role of structural or functional diversity is still less emphasized than the importance of species diversity.

In experiments including functional diversity as well as species diversity, autocorrelation between the different measures may occur (Fig. 5; Tilman et al. 1997a; Naeem 2002a; Petchey and Gaston 2002; Naeem and Wright 2003). Depending on the design of the experiment, it can be impossible to separate their effects (Schmid et al. 2002a). Functional diversity and species diversity is tightly linked due to evolutionary constraints (Reich et al. 2003). Species diversity might be, as well as functional type diversity, a rather good surrogate for the more complex diversity of functional traits.

Additionally, it has to be asked whether it is applicable to separate effects of different richness components, such as richness of species and plant functional

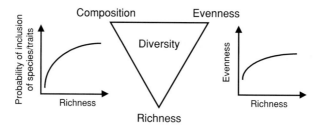

Fig. 5. Scheme of general relationship between different diversity parameters. The diversity of elements can be seen as a combination of their richness (number of elements), their composition (if certain elements are included or not) and their evenness (their abundance in the community). All three parameters will be more or less tightly correlated (graphs on the *right* and the *left*), leading to the inevitable consequence of autocorrelation in experiments

types, and to ignore other diversity components (Schmid et al. 2002b)? On an abstract level, separation of effects should be attempted between richness, evenness and composition effects. Richness will in most cases be correlated with evenness – the more species, the more even distribution of abundance might be expected (Fig. 6, right) – and compositional effects; the probability of a species included in a mixture will increase with higher richness, the sampling or selection effect. Instead of trying to statistically separate effects of single diversity parameters within one analysis, e.g. species and functional type richness and mixture effects in an ANOVA of a multiple regression, it might

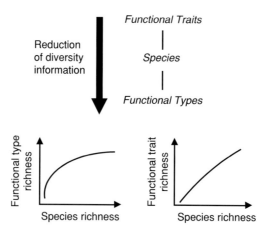

Fig. 6. Different diversity levels used in diversity experiments. From the functional trait to the species and the functional type level, the information within the level is reduced, leading to specific correlations between the levels (*bottom graphs*)

be more applicable to accept that the different parameters have to be considered jointly, e.g. by using them in a principal component analysis to derive independent orthogonal factors from them. These factors can help us to analyse the relationship between diversity and functioning (Naeem 2002a; Nesshöver et al., in review). However, one still has to choose which diversity parameters should be included in such an analysis (Nesshöver 2004) or use randomised bootstrapping techniques for analysis (Petchey et al. 2004).

6.2 Individuality, Assembly Rules and Non-Stochastic Extinction

In order to statistically test hypotheses, data have to be as appropriate as possible, and tests make only sense when replicates are included. However, after some years of a study, it becomes obvious that even if there are still similarities, the development of each plot is unique. The quality of replicates thus depends on time and is better in the beginning of an experiment. In addition, at the beginning of an experiment, stochastic events, such as the effects of pioneer species, and the role of slight differences in the mechanical disturbance during the setting up of the experiment, are relevant. They become less important during the development of the stands, when biotic interaction and space-filling takes place.

Due to the target of identifying general tendencies or rules, biodiversity experiments use a specific species pool. They mix species for certain diversity levels. This mechanistic approach does not reflect the assembly rules and population dynamics in natural communities (see Lepš 2004; White and Jentsch 2004). Biodiversity experiments tend to ignore the coevolved and non-stochastic inter-relationships between species. When the consequences of extinction are derived from assembly experiments, it cannot be neglected that simulated random loss of species is not realistic (Fonseca and Ganade 2001; Schläpfer et al. 2005). Causing species loss by highly intensive land use and then allowing regeneration, Schläpfer et al. (2005) show that highly productive monoculture species were most persistent when subjected to community disturbance. However, these species were not superior to others during the regeneration phase. Species do not react equally to environmental stress. Their identity plays a role in their resilience. Predictions of consequences of plant species loss have to consider both specific extinction probabilities and post-extinction development (Schläpfer et al. 2005). When knowledge about species traits, such as extinction risk, is available (Pimm et al. 1988; Stöcklin and Fischer 1999; Thomas et al. 2004), this can be integrated into the design of diversity experiments. Non-random extinctions can disproportionately affect the invasibility of communities (Zavaleta and

Hulvey 2004). Ives and Cardinale (2004) stress the fact that the community performance after non-random extinctions depends strongly on food web interactions. When simulated extinctions are random and conclusions are based on mean values, the findings will deviate strongly from observation. Models indicate that the high importance of competition increases differences between random and ordered extinctions, but mutualisms reduce the differences (Gross and Cardinale 2005).

6.3 The Importance of Temporal and Spatial Scales

According to temporal scales, there are some fundamental problems inherent in experimental approaches. Compared to the development of natural ecosystems, biodiversity experiments aim to produce results within a very short time period. Thus, they are composed of rather short-lived or pioneer species. For many of them, a 3-year period is rather short to attain optimal performance. In addition, such short periods can hardly ensure that adapted mutualistic communities of soil biota can develop. Another temporal problem is synchrony. Experiments start on the same date with species characterized by different life cycles. It is technically almost impossible to ensure from the beginning a diversified age structure within the populations of different species in mixtures: juvenile, adult and senescent individuals in comparable proportions for different species with different potential age and development. As a consequence, the dominance structures change as experiments continue. It may take years until late successional species, such as *Arrhenatherum elatius* in European grasslands, gain their natural importance (Fig. 4b). Organisms which respond slowly, such as most characteristic competitive species of natural communities, which follow the strategy of resource allocation to storage organs, need some time to play the role that is attributed to them in nature. This can be ignored if a comparable proportion of species is aimed at. However, this target can hardly be achieved at any time of an experiment. Extrapolations of short-term experiments to natural processes and responses are naïve if they ignore this. However, there seems to be an influence of diversity on temporal variability of the community (Cottingham et al. 2001). At the very least, hysteresis can occur. The loss and the addition of species can produce contradictory results. Starting from low diversity and complexity this can slowly contribute to increased community performance, and the loss of species, starting from diverse communities can be compensated for (Fig. 7). If such mechanisms are effective, additive experiments have to be tested against removal experiments.

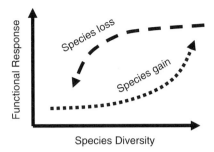

Fig. 7. Experiments manipulate species diversity in order to simulate the consequences of species loss. The better experimental reliability (replicates, controlled communities) of an additive set up compared to species removal is the reason why most experiments rely on additive approaches. However, there can be hysteresis in the system. The direction of the development, species gain or loss, can be related to a delayed response or functional persistence. It takes time until facilitative interactions are built up and the loss of complexity may be temporarily compensated for (after Beierkuhnlein and Jentsch 2005)

Bommarco (2003) demonstrates in a meta-analysis that herbivore and predator abundance depends on available space. For many reasons, the scale of experimental plots will influence results where biocoenotic interferences are concerned. Cardinale et al. (2004) simulated changes in spatial scale and could not find an effect on diversity–productivity relationships. However, as discussed earlier, such simplified models cannot replace experiments because they are based on unrealistic simplification. Whittaker and Heegard (2003) found scale-dependency of the species richness–productivity relationship in observational studies. Up until now, only few experimental studies have focussed on spatial aspects. Varying soil depth and rooting space yielded a remarkable influence on the performance of plant species diversity–productivity effects (Dimitrakopoulos and Schmid 2004). These authors found a linear increase in the magnitude of positive diversity effects on aboveground as well as belowground biomass with increasing availability of space and access to resources.

Biodiversity experiments are not comparable in size to natural communities due to logistic restrictions. This affects not only the population dynamics within the plots and treatments but has also an influence on microclimate and on biotic interactions with herbivores. This is because there will be side-effects by neighbouring and mostly structurally different plots, or side-effects at the edge of the experiment. This problem has been broached in the Jena experiment, but still, the size of the plots is not comparable to grassland communities found in the cultural landscape. One approach used to take into account the heterogeneity within communities of differing diversity is to look at evenness (Mulder et al. 2004). However, if the scale of investigation is increased to the landscape scale, the role of diversity in the system can be expected to change completely with respect to increased abiotic heterogeneity (Bengtsson et al. 2002).

6.4 Separation of External Factors and Intrinsic Factors – Multi-Site Experiments and the Connection Between Experimental and Observational Studies

Site productivity and resource availability control strongly the performance of entire communities. Many experiments (e.g. Cedar Creek) were carried out on sites where nutrients or other resources were limited. Consequently, small differences in habitat carrying capacity may alter the relationship between diversity and functioning (Huston and McBride 2002; Aarssen et al. 2003). On a gradient of natural communities with different habitat carrying capacity (or site productivity), this leads to the observed hump-shaped curve first identified by Grime (1973). The results from the experiments using a logarithmic relationship between diversity and productivity seemed to contradict these findings (Mittelbach et al. 2001). The BIODEPTH experiment tried to identify general mechanisms across a wide spatial range in Europe with different environmental conditions. Differences in climate and soil conditions between the sites were part of the design that considered transects from oceanic to continental and from boreal to Mediterranean climates.

Schmid (2002) approached this by applying a three-dimensional model to separate habitat carrying capacity from community productivity (Fig. 8a). With differences in resource supply, the plane of an observed relationship moves from plane A to plane B. According to the hump-shaped model, which restricts the possible combinations, the planes change their extension. The results from different experiments with given species numbers are reflected by trajectories within the planes A and B (Schmid 2002). Projected back into the two-dimensional diversity–productivity model, this predicts specific trajectories within the hump under specific environmental conditions (Fig. 8b, curves 1, 2, 3). This is found when comparing different BIODEPTH sites (Schmid and Hector 2004). The question in this context is, how can the observed experimental relationship be related to natural conditions? Schmid and Hector (2004) argue that, if the weeding is stopped, diversity as well as productivity will shift along the observed trajectories towards an optimised state within the hump (arrows, Fig. 8b), as was observed at the Swiss BIODEPTH site (Pfisterer et al. 2004).

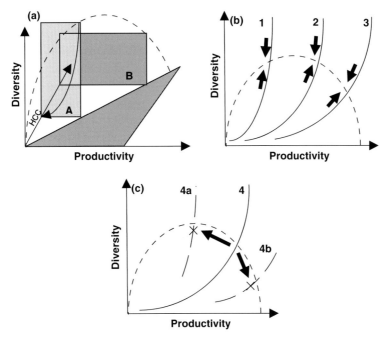

Fig. 8a–c. Schematic relationship between diversity, stand productivity and habitat carrying capacity (*HCC*). **a** Three-dimensional relation of diversity, productivity and capacity, **b** schematic visualisation of the logarithmic relationship found in experiments under different HCC (*graphs 1–3*, e.g. the different sites of the BIODEPTH experiment). The *arrows* indicate the potential development of productivity and diversity towards a convergence point along the hump (*dotted line*) if the diversity gradient is not maintained (compare Schmid and Hector 2004). **c** If the HCC is altered, alternative convergence points might be approached, leading to either increased or decreased diversity. See text for further explanation

Nevertheless, this effect will only appear if the carrying capacity of the site remains constant over time. For the German BIODEPTH site with the strong influence of legumes on the nutrient status of the stands, an additional shift was observed, since carrying capacity decreased with the decreasing abundance of legumes. Thus, an unexpected additional shift towards less productive and more species-rich stands was found (Fig. 8c, curves 4, 4a). Since the gradient of diversity is shortened as a consequence of immigration and succession, the hypothetical relationship (dashed line, Fig. 8c) is no longer found. All plots converge towards one point within the hump (cross, Fig. 8c). Another hypothetical question is: what would happen if the carrying capacity were strongly increased (e.g. by fertilization; Schmid 2002)? Then, diversity can decrease, since dominant species out-perform others, and productivity can rise at the same time (Fig. 8c, curve 4b).

7 Outlook

Biodiversity experiments are obviously both artificial constructs and heuristic tools, and there is compelling evidence that these qualities are tightly linked. The particular value of experiments lies in their contribution to an examination of causal effects. They can be designed in a way which aims to falsify specific hypotheses. Natural communities do not offer this opportunity. They share unobservable and immeasurable factors that have previously influenced their present-day condition. In addition, there are many measurable factors which interact at each site at different intensities.

Biodiversity experiments mainly involve rather simple or simplified communities – and if only annuals are concerned, they are often far too simplified to yield valuable results. Up until today, most approaches have used the community scale. Only the ECOTRON chamber experiment integrated higher trophic levels. The realization of more natural experiments, meaning systems that show a higher degree of complexity, is a major challenge in ecology (Belovsky et al. 2004). Large-scale experiments with higher complexity and with higher longevity of organisms are required – perhaps even across different ecosystems. Such experiments would be more representative of natural mechanisms and give us more information about the real effects of biodiversity loss. However, it would take years until relevant results could be achieved. Plantation experiments have been installed recently in forests around the world (e.g. Finland, Germany, Panama, and Borneo) with a varying diversity of tree species and genera (Scherer-Lorenzen et al. 2005b). At the global scale, the contribution of forests to ecosystem functioning and biodiversity loss is outstanding. Their structural diversity but also the integration of long-term site conditions, management, and natural disturbances have strong impacts on data (Pretzsch 2005). Up until now, silvicultural experiments have concentrated on tree species and ignored shrubs, dwarf shrubs and grasses, which might benefit from tree species diversity. This is reasonable from an economic point of view but limits their relevance for ecology. In the future, entire forest communities should be investigated. This means also looking, for instance, at epiphytes and cryptogams. Nevertheless, manipulating diversity in forests is an enormous step towards experiments which are of a more realistic scale and representative of nature.

Despite partly contradictory results from different case studies, there is more and more consensus in the scientific community regarding the effects of biodiversity loss (Hooper et al. 2005). Obviously, it depends on the system under consideration whether effects of diversity per se or of certain parts of the community, such as productive species, emerge in additive experiments (Tilman 1997). There is some indication of the higher probability of selection of highly productive species with increasing species richness, but a sophisticated experimental design allows us to cope with this. Not forgetting the initial motivation of combating species loss, a more decisive question is, which species are prone to extinction? In nature, it is obvious that the risk of becoming extinct is not randomly distributed. In this case, non-random selection means selecting those which have the highest extinction probability, e.g. short-lived seeds, low competitive ability (Stöcklin and Fischer 1999), not the ones with the greatest contribution to ecosystem functioning, to simulate biodiversity loss (Lepš 2004). As there is a reason for their being under threat, such species are often difficult to cultivate in experiments.

Productivity is a prominent ecosystem function and the most widely considered variable in many experiments. However, a large proportion of the diverse ecosystems of the world is not very productive (e.g. South African Fynbos vegetation). Thus ecosystem functions other than productivity might be of more relevance in these cases. The role of structural diversity as habitat quality or as a factor to prevent soil erosion can matter. Until now, most biodiversity experiments have been carried out in the temperate region, and have concentrated on grasslands. However, biodiversity at the global scale is mainly threatened in tropical and subtropical areas. There, many structural and logistic constraints have to be considered when planning experiments. The first biodiversity experiments have only recently been installed in tropical areas (Scherer-Lorenzen et al. 2005b). More of them are urgently needed.

In order to evaluate the results that have been obtained experimentally, it seems inevitable that biodiversity models will become more complex on the one hand and that these data will need to be related to natural ecosystems on the other. Up until now, convincing concepts devised to calibrate experimental data according to real world data, do not exist. In our opinion, only the combination of various methodological approaches will deliver the insights required into the functioning of ecosystems. Any approach has its methodological restrictions and constructs. To broaden the validity of results from within experiments to within natural communities, data from various sources have to be adjusted. It would be most welcome if it were planned from the beginning that experimental results should be integrated into meta-analyses, in order to separate site-specific effects from general trends. Single-site experiments just add another unit to a large puzzle. What has been found for northern European grasslands does not necessarily hold true for the Mediterranean macchia (see Wardle 2001b; but also Troumbis and Memtsas 2000). In the case of biodiversity studies, this affords not only comparable levels of diversity but also a standardized experimental design. This was applied in the scope of the BIODEPTH project, but also independent experiments could follow such a rationale.

One major reason for confusion in biodiversity experiments is the multi-dimensional nature of biodiversity. There is no concept of biodiversity that concentrates merely on one aspect, such as species diversity. Diversity includes richness, evenness and composition components. Most experiments still concentrate on species richness as it is easily established. Nevertheless, community heterogeneity is of increasing importance, if larger plot sizes are used, as clumped aggregations (Stoll and Prati 2001) and small-scale hot spots of functional effects become more probable.

Experiments on biodiversity and its changes in ecosystems cannot be solely restricted to species diversity and its decline. Changes of diversity occur at various scales as do qualities. The total extinction of a genetic pool of a species is only one facet. The local or regional loss of populations may be another one and can result in a decrease in local alpha diversity. However, this may locally be (over-) compensated for in numbers by alien or even invasive species (Sax and Gaines 2003). Present approaches consider invasibility as an effect of declining diversity. Such experiments are not designed to work out the effects of invasion and of a conceivable increase in diversity on ecosystem functioning.

A local increase in species richness, nevertheless, can be accompanied by extinction of rare or specialized species. Again, we do have to consider scale and ask which communities profit and which suffer. As an effect of such changes, the dissimilarity (beta diversity) between communities (see also Fukami et al. 2001) and between areas may decrease. This will lead to a loss of biodiversity at the landscape level that cannot be characterized by numbers of species. Many biodiversity experiments focus on alpha diversity – no matter if referring to species or functional groups. The loss of dissimilarity between stands has not been identified as an experimental research objective. One of the probable effects is a lower colonization rate by new species from neighbouring communities, as these are more similar, and thus less temporal turnover in species composition. In the face of global changes, this can mean lower resilience and adaptational capability. Even less is known about the consequences of modified turnover and age structure diversity within given communities or levels of species diversity.

Functional traits have been found to be a good tool with which to characterize the functioning of communities. However, according to temporal traits, observational knowledge that can be used in the design of experiments is only satisfactory for tree species. The population dynamics and life cycles are only understood for few species. It matters for metabolism and functioning, whether the individuals within an experiment are juveniles or adults, but as long as we cannot differentiate them, which is the case for many grasses and herbs, this cannot be considered. Up until now, many mechanisms which are known to occur according to interactions between organisms, have not been applied to diversity experiments. The diversity of interactions is still largely unknown. The role of allelopathy, mutualisms of various kinds, interactions via promoted or suppressed soil organisms, are widely unknown. Many of these important functional processes will be species specific. Theoretical considerations help us to develop hypotheses; however, these have to be proven or withdrawn. The intrinsic epistemological problem of biodiversity research is that it is almost impossible to carry out in the vast majority of natural ecosystems. One reason for this restriction is to be found in the complexity of biotic interactions across and within trophic levels and functional groups. Species play a variety of roles, and there are many of these, including competition, facilitation and mutualism.

One factor limiting our ability to analyse biodiversity loss is the contribution of historic factors to ecosystem patterns and traits which have emerged. Recent aspects and structures are often attributed to past processes (see Davis and Pelsor 2001). Historical site conditions and ecosystem performance can be documented by storage matter or informational qualities such as patterns of behaviour. Connell (1980) stressed that the "ghost of competition past", evolutionary forces that are not necessarily effective today, have often had repercussions not only on the behaviour, distribution or morphology of species but also on their performance. Past interferences and facilitation stimulated coevolutive adaptation. If this is reflected in genotypes, which are not detectable by morphologic means, only experiments can prove its effect.

Various restrictions of biodiversity experiments have been discussed in this paper. What has been found to be true in one experiment (e.g. the positive correlation between productivity and species diversity by Hector et al. 1999) may not be confirmed when the scale of the investigation is changed (Waide et al. 1999; Bengtsson et al. 2002; Chase and Leibold 2002). This may be attributed to the fact that the emergence of biodiversity and of ecological complexity is strongly related to spatial scale. The size of experimental plots differs and larger plots are regarded as delivering more promising results. Experiments with varying plot size could help us to find answers to this technical problem (Roscher et al. 2005).

The aim to search for generality was an initial impetus of biodiversity–ecosystem functioning research. However, generality has still not been found here, and maybe this is not possible at all. Both of the terms, biodiversity and ecosystem functioning, turn out to be just too wide for testing by hypotheses. Functions, species and ecosystems integrate a huge amount of individuality and complexity. First conclusions based on single case studies had to be revised or at least restricted. Today, the focus is on more specific research questions. It would be a commonly made mistake to accept or reject theories as units (Loehle 1988). Nevertheless, ecological experiments have contributed tremendously to a better understanding of the effects of biodiversity loss. Valuable insights have been gained, which are of great relevance for society and policy. The recognition of their limited validity and partly controversial results indicate the ongoing development of ecological insight.

References

Aarssen LW (1997) High productivity in grassland ecosystems: effected by species diversity or productive species. Oikos 80:183–184

Aarssen LW, Laird, RA, Pither J (2003) Is the productivity of vegetation plots higher or lower when there are more species? Variable predictions from interactions of the "sampling effect" and "competitive dominance effect" on the habitat template. Oikos 102:427–431

Allison GW (1999) The implications of experimental design for biodiversity manipulations. Am Nat 153:26–45

Altieri MA (1994) Biodiversity and pest management in agroecosystems. Food Products Press

Andow DA (1991) Vegetational diversity and arthropod population response. Annu Rev Entomol 36:561–586

Barbault R, Colwell RK, Dias B, et al. (1991) Conceptual framework and research issues for species diversity at the community level. In: Solbrig OT (ed) From genes to ecosystems: a research agenda for biodiversity. IUBS, Cambridge, Mass., pp 37–71

Bardgett RD, Shine A (1999) Linkages between plant litter diversity, soil microbial biomass and ecosystem function in temperate grasslands. Soil Biol Biochem 31:317–321

Beierkuhnlein C, Jentsch A (2005) Ecological importance of species diversity. In: Henry R (ed) Plant diversity and evolution: genotypic and phenotypic variation in higher plants. CABI, Wallingford, pp 249–285

Beierkuhnlein C, Schulte A (2001) Plant functional types – Einschränkungen und Möglichkeiten funktionaler Klassifikationsansätze in der Vegetationsökologie. Jax K (ed) Funktionsbegriff und Unsicherheit in der Ökologie. Lang, Germany, pp 45–64

Belovsky GE, Botkin DB, Crowl TA, et al. (2004) Ten suggestions to strengthen the science of ecology. Bioscience 54:345–351

Benedett-Cecchi L (2004) Increasing accuracy of causal inference in experimental analysis of biodiversity. Funct Ecol 18:761–768

Bengtsson J, Engelhardt K, Giller P, et al. (2002) Slippin' and slidin' between the scales: the scaling components of biodiversity–ecosystem functioning relations. In: Loreau M, Naeem S, Inchausti P (eds) Biodiversity and ecosystem functioning – synthesis and perspectives. Oxford University Press, Oxford, pp 209–220

Berlow EL, D'Antonio CM, Swartz H (2003) Response of herbs to shrub removal across natural and experimental variation in soil moisture. Ecol Appl 13:1375–1387

Bertness MD, Callaway R (1994) Positive interactions in communities. Trends Ecol Evol 9:191–193

Bommarco R (2003) Scale as modifier in vegetation diversity experiments: effects on herbivores and predators. Oikos 102:440–448

Bret-Harte MS, Garcia EA, Sacre VM, et al. (2004) Plant and soil responses to neighbour removal and fertilization in Alaskan tussock tundra. J Ecol 92:635–647

Bruno JF, Stachowicz JJ, Bertness MD (2003) Inclusion of facilitation into ecological theory. Trends Ecol Evol 18:119–125

Callaway RM, DeLucia EH, Moore D, et al. (1996) Competition and facilitation: contrasting effects of *Artemisia tridentata* on *Pinus ponderosa* and *P. monophylla*. Ecology 77:2130–2141

Cardinale BJ, Palmer MA, Collins SL (2002) Species diversity enhances ecosystem functioning through interspecific facilitation. Nature 415:426–429

Cardinale BJ, Ives AR, Inchausti P (2004) Effects of species diversity on the primary productivity of ecosystems: extending our spatial and temporal scale of interference. Oikos 104:437–450

Chapin FS, Sala OE, Burke IC, et al. (1998) Ecosystem consequences of changing biodiversity. BioScience 48:45–52

Chapin FS III, Zavaleta ES, Eviner VT, et al. (2000) Consequences of changing biodiversity. Nature 405:234–242

Chase JM, Leibold MA (2002) Spatial scale dictates the productivity–biodiversity relationship. Nature 416:427–430

Chetty CKR, Reddy MN (1984) Staple land equivalent ratio for assessing yield advantage from intercropping. Exp Agric 20:171–177

Connell JH (1980) Diversity, and the coevolution of competitors, or the ghost of competition past. Oikos 35:131–138

Connell JH (1983) Interpreting the results of field experiments: effects of indirect interactions. Oikos 41:290–291

Cottingham KL, Brown BL, Lennon JT (2001) Biodiversity may regulate the temporal variability of ecological systems. Ecol Lett 4:72–85

Craine JM, Tilman D, Wedin D, et al. (2002) Functional traits, productivity and effects of nitrogen cycling of 33 grassland species. Funct Ecol 16:563–574

Crawley MJ, Brown SL, Heard MS, Edwards GR (1999) Invasion-resistance in experimental grassland communities: species richness or species identity? Ecol Lett 2:140–148

Davis MA, Pelsor M (2001) Experimental support for a resource-based mechanistic model of invasibility. Ecol Lett 4:421–428

Deutschman DH (2001) Design and analysis of biodiversity field experiments. Ecol Res 16:833–843

Díaz S, Cabido M (2001) Vive la difference: plant functional diversity matters to ecosystem processes. Trends Ecol Evol 16:646–655

Díaz S, Symstad AJ, Chapin FSI, Wardle DA, Huenneke LF (2003) Functional diversity revealed by removal experiments. Trends Ecol Evol 18:140–146

Diemer M, Joshi J, Schmid B, et al. (1997) An experimental protocol to assess the effects of plant diversity on ecosystem functioning utilised by a European research network. Bull Geobot Inst ETH Zürich 63:95–10

Dietz H, Steinlein T (2001) Ecological aspects of clonal growth in plants. Progr Bot 62:511–530

Dimitrakopoulos PG, Schmid B (2004) Biodiversity effects increase linearly with biotope space. Ecol Lett 7:574–583

Doak DF, Bigger D, Harding EK, et al. (1998) The statistical inevitability of stability–diversity relationships in community ecology. Am Nat 151:264–276

Drake JM (2003) Why does grassland productivity increase with species richness? Disentangling species richness and composition with tests for overyielding and superyielding in biodiversity experiments. Proc R Soc Lond B 270:1713–1719

Dukes JS (2001a) Biodiversity and invasibility in grassland microcosms. Oecologia 126:563–568

Dukes JS (2001b) Productivity and complementarity in grassland microcosms of varying diversity. Oikos 94:468–480

Ehrlich PR, Ehrlich AH (1981) Extinction: the causes and consequences of the disappearance of species. Random House, New York

Fargione J, Brown CS, Tilman D (2003) Community assembly and invasion: an experimental test of neutral versus niche processes. PNAS 100:8916–8920

Fay PA, Carlisle JD, Knapp AK, et al. (2000) Altering rainfall timing and quantity in a mesic grassland ecosystem: design and performance of rainfall manipulation shelters. Ecosystems 3:308–319

Fonseca CR, Ganade G (2001) Species functional redundancy, random extinctions and the stability of ecosystems. J Ecol 89:118–125

Fox BJ (1987) Species assembly and the evolution of community structure. Evol Ecol 1:201–213

Fridley JD (2001) The influence of species diversity on ecosystem productivity: how, where, and why? Oikos 93:514–526

Fridley JD (2002) Resource availability dominates and alters the relationship between species diversity and ecosystem productivity in experimental plant communities. Oecologia 132:271–277

Fridley JD (2003) Diversity effects on production in different light and fertility environments: an experiment with communities of annual plants. J Ecol 91:396–406

Fukami TS, Naeem S, Wardle DA (2001) On similarity among local communities in biodiversity experiments. Oikos 95:340–348

Garnier E, Navas ML, Austin MP, et al. (1997) A problem for biodiversity–productivity studies: how to compare the productivity of multispecific plant mixtures to that of monocultures? Acta Oecol 18:657–670

Ghilarov AM (2000) Ecosystem functioning and intrinsic value of biodiversity. Oikos 90:408–412

Gitay H, Noble IR (1997) What are functional types and how should we seek them? In: Smith TM, Shugart HH, Woodward FI (eds) Plant functional types: their relevance to ecosystem properties. Cambridge University Press, Cambridge, pp 3–19

Gitay H, Wilson JB , Lee WG (1996) Species redundancy: a redundant concept? J Ecol 84:121124

Gleason HA (1926) The individualistic concept of the plant association. Bull Torrey Bot Club 53:7–26

Gowdy JM, Daniel CN (1995) One world, one experiment: addressing the biodiversity–economics conflict. Ecol Econ 15:181–192

Grime JP (1973) Competitive exclusion in herbaceous vegetation. Nature 242:344–347

Grime JP (1997) Biodiversity and ecosystem function: the debate deepens. Science 277:1260–1261

Grime JP (1998) Benefits of plant diversity to ecosystems: immediate, filter and founder effects. J Ecol 86:902–910

Grime JP (2002) Declining plant diversity: empty niches or functional shifts? J Veg Sci 13:457–460

Grimm V, Wissek C (1997) Babel, or the ecological stability discussion: an inventory and analysis of terminology and a guide for avoiding confusion. Oecologia 109:323–334

Gross K, Cardinale BJ (2005) The functional consequences of random vs. ordered species extinctions. Ecol Lett 8:409–418

Hall SJ, Gray SA, Hammett ZL (2000) Biodiversity–productivity relations: an experimental evaluation of mechanisms. Oecologia 122:545–555

Hanley ME (2004) Seedling herbivory and the influence of plant species richness in seedling neighbourhoods. Plant Ecol 170:35–41

Hartley SE, Jones TH (2003) Plant diversity and insect herbivores: effects of environmental change in contrasting model systems. Oikos 101:6–17

Hättenschwiler S (2005) Effects of tree species diversity on litter quality and decomposition. In: Scherer-Lorenzen M, Körner C, Schulze ED (eds) Forest diversity and function. Springer, Berlin Heidelberg New York, pp 149–164

He JS, Bazzaz FA, Schmid B (2002) Interactive effects of diversity, nutrients and elevated CO_2 on experimental plant communities. Oikos 97:337–348

Hector A (1998) The effect of diversity on productivity: detecting the role of species complementarity. Oikos 82:597–599

Hector A (2002) Biodiversity and the functioning of grassland ecosystems: multi-site comparisons. In: Kinzig AP, Pacala SW, Tilman D (eds) The functional consequences of biodiversity. Princeton University Press, Princeton, N.J., pp 71–95

Hector A, Hooper R (2002) Darwin and the first ecological experiment. Science 295:639–640

Hector A, Schmid B, Beierkuhnlein C, et al. (1999) Plant diversity and productivity experiments in European grasslands. Science 286:1123–1127

Hector A, Beale AJ, Minns A, et al. (2000) Consequences of the reduction of plant diversity for litter decomposition: effects through litter quality and microenvironment. Oikos 90:357–371

Hector A, Dobson K, Minns A, et al. (2001a) Community diversity and invasion resistance: an experimental test in a grassland ecosystem and a review of comparable studies. Ecol Res 16:819–131

Hector A, Joshi J, Lawler SP, et al. (2001b) Conservation implications of the link between bio diversity and ecosystem functioning. Oecologia 129:624–628

Hector A, Bazeley-White E, Loreau M, et al. (2002a) Overyielding in grassland communities: testing the sampling effect hypothesis with replicated biodiversity experiments. Ecol Lett 5:502–511

Hector A, Loreau M, Schmid B, et al. (2002b) Biodiversity manipulations experiments: studies replicated at multiple sites. In: Loreau M, Naeem S, Inchausti P (eds) Biodiversity and ecosystem functioning: synthesis and perspectives. Oxford University Press, Oxford, pp 36–46

Hobbie SE, Shevtsova A, Chapin FS I (1999) Plant responses to species removal and experimental warming in Alaskan tussock tundra. Oikos 84:417–434

Holmgren M, Scheffer M, Huston MA (1997) The interplay of facilitation and competition in plant communities. Ecology 78:1966–1975

Holt RD Loreau M (2002) Biodiversity and ecosytem functioning: the role of trophic interactions and the importance of system openess. In: Kinzig AP, Pacala SW , Tilman D (eds) The functional consequences of biodiversity. Princeton University Press, Princeton, N.J., pp 246–262

Hooper DU (1998) The role of complementarity and competition in ecosystem responses to variation in plant diversity. Ecology 79:704–719

Hooper DU, Dukes JS (2004) Overyielding among plant functional groups in a long-term experiment. Ecol Lett 7:95–105

Hooper DU, Vitousek PM (1997) The effect of plant composition and diversity on ecosystem processes. Science 277:1302–1305

Hooper DU, Vitousek PM (1998) Effects of plant composition and diversity on nutrient cycling. Ecol Monogr 68:121–149

Hooper DU, Solan M, Symstad A, et al. (2002) Species diversity, functional diversity and ecosystem functioning. In: Loreau M, Naeem S, Inchausti P (eds) Biodiversity and ecosystem functioning: synthesis and perspectives. Oxford University Press, Oxford, pp 195–208

Hooper DU, Chapin FS, Ewel JJ, et al. (2005) Effects of biodiversity on ecosystem functioning: a consensus of current knowledge. Ecol Monogr 75:3–35

Hurlbert SH (1984) Pseudoreplication and the design of ecological field experiments. Ecol Monogr 54:187–211

Hurlbert SH (2004) On misinterpretations of pseudoreplication and related matters: a reply to Oksanen. Oikos 104:591–597

Huston MA (1997) Hidden treatments in ecological experiments: re-evaluating the ecosystem function of biodiversity. Oecologia 110:449–460

Huston MA, McBride AC (2002) Evaluating the relative strengths of biotic versus abiotic controls on ecosystem processes. In: Loreau M, Naeem S, Inchausti P (eds) Biodiversity and ecosystem functioning: synthesis and perspectives. Oxford University Press, Oxford , pp 47–60

Huston MA, Aarssen LW, Austin MP, et al. (2000) No consistent effect of plant diversity on productivity. Science 289:1255a

Ives AR, Cardinale BJ (2004) Food-web interactions govern the resistance of communities after non-random extinctions. Nature 429:174–177

Ives AR, Klug JL, Gross K (2000) Stability and species richness in complex communities. Ecol Lett 3:399–411

Jentsch A, Wittmer H, Jax K, et al. (2003) Biodiversity – emerging issues for linking natural and social sciences. Gaia 2:121–128

Johnson KH, Vogt KA, Clark HJ, et al. (1996) Biodiversity and the productivity and stability of ecosystems. Trends Ecol Evol 11:372–377

Jones CG, Lawton JH, Shachak M (1994) Organisms as ecosystem engineers. Oikos 69:376–386

Kaiser C (2000) Rift over biodiversity divides ecologists. Science 289:1282–1283

Kenkel NC, Peltzer DA, Baluta D, Pirie D (2000) Increasing plant diversity does not influence productivity: empirical evidence and potential mechanisms. Community Ecol 1:165–170

Kennedy TA, Naeem S, Howe KM, Knops JMH, Tilman D, Reich P (2002) Biodiversity as a barrier to ecological invasion. Nature 417:636–638

Knops JMH, Tilman D, Haddad NM, et al. (1999) Effects of plant species richness on invasion dynamics, disease outbreaks, insect abundances and diversity. Ecol Lett 2:286–293

Knops JMH, Wedin D, Tilman D (2001) Biodiversity and decomposition in experimental grassland ecosystems. Oecologia 126:429–433

Knops JMH, Bradley KL, Wedin DA (2002) Mechanisms of plant species impacts on ecosystem nitrogen cycling. Ecol Lett 5:454–466

Koricheva J, Mulder CPH, Schmid B, et al. (2000) Numerical responses of different trophic groups of invertebrates to manipulation of plant diversity in grassland. Oecologia 125:271–282

Kutschera L, Lichtenegger E, Sobotik M (1992a) Wurzelatlas Mitteleuropäischer Grünlandpflanzen, vol 1. Fischer, Stuttgart

Kutschera L, Lichtenegger E, Sobotik M (1992b) Wurzelatlas Mitteleuropäischer Grünlandpflanzen, vol 2. Fischer, Stuttgart

Lambers JHR, Harpole WS, Tilman D, et al. (2004) Mechanisms responsible for the positive diversity–productivity relationship in Minnesota grasslands. Ecol Lett 7:661–668

Lamont BB (1995) Testing the effect of ecosystem composition/structure on its functioning. Oikos 74:283–295

Lavorel S, Garnier E (2002) Predicting changes in community composition and ecosystem functioning from plant traits: revisiting the holy grail. Funct Ecol 16:545–556

Lavorel S, Prieur-Richard AH, Grigulis K (1999) Invasibility and diversity of plant communities: from pattern to process. Divers Distrib 5:41–49

Lawton JH (1995) Ecological experiments with model systems. Science 269:328–331

Lawton JH, Brown VK (1993) Redundancy in ecosystems. In: Schulze ED, Mooney HA (eds) Biodiversity and ecosystem function. (Ecological studies no. 99) Springer, Berlin Heidelberg New York, pp 255–270

Lawton JH, Naeem S, Thompson LJ, et al. (1998) Biodiversity and ecosystem function: getting the Ecotron experiment in its correct context. Funct Ecol 12:848–852

Lehman CL, Tilman D (2000) Biodiversity, stability, and productivity in competitive communities. Am Nat 156:534–552

Leigh EG (1965) On the relationship between productivity, biomass, diversity and stability of a community. Proc Natl Acad Sci USA 53:777–783

Lepš J (2004) What do the biodiversity experiments tell us about the consequences of plant species loss in the real world? Basic Appl Ecol 5:529–534

Lepš, J, Brown VK, Diaz Len TA, et al. (2001) Separating the chance effect from other diversity effects in the functioning of plant communities. Oikos 95:123–134

Levine JM, D'Antonio CM (1999) Elton revisited: a review of evidence linking diversity and invisibility. Oikos 87:15–26

Lloret F, Peñuelas J, Estiarte M (2004) Experimental evidence of reduced diversity of seedlings due to climate modification in a Mediterranean-type community. Global Change Biol 10:248–258

Loehle C (1987) Hypothesis testing in ecology: psychological effects and the importance of theory maturation. Q Rev Biol 62:397–409

Loehle C (1988) Philosophical tools: potential contributions to ecology. Oikos 51:97–104

Loreau M (1998a) Biodiversity and ecosystem functioning: a mechanistic model. Proc Natl Acad Sci USA 95:5632–5636

Loreau M (1998b) Separating sampling and other effects in biodiversity experiments. Oikos 82:600–602

Loreau M (2000) Biodiversity and ecosystem functioning: recent theoretical advances. Oikos 91:3–17

Loreau M, Behara N (1999) Phenotypic diversity and stability of ecosystem processes. Theor Popul Biol 56:29–47

Loreau M, Hector A (2001) Partitioning selection and complementarity in biodiversity experiments. Nature 412:72–76

Loreau M, Naeem S , Inchausti P, et al. (2001) Biodiversity and ecosystem functioning: current knowledge and future challenges. Science 295:804–808

Loreau M, Naeem S, Inchausti P (eds) (2002a) Biodiversity and ecosystem functioning: current knowledge and future challenges. Science 294:804–808

Loreau M, Downing A, Emmerson M, et al. (2002b) A new look at the relationship between diversity and stability. In: Loreau M, Naeem S, Inchausti P (eds) Biodiversity and ecosystem functioning: synthesis and perspectives. Oxford University Press, Oxford, pp 79–91

Lyons KG, Schwartz MW (2001) Rare species loss alters ecosystem function: invasion resistance. Ecol Lett 4:358–365

MacArthur R (1955) Fluctuations of animal populations and a measure of community stability. Ecology 36:533–536

Mackey RL, Currie DJ (2001) The diversity–disturbance relationship: is it generally strong and peaked? Ecology 82:3479–3492

May RM (1972) Will a large complex system be stable? Nature 238:413–414

McCann KS (2000) The diversity–stability debate. Nature 405:228–233

McGrady-Steed J, Harris PM, Morin PJ (1997) Biodiversity regulates ecosystem predictability. Nature 390:162–165

McLellan AJ, Fitter AH, Law R (1995) On decaying roots, mycorrhizal colonization and the design of removal experiments. J Ecol 83:225–230

Mead R, Willey RW (1980) The concept of a "land equivalent ratio" and advantages on yields from intercropping. Exp Agric 16:217–228

Meehl GA, Karl T, Easterling DR, et al. (2000) An Introduction to trends in extreme weather and climate events: Observations, socioeconomic impacts, terrestrial ecological impacts, and model projections. Bull Am Meteorol Soc 81:413–416

Meiners SJ, Cadenasso ML, Pickett STA (2004) Beyond biodiversity: individualistic controls of invasion in a self-assembled community. Ecol Lett 7:121–126

Minns A, Finn J, Hector A, et al. (2001) The functioning of European grassland ecosystems: potential benefits of biodiversity to agriculture. Outlook Agric 30:179–185

Mitchell CE, Tilman D, Groth JV (2002) Effects of grassland plant species diversity, abundance, and composition on foliar fungal disease. Ecology 83:1713–1726

Mittelbach GSM, Steiner CF, Scheiner SM, et al. (2001) What is the observed relationship between species richness and productivity? Ecology 82:2381–2396

Montgomery CA, Pollack RA (1996) Economics and biodiversity: weighing the benefits and costs of conservation. J For 94:34–38

Mooney HA (2002) The debate on the role of biodiversity in ecosystem functioning. In: Loreau M, Naeem S, Inchausti P (eds) Biodiversity and ecosystem functioning: synthesis and perspectives. Oxford University Press, Oxford, pp 12–17

Mooney HA, Cushman JH, Medina E, Sala OE, Schulze ED (eds) (1996) Functional roles of biodiversity – a global perspective. Wiley, New York

Mouquet N, Leadley P, Meriguet J, Loreau M (2004) Immigration and local competition in herbaceous plant communities: a three-year seed-sowing experiment. Oikos 104:77–90

Mulder CPH, Koricheva J, Huss-Danell K, et al. (1999) Insects affect relationship between plant species richness and ecosystem processes. Ecol Lett 2:237–246

Mulder CPH, Jumpponen A, Hogberg P, Huss-Danell K (2002) How plant diversity and legumes affect nitrogen dynamics in experimental grassland communities. Oecologia 133:412–421

Mulder CPH, Bazeley-White E, Dimitrakopoulos PG, et al. (2004) Species eveness and productivity in experimental plant communities. Oikos 107:50

Naeem S (1998) Species redundancy and ecosystem reliability. Conserv Biol 12:39–45

Naeem S (2002a) Disentangling the impacts of diversity on ecosystem functioning in combinatorial experiments. Ecology 83:2925–2935

Naeem S (2002b) Ecosystem consequences of biodiversity loss: the evolution of a paradigm. Ecology 83:1537–1552

Naeem S, Li S (1997) Biodiversity enhances ecosystem reliability. Nature 390:507–509

Naeem S, Wright JP (2003) Disentangling biodiversity effects on ecosystem functioning: deriving solutions to a seemingly insurmountable problem. Ecol Lett 6:567–579

Naeem S, Thompson LJ, Lawler SP, et al. (1994) Declining biodiversity can alter the performance of ecosystems. Nature 368:734–737

Naeem S, Thompson LJ, Lawler SP, et al. (1995) Empirical evidence that declining species diversity may alter the performance of terrestrial ecosystems. Phil Trans R Soc Lond B 347:249–262

Naeem S, Håkansson K, Lawton JH, et al. (1996) Biodiversity and plant productivity in a model assemblage of plant species. Oikos 76:259–264

Naeem S, Hahn DR, Schurmann G (2000a) Producer–decomposer co-dependency influences biodiversity effects. Nature 403:762–764

Naeem S, Knops JHM, Tilman D, et al. (2000b) Plant diversity increases resistance to invasion in the absence of covarying extrinsic factors. Oikos 91:97–108

Naeem S, Loreau M, Inchausti P (2002) Biodiversity and ecosystem functioning: the emergence of a synthetic ecological framework. In: Loreau M, Naeem S, Inchausti P (eds) Biodiversity and ecosystem functioning: synthesis and perspectives. Oxford University Press, Oxford, pp 3–11

Nesshöver C (2004) The role of plant functional diversity in Central European grasslands for ecosystem functioning – a conceptual framework and evidence from a long-term experiment. PhD thesis. University of Bayreuth, Germany

Nesshöver C, Scherer-Lorenzen M, Beierkuhnlein C (in review) The two facets of the diversity–productivity relationship in a long-term grassland experiment. Oecologia

Norberg J, Swaney DP, Dushoff J, Lin J, et al. (2001) Phenotypic diversity and ecosystem functioning in changing environments: a theoretical framework. Proc Natl Acad Sci USA 98:11376–11381

Oksanen L (2001) Logic of experiments in ecology: is pseudoreplication a pseudoissue? Oikos 94:27–38

Palmer MW, Maurer T (1997) Does diversity beget diversity? A case study of crops and weeds. J Veg Sci 8:235–240

Perrings C (1995) Biodiversity loss – economic and ecological issues. Cambridge University Press, Cambridge

Petchey OL, Gaston KJ (2002) Extinction and the loss of functional diversity. Proc R Soc Lond B 269:1721–1727

Petchey OL, Hector A, Gaston KJ (2004) How do different measures of functional diversity perform? Ecology 85:847–857

Peters RL (1994) Conserving biological diversity in the face of climate change. In: Kim KC, Weaver RD (eds) Biodiversity and landscapes. Cambridge University Press, New York, pp 105–132

Peterson G, Allen CR, Holling CS (1998) Ecological resilience, biodiversity and scale. Ecosystems 1:6–18

Pfisterer AB, Schmid B (2002) Diversity–dependent production can decrease the stability of ecosystem functioning. Nature 416:84–86

Pfisterer AB, Joshi J, Schmid B, Fischer M (2004) Rapid decay of diversity–productivity relationships after invasion of experimental plant communities. Basic Appl Ecol 5:5–14

Pimm SL, Jones HL, Diamond J (1988) On the risk of extinction. Am Nat 132:757–785

Polley HW, Wilsey BJ, Derner JD (2003) Do species evenness and plant diversity influence the magnitude of selection and complementarity effects in annual plant species mixtures. Ecol Lett 6:248–256

Popper KR (1959) The logic of scientific discovery. Hutchinson, London

Popper KR (1963) Conjectures and refutations: the growth of scientific knowledge. Harper & Row, New York

Poschlod P, Kleyer M, Tackenberg O (2000) Databases on life history traits as a tool for risk assessment in plant species. Zeitschr. Okol Natursch 9:3–18

Pretzsch H (2005) Diversity and productivity in forests: evidence from long-term experimental plots. In: Scherer-Lorenzen M, Schulze ED, Körner C (eds) Forest diversity and function. Springer, Berlin Heidelberg New York, pp 41–64

Prieur-Richard AH, Lavorel S (2000) Invasions: the perspective of diverse plant communities. Aust J Ecol 25:1–7

Prieur-Richard AH, Lavorel S, Grigulis K, Dos Santos A (2000) Plant community diversity and invasibility by exotics: invasion of Mediterranean old fields by *Conyza bonariensis* and *Conyza canadiensis*. Ecol Lett 3:412–422

Reich PB, Knops J, Tilman D, et al. (2001) Plant diversity enhances ecosystem responses to elevated CO2 and nitrogen deposition. Nature 410:809–812

Reich PB, Wright IJ, Cavender-Bares J, et al. (2003) The evolution of plant functional variation: traits, spectra, and strategies. Int J Plant Sci 164 [Suppl]:143–164

Robinson GR, Quinn JF, Stanton ML (1995) Invasibility of experimental habitat islands in a California winter annual grassland. Ecology 76:786–794

Roscher C, Schumacher J, Baade J, et al. (2004) The role of biodiversity for element cycling and trophic interactions: an experimental approach in a grassland community. Basic Appl Ecol 5:107–121

Roscher C, Temperton VM, Scherer-Lorenzen M, et al. (2005) Overyielding in experimental grassland communities irrespective of species pool or spatial scale. Ecol Lett 8:419–429

Sala OE, Chapin FS III, Armesto JJ, et al. (2000) Global biodiversity scenarios for the year 2100. Science 287:1770–1774

Sankaran M, McNaughton SJ (1999) Determinants of biodiversity regulate compositional stability of communities. Nature 401:691–693

Sax DF, Gaines SD (2003) Species diversity: from global decreases to local increases. Trends Ecol Evol 18:561–566

Scheiner SM, Gurevitch J (eds) (1993) Design and analysis of ecological experiments. Chapman and Hall, New York

Scherer-Lorenzen M (1999) Effects of plant diversity on ecosystem processes in experimental grassland communities. Bayr Forum Okol 75:195

Scherer-Lorenzen M, Palmborg C, Prinz A, Schulze ED (2003) The role of plant diversity and composition for nitrate leaching in grasslands. Ecology 84:1539–1552

Scherer-Lorenzen M, Körner C, Schulze ED (2005a) The functional significance of forest diversity: the starting point. In: Scherer-Lorenzen M, Körner C, Schulze ED (eds) Forest diversity and function. Springer, Berlin Heidelberg New York, pp 3–12

Scherer-Lorenzen M, Potvin C, Koricheva J, et al. (2005b) The design of experimental tree plantations for functional biodiversity research. In: Scherer-Lorenzen M, Körner C, Schulze ED (eds) Forest diversity and function. Springer, Berlin Heidelberg New York, pp 347–376

Schläpfer F, Schmid B (1999) Ecosystem effects of biodiversity: a classification of hypothesis and exploration of empirical results. Ecol Appl 9:893–912

Schläpfer F, Pfisterer AB, Schmid B (2005) Non-random species extinction and plant production: implications for ecosystem functioning. J Appl Ecol 42:13–24
Schmid B (2002) The species richness–productivity controversy. Trends Ecol Evol 17:113–114
Schmid B, Hector A (2004) The value of biodiversity experiments. Basic Appl Ecol 5:535–542
Schmid B, Hector A, Huston MA, et al. (2002a) The design and analysis of biodiversity experiments. In: Loreau M, Naeem S, Inchausti P (eds) Biodiversity and ecosystem functioning:
  synthesis and perspectives. Oxford University Press, Oxford, pp 61–75
Schmid B, Joshi J, Schläpfer F (2002b) Empirical evidence for biodiversity–ecosytem func-
  tioning relationships. In: Kinzig AP, Pacala SW, Tilman D (eds) The functional conse-
  quences of biodiversity. Princeton University Press, Princeton, N.J., pp 120–150
Schroth G (1999) A review of belowground interactions in agroforestry, focussing on mech-
  anisms of management options. Agrofor Syst 43:5–34
Schulze ED, Mooney HA (1993) (eds) Biodiversity and ecosystem function. Springer, Berlin
  Heidelberg New York
Schwartz MW, Brigham CA, Hoeksema JD, et al. (2000) Linking biodiversity to ecosystem
  function: implications for conservation ecology. Oecologia 122:297–305
Shmida A, Ellner S (1984) Coexistence of plant species with similar niches. Vegetatio 58:29–55
Simberloff D (1980) A succession of paradigms in ecology: essentialism to materialism and
  probabilism. Synthesis 43:3–39
Smith MD, Knapp AK (2003) Dominant species maintain ecosystem function with non-ran-
  dom species loss. Ecol Lett 6:509–517
Smith MD, Wilcox JD, Kelly T, Knapp AK (2004) Dominance not richness determines invasi-
  bility of tallgrass prairie. Oikos 106:253–262
Smith TM, Shugart HH, Woodward FI (1997) Plant functional types. Cambridge University
  Press, Cambridge
Spaèková I, Lepš J (2001) Procedure for separating the selection effect from other effects in
  diversity–productivity relationship. Ecol Lett 4:585–594
Spehn EM, Joshi J, Schmid B, et al. (2000) Above-ground resource use increases with plant
  species richness in experimental grassland ecosystems. Funct Ecol 14:326–337
Spehn EM, Scherer-Lorenzen M, Schmid B, et al. (2002) The role of legumes as a component of
  biodiversity in a cross-European study of grassland biomass nitrogen. Oikos 98:205–218
Spehn EM, Hector A, Joshi J, et al. (2005) Ecosystem effects of biodiversity manipulations in
  European grasslands. Ecol Monogr 75:37–63
Sprugel DG (1991) Disturbance, equilibrium, and environmental variability. Biol Conserv
  58:1–18
Stephan A, Meyer AH, Schmid B (2000) Plant diversity positively affects soil bacterial diver-
  sity in experimental grassland ecosystems. J Ecol 88:988–998
Stevens MHH, Carson WP (2001) Phenological complementarity, species diversity, and
  ecosystem function. Oikos 92:291–296
Stocker R, Körner C, Schmid B, et al. (1999) A field study of the effects of elevated CO2 and
  plant species diversity on ecosystem-level gas exchange in a planted calcareous grassland.
  Global Change Biol 5:95–105
Stöcklin J, Fischer M (1999) Plants with longer-lived seeds have lower local extinction rates in
  grassland remnants 1950–1985. Oecologia 120:539–543
Stoll P, Prati D (2001) Intraspecific aggregation alters competitive interactions in experimen-
  tal plant communities. Ecology 82:319–327
Sullivan G, Zedler JG (1999) Functional redundancy among tidal mash halophytes: a test.
  Oikos 84:246–260
Swift MJ, Anderson JM (1993) Biodiversity and ecosystem functioning in agricultural sys-
  tems. In: Schulze ED, Mooney HA (eds) Biodiversity and ecosystem function. (Ecological
  studies no. 99) Springer, Berlin Heidelberg New York, pp 15–41
Symstad AJ (2000) A test of the effects of functional group richness and composition on
  grassland invasibility. Ecology 81:99–109

Symstad AJ, Tilman D (2001) Diversity loss, recruitment limitation, and ecosystem functioning: lessons learned from a removal experiment. Oikos 92:424–435

Symstad AJ, Tilman D, Willson J, Knops J (1998) Species loss and ecosystem functioning: effects of species identity and community composition. Oikos 81:389–397

Symstad AJ, Chapin FD, Wall DH, et al. (2003) Long-term and large-scale perspectives on the relationship between biodiversity and ecosystem functioning. Bioscience 53:89–98

Thomas CD, Cameron A, Green RF, et al. (2004) Extinction risk from climate change. Nature 427:145–148

Thompson K, Hodgson JG, Grime JP, Burke MJ (2001) Plant traits and temporal scale: evidence from a 5-year invasion experiment using native species. J Ecol 89:1054–1060

Tilman D (1996) Biodiversity: population versus ecosystem stability. Ecology 77:350–363

Tilman D (1997) Distinguishing between the effects of species diversity and species composition. Oikos 80:185–185

Tilman D (1999) The ecological consequences of changes in biodiversity: a search for general principles. Ecology 80:1455–1474

Tilman D, Downing JA (1994) Biodiversity and stability in grasslands. Nature 367:363–365

Tilman D, Lehman C (2002) Biodiversity, composition, and ecosystem processes: theory and concepts. In: Kinzig AP, Pacala SW, Tilman D (eds) The functional consequences of biodiversity. Princeton University Press, Princeton, N.J., pp 9–41

Tilman D, Wedin D, Knops J (1996) Productivity and sustainability influenced by biodiversity in grassland ecosystems. Nature 379:718–720

Tilman D, Knops J, Wedin D, et al. (1997a) The influence of functional diversity and composition on ecosystem processes. Science 277:1300–1302

Tilman D, Naeem S, Knops J, et al. (1997b) Biodiversity and ecosystem properties. Science 278:1866–1867

Tilman D, Lehman CL, Bristow CE (1998) Diversity–stability relationships: statistical inevitability or ecological consequence? Am Nat 151:277–282

Tilman D, Reich PB, Knops JH, et al. (2001) Diversity and productivity in a long-term grassland experiment. Science 294:843–845

Tilman D, Knops J, Wedin D, Reich P (2002a) Experimental and observational studies of diversity, productivity, and stability. In: Kinzig AP, Pacala SW, Tilman D (eds) The functional consequences of biodiversity, Princeton University Press, Princeton, N.J., pp 42–70

Tilman D, Knops J, Wedin D, Reich P (2002b) Plant diversity and composition: effects on productivity and nutrient dynamics of experimental grasslands. In: Loreau M, Naeem S, Inchausti P (eds) Biodiversity and ecosystem functioning: synthesis and perspectives. Oxford University Press, Oxford, pp 21–35

Trenbath BR (1974) Biomass productivity of mixtures. Adv Agron 26:117–210

Trenbath BR (1976) Plant interactions in mixed crop communities. In: Papendick RI, Sanchez PA, Triplett GB (eds) Multiple cropping. ASA special publication no. 27. American Society of Agronomy, Madison, Wis., pp 129–170

Troumbis AY, Memtsas D (2000) Observational evidence that diversity may increase productivity in Mediterranean shrublands. Oecologia 125:101–108

Underwood AJ (1997) Experiments in ecology. Cambridge University Press, Cambridge

Van der Heijden MGA, Cornelissen JHC (2002) The critical role of plant–microbe interactions on biodiversity and ecosystem functioning: arbuscular mycorrhizal associations as an example. In: Loreau M, Naeem S, Inchausti P (eds) Biodiversity and ecosystem functioning: synthesis and perspectives. Oxford University Press, Oxford, pp 181–192

Van der Heijden MGA, Klironomos JN, Ursic M, et al. (1998) Mycorrhizal fungal diversity determines plant biodiversity, ecosystem variability and productivity. Nature 396:69–72

Van der Maarel E (1993) Some remarks on disturbance and its relation to diversity and stability. J Veg Sci 4:733–736

Van der Putten WH, Mortimer SR, Hedlund K, et al. (2000) Plant species diversity as a driver of early succession in abandoned fields: a multi-site approach. Oecologia 124:91–99

Van Ruijven J, Berendse F (2003) Positive effects of plant species diversity on productivity in the absence of legumes. Ecol Lett 6:170–175

Van Ruijven, J, Berendse F (2005) Diversity productivity relationships: initial effects, long-term patterns, and underlying mechanisms. Proc Natl Acad Sci USA 102:695–700

Van Ruijven, J, de Deyn GB, Berendse F (2003) Diversity reduces invasibility in experimental plant communities: the role of plant species. Ecol Lett 6:910–918

Vandermeer J (1989) The ecology of intercropping. Cambridge University Press, Cambridge

Vandermeer J, Lawrence D, Symstad AJ, Hobbie S (2002) Effect of biodiversity on ecosystem functioning in managed ecosystems. In: Loreau M, Naeem S, Inchausti P (eds) Biodiversity and ecosystem functioning: synthesis and perspectives. Oxford University Press, Oxford, pp 221–233

Waide RB, Willig MR, Steiner CF, et al. (1999) The relationship between productivity and species richness. Annu Rev Ecol Syst 30:257–300

Walker BH (1992) Biodiversity and ecological redundancy. Conserv Biol 6:18–23

Walker BH, Kinzig A, Langridge J (1999) Plant attribute diversity, resilience, and ecosystem function: the nature and significance of dominant and minor species. Ecosystems 2:95–113

Walker LR, Vitousek PM (1991) An invader alters germination and growth of a native dominant tree in Hawaii. Ecology 72:1449–1455

Wall DH (1999) Biodiversity and ecosystem functioning. BioScience 49:107–108

Wardle DA (1999) Is sampling effect a problem for experiments investigating biodiversity–ecosystem function relationships? Oikos 87:403–407

Wardle DA (2001a) Experimental demonstration that plant diversity reduces invasibility – evidence of a biological mechanism or a consequence of sampling effect? Oikos 95:161–170

Wardle DA (2001b) No observational evidence for diversity enhancing productivity in Mediterranean shrublands. Oecologia 129:620–621

Wardle DA, Grime JP (2003) Biodiversity and stability of grassland ecosystem functioning. Oikos 100:622–623

Wardle DA, Bonner KI, Nicholson KS (1997) Biodiversity and plant litter: experimental evidence which does not support the view that enhanced species richness improves ecosystem function. Oikos 79:247–258

Wardle DA, Bonner KI, Barker GM, et al. (1999) Plant removals in perennial grassland: vegetation dynamics, decomposers, soil biodiversity, and ecosystem properties. Ecol Monogr 69:535–568

Wardle DA, Bonner KI, Barker GM (2000a) Stability of ecosystem properties in response to above-ground functional group richness and composition. Oikos 89:11–23

Wardle DA, Huston MA, Grime JP, et al. (2000b) Biodiversity and ecosystem function: an issue in ecology. Bull Ecol Soc Am 81:235–239

Wardle DA, Van der Putten WH (2002) Biodiversity, ecosystem functioning and above-ground–belowground linkages. In: Loreau M, Naeem S, Inchausti P (eds) Biodiversity and ecosystem functioning: synthesis and perspectives. Oxford University Press, Oxford, pp155–168

Weiher E, Van der Werf A, Thompson K, et al. (1999) Challenging Theophrastus: a common core list of plant traits for functional ecology. J Veg Sci 10:609–620

Weltzin JF, Muth NZ, Von Holle B, Cole PG (2003) Genetic diversity and invasibility: a test using a model system with a novel experimental design. Oikos 103:505–518

Westoby M, Leishman M (1997) Categorizing plant species into functional types. In: Smith TM, Shugart HH, Woodward FI (eds) Plant functional types. Cambridge University Press, Cambridge, pp 104–121

White PS, Jentsch A (2001) The search for generality in studies of disturbance and ecosystem dynamics. Prog Bot 63:349–399

White PS, Jentsch A (2004) Disturbance, succession and community assembly in terrestrial plant communities. In: Temperton VM, Hobbs R, Fattorini M, Halle S (eds) Assembly rules in restoration ecology – bridging the gap between theory and practise. Island Press, Washington, D.C., pp 342–366

Whittaker RJ, Heegaard E (2003) What is the observed relationship between species richness and productivity? Ecology 84:3384–3389

Willey RW (1985) Evaluation and presentation of intercropping advantages. Exp Agric 21:119–133

Woodward FI, Kelly CK (1997) Plant functional types: towards a definition by environmental constraints. In: Smith TM, Shugart HH, Woodward FI (eds) Plant functional types: their relevance to ecosystem properties and global change. Cambridge University Press, Cambridge, pp 47–65

Yachi S, Loreau M (1999) Biodiversity and ecosystem productivity in a fluctuating environment: the insurance hypothesis. Proc Natl Acad Sci USA 96:1463–1468

Zak DR, Holmes WE, White DC, et al. (2003) Plant diversity, soil microbial communities, and ecosystem function: are there any links? Ecology 84:2042–2050

Zaller JG, Arnone JA III (1999) Earthworm responses to plant species' loss and elevated CO_2 in calcareous grassland. Plant Soil 208:1–8

Zavaleta ES, Hulvey KB (2004) Realistic species losses disproportionately reduce grassland resistance to biological invaders. Science 306:1175–1177

Carl Beierkuhnlein
Department of Biogeography
University of Bayreuth
95440 Bayreuth
Germany
Tel.: +49-921-552270
Fax: +49-921-552315
e-mail: Carl.Beierkuhnlein@uni-bayreuth.de

Carsten Nesshoever
Centre for Environmental Research, Halle-Leipzig
Permoserstrasse 15
04318 Leipzig
Germany
Tel.: +49-341-2353282
Fax: +49-341-2353191

Resource allocation in clonal plants

Markus Lötscher

1 Introduction

Vegetative growth is one of the species traits which has become increasingly important in determining the success of species in grassland communities (Pywell et al. 2003). In fact, the majority of species in grassland are clonal plants. Clonal growth (vegetative propagation) results in the production of new, genetically identical descendants (ramets) with the potential to become independent of the mother organism. Clonal growth modes have been classified based on the origin of the connections (root and stem derived), placement (aboveground, belowground), distance (short and long spacers) and persistence (long-lived, short-lived) of the clonal growth organs (Klimeš et al. 1997). The CLO-PLA2 database includes 2,749 species and focuses on the architectural aspects of clonal growth in vascular plants of central Europe (Klimeš and Klimešová 1999). In cultivated meadows and pastures about 85% of the species are clonal plants, mainly rhizome-(belowground) and stolon-(aboveground) forming types (Klimeš et al. 1997). The success of these species in grassland may be explained by:

1. The position of the meristems near or below the soil surface avoiding their loss during defoliation (Briske 1996).
2. The ability to spread physically and physiologically integrated ramets through heterogeneously distributed resource patches.
3. The potential to store resources in plagiotropic stems which are not removed by grazers.

Ecological aspects of clonal growth have been reviewed by Dietz and Steinlein (2001). The present review focuses on the physiological aspects of clonal growth to elucidate the cost/benefit relationships in maintaining physical connections between foraging ramets.

Progress in Botany, Vol. 67
© Springer-Verlag Berlin Heidelberg 2006

2 Foraging Behaviour

It has been suggested that the growth pattern of a stoloniferous or rhizomatous species is analogous to the search path of a foraging animal (Sutherland 1987 and references therein). Hutchings and De Kroon (1994) defined foraging modified after Slade and Hutchings (1987) as "the processes whereby an organism searches, or ramifies within its habitat, which enhance its acquisition of essential resources". The foraging theory predicts a decrease in internode length and/or an increase in branching frequencies in response to high local resource availability, because such a pattern would enable the plant to accumulate ramets in favourable microsites. Maximal rates of resource uptake on favourable sites and low costs on resource-poor sites due to low growth investments should maximise growth in a heterogeneous environment. The crucial variables for such behaviour were suggested to be the branching probability, internode length and branching angle (Sutherland 1987). Sutherland and Stillman (1988) and De Kroon and Hutchings (1995) summarized the responses of 26 species. Branching probability was most consistent in that high photon flux density and nutrient uptake enhanced branching, except for some rhizomatous species, whereas internode length was less consistent. Branching angle was not affected, but only three studies analysed this trait (however, see Stoll et al. 1998).

2.1 Vertical versus Horizontal Growth

Reductions in the red:far-red (R:FR) ratio of the canopy-filtered or plant-reflected light increased significantly internode length in many orthotropic plants. This response has been termed "foraging for light" (Ballaré et al. 1997) or "shade-avoidance response". The links between an environmental cue, gene expression and resultant phenotype have been intensively investigated (Nagy and Schäfer 2000; Schlichting and Smith 2002; Morelli and Ruberti 2002), and the effect of the phenotypic plasticity on relative fitness has been analysed by the introduction of transgenes which alter the plastic response of the plants to the environmental cue (Schmitt et al. 1999). When plants grow in close proximity, a vertical light gradient develops. In this system the chance of survival depends on the potential to position leaves in the uppermost canopy layer. Thus, plants which can adjust shoot height to canopy height have a selective advantage. In relatively nutrient-rich habitats the probability of the existence of a vertical light gradient, and its duration, might be higher than that of horizontally distributed light gaps. As plasticity is favoured when environmental cues are reliable and operate long enough for plants to show a response (Schlichting and Smith 2002; Alpert and Simms 2002) plasticity in shoot height rather than vertical spread might be advantageous in many habitats. In fact, horizontal foraging traits (internode length) increased by 0–74% and vertical foraging traits (mainly increased petiole length) by 20–300% in response to shading in stoloniferous species (Dong 1993, 1995; Price and Hutchings 1996; Huber and Wiggerman 1997; Lötscher and Nösberger 1997; Stuefer and Huber 1998; Marcuvitz and Turkington 2000; Van Kleunen and Fischer 2001). This led to a reworking of the foraging concept. De Kroon and Hutchings (1995) suggested that morphological responses of shoots and roots are much more important than the morphological plasticity of rhizomes and stolons for the placement of leaves and root tips in patches of high resource availability. Campbell et al. (1991) concluded that subordinate plants appear to have a greater potential for precise vertical foraging whereas dominant plants are relatively insensitive to fine-grained patchiness. Precise foraging for light was mainly achieved in species with orthotropic stems where leaves arise in the upper layer of the canopy, whereas for turf grasses and rosette forbs it has been assumed that basal meristems are too remote from leaf tips to allow fine adjustments in position (Grime and Mackey 2002). However, some forbs with plagiotropic stems position their leaves by growth zones in the petioles while photoperception seems to occur in the tips rather than the base (Thompson 1995), thus allowing them to precisely position their foliage in the upper canopy layer (Leeflang et al. 1998).

2.2 Lateral Spread

Although plasticity in internode length (distance between ramets) may contribute little to the success in locating and occupying resource-rich patches, the internode length defines the growth mode. Within a given time period species with longer internodes explore more space than species with short internodes, which increases the probability of locating favourable patches. Thus, a higher proportion of fast and widely lateral spreading species would be expected in pastures rather than in meadows, because animal grazing creates a more patchy distribution of resources than mowing. In fact, in species-rich grasslands the proportion of species with a fast spreading mode (Stammel et al. 2003) and a wide lateral spread (Kahmen et al. 2002) was statistically higher in pastures than in meadows. However, in these studies pastures were defoliated twice a year or even during the whole vegetation period whereas the meadows were defoliated once in late summer or autumn. Thus, canopy height, frequency of defoliation and date of defoliation rather than patchiness of the grassland might be other reasons for the shift in species composition.

2.3 Branching

Increased branching when resources are high is part of the foraging behaviour of plants. Branching of clonal herbs and grasses has been investigated in terms of changes in light intensity (photon fluence density; PFD) and the R:FR ratio at the whole plant level (Casal et al. 1985, 1987, 1990; Solangaarachchi and Harper 1987; Robin et al. 1992; Huber and Stuefer 1997; Skálová et al. 1997; Stuefer and Huber 1998; Marcuvitz and Turkington 2000), changes in blue light (Gautier et al. 1998), and localised response to a changed R:FR ratio (Deregibus et al. 1985; Skálová and Krahulec 1992; Robin et al. 1994a,b; Teuber and Laidlaw 1996; Lötscher and Nösberger 1997; Leeflang 1999; Hay et al. 2001; Héraut-Bron et al. 2001). The effect of light intensity and quality on branching has been analysed in some detail in *Trifolium repens*. Lowering PFD on the whole plant or a single branch delayed outgrow of branches (Solangaarachchi and Harper 1987; Thompson and Harper 1988; Kemball et al. 1992; Lötscher and Nösberger 1997). It could be argued that the effects of low PFD were due to the low fluence rate of blue light. However, blue light did not affect branching, although plants grown without blue light had longer petioles, larger leaves and shorter internodes on branches (Gautier et al. 1997, 1998). There is an increasing interest in how light quantity affects the morphological responses of plants (Vandenbussche et al. 2003).

Exposing whole plants to low R:FR ratios delayed branching significantly. However, there is still controversy concerning whether the light at the axillary bud is the sole signal affecting branch outgrowth. It might be advantageous for plagiotropic plants growing in a vertical light gradient if branching is not suppressed as long the leaves are positioned at the top of the canopy although the stem and its axillary buds perceive a low R:FR ratio. In fact, there is some evidence that the light perceived by leaves affects branching. Firstly, a delay in branching was induced when the developing leaf received low R:FR ratios but not when the light treatment was applied to the axillary bud (Robin et al. 1994a). Hay et al. (2001) suggested that there is a signalling substance which is produced in the developing leaf and transported to branches delaying outgrowth and the production of ramets only when the leaf is shaded. Secondly, shading of the plagiotropic stem did not reduce branching when the unfolded leaves were exposed to high light (Lötscher and Nösberger 1997). Furthermore, in a natural grassland stand all plants except *Trifolium fragiferum* were cut every other week at a height of 0, 5, 10 and 20 cm, respectively (Huber and Wiggerman 1997). In this experiment petiole length adapted to the stand height and leaves were positioned in the upper canopy layers whereas the stem remained shaded. Branching on primary ramets was not affected, but the number of secondary ramets was reduced. Thirdly, light with a low R:FR ratio reflected from mirrors or neighbouring plants without significantly reducing PFD on the plant did not affect branching in several clonal species (Leeflang 1999; Marcuvitz and Turkington 2000). Furthermore, Barthram (1997) showed that when *T. repens* was grown in association with grasses, increasing height of the grazed sward increased petiole length, but did not affect branching rates. The number of sheep was adjusted twice a week to maintain the swards at their target heights of 3, 5 and 10 cm, respectively. In a lightly grazed system not all unfolded leaves were removed, so that at least some leaves experienced favourable light conditions which might have facilitated branching. In contrast, when perennial ryegrass–white clover swards were cut at 4-week intervals to heights of 2, 4, 6, 8 and 10 cm, increasing the height at which they were cut reduced the number of branches per nodes and the proportion of *T. repens* in the sward (Acuña and Wilman 1993; Wilman and Acuña 1993). In this experiment, unfolded leaves were frequently defoliated and light was mainly intercepted by the developing leaves.

It is often argued that the lower number of branches in less frequently defoliated stands is due to the stronger shading of the plant, which inhibits branch development, e.g. Simon et al. (2004). As severe defoliation results in a loss of vegetative bud viability (Hay and Newton 1996) and defoliation increases the time until a branch appears (Lötscher and Nösberger 1997) frequent defoliation should reduce rather than increase the number of growing points per area. However, frequent defoliation increased the residual leaf area, probably as a result of the formation of shorter petioles (Hay and Newton 1996) and there was a strong correlation between residual leaf area and number of growing points (Simon et al. 2004). There might be a trade-off between assimilates invested into growth of petioles and growth of secondary ramets. As maintenance of a short sward by grazing and a high frequency of mowing reduce petiole length, relatively more assimilates might be available for branch development. Héraut-Bron et al. (2001) analysed the allocation of currently assimilated ^{14}C in *T. repens* plants whose stems were either shaded or unshaded. The treatment effects were rather small but the shading treatment, which favoured petiole growth, slightly increased the allocation of ^{14}C to the developing ramets and petioles at the main axis and this was accompanied by reduced branch production.

3 Clonal Integration

The benefits of physiological integration may include extended support of new ramets, recycling and sharing of resources among ramets, buffering of environmental heterogeneity and local stress, and regulation of intra-clonal competition among ramets through developmental control of ramet production (Jónsdóttir and Watson 1997). On the other hand, modular independence may be favourable when respiratory costs of maintaining rhizome or stolon tissue are high, or dilution of internal resources as the result of resource sharing among ramets leads to a higher risk of genet mortality (Jónsdóttir and Watson 1997). Kelly (1995) argues that the possible costs of maintenance and transport incurred by inter-ramet connections may be seen most reasonably to be a factor selecting for disintegration. However, costs to maintain interconnections in clonal plants have not yet been quantified, and the use of these structures as storage organs may exceed the costs of maintenance. Nevertheless, physiological independence of ramets could prevent the accumulation of somatic mutations, viral infections and plant pathogens in the clone (Hutchings and Bradbury 1986).

There have been different approaches to studying the extent of integration in clonal plants: intact plants growing in environments of different resource availability, comparison of growth of plants with intact connections between ramets and plants with severed ramets, and analysis of nutrient transport by labelling of nutrients.

3.1 Extended Support of New Ramets

One of the most important traits of clonal growth is the support of young ramets by resources from parent ramets. This allows the ramet to colonise new space and potentially outcompete neighbours before it becomes an autonomous individual. Thus, connected ramets may have a higher probability of survival in comparison with seedlings (Hutchings and Bradbury 1986). Dependence on photoassimilates may be more temporary than on water and mineral nutrients as a positive net export rate of photoassimilates is reached when the leaf has about 75% of its final area (e.g. Chapman et al. 1990), whereas initiation of root growth and therefore water and nutrient uptake generally occur at a later stage of ramet development. The dependence of a ramet on its connection to a parent ramet has been studied by comparing the survival and growth of severed apical ramets to those of connected ramets. The dependence of a ramet depended on its species, genotype and environmental conditions.

The survival rate of severed ramets in *Carex bigelowii* increased significantly with the number of apical ramets which remained connected after severance. In one genotype two connected apical ramets were sufficient to ensure survival, whereas in another genotype at least four ramets had to be connected to ensure a 100% survival rate (Jónsdóttir and Callaghan 1988). A significant genetic variation in survival of apical ramets after severing was also reported for *Ranunculus reptans* (Van Kleunen et al. 2000). In *Glechoma hederacea* growth and development after severance was limited in all fragments which had less than five connected ramets at the time of severance (Birch and Hutchings 1999). In *Holcus lanatus* survival and growth of tillers, which were less than 2 weeks old, decreased after severing (Bullock et al. 1994). As root development lags behind shoot development it would be interesting to know whether the root:shoot ratio of the disconnected fragments determines the developmental stage at which a severed fragment develops without restriction after severance. Although it seems to be an important factor, root growth was not reported in these studies. Survival might be improved when water loss is reduced after severance, for instance by shading (Lau and Young 1988) or defoliation (Birch and Hutchings 1999; Stuefer and Huber 1999).

3.2 Cost/Benefit

Many studies analysed sharing of resources among ramets by exposing fragments of connected ramets to differential conditions of resource availability such as light, N, water (Alpert and Mooney 1986; Alpert 1991, 1999a,b; Evans 1991, 1992; Friedman and Alpert 1991; Stuefer and Hutchings 1994; Fransen et al. 1996), salinity (Hester et al. 1994; Shumway 1995; Pennings and Callaway 2000), and burial (Yu et al. 2002). Differences in biomass accumulation of connected and severed fragments were then used to estimate the direction of resource transport and degree of clonal integration. Fragments from resource-poor environments accumulated more biomass when they were connected to fragments from resource-rich environments in comparison with severed fragments. Biomass gain was more significant when the fragment of the resource-poor environment was younger than the fragment of the resource-rich environment, and less distinct in the opposite direction. Such enhanced growth has been interpreted to reflect the transport of resources from donor to recipient fragments. In comparison with severed fragments, biomass gain was significantly higher in connected recipient fragments, whereas biomass accumulation of the donor fragments was not always significantly lower. From this it has been concluded that gain in recipient fragments was achieved at little or no cost to the donor fragment (Van Kleunen et al. 2000; Yu et al. 2002). However, this may depend on the size differences between the donor and recipient fragment, particularly when the recipient fragment is the younger part of the plant. A small amount of resources spent by a large donor fragment may significantly enhance growth of a small recipient fragment without provoking a significant reduction in growth of the donor fragment. By comparing the biomass gain (G, positive value) of recipient fragments with the biomass reduction (R, negative value) of donor fragments in ten comparisons of six studies (Friedman and Alpert 1991; Birch and Hutchings 1999; Alpert 1999a,b; Van Kleunen et al. 2000; Yu et al. 2002) a correlation between benefit for the recipient and costs for the donor was shown: G (g) $= -1.15R+0.28$, $r^2 = 0.65$ ($P<0.001$). This indicates that the benefit in terms of the biomass of recipients involves approximately the same costs in the form of reduced biomass production in donors. Thus, total biomass production (M) was similar for severed and intact plants: $\log(M_{severed})=1.04\log(M_{intact})-0.06$, $r^2=0.94$ ($P<0.001$).

However, severing of a clone may break apical dominance and initiate the growth of lateral branches. This may result in a pattern of resource allocation which differs from that of intact plants. Furthermore, in many studies heterogeneous and homogeneous treatments provided different total amounts of resources (Hutchings 1999). There are only a few studies which compared biomass production in homogeneous and heterogeneous environments in intact plants with similar amounts of resource supply. These studies indicated that biomass production is higher when resources are heterogeneously distributed. When *G. hederacea* was grown over an area of patchily distributed nutrient availability, total biomass production was higher than for plants grown with the same amount of nutrient supply but homogeneously distributed (Birch and Hutchings 1994, reviewed by Hutchings et al. 2000).

Besides sharing of resources between fragments there is some evidence that "signals" are transported between ramets. Physical connections between fragments of a *Fragaria chiloensis* clone known to share resources induced them to segregate their roots, thus increasing clonal performance (Holzapfel and Alpert 2003). The authors suggested that the mechanism for root segregation might depend upon transport of a signal between connected fragments.

In many studies it has been assumed that each ramet has the same function in terms of resource uptake and production of vegetative or generative offspring. However, it has been recognised that within an aggregation of physically connected ramets the function of ramets may differ, and the hypothesis of division of labour among ramets has been formulated.

3.3 Division of Labour

Division of labour in clonal plants has been defined as the coordinated specialisation of individual ramets for the acquisition of different resources, combined with the reciprocal exchange of these resources between ramets (Alpert and Stuefer 1997; Hutchings and Wijesinghe 1997). Coordinated specialisation appears to be developmentally programmed and environmentally induced. The first case is typical for species which consist of a long-lived rhizome and root systems with aerial shoots of a relatively short lifespan. In these species older leafless ramets import carbohydrates from younger leafy ramets whereas the leafy ramets import water and nutrients from the leafless ones (Jónsdóttir and Callaghan 1988; Jónsdóttir and Watson 1997; D'Hertefeldt and Falkengren-Grerup 2002). The characteristics of this pattern of specialisation are that it develops independently of resource availabilities. However, the ratio of differently specialized ramets can change within clones according to environmental conditions as the formation of shootless ramets increased at higher levels of shoot density (Charpentier and Stuefer 1999). In the second type of division of labour specialisation is induced by disturbance and environmental heterogeneity.

Defoliation by grazing and cutting reduces significantly the fraction of foliated ramets. For instance, under continuous grazing, only the three youngest ramets of a stem axis are foliated (Chapman 1986) so that an individual plant consists of many leafless, partially rooted ramets connected to young, leafy ramets with (if at all) small roots. In this system each ramet has its specific function like assimilation of photosynthates, uptake of water and nutrients, storage of reserves and maintenance of resource pathways and connections between ramets. In situations where defoliation occurs the functions of storage and connecting ramets are likely to be the most common tasks of a ramet in stoloniferous species because often less than half of the ramets bear a root system depending on season and defoliation management (Newton and Hay 1994).

It has been hypothesized that if clonal fragments were placed in an heterogeneous environment each ramet should become specialised in order to acquire a resource that was relatively abundant in its microsite and relatively scarce in the microsite of connected ramets (Alpert and Stuefer 1997). Ramets in both patch types should adjust their root–shoot allocation in a way to increase the uptake of a locally abundant resource. Evidence for environmentally induced division of labour was mainly found when light and water were the reciprocal resources (Salzman and Parker 1985; Alpert and Mooney 1986; Evans and Whitney 1992; Stuefer 1995; Stuefer et al. 1994, 1996). A simulation model suggests that the conductivities for water transport in stolon internodes are crucial in determining the degree of specialisation and co-operation in clonal plants (Stuefer et al. 1998). Experiments that varied N levels instead of water did not always find that resource sharing induced specialisation of resource uptake (Stuefer and Hutchings 1994; but see Alpert and Stuefer 1997). A reason for this might be that transfer of N tends to be directed from older to younger ramets and may depend on the bulk flow of water in xylem. As pointed out by Alpert and Stuefer (1997), a major drawback of the specialisation of resource uptake is that when fragments fall apart specialised ramets may suffer higher mortality (if the specialisation goes in the wrong direction) than unspecialised ramets. High risks of fragmentation could therefore select against a strong intra-clonal specialisation and environmentally induced division of labour in clonal plants (Stuefer et al. 1998). Many of the aforementioned studies analysed growth and development of clonal fragments. However, sharing of resources in intact plants is more complex and often limited to sections of the plant.

3.4 Resource Transport and Sectoriality

The distribution of resources is often restricted to certain regions of the plant. Such sectorial patterns of distribution are closely related to shoot phyllotaxy. Thus, sectoriality can be regarded as a structural constraint on the transport of resources in both the phloem and xylem. The phyllotaxy of the shoot axis determines which plant parts will be physiologically integrated, i.e. it determines the physiological link between source organs and particular sinks (Marshall and Price 1997). These integrated physiological units are commonly expressed as sectorial compartments within which the movement of metabolites and dyes is largely confined (Watson and Caspar 1984). Studies on the distribution of resources amongst ramets and physiological organisation in clonal systems cover a wide range of species (Marshall 1990, 1996). However, there are only a few clonal species in which patterns of ramet integration have been systematically investigated as a function of the vascular anatomy (Stuefer 1996). In the following it will be shown that knowledge of the vascular anatomy is crucial for the interpretation of physiological integration.

3.4.1 *The Model Plants* Glechoma hederacea *and* Trifolium repens

The vascular anatomy of the two clonal species *G. hederacea* and *T. repens* has been studied in detail (Thomas 1987; Price et al. 1992; Sackville Hamilton and Hay 1998). The two species differ in that a ramet of *G. hederacea* consists of two leaves which emerge from opposite sides of the stolon whereas ramets of *T. repens* consist of one leaf alternately arising at successive ramets from the left and right side of the stolon. Buds in the axil of each leaf may form an inflorescence or branch. In both species the vascular architecture of a stolon is characterised by four main axial vascular bundles, an upper and a lower one on each side of the stolon, that run the length of the stolon. There are no cross-connections between bundles of the left and right side of the stolon. In *G. hederacea* leaf traces connect each leaf of a ramet to the proximal pair of the vascular bundles, which also provide the vascularisation for the secondary stolon developing from the axil of that leaf. In *T. repens* three leaf traces connect each leaf to the axial vascular bundles. One trace connects with the near-side axial bundles (on the same side of the stolon as the connected leaf) at the second node basal to the node bearing the leaf. The two other traces connect with the far-side axial bundles one and two nodes basal to the leaf. Thus, *G. hederacea* shows a more restricted distribution of assimilates exported from the leaf in that a leaf supplies the branches of the near-side of the stolon, whereas in *T. repens* a leaf supplies branches of both sides of the stolon. ^{14}C exported from a young leaf supports the apex, but the bulk of transport is mainly basipetal (Chapman et al. 1991; Kemball and Marshall 1995). On older sections of the stolon the main sinks are the two branches proximal to the node with the source leaf and the branch at the node with the source leaf reflecting the leaf traces (Chapman et al. 1991; Kemball and Marshall 1995), and indicating a preferential acropetal transport of assimilates after entering the axial bundles. Manipulation of the plant by severing the stolon may strongly affect the distribution pattern in that new pathways are formed which do not occur in intact plants. For example, the leaf of the most basal node of a stolon fragment supplied the fragment with ^{14}C although cutting the stolon should have removed any of the leaf traces connected to the vascular bundles of the stolon (Kemball and Marshall 1995). Furthermore, local shading and partial defoliation may induce the establishment of new inter-orthostichy pathways and assimilate transport becomes more closely related to the metabolic demands of sinks rather than to their phyllotactic relations (Wardlaw 1990; Marshall 1996).

The missing cross-connections between near- and far-side axial bundles are reflected in a strong sectoriality of assimilate distribution. There is little movement of ^{14}C between two branches at a ramet in *G. hederacea* (Price et al. 1992) and a defoliated branch is not supported by its opposite non-defoliated branch (Price and Hutchings 1992; Price et al. 1996). This strong sectoriality in *G. hederacea* compared with the rather widespread pattern of assimilates distributed from the leaf in *T. repens* led to the conclusion that the formation of physiological integrated units is more distinct in *G. hederacea* than in *T. repens* (Stuefer 1996). However, the distribution of assimilates among branches has been rarely studied in detail. When the assimilate distribution from a source leaf of a branch was analysed, other branches were severed (Robin et al. 1989) or pooled so that assimilate distribution from near-side branches to far-side branches could not be analysed (Chapman et al. 1991, 1992a,b; Kemball and Marshall 1995). In the study of Kemball et al. (1992) only little ^{14}C moved from a near-side branch to a far-side branch. Thus, there is little evidence that opposite branches in *T. repens* are more integrated than those of *G. hederacea*.

Distribution patterns of nutrients also suggest that the vascular architecture constrains the physiological integration among branches in *T. repens*. Roots have vascular connections that supply the near-side axial bundle of the stolon and axial bundles of the branch arising at the node. ^{32}P and ^{45}Ca exported from a nodal root move preferentially to the branch at the rooted node and to distal near-side leaves and branches (Chapman and Hay 1993; Kemball and Marshall 1994; Lötscher and Hay 1996a,b, 1997). The only marginal supply from far-side branches continued over a 3-day interval indicating that potential redistribution of ^{32}P via leaves did not remove the sectorial pattern (Lötscher and Hay 1996b). However, manipulation of intra-plant source–sink relationships by severance of far-side roots and defoliation of parent stolon leaves promoted lateral transport to the far-side branches (Lötscher and Hay 1996a).

The strong sectoriality is also evident when the development of branches along irregularly rooted stolons is studied. Branches distal to the rooted node showed a delayed outgrowth of the axillary bud, a slower elongation rate and a slower dry matter accumulation when they where positioned on the far-side of the stolon (Lötscher and Nösberger 1996; Thomas et al. 2002). Recently Thomas et al. (2003, 2004) showed that bud outgrowth and branch development on non-rooted stolon fragments are regulated by acropetal transport of root-supplied resources and competition among sinks for root-derived resources.

3.4.2 Complexity of the Clone

The degree of physiological integration within a clone may largely depend on the structural complexity of the clone (Marshall and Price 1997), and sectoriality might be as common in most grasses as it is in herbaceous species (Derner and Briske 1998). For instance, the detailed studies on *G. hederacea* and *T. repens* revealed that both species develop physiologically independent units. Still, constraints in resource sharing occur mainly between branches which develop on opposite sides of the parent axis, whereas resources are more freely distributed along an axis of sequentially developed ramets and between the axis and its branches. Differences in the degree of resource sharing within a clone were also found in rhizomatous and caespitose grasses (Derner and Briske 1998). Resource sharing among ramet generations was high but at the level of the entire clone there was a high degree of ramet independence. Extensive physiological integration reported for *Carex* species and *Podophyllum peltatum* (Jónsdóttir and Watson 1997; D'Hertefeldt and Falkengren-Grerup 2002) does not conflict with the findings of Derner and Briske (1998) as the clones consisted only of a single axis of sequentially developed unbranched ramets.

3.4.3 Genetic Variation

Evaluation of the importance of physiological integration on the benefits of resource capture within a heterogeneous environment would be most promising if the behaviour of genotypes with a differential degree of physiological integration could be analysed. However, there are only a few studies which analysed genetic variation in clonal integration and none of them compared growth of genotypes under realistic field conditions. Genetic variation in clonal integration was found in *Fragaria chiloensis* (Alpert 1999a). Clones from grassland where resources are relatively uniform had low rates of resource sharing compared to clones from sand dunes, where resources are more patchy. The authors concluded that resource sharing between fragments within clones is likely to be disadvantageous in uniform habitats and advantageous in patchy ones. However, the mechanisms which explain the variation in resource transport remain to be elucidated (see also Holzapfel and Alpert 2003). Van Kleunen et al. (2000) compared *Ranunculus reptans* genotypes from heterogeneous microhabitats with those from homogeneous ones and found significant variation among genotypes within populations rather than among populations.

In *T. repens* distribution of resources within a clone was more restricted in genotypes with a high number of vascular bundles within stolons than in those with a low number. Although the functional mechanisms were not studied it seems likely that a high number of vascular bundles facilitates a more precise allocation pattern of resources. The number of vascular bundles correlated negatively with leaf size and stolon diameter (Lötscher and Hay 1995, 1996a) indicating that small-leaved genotypes were physiologically more integrated than large-leaved genotypes. In contrast, Welham et al. (2002) used four *T. repens* genotypes differing in leaf size but found no evidence for genotypic variance of physiological variation.

4 Importance of Storage Pools

In the context of clonal growth studies the function of inter-ramet connections were mainly discussed in terms of intra-plant resource distribution and costs which arise in maintaining these structures. Only recently has it been recognised that ramet-connecting structures such as rhizomes and stolons serve as storage pools (Stuefer and Huber 1999; Suzuki and Stuefer 1999; Klimeš and Klimešová 2002; Price et al. 2002). Still, there is little information on allocation patterns of remobilised reserves among ramets and on the importance of reserves in situations of spatially heterogeneous resource availability. Although there is a vast body of evidence that recently assimilated resources were transported from mother ramets to daughter ramets and from well-supplied ramets to starving ramets, much less is known about the role of stored resources.

Resources used to support starving ramets may originate from current assimilation or from remobilisation of stored resources. In both scenarios support of starving ramets reduces the potential of building up a storage pool, i.e. resources which might be essential in future growth and winter survival (Bouchart et al. 1998; Lüscher et al. 2001; Frankow-Lindberg 2001). Especially in situations of defoliation, stored resources are crucial for fast regrowth to sustain a plant's position among competing neighbours. The importance of stored resources for regrowth has been extensively investigated in some of the agriculturally most important clonal plants.

In *T. repens* defoliation immediately reduces the concentration of water-soluble carbohydrates and starch in shoots and roots (Kang and Brink 1995; Singh and Sale 1997; Lawson et al. 2000; Simon et al. 2004). The time taken to replenish the reserve pools was approximately 4 weeks. More frequent defoliation reduced the reserve pools to lower levels, reduced leaf appearance rate and stolon elongation rate and reduced survival of stolons (Lawson et al. 2000; Simon et al. 2004). In grasses frequent defoliation decreased starch and sucrose levels but had little effect on glucose, fructose and fructan levels (Klimeš and Klimešová 2002; Donaghy and Fulkerson 2002). The negative effect of a higher defoliation frequency on carbohydrate reserves in stubble was more pronounced in the summer–autumn period than in the spring–summer period (Boschma et al. 2003). Carbohydrate reserves were estimated by determining etiolated regrowth to evaluate the suitability of the species for grazing (Lardner et al. 2003). Etiolated regrowth appeared to be more sensitive in detecting plants near death than concentrations of water-soluble carbohydrates and fructan (Boschma et al. 2003), indicating that total amount rather than concentration of reserves determines regrowth. However, the pronounced decrease in stored carbohydrates after defoliation may not fully account for investment in regrowing leaves, but might rather be the result of the use of carbohydrates in respiration. Thus, concentrations of stored carbohydrates showed rather low correlations with the regrowth potential (Volenec et al. 1996). Recent investigations of N reserves suggest that remobilisation of both N and carbohydrate might be crucial for fast regrowth.

Shoot dry matter production during regrowth is related to the concentration of soluble proteins rather than to that of total N (Avice et al. 1997). Specific proteins have been characterised in forage legumes. These proteins act as vegetative storage proteins (VSP) (Hendershot and Volenec 1993; Corre et al. 1996; Cunningham and Volenec 1996). VSPs are defined as proteins which are preferentially synthesised during the development of storage organs and depleted from storage organs during reactivation of meristems. Their abundance should greatly exceed that of other proteins in perenniating organs (Cyr and Bewley 1990; Staswick 1994; Volenec et al. 1996). In *T. repens* VSPs account for at least 20% of total soluble proteins (Corre et al. 1996). The relative abundance of the VSPs was found to decrease during early regrowth and increased in later stages of regrowth (Goulas et al. 2001, 2002). Recently some polypeptides were identified in *Lolium perenne* which satisfied some of the criteria for classification as VSPs. However, they do not satisfy the VSP criterion of high abundance (Cunningham and Volenec 1996; Li et al. 1996; Louahlia et al. 1999).

Leaf growth for the first 2 days after defoliation was found to a large extent to be dependent on the use of C and N reserves. The quantitative contribution of current photosynthates to the C required for shoot regrowth starts to outweigh that of the stored carbohydrate by the second day following defoliation, whereas N reserves were the dominant source until 3 days after defoliation (Schnyder and De Visser 1999; Morvan-Bertrand et al. 1999a,b; Lattanzi et al. 2004a,b). Thus, the dependence of growth zones on reserves was stronger for N than for C (De Visser et al. 1997). Mobilisation of N compounds from tissues remaining after defoliation play an important role in providing N for subsequent leaf growth since both N uptake from the soil and N_2 fixation are severely down-regulated for several days (Ourry et al. 1988, 1989; Kim et al. 1993; Thornton and Millard 1993, 1996; Thornton et al. 1993, 1994; Volenec et al. 1996).

The relative contribution of each N source – root uptake and remobilisation – depends on the N status of the plant and varies among species and cultivars (Thornton et al. 1994; Louahlia et al. 1999). Thornton et al. (1993) suggested that species which tend more to the C strategy (competitor, after Grime 1977) are species which rely relatively more on current root uptake to supply N for regrowth of laminae. In turn, the species which tend to the

S strategy (stress tolerator) are species which use remobilisation predominantly to supply N for regrowth. Differences in concentrations of carbohydrates and vegetative storage proteins among cultivars were also documented for *T. repens* (Kang and Brink 1995). Lawson et al. (2000) showed that cultivars with higher levels of starch and carbohydrates and higher concentrations in the stolons were less affected by defoliation and showed higher growth rate and higher survival of stolons. In contrast, Simon et al. (2004) compared a giant-type and a dwarf-type cultivar. Although the giant-type had higher starch and VSP concentrations the cutting-induced decreases in both reserves were similar for each cultivar.

5 Simulation Models

The foraging theory predicted decreased internode length and increased branching rate in favourable conditions. The consequences of these responses on the success of ramets placed in favourable conditions were determined by means of simulation models. The simulation models of Sutherland and Stillman (1988, 1990) showed that aggregation in the better sites resulted from ramets branching more and producing shorter internodes. Modifying the branching angle had no effect and aggregation depended on the internode length relative to the patch size. Cain (1994) and Cain et al. (1996) suggested from their model that when plants were able to concentrate in favourable habitats, this was usually caused by increased daughter ramet production. Variation in growth angles had little impact on the ability of ramets to locate favourable patches, but increased the ability of clones to remain in favourable patches once found. Oborny (1994) analysed the effect of growth rules (e.g. random search, branching probability, internode length, formation of meristem banks) on the effectiveness of space occupation. The results of the simulation suggested that plastic growth forms are not necessarily more advantageous than rigid structures. However, the success of a growth form in resource acquisition depended on the predictability of resource availability in space and time (Oborny and Cain 1997). Patch size and predictability of spatio-temporal resource availability were also important factors in the simulation model of Piqueras et al. (1999). In this model *Trientalis europaea* was effective at concentrating its ramets in favourable patches, but this process was strongly influenced by patch size. The larger patches resulted in the highest levels of ramet accumulation. Ramet aggregation was mainly due to the enhanced performance of clones located initially in the favourable patches, or clones that were located in favourable patches by chance. Winkler and Stöcklin (2000) and Winkler and Fischer (2002) predicted optimal clonal plant life histories by introducing relative allocation to sexual and vegetative reproduction. Simultaneous trade-offs between sexual and vegetative reproduction and between the length and number of spacers led to fitness optima at intermediate parameter values, depending on the success of seedling establishment. Herben and Suzuki (2002) introduced resource translocation and considered maintenance costs within a set of physiologically integrated ramets. Resource translocation increased ramet density and made the plant more competitive. If translocation was involved, some of the plants in a highly competitive area were likely to get support from outside the high-competition patch.

6 Conclusions

The stimulating thoughts published on foraging behaviour and functional integration in the late 1980s (Hutchings and Bradbury 1986; Sutherland 1987) initiated numerous studies on resource translocation and biomass accumulation in clonal plants growing in a heterogeneous environment. Accumulation of biomass in favourable patches and accelerated unidirectional growth in unfavourable patches have been defined as foraging behaviour in clonal plants. Three variables have been suggested to be decisive for this behaviour: internode length, branching angle and branching probability. A relatively small variability in internode length induced by nutrients and light was already recognised in early studies and has been validated by numerous recent studies. On the other hand, it has been shown for many clonal plants that plasticity in vertical expansion is much higher than that of horizontal expansion. Thus, it is suggested that in scenarios with competing neighbours vertical expansion is much more crucial for survival than horizontal expansion, i.e. the search for favourable patches. This needs to be validated in more detail. Little attention has been paid to branching angle in clonal growth studies so that there is still little knowledge about the variability in this variable. Branching probability seems to be the most consistent variable in foraging behaviour. However, there is still lack of knowledge about the link between the site of light signal perception and site of reaction. There is some evidence that the spectral composition at the branch bud is not the sole signal affecting branch outgrowth but that the light situation of the whole plant or ramet might be involved in controlling branching. This suggests that branching probability is linked with the capacity of the ramet not only to spread horizontally, but also vertically.

Extensive research on their vascular architecture and labelling with radioactive tracers in *G. hederacea* and *T. repens* gave insight into the constraints and potential of the translocation of currently assimilated resources among connected ramets. The outcome of resource translocation in heterogeneous environments has been mainly studied in plant fragments in terms of biomass production. Although cost/benefit analyses have often been claimed, there is virtually no study which quantified mobilisation and translocation of the key resources including stored resources and their utilisation in growth and maintenance. Thus, there is still a great need for knowledge about stored resources in order to quantify the long-term benefits of physiological integration in clonal plants. To achieve an understanding of clonal growth requires the development of integrated research concepts by plant ecologists and physiologists. Labelling technology employing stable isotopes is a promising tool with which to analyse cost/benefit relationships of clonal plants foraging in heterogeneous environments. The results of such studies could be incorporated into simulation models, which aim to quantify cost/benefit relationships in clonal growth strategies.

References

Acuña GH, Wilman D (1993) Effects of cutting height on the productivity and composition of perennial ryegrass–white clover swards. J Agric Sci 121:29–37

Alpert P (1991) Nitrogen sharing among ramets increases clonal growth in *Fragaria chiloensis*. Ecology 72:69–80

Alpert P (1999a) Clonal integration in *Fragaria chiloensis* differs between populations: ramets from grassland are selfish. Oecologia 120:69–76

Alpert P (1999b) Effects of clonal integration on plant plasticity in *Fragaria chiloensis*. Plant Ecol 141:99–106

Alpert P, Mooney HA (1986) Resource sharing among ramets in the clonal herb, *Fragaria chiloensis*. Oecologia 70:227–233

Alpert P, Simms EL (2002) The relative advantages of plasticity and fixity in different environments: when is it good for a plant to adjust? Evol Ecol 16:285–297

Alpert P, Stuefer JF (1997) Division of labour in clonal plants. In: De Kroon H, Van Groenendal J (eds) The ecology and evolution of clonal plants. Backhuys, Leiden, pp 137–154

Avice JC, Ourry A, Lemaire G, Volenec JJ, Boucaud J (1997) Root protein and vegetative storage protein are key organic nutrients for alfalfa shoot regrowth. Crop Sci 37:1187–1193

Ballaré CL, Scopel AL, Sánchez RA (1997) Foraging for light: photosensory ecology and agricultural implications. Plant Cell Environ 20:820–825

Barthram GT (1997) Shoot characteristics of *Trifolium repens* grown in association with *Lolium perenne* or *Holcus lanatus* in pastures grazed by sheep. Grass Forage Sci 52:336–339

Birch CPD, Hutchings MJ (1994) Exploitation of patchily distributed soil resources by the clonal herb *Glechoma hederacea*. J Ecol 82:653–664

Birch CPD, Hutchings MJ (1999) Clonal segmentation. Plant Ecol 141:21–31

Boschma SP, Scott JM, Hill MJ, King JR, Lutton JJ (2003) Plant reserves of perennial grasses subjected to drought and defoliation stresses on the Northern Tablelands of New South Wales, Australia. Aust J Agric Res 54:819–828

Bouchart V, Macduff JH, Ourry A, Svenning MM, Gay AP, Simon JC, Boucaud J (1998) Seasonal pattern of accumulation and effects of low temperatures on storage compounds in *Trifolium repens*. Physiol Plant 104:65–74

Briske DD (1996) Strategies of plant survival in grazed systems: a functional interpretation. In: Lemaire G, Chapman D (eds) The ecology and management of grazing systems. CAB, Wallingford, pp 37–67

Bullock JM, Mortimer AM, Begon M (1994) Physiological integration among tillers of *Holcus lanatus*: age-dependence and responses to clipping and competition. New Phytol 128:737–747

Cain ML (1994) Consequences of foraging in clonal plant species. Ecology 75:933–944

Cain ML, Dudle DA, Evans JP (1996) Spatial models of foraging in clonal plant species. Am J Bot 83:76–85

Campbell BD, Grime JP, Mackey JML (1991) A trade-off between scale and precision in resource foraging. Oecologia 87:532–538

Casal JJ, Deregibus VA, Sánchez RA (1985) Variations in tiller dynamics and morphology in *Lolium multiflorum* Lam. vegetative and reproductive plants as affected by differences in red/far-red irradiation. Ann Bot 56:553–559

Casal JJ, Sánchez RA, Deregibus VA (1987) Tillering responses of *Lolium multiflorum* plants to changes of red/far-red ratio typical of sparse canopies. J Exp Bot 38:1432–1439

Casal JJ, Sánchez RA, Gibson D (1990) The significance of changes in the red/far-red ratio, associated with either neighbour plants or twilight, for tillering in *Lolium multiflorum* Lam. New Phytol 116:565–572

Chapman DF (1986) Development, removal, and death of white clover leaves under three grazing managements in hill country. N Z J Agric Res 29:39–47

Chapman DF, Hay MJM (1993) Translocation of phosphorus from nodal roots in two contrasting genotypes of white clover (*Trifolium repens*). Physiol Plant 89:323–330

Chapman DF, Robson MJ, Snaydon RW (1990) The carbon economy of developing leaves of white clover (*Trifolium repens* L.). Ann Bot 66:623–628

Chapman DF, Robson MJ, Snaydon RW (1991) The influence of leaf position and defoliation on the assimilation and translocation of carbon in white clover (*Trifolium repens* L). 1. Carbon distribution patterns. Ann Bot 67:295–302

Chapman DF, Robson MJ, Snaydon RW (1992a) The carbon economy of clonal plants of *Trifolium repens* L. J Exp Bot 43:427–434

Chapman DH, Robson MJ, Snaydon RW (1992b) Physiological integration in the clonal perennial herb *Trifolium repens* L. Oecologia 89:338–347

Charpentier A, Stuefer JF (1999) Functional specialization of ramets in *Scirpus maritimus*. Plant Ecol 141:129–136

Corre N, Bouchart V, Ourry A, Boucaud J (1996) Mobilization of nitrogen reserves during regrowth of defoliated *Trifolium repens* L. and identification of potential vegetative storage proteins. J Exp Bot 47:1111–1118

Cunningham SM, Volenec JJ (1996) Purification and characterization of vegetative storage proteins from alfalfa (*Medicago sativa* L.) taproots. J Plant Physiol 147:625–632

Cyr DR, Bewley JD (1990) Proteins in the roots of the perennial weeds chicory (*Cichorium intybus* L.) and dandelion (*Taraxacum officinale* Weber) are associated with overwintering. Planta 182:370–374

D'Hertefeldt T, Falkengren-Grerup U (2002) Extensive physiological integration in *Carex arenaria* and *Carex disticha* in relation to potassium and water availability. New Phytol 156:469–477

De Kroon H, Hutchings MJ (1995) Morphological plasticity in clonal plants: the foraging concept reconsidered. J Ecol 83:143–152

De Visser R, Vianden H, Schnyder H (1997) Kinetics and relative significance of remobilized and current C and N incorporation in leaf and root growth zones of *Lolium perenne* after defoliation: assessment by ^{13}C and ^{15}N steady-state labelling. Plant Cell Environ 20:37–46

Deregibus VA, Sánchez RA, Casal JJ, Trlica MJ (1985) Tillering responses to enrichment of red light beneath the canopy in a humid natural grassland. J Appl Ecol 22:199–206

Derner JD, Briske DD (1998) An isotopic (^{15}N) assessment of intraclonal regulation in C4 perennial grasses: ramet interdependence, independence or both. J Ecol 86:305–314

Dietz H, Steinlein T (2001) Ecological aspects of clonal growth in plants. Prog Bot 62:511–529

Donaghy DJ, Fulkerson WJ (2002) The impact of defoliation frequency and nitrogen fertilizer application in spring on summer survival of perennial ryegrass under grazing in subtropical Australia. Grass Forage Sci 57:351–359

Dong M (1993) Morphological plasticity of the clonal herb *Lamiastrum galeobdolon* (L.) Ehrend. & Polatschek in response to partial shading. New Phytol 124:291–300

Dong M (1995) Morphological responses to local light conditions in clonal herbs from contrasting habitats, and their modification due to physiological integration. Oecologia 101:282–288

Evans JP (1991) The effect of resource integration on fitness-related traits in a clonal dune perennial, *Hydrocotyle bonariensis*. Oecologia 86:268–275

Evans JP (1992) The effect of local resource availability and clonal integration on ramet functional morphology in *Hydrocotyle bonariensis*. Oecologia 89:265–276

Evans JP, Whitney S (1992) Clonal integration across a salt gradient by a nonhalophyte, *Hydrocotyle bonariensis* (Apiaceae). Am J Bot 79:1344–1347

Frankow-Lindberg BE (2001) Adaptation to winter stress in nine white clover populations: Changes in non-structural carbohydrates during exposure to simulated winter conditions and "spring" regrowth potential. Ann Bot 88:745–751

Fransen B, Van Rheenen JWA, Van Dijk A, Kreulen RDK (1996) High levels of inter-ramet water translocation in two rhizomatous *Carex* species, as quantified by deuterium labelling. Oecologia 106:73–84

Friedman D, Alpert P (1991) Reciprocal transport between ramets increases growth of *Fragaria chiloensis* when light and nitrogen occur in separate patches but only if patches are rich. Oecologia 86:76–80

Gautier H, Varlet-Grancher C, Baudry N (1997) Effects of blue light on the vertical colonization of space by white clover and their consequences for dry matter distribution. Ann Bot 80:665–671

Gautier H, Varlet-Grancher C, Baudry N (1998) Comparison of horizontal spread of white clover (*Trifolium repens* L.) grown under two artificial light sources differing in their content of blue light. Ann Bot 82:41–48

Goulas E, Le Dily F, Teissedre L, Corbel C, Robin C, Ourry A (2001) Vegetative storage proteins in white clover (*Trifolium repens* L.): quantitative and qualitative features. Ann Bot 88:789–795

Goulas E, Le Dily F, Simon JC, Ourry A (2002) Morphological pattern of development affects the contribution of nitrogen reserves to regrowth of defoliated white clover (*Trifolium repens* L.). J Exp Bot 53:1941–1948

Grime JP (1977) Evidence for the existence of three primary strategies in plants and their relevance to ecological and evolutionary theory. Am Nat 111:1169–1194

Grime JP, Mackey JML (2002) The role of plasticity in resource capture by plants. Evol Ecol 16:299–307

Hay MJM, Newton PCD (1996) Effect of severity of defoliation on the viability of reproductive and vegetative axillary buds of *Trifolium repens* L. Ann Bot 78:117–123

Hay MJM, Newton PCD, Robin C, Cresswell A (2001) Branching responses of a plagiotropic clonal herb to localised incidence of light simulating that reflected from vegetation. Oecologia 127:185–190

Hendershot KL, Volenec JJ (1993) Taproot nitrogen accumulation and use in overwintering alfalfa (*Medicago sativa* L.). J Plant Physiol 141:68–74

Héraut-Bron V, Robin C, Verlet-Grancher C, Guckert A (2001) Phytochrome-mediated effects on leaves of white clover: consequences for light interception by the plant under competition for light. Ann Bot 88:737–743

Herben T, Suzuki J (2002) A simulation study of the effects of architectural constraints and resource translocation on population structure and competition in clonal plants. Evol Ecol 15:403–423

Hester MW, McKee KL, Burdick DM, Koch MS, Flynn KM, Patterson S, Mendelssohn IA (1994) Clonal integration in *Spartina patens* across a nitrogen and salinity gradient. Can J Bot 72:767–771

Holzapfel C, Alpert P (2003) Root cooperation in a clonal plant: connected strawberries segregate roots. Oecologia 134:72–77

Huber H, Stuefer JF (1997) Shade-induced changes in the branching pattern of a stoloniferous herb: functional response or allometric effect? Oecologia 110:478–486

Huber H, Wiggerman L (1997) Shade avoidance in the clonal herb *Trifolium fragiferum*: a field study with experimentally manipulated vegetation height. Plant Ecol 130:53–62

Hutchings M, Bradbury IK (1986) Ecological perspectives on clonal perennial herbs. BioScience 36:178–182

Hutchings MJ (1999) Clonal plants as cooperative systems: benefits in heterogeneous environments. Plant Species Biol 14:1–10

Hutchings MJ, De Kroon H (1994) Foraging in plants: the role of morphological plasticity in resource acquisition. Adv Ecol Res 25:159–238

Hutchings MJ, Wijesinghe DK (1997) Patchy habitats, division of labour and growth dividends in clonal plants. Trends Ecol Evol 12:390–394

Hutchings MJ, Wijesinghe DK, John EA (2000) The effects of heterogeneous nutrient supply on plant performance: a survey of responses, with special reference to clonal herbs. In: Hutchings MJ, John EA, Stewart AJA (eds) The ecological consequences of environmental heterogeneity. Blackwell, Oxford, pp 91–130

Jónsdóttir IS, Callaghan TV (1988) Interrelationships between different generations of interconnected tillers of *Carex bigelowii*. Oikos 52:120–128

Jónsdóttir IS, Watson MA (1997) Extensive physiological integration: an adaptive trait in resource-poor environments? In: De Kroon H, Van Groenendal J (eds) The ecology and evolution of clonal plants. Backhuys, Leiden, pp 109–136

Kahmen S, Poschlod P, Schreiber K-F (2002) Conservation management of calcareous grasslands. Changes in plant species composition and response of functional traits during 25 years. Biol Conserv 104:319–328

Kang JH, Brink GE (1995) White clover morphology and physiology in response to defoliation interval. Crop Sci 35:264–269

Kelly CK (1995) Thoughts on clonal integration: facing the evolutionary context. Evol Ecol 9:575–585

Kemball WD, Marshall C (1994) The significance of nodal rooting in *Trifolium repens*: ^{32}P distribution and local growth responses. New Phytol 127:83–91

Kemball WD, Marshall C (1995) Clonal integration between parent and branch stolons in white clover: a developmental study. New Phytol 129:513–521

Kemball WD, Palmer MJ, Marshall C (1992) The effect of local shading and darkening on branch growth, development and survival in *Trifolium repens* and *Galium aparine*. Oikos 63:366–375

Kim TH, Ourry A, Boucaud J, Lemaire G (1993) Partitioning of nitrogen derived from N$_2$ fixation and reserves in nodulated *Medicago sativa* L. during regrowth. J Exp Bot 44:555–562

Klimeš L, Klimešová J (1999) CLO-PLA2 – a database of clonal plants in central Europe. Plant Ecol 141:9–19

Klimeš L, Klimešová J (2002) The effects of mowing and fertilization on carbohydrate reserves and regrowth of grass: do they promote plant coexistence in species-rich meadows? Evol Ecol 15:363–382

Klimeš L, Klimešová J, Hendriks R, Van Groenendael J (1997) Clonal plant architecture: a comparative analysis of form and function. In: De Kroon H, Van Groenendal J (eds) The ecology and evolution of clonal plants. Backhuys, Leiden, pp 1–29

Lardner HA, Wright SBM, Cohen RDH (2003) Assessing eight grass species for pasture by measuring etiolated spring regrowth. Can J Plant Sci 83:551–554

Lattanzi FA, Schnyder H, Thornton B (2004a) Defoliation effects on carbon and nitrogen substrate import and tissue-bound efflux in leaf growth zones of grasses. Plant Cell Environ 27:347–356

Lattanzi FA, Schnyder H, Thornton B (2004b) The sources of carbon and nitrogen supplying leaf growth. Assessment of the role of stores with compartmental models. Plant Physiol 137:383–395

Lau RR, Young DR (1988) Influence of physiological integration on survivorship and water relations in a clonal herb. Ecology 69:215–219

Lawson AR, Kelly KB, Sale PWG (2000) Defoliation frequency and cultivar effects on the storage and utilisation of stolon and root reserves in white clover. Aust J Agric Res 51:1039–1046

Leeflang L (1999) Are stoloniferous plants able to avoid neighbours in response to low R:FR ratios in reflected light? Plant Ecol 141:59–65

Leeflang L, During HJ, Werger MJA (1998) The role of petioles in light acquisition by Hydrocotyle vulgaris L. in a vertical light gradient. Oecologia 117:235–238

Li R, Volenec JJ, Joern BC, Cunningham SM (1996) Seasonal changes in nonstructural carbohydrates, protein, and macronutrients in roots of alfalfa, red clover, sweetclover, and birdsfoot trefoil. Crop Sci 36:617–623

Lötscher M, Hay MJM (1995) Differences in resource allocation to stolon branches of Kopu white clover genotypes induced by manipulation of rooting. Agric Soc N Z Special Publ 11:145–148

Lötscher M, Hay MJM (1996a) Distribution of mineral nutrient from nodal roots of Trifolium repens: genotypic variation in intra-plant allocation of ^{32}P and ^{45}Ca. Physiol Plant 97:269–276

Lötscher M, Hay MJM (1996b) Distribution of phosphorus and calcium from nodal roots of Trifolium repens: the relative importance of transport via xylem or phloem. New Phytol 133:445–452

Lötscher M, Hay MJM (1997) Genotypic differences in physiological integration, morphological plasticity and utilization of phosphorus induced by variation in phosphate supply in Trifolium repens. J Ecol 85:341–350

Lötscher M, Nösberger J (1996) Influence of position and number of nodal roots on outgrowth of axillary buds and development of branches in Trifolium repens (L.). Ann Bot 78:459–465

Lötscher M, Nösberger J (1997) Branch and root formation in Trifolium repens is influenced by the light environment of unfolded leaves. Oecologia 111:499–504

Louahlia S, Macduff JH, Ourry A, Humphreys M, Boucaud J (1999) Nitrogen reserve status affects the dynamics of nitrogen remobilization and mineral nitrogen uptake during recovery of contrasting cultivars of Lolium perenne from defoliation. New Phytol 142:451–462

Lüscher A, Stäheli B, Braun R, Nösberger J (2001) Leaf area, competition with grass, and clover cultivar: key factors for successful overwintering and fast regrowth of white clover (Trifolium repens L.) in spring. Ann Bot 88:725–735

Marcuvitz S, Turkington R (2000) Differential effects of light quality, provided by different grass neighbours, on the growth and morphology of Trifolium repens L. (white clover). Oecologia 125:293–300

Marshall C (1990) Source-sink relations of interconnected ramets. In: Van Groenendael J, De Kroon H (eds) Clonal growth in plants: regulation and function. SPB, The Hague, pp 23–41

Marshall C (1996) Sectoriality and physiological organisation in herbaceous plants: an overview. Vegetatio 127:9–16

Marshall C, Price EAC (1997) Sectoriality and its implications for physiological integration. In: De Kroon H, Van Groenendael J (eds) The ecology and evolution of clonal plants. Backhuys, Leiden, pp 79–107

Morelli G, Ruberti I (2002) Light and shade in the photocontrol of *Arabidopsis* growth. Trends Plant Sci 7:399–404

Morvan-Bertrand A, Boucaud J, Prud'Homme M-P (1999a) Influence of initial levels of carbohydrates, fructans, nitrogen, and soluble proteins on regrowth of *Lolium perenne* L. cv. Bravo following defoliation. J Exp Bot 50:1817–1826

Morvan-Bertrand A, Pavis N, Boucaud J, Prud'Homme M-P (1999b) Partitioning of reserve and newly assimilated carbon in roots and leaf tissues of *Lolium perenne* during regrowth after defoliation: assessment by ^{13}C steady-state labelling and carbohydrate analysis. Plant Cell Environ 22:1097–1108

Nagy F, Schäfer E (2000) Control of nuclear import and phytochromes. Curr Opin Plant Biol 3:450–454

Newton PCD, Hay MJM (1994) Patterns of nodal rooting in *Trifolium repens* (L.) and correlations with stages in the development of axillary buds. Grass Forage Sci 49:270–276

Oborny B (1994) Growth rules in clonal plants and environmental predictability – a simulation study. J Ecol 82:341–351

Oborny B, Cain ML (1997) Models of spatial spread and foraging in clonal plants. In: De Kroon H, Van Groenendal J (eds) The ecology and evolution of clonal plants. Backhuys, Leiden pp 155–183

Ourry A, Boucaud J, Salette J (1988) Nitrogen mobilization from stubble and roots during re-growth of defoliated perennial ryegrass. J Exp Bot 39:803–809

Ourry A, Bigot J, Boucaud J (1989) Protein mobilization from stubble and roots, and proteolytic activities during post-clipping re-growth of perennial ryegrass. J Plant Physiol 134:298–303

Pennings S, Callaway RM (2000) The advantages of clonal integration under different ecological conditions: a community-wide test. Ecology 81:709–716

Piqueras J, Klimeš L, Redbo-Torstensson P (1999) Modelling the morphological response to nutrient availability in the clonal plant *Trientalis europaea* L. Plant Ecol 141:117–127

Price EAC, Hutchings MJ (1992) Studies of growth in the clonal herb *Glechoma hederacea*. II. The effects of selective defoliation. J Ecol 80:39–47

Price EAC, Hutchings MJ (1996) The effects of competition on growth and form in *Glechoma hederacea*. Oikos 75:279–290

Price EAC, Marshall C, Hutchings MJ (1992) Studies of growth in the clonal herb *Glechoma hederacea*. I. Patterns of physiological integration. J Ecol 80:25–38

Price EAC, Hutchings MJ, Marshall C (1996) Causes and consequences of sectoriality in the clonal herb *Glechoma hederacea*. Vegetatio 127:41–54

Price EAC, Gamble R, Williams GG, Marshall C (2002) Seasonal patterns of partitioning and remobilization of ^{14}C in the invasive rhizomatous perennial Japanese knotweed [*Fallopia japonica* (Houtt.) Ronse Decraene]. Evol Ecol 15:347–362

Pywell RF, Bullock JM, Roy DB, Warman L, Walker KJ, Rothery P (2003) Plant traits as predictors of performance in ecological restoration. J Appl Ecol 40:65–77

Robin C, Guerin V, Guckert A (1989) Effect of ramification of the stolon on the assimilates distribution in the white clover. Agronomie 9:849–857

Robin C, Varlet-Grancher C, Gastal F, Flenet F, Guckert A (1992) Photomorphogenesis of white clover (*Trifolium repens* L.): phytochrome mediated effects on ^{14}C-assimilate partitioning. Eur J Agron 1:235–240

Robin C, Hay MJM, Newton PCD (1994a) Effect of light quality (red:far-red ratio) and defoliation treatments applied at a single phytomer on axillary bud outgrowth in *Trifolium repens* L. Oecologia 100:236–242

Robin C, Hay MJM, Newton PCD, Greer DH (1994b) Effect of light quality (red:far-red ratio) at the apical bud of the main stolon on morphogenesis of *Trifolium repens* L. Ann Bot 74:119–123

Sackville Hamilton NR, Hay MJM (1998) Vascular architecture of a large-leafed genotype of *Trifolium repens*. Ann Bot 81:441–448

Salzman AG, Parker MA. (1985) Neighbors ameliorate local salinity stress for a rhizomatous plant in a heterogeneous environment. Oecologia 65:273–277

Schlichting CD, Smith H (2002) Phenotypic plasticity: linking molecular mechanisms with evolutionary outcomes. Evol Ecol 16:189–211

Schmitt J, Dudley SA, Pigliucci M (1999) Manipulative approaches to testing adaptive plasticity: phytochrome-mediated shade-avoidance responses in plants. Am Nat 154:S43–S54

Schnyder H, De Visser R (1999) Fluxes of reserve-derived and currently assimilated carbon and nitrogen in perennial ryegrass recovering from defoliation. The regrowing tiller and its component functionally distinct zones. Plant Physiol 119:1423–1435

Shumway SW (1995) Physiological integration among clonal ramets during invasion of disturbance patches in a New England salt marsh. Ann Bot 76:225–233

Simon JC, Jacquet A, Decau ML, Goulas E, Le Dily F (2004) Influence of cutting frequency on the morphology and C and N reserve status of two cultivars of white clover (*Trifolium repens* L.). Eur J Agron 20:341–350

Singh DK, Sale PWG (1997) Defoliation frequency and the response by white clover to increasing phosphorus supply. 2. Non-structural carbohydrate concentrations in plant parts. Aust J Agric Res 48:119–124

Skálová H, Krahulec F (1992) The response of three *Festuca rubra* clones to changes in light quality and plant density. Funct Ecol 6:282–290

Skálová H, Pecháčková S, Suzuki J, Herben T, Hara T, Hadincová V, Krahulec F (1997) Within-population genetic differentiation in traits affecting clonal growth: *Festuca rubra* in a mountain grassland. J Evol Biol 10:383–406

Slade AJ, Hutchings MJ (1987) The effects of nutrient availability on foraging in the clonal herb *Glechoma hederacea*. J Ecol 75:95–112

Solangaarachchi SM, Harper JL (1987) The effect of canopy-filtered light on the growth of white clover *Trifolium repens*. Oecologia 72:372–376

Stammel B, Kiehl K, Pfadenhauer J (2003) Alternative management on fens: response of vegetation to grazing and mowing. Appl Veg Sci 6:245–254

Staswick PE (1994) Storage proteins of vegetative tissues. Annu Rev Plant Physiol Plant Mol Biol 45:303–322

Stoll P, Egli P, Schmid B (1998) Plant foraging and rhizome growth pattern of *Solidago altissima* in response to mowing and fertilizer application. J Ecol 86:341–354

Stuefer JF (1995) Separating the effects of assimilate and water integration in clonal fragments by the use of steam-girdling. Abstr Bot 19:75–81

Stuefer JF (1996) Potential and limitations of current concepts regarding the response of clonal plants to environmental heterogeneity. Vegetatio 127:55–70

Stuefer JF, Hutchings MJ (1994) Environmental heterogeneity and clonal growth: a study of the capacity for reciprocal translocation in *Glechoma hederacea* L. Oecologia 100:302–308

Stuefer JF, Huber H (1998) Differential effects of light quantity and spectral light quality on growth, morphology and development of two stoloniferous *Potentilla* species. Oecologia 117:1–8

Stuefer JF, Huber H (1999) The role of stolon internodes for ramet survival after clone fragmentation in *Potentilla anserina*. Ecol Lett 2:135–139

Stuefer JF, During HJ, De Kroon H (1994) High benefits of clonal integration in two stoloniferous species, in response to heterogeneous light environments. J Ecol 82:511–518

Stuefer JF, De Kroon H, During HJ (1996) Exploitation of environmental heterogeneity by spatial division of labour in a clonal plant. Funct Ecol 10:328–334

Stuefer JF, During HJ, Schieving F (1998) A model on optimal root–shoot allocation and water transport in clonal plants. Ecol Model 111:171–186

Sutherland WJ (1987) Growth and foraging behaviour. Nature 330:18–19

Sutherland WJ, Stillman RA (1988) The foraging tactics of plants. Oikos 52:239–244

Sutherland WJ, Stillman RA (1990) Clonal growth: insights from models. In: Van Groenendael J, De Kroon H (eds) Clonal growth in plants: regulation and function. SPB, The Hague, pp 95–111

Suzuki JI, Stuefer JF (1999) On the ecological and evolutionary significance of storage in clonal plants. Plant Species Biol 14:11–17

Teuber N, Laidlaw AS (1996) Influence of irradiance on branch growth of white clover stolons in rejected areas within grazed swards. Grass Forage Sci 51:73–80

Thomas RG (1987) The structure of a mature plant. In: Baker MJ, Williams WM (eds) White clover. CAB, Wallingford, pp 1–29

Thomas RG, Hay MJM, Newton PCD (2002) A developmentally based categorization of branching in *Trifolium repens* L.: influence of nodal roots. Ann Bot 90:379–389

Thomas RG, Hay MJM, Newton PCD (2003) Relationships among shoot sinks for resources exported from nodal roots regulated branch development of distal non-rooted portions of *Trifolium repens* L. J Exp Bot 54:2091–2104

Thomas RG, Hay MJM, Newton PCD, Tilbrook JC (2004) Relative importance of nodal roots and apical buds in the control of branching in *Trifolium repens* L. Plant Soil 255:55–66

Thompson L (1995) Sites of photoperception in white clover. Grass Forage Sci 50:259–262

Thompson L, Harper JL (1988) The effect of grasses on the quality of transmitted radiation and its influence on the growth of white clover *Trifolium repens*. Oecologia 75:343–347

Thornton B, Millard P (1993) The effects of nitrogen supply and defoliation on the seasonal internal cycling of nitrogen in *Molinia caerulea*. J Exp Bot 44:531–536

Thornton B, Millard P (1996) Effects of severity of defoliation on root functioning in grasses. J Range Manage 49:443–447

Thornton B, Millard P, Duff EI, Buckland ST (1993) The relative contribution of remobilization and root uptake in supplying nitrogen after defoliation for regrowth of laminae in four grass species. New Phytol 124:689–694

Thornton B, Millard P, Duff EI (1994) Effects of nitrogen supply on the source of nitrogen used for regrowth of laminae after defoliation of four grass species. New Phytol 128:615–620

Van Kleunen M, Fischer M (2001) Adaptive evolution of plastic foraging responses in a clonal plant. Ecology 82:3309–3319

Van Kleunen M, Fischer M, Schmid B (2000) Clonal integration in *Ranunculus reptans*: by-product or adaptation? J Evol Biol 13:237–248

Vandenbussche F, Vriezen WH, Smalle J, Laarhoven LJJ, Harren FJM, Van Der Straeten D (2003) Ethylene and auxin control the *Arabidopsis* response to decreased light intensity. Plant Physiol 133:517–527

Volenec JJ, Ourry A, Joern BC (1996) A role for nitrogen reserves in forage regrowth and stress tolerance. Physiol Plant 97:185–193

Wardlaw IF (1990) The control of carbon partitioning in plants. New Phytol 116:341–381

Watson MA, Caspar BB (1984) Morphogenetic constraints on patterns of carbon distribution in plants. Annu Rev Ecol Syst 15:233–258

Welham CVJ, Turkington R, Sayre C (2002) Morphological plasticity of white clover (*Trifolium repens* L.) in response to spatial and temporal resource heterogeneity. Oecologia 130:231–238

Wilman D, Acuña GH (1993) Effects of cutting height on the growth of leaves and stolons in perennial ryegrass-white clover swards. J Agric Sci 121:39–46

Winkler E, Fischer M (2002) The role of vegetative spread and seed dispersal for optimal life histories of clonal plants: a simulation study. Evol Ecol 15:281–301

Winkler E, Stöcklin J (2000) Sexual and vegetative reproduction of *Hieracium pilosella* L. under competition and disturbance: a grid-based simulation model. Ann Bot 89:525–536

Yu F, Chen Y, Dong M (2002) Clonal integration enhances survival and performance of *Potentilla anserina*, suffering from partial sand burial on the Ordos Plateau, China. Evol Ecol 15:303–318

Markus Lötscher
Lehrstuhl für Grünlandlehre
Technische Universität München
Am Hochanger 1
Germany 70-85350 Freising
Tel.: +49-8161-713722
Fax: +49-8161-713243
e-mail: loetscher@wzw.tum.de

Subject Index

Printing: Krips bv, Meppel
Binding: Stürtz, Würzburg